Springer Proceedings in Mathematics & Statistics

Volume 186

Springer Proceedings in Mathematics & Statistics

This book series features volumes composed of selected contributions from workshops and conferences in all areas of current research in mathematics and statistics, including operation research and optimization. In addition to an overall evaluation of the interest, scientific quality, and timeliness of each proposal at the hands of the publisher, individual contributions are all refereed to the high quality standards of leading journals in the field. Thus, this series provides the research community with well-edited, authoritative reports on developments in the most exciting areas of mathematical and statistical research today.

More information about this series at http://www.springer.com/series/10533

Jim M. Cushing · M. Saleem
H.M. Srivastava · Mumtaz Ahmad Khan
M. Merajuddin
Editors

Applied Analysis in Biological and Physical Sciences

ICMBAA, Aligarh, India, June 2015

Springer

Editors
Jim M. Cushing
Department of Mathematics and Program in Applied Mathematics
University of Arizona
Tucson, USA

Mumtaz Ahmad Khan
Department of Applied Mathematics
Aligarh Muslim University
Aligarh, Uttar Pradesh
India

M. Saleem
Department of Applied Mathematics
Aligarh Muslim University
Aligarh, Uttar Pradesh
India

M. Merajuddin
Department of Applied Mathematics
Aligarh Muslim University
Aligarh, Uttar Pradesh
India

H.M. Srivastava
Department of Mathematics and Statistics
University of Victoria
Victoria, BC
Canada

ISSN 2194-1009 ISSN 2194-1017 (electronic)
Springer Proceedings in Mathematics & Statistics
ISBN 978-81-322-3638-2 ISBN 978-81-322-3640-5 (eBook)
DOI 10.1007/978-81-322-3640-5

Library of Congress Control Number: 2016951965

Mathematics Subject Classification (2010): 92B05, 92D25, 92C60, 35Q35, 46T99

Printed on acid-free paper

This Springer imprint is published by Springer Nature
The registered company is Springer (India) Pvt. Ltd.
The registered company address is: 7th Floor, Vijaya Building, 17 Barakhamba Road, New Delhi 110 001, India

Preface

This volume comprises three parts. Part I contains the contributions from the discipline of nonlinear dynamics and its applications to the biological sciences, Part II presents the research papers on nonlinear analysis and applications to a variety of problems in science, engineering and industry and lastly Part III focuses on contributions concerning applied analysis. The authors were speakers and participants at the conference. Their papers touch upon a variety of important contemporary topics including linear/nonlinear analysis, mathematical biology/ecology, dynamical systems, graph theory, variational inequalities/functional analysis, differential and difference equations, partial differential equations, numerical analysis/ techniques, chaos and wavelet analysis. The emphasis is on both the mathematical and the applied aspects of these topics. All contributions were peer reviewed.

This volume mainly focuses on current research in fields of mathematical analysis that can be used as sophisticated tools for the study of scientific problems. The reader will find a variety of applications and different facets of nonlinear analysis, with an interdisciplinary flavor that ranges from model development and formulation to the mathematical and theoretical analysis of the models. We hope that the work presented in this volume will appeal to and benefit researchers, academicians and engineers equally. We further hope that the research embodied in this volume will stimulate the formation of interdisciplinary groups for fruitful collaborative research.

The Department of Applied Mathematics at Aligarh Muslim University, Aligarh, India, offers a Ph.D. research program in areas ranging from applied disciplines such as Mathematical Biology and Graph Theory, to purely mathematical disciplines, like Functional Analysis, Special Functions and Algebra. This department is part of the Faculty of Engineering and Technology and actively interacts with engineering faculty and students. When in June 2014 the department was entrusted with the task of organizing and hosting an international conference, it was natural to include applied analysis and applications to the physical sciences as major themes. Owing to the increasing interest and importance of applications in the biosciences, it was decided also to include them as a major theme of the conference.

We are very grateful to the members of the Editorial Board for their guidance. Our sincere thanks are due to many experts, including those on the Editorial Board, who have acted as peer reviewers/referees.

The faculty members and the ministerial staff of the Department of Applied Mathematics provided invaluable help with the organization of ICMBAA-2015. We would particularly like to thank Lt. Gen. Zameer Uddin Shah (Vice Chancellor, AMU) and Brigadier S. Ahmad Ali (Pro Vice Chancellor, AMU) for their support and patronage. Dr. Sabiha Tabassum, Dr. Ghazala Yasmin, Mr. Mohammad Malik, and Dr. Abdullah Bin Abu Baker handled innumerable tasks during the organization of the conference and also during the development of this volume with style and grace under pressure. Financial support for the conference from DST, New Delhi; CSIR, New Delhi; TEQIP-II and AMU, Aligarh is greatly acknowledged. TEQIP-II Program in the Z.H. College of Engineering and Technology needs a special mention here for its generous support towards the preparation of this volume.

We place on record our deep appreciation and thanks to all the contributors of this volume. This volume was made possible only by their efforts and cooperation. Finally, we gratefully acknowledge our indebtedness to Springer and their support team for their cooperation and patience. Our special thanks are given to Mr. Shamim Ahmad, Ms. Ignasy Devi and Mr. V. Praveenkumar from Springer for their cooperation and encouragement.

Arizona, USA — Jim M. Cushing
Aligarh, India — M. Saleem
Victoria, Canada — H.M. Srivastava
Aligarh, India — Mumtaz Ahmad Khan
Aligarh, India — M. Merajuddin
March 2016

Editorial Board

Contents

Editors and Contributors

About the Editors

Jim M. Cushing is Professor in the Department of Mathematics and member of the Interdisciplinary Program in Applied Mathematics at the University of Arizona, Tucson, U.S.A. His professional interests include discrete and continuous dynamic systems and applications in population, ecological and evolutionary biology. He is a fellow of the American Mathematical Society, editor-in-chief of the Journal of Biological Dynamics and past president of the International Society of Difference Equations. He is also the Alexander von Humboldt Fellow.

M. Saleem is Professor in the Department of Applied Mathematics, Aligarh Muslim University, India. He did his Ph.D. from the Indian Institute of Technology Kanpur, India, in 1980. His main area of research is mathematical modelling of ecological problems. He has also published numerous papers on elasticity in several journals of international repute. His current research interests include cancer growth models. He is a life member of the Indian Society for Mathematical Modelling & Computer Simulation.

H.M. Srivastava has held the position of Professor Emeritus in the Department of Mathematics and Statistics at the University of Victoria in Canada since 2006. Prior to this, he was a professor in the same department from 1974 to 2006 and associate professor in the same department from 1969 to 1974. He has held numerous visiting research and honorary chair positions at many universities and research Institutes in different parts of the world. He is also actively editorially associated with numerous international scientific research journals. His current research interests include several areas of pure and applied mathematical sciences such as real and complex analysis, fractional calculus and its applications, integral equations and transforms, higher transcendental functions and their applications, q-series and q-polynomials, analytic number theory, analytic and geometric inequalities, probability and statistics, and inventory modelling and optimization. He has published 25 books, monographs and edited volumes, 30 book and encyclopaedia chapters, 45 papers in international conference proceedings, and more than 1,000 scientific research journal articles, as well as written forewords and prefaces to many books and journals. He is currently listed as the 2015 Thomson-Reuters Highly Cited Researcher.

Mumtaz Ahmad Khan is Dean of the Faculty of Engineering & Technology and Professor in the Department of Applied Mathematics, Aligarh Muslim University, India. He did both his M.Sc. and Ph.D. from Lucknow University in 1970 and 1978, respectively. He has 150 research papers to his credit published in several international journals of repute. He has successfully guided 17 Ph.D. and 12 M.Phil students. He is in the editorial board of International Transactions in Mathematical Sciences and Computer. He is a reviewer for American Mathematical Review. His current research interests include discrete complex analysis, fractional calculus and its application, integral equations and transforms, higher transcendental functions and their applications, q-series, q-polynomials and umbral calculus.

M. Merajuddin is Professor in the Department of Applied Mathematics, Aligarh Muslim University, India. He did both his M.Sc. and Ph.D. from the Indian Institute of Technology Kanpur, India, in 1978 and 1985, respectively. His research interests include graph theory and discrete mathematics.

Contributors

Tazid Ali Department of Mathematics, Dibrugarh University, Dibrugarh, India

S. Arumugam National Centre for Advanced Research in Discrete Mathematics, Kalasalingam University, Krishnankoil, Tamil Nadu, India; Department of Computer Science, Liverpool Hope University, Liverpool, UK; Department of Computer Science, Ball State University, Muncie, IN, USA

M.A. Aziz-Alaoui UniHavre, LMAH, FR CNRS 3335, ISCN, Normandie University, Le Havre, France

D. Bahuguna Department of Mathematics & Statistics, Indian Institute of Technology Kanpur, Kanpur, India

Malay Banerjee Department of Mathematics and Statistics, IIT Kanpur, Kanpur, India

Anupma Bansal Department of Mathematics, DAV College for Women, Ferozepur Cantt, Punjab, India

Edy Tri Baskoro Combinatorial Mathematics Research Group, Faculty of Mathematics and Natural Sciences, Institut Teknologi Bandung, Bandung, Indonesia

Veeky Baths Department of Biological Sciences, Birla Institute of Technology and Science Pilani, Zuarinagar, Goa, India

Mohammad Belal Department of Mechanical Engineering, Aligarh Muslim University, Aligarh, India

Peeyush Chandra Department of Mathematics and Statistics, IIT Kanpur, Kanpur, India

Yuqing Chen College of Applied Mathematics Guangdong University of Technology, Guangzhou, Guangdong, People's Republic of China

Yeol Je Cho Department of Mathematics Education, The Research Institute of Natural Sciences, Gyongsang National University, Jinju, Korea

B.S. Choudhury Department of Mathematics, IIEST, Shibpur, Howrah, West Bengal, India

Jim M. Cushing Department of Mathematics and Program in Applied Mathematics, University of Arizona, Tucson, AZ, USA

M. Darabi Department of Basic Sciences, Golpayegan University of Technology, Golpayegan, Isfahan, Iran

Gunamani Deheri Department of Mathematics, Sardar Patel University, Anand, Gujarat, India

Balram Dubey Department of Mathematics, BITS Pilani-Pilani Campus, Pilani, Rajasthan, India

Preeti Dubey Department of Mathematics, BITS Pilani-Pilani Campus, Pilani, Rajasthan, India

Uma S. Dubey Department of Biological Sciences, BITS Pilani-Pilani Campus, Pilani, Rajasthan, India

Arnaud Ducrot IMB, UMR CNRS 5251, University of Bordeaux, Bordeaux, France

Samira El Yacoubi IMAGES-Espace-Dev, UMR 228 IRD UM UR UG, University of Perpignan Via Domitia, Perpignan, France

Sunita Gakkhar Department of Mathematics, Indian Institute of Technology Roorkee (IITR), Roorkee, Uttarakhand, India

Anju Goel Department of Mathematics, Indian Institute of Technology Roorkee (IITR), Roorkee, Uttarakhand, India

Nisha Gohain Department of Mathematics, Dibrugarh University, Dibrugarh, India

R.K. Gupta School of Mathematics and Computer Applications, Thapar University, Patiala, Punjab, India

R.P. Gupta Department of Mathematics, VNIT, Nagpur, India

Sahil Gupta Guru Nanak Dev University, Amritsar, India

Sanjeev Gupta Department of Humanities and Social Sciences, Indian Institute of Technology, Kanpur, India

Nadeem Hasan Department of Mechanical Engineering, Aligarh Muslim University, Aligarh, India

Shamshad Husain Department of Applied Mathematics, Aligarh Muslim University, Aligarh, India

M. Indhumathy National Centre for Advanced Research in Discrete Mathematics, Kalasalingam University, Krishnankoil, Tamil Nadu, India

Akhlad Iqbal Department of Mathematics, Aligarh Muslim University, Aligarh, India

Kamaljeet Department of Mathematics & Statistics, Indian Institute of Technology Kanpur, Kanpur, India

Kanika Khatter Department of Applied Mathematics, Delhi Technological University, Delhi, India

Jong Kyu Kim Department of Mathematics Education, Kyungnam University, Changwon, Korea

Daniel A. Korytowski School of Mathematical and Statistical Sciences, Arizona State University, Tempe, AZ, USA

Rajeev Kumar Department of Mathematics, Maharishi Markandeshwar University, Mullana, Ambala, Haryana, India

A. Kundu Department of Mathematics, Siliguri Institute of Technology, Darjeeling, West Bengal, India

Hiroshi Matano Graduate School of Mathematical Sciences, University of Tokyo, Komaba, Tokyo, Japan

Arti Mishra Department of Mathematics, Indian Institute of Technology, Roorkee, Uttarakhand, India

P. Muthu Department of Mathematics, National Institute of Technology, Warangal, Telangana, India

T.D. Narang Guru Nanak Dev University, Amritsar, India

V. Ravichandran Department of Mathematics, University of Delhi, Delhi, India

Huma Sahper Department of Applied Mathematics, Aligarh Muslim University, Aligarh, India

V. Samhita Department of Mathematics, BITS Pilani Hyderabad Campus, Hyderabad, Telangana, India

Moitri Sen SERC, IISc Bangalore, Bangalore, India

R.K. Shukla Invertis University, Bareilly, India

Snehal Shukla Department of Mathematics, Shri R.K. Parikh Arts and Science College, Petlad, Gujarat, India

Anuraj Singh Graphic Era University, Dehrdaun, India

Tarkeshwar Singh Department of Mathematics, Birla Institute of Technology and Science Pilani, Zuarinagar, Goa, India

S. Sivaprasad Kumar Department of Applied Mathematics, Delhi Technological University, Delhi, India

Hal L. Smith School of Mathematical and Statistical Sciences, Arizona State University, Tempe, AZ, USA

M.A. Sofi Department of Mathematics, University of Kashmir, Srinagar, India

T. Som Department of Mathematical Sciences, IIT (BHU), Varanasi, India

Pratik Suchde TIFR Centre for Applicable Mathematics, Bangalore, India

M. Varunkumar Department of Mathematics, National Institute of Technology, Warangal, Telangana, India

A.S. Vasudeva Murthy TIFR Centre for Applicable Mathematics, Bangalore, India

Vitaly Volpert Institut Camille Jordan, UMR 5208 CNRS, University Lyon 1, Villeurbanne, France

Kristiana Wijaya Combinatorial Mathematics Research Group, Faculty of Mathematics and Natural Sciences, Institut Teknologi Bandung, Bandung, Indonesia

Radouane Yafia Ibn Zohr University, Campus Universitaire Ait Melloul, Agadir, Morocco

J. Zafarani Department of Mathematics, Sheikhbahaee University, Isfahan, Iran; University of Isfahan, Isfahan, Iran

Part I
Applications to Biological Sciences

Modeling and Dynamics of Predator Prey Systems on a Circular Domain

Radouane Yafia, M.A. Aziz-Alaoui and Samira El Yacoubi

Abstract The present chapter is devoted to the mathematical modeling and the analysis of the dynamics of predator prey systems on a circular domain. We first give some reminders on the Laplace operator and spectral theory on a disc. Then, we analyze the dynamics of two mathematical models with two or three reaction diffusion equations, defined on a circular domain. The results are given in terms of local/global stability and of emergence of spatio-temporal patterns due to symmetry-breaking bifurcations. One basic type of such a phenomenon is Turing bifurcation which gives rise to pattern formation, a process by which a spatially uniform state loses stability to a non-uniform state. We derive, theoretically, the conditions for Turing diffusion driven instability to occur, and perform numerical simulations to illustrate how biological processes can affect spatiotemporal pattern formation in a spatial domain.

Keywords Dynamics · Predator prey · Spatio-temporal · Circular domain · Patterns · Turing instability

R. Yafia
Ibn Zohr University, Campus Universitaire Ait Melloul,
Route Nationale N°10, Agadir, Morocco
e-mail: yafia1@yahoo.fr

M.A. Aziz-Alaoui (✉)
UniHavre, LMAH, FR CNRS 3335, ISCN, Normandie University, 76600 Le Havre, France
e-mail: aziz.alaoui@univ-lehavre.fr

S. El Yacoubi
IMAGES-Espace-Dev, UMR 228 IRD UM UR UG,
University of Perpignan Via Domitia,
52, rue Paul Alduy, Perpignan, France
e-mail: yacoubi@univ-perp.fr

J.M. Cushing et al. (eds.), *Applied Analysis in Biological and Physical Sciences*,
Springer Proceedings in Mathematics & Statistics 186,
DOI 10.1007/978-81-322-3640-5_1

1 Introduction

In our knowledge, the first mathematical model of predator prey interaction is given by A. Lotka [16] and V. Volterra [20]. This model is a simplified system of two ordinary differential equations which does not take into account the space variable and supposes that every individual is accessible to every other individual and produces the so-called "mean-field description of the system". One of the oldest spatio-temporal model which takes into account the movement of individuals/organisms/particules is the standard reaction diffusion system (Fisher [13], Kolmogorov et al. [15], Murray [17]):

$$\frac{\partial N(X,t)}{\partial t} = D\Delta N(X,t) + f(N(X,t)), \qquad (X,t) \in \Omega \times \mathbb{R}^+, \Omega \subseteq \mathbb{R}^n, \quad (1)$$

where N is a p components vector, Δ is the Laplacian operator, D is the diffusion matrix and f is a nonlinear term (reaction term) representing the interactions between species N (individuals/organisms/particules).

From the mathematical modeling point of view, if $N(x,t)$ is the concentration of individuals/organisms/particules at time $t > 0$ and the position x. Then the diffusion term can be regarded as:

$$\frac{\partial N(X,t)}{\partial t} = D\Delta N(X,t)$$

where D (which can depend on x) is a positive definite symmetric diffusion matrix which describes the non-homogeneous diffusion. Therefore, the local reaction process is modeled by a local dynamical system as follows:

$$\frac{\partial N(X,t)}{\partial t} = f(N(X,t))$$

To describe the interaction of both types of processes (diffusion and reaction), we suppose that they happen on a small time interval. If we let this interval to tend to zero, then this time-splitting scheme turns into the so-called reaction-diffusion system, given by system (1).

If the reaction diffusion processes occur in a spatially confined domain Ω, then boundary conditions have to be imposed, for example the Dirichlet condition when specifying the values that the solution must check on the boundaries of the field:

$$N(X,t) = \varphi(X), \;\; X \in \partial\Omega$$

or the Neumann condition when specifying the values the derivative of the solution must satisfy on the boundaries of the field :

$$\frac{\partial N}{\partial n}(X,t) = \psi(X), \;\; X \in \partial\Omega; \;\; n \text{ is outflow through the boundary of } \Omega.$$

If $\psi(X) = 0$, then, for the dynamic of the populations, there is no immigration nor emigration.

There are other possible boundary conditions. For example the Robin boundary conditions, which are a combination of Dirichlet and Neumann conditions. The dynamic boundary conditions, or the mixed boundary conditions which correspond to the juxtaposition of different boundary conditions on different parts of the border of the domain.

A lot of mathematical problems arise from reaction diffusion theory such as: existence and regularity of solutions, boundedness of solutions, stability, traveling waves etc. [3–5, 7–10, 14, 23, 24]. One of these questions is: how the diffusion term can affect the asymptotic behavior of the corresponding system without diffusion term? In 1952, Turing prove that, under certain conditions, chemical products react and diffuse to produce non constant steady state and induce spatial patterns. This property can be explained as follows: In the absence of diffusion, the stable uniform steady state of the corresponding ordinary differential equation becomes unstable in the presence of diffusion (which called diffusion driven instability or Turing instability) and spatial patterns can evolve through bifurcations [17].

2 Spectral Theory on a Circular Domain

In this section, since there exists a difference between the analysis in a rectangle domain and a circular domain (disc), we give some results on the Laplace operator on a circular domain (see, [17]).

Let us consider a disc with a radius R as follows:

$$\mathscr{D} = \{(r, \theta) : 0 \leq r < R\}.$$

Then the Laplace operator is defined in cartesian coordinates as $\Delta\varphi = \frac{\partial^2}{\partial x^2}\varphi + \frac{\partial^2}{\partial y^2}\varphi$ and in polar coordinates (r, θ) as $\Delta_{r\theta}\varphi = \frac{\partial^2}{\partial r^2}\varphi + \frac{1}{r}\frac{\partial}{\partial r}\varphi + \frac{1}{r^2}\frac{\partial^2}{\partial \theta^2}\varphi$, with $x = r\cos(\theta)$, $y = r\sin(\theta)$ and $r = \sqrt{x^2 + y^2}$ and $\tan(\theta) = \frac{y}{x}$.

To compute the eigenvectors on the circular domain, one needs to separate variables using polar coordinates. Considering the eigenvalue problem

$$\begin{cases} \Delta_{r\theta}\varphi = -\lambda\varphi \\ \varphi(R, \theta) = 0, \ \theta \in [0, 2\pi] \\ \frac{\partial\varphi}{\partial\eta} = 0, \ \text{on } r = R \text{ and } \theta \in [0, 2\pi] \end{cases} \tag{2}$$

and looking for solutions of the form $\varphi(r, \theta) = P(r)\Phi(\theta)$. By differentiation and from the Eq. (2) we have:

$$P''(r)\Phi(\theta) + \frac{1}{r}P'(r)\Phi(\theta) + \frac{1}{r^2}P(r)\Phi''(\theta) = -\lambda P(r)\Phi(\theta) \tag{3}$$

Therefore

$$\frac{r^2}{P(r)}\{P''(r)+\frac{1}{r}P'(r)+\lambda P(r)\}=-\frac{\Phi''(\theta)}{\Phi(\theta)} \tag{4}$$

The only way for these two expressions to equal for all possible values of r and θ is to have them both equal a constant. Therefor, there exists k such that $-\Phi''(\theta)=k^2\Phi(\theta)$

The appropriate boundary conditions to apply to this problem state that the function $\Phi(\theta)$ and its first derivative with respect to θ are periodic in θ.

Then, the solution is given by:

$$\Phi_n(\theta)=a_n\sin(n\theta)+b_n\cos(n\theta) \text{ for integers } k=n\geq 1$$

where a_n and b_n are constants.

Then we have the following second order differential equation of

$$P''(r)+\frac{1}{r}P'(r)+\left(\lambda-\frac{k^2}{r^2}\right)P(r)=0, \text{ such that } P'(R)=0 \tag{5}$$

Let $x=\sqrt{\lambda}r$ and $P(x)=J(\frac{x}{\sqrt{\lambda}})$. Then, we have

$$J''(x)+\frac{1}{x}J'(x)+\left(1-\frac{k^2}{x^2}\right)J(x)=0 \text{ (called Bessel equation)} \tag{6}$$

The solution for it is the nth Bessel function

$$J_n(x)=\sum_{l=0}^{+\infty}\frac{(-1)^l}{l!(n+l)!}\left(\frac{x}{2}\right)^{n+2l}$$

Since $P(r)=J_n(\sqrt{\lambda}r)$, we get:

$$\phi_n^\lambda(r,\theta)=\Phi_n(\theta)J_n(\sqrt{\lambda}r) \tag{7}$$

which are eigenfunctions of the Laplacian operator in polar coordinates.

The eigenvalues λ associated to the eigenvector ϕ_n^λ are determined from the boundary conditions.

From Dirichlet boundary conditions defined as follows $\phi_n^\lambda(R,\theta)=0, \forall\theta\in[0,2\pi]$ we get $J_n(\sqrt{\lambda}R)=0$. This means that $\sqrt{\lambda}R$ is a root of J_n.

From the Neumann boundary conditions: $\partial_r\phi_n^\lambda(R,\theta)=0, \forall\theta\in[0,2\pi]$ we get $J_n'(\sqrt{\lambda}R)=0$. This means that $\sqrt{\lambda}R$ is a root of J_n'.

We denote these roots by α_{nm} and assume they are indexed in increasing order:

$$J_n(\alpha_{nm})=0, \alpha_{n1}<\alpha_{n2}<\alpha_{n3}<....$$

Therefore $\sqrt{\lambda}R = \alpha_{nm}$ for some index m and the eigenvalues will be written in the following form:

$$\lambda_{nm} = \left(\frac{\alpha_{nm}}{R}\right)^2$$

where n is the index of nth Bessel function and m is the index number of their roots. If $R = 1$, then the eigenvalues of the equations $\Delta\varphi = -\lambda\varphi$ are the square of zero solution of Bessel functions.

3 Mathematical Model of Two Species

In this section, we consider a 2-D reaction diffusion model which is based on the modified Leslie-Gower model with Beddington-DeAngelis functional responses [4–6, 11, 12, 18, 19, 21, 22]:

$$\begin{cases} \frac{\partial u(t,X)}{\partial t} = D_1 \Delta u(t,X) + \left(a_1 - b_1 u(t,X) - \frac{c_1 v(t,X)}{d_1 u(t,X)+d_2 v(t,X)+k_1}\right) u(t,X) \\ \frac{\partial v(t,X)}{\partial t} = D_2 \Delta v(t,X) + \left(a_2 - \frac{c_2 v(t,X)}{u(t,X)+k_2}\right) v(t,X) \end{cases} \tag{8}$$

$u(t, X)$ and $v(t, X)$ represent population densities at time t and space $X = (x, y)$ defined on a circular domain (or disc domain) with radius R (i.e. $\Omega = \{X = (x, y) \in \mathbb{R}^2, x^2 + y^2 < R^2\}$), $r_1, a_1, b_1, k_1, r_2, a_2,$ and k_2 are model parameters assuming only positive values, a_1 is the growth rate of preys u, a_2 describes the growth rate of predators v, b_1 measures the strength of competition among individuals of species u, c_1 is the maximum value of the per capita reduction of u due to v, c_2 has a similar meaning to c_1, k_1 measures the extent to which environment provides protection to prey u, k_2 has a similar meaning to k_1 relatively to the predator v, d_1 and d_2 are two positive constants, D_1 and D_2 are the terms diffusions of the preys and the predators.

Steady States and Stability

We consider the reaction diffusion system of two species (8) defined on a circular domain with Neumann boundary conditions (which means that there are no flux of species of both predator and prey on the boundary of the circular domain Ω), where $\Omega = \{(x, y) : x^2 + y^2 < R^2\}$. We can write x and y in polar coordinates as follow $x = rcos\theta$ and $y = rsin\theta$, applying the polar coordinate transformation we find $\Gamma = \{(r, \theta) : 0 < r < R,\ 0 \leq \theta < 2\pi\}$, R the radius of the disk Ω; $r = \sqrt{x^2 + y^2}$, and $\theta = \tan^{-1}(\frac{y}{x})$.

Without loss of generalities we denote also $u(t, x, y) = u(t, rcos(\theta), rsin(\theta)) = u(t, r, \theta)$ and $v(t, x, y) = v(t, rcos(\theta), rsin(\theta)) = v(t, r, \theta)$ are thc densities of prey and predators respectively in polar coordinates, at $t = 0$, $u(0, r, \theta) = u_0(r, \theta) \geq 0$, $v(0, r, \theta) = v_0(r, \theta) \geq 0$. Therefore the Laplacian operator in polar coordinates is given by:

$$\Delta_{r\theta}u = \frac{\partial^2 u}{\partial r^2} + \frac{1}{r}\frac{\partial u}{\partial r} + \frac{1}{r^2}\frac{\partial^2 u}{\partial \theta^2}, \tag{9}$$

Then, the spatio-temporal system (8) in polar coordinates is written as follows:

$$\begin{cases} \frac{\partial u(t,r,\theta)}{\partial t} = D_1\Delta_{r\theta}u(t,r,\theta) + f(u(t,r,\theta), v(t,r,\theta))\ \forall (r,\theta) \in \Gamma, t > 0 \\ \frac{\partial v(t,r,\theta)}{\partial t} = D_2\Delta_{r\theta}v(t,r,\theta) + g(u(t,r,\theta), v(t,r,\theta))\ \forall (r,\theta) \in \Gamma, t > 0 \\ \frac{\partial u(t,r,\theta)}{\partial n} = \frac{\partial v(t,r,\theta)}{\partial n} = 0, \qquad \forall (r,\theta) \in \partial\Gamma \end{cases} \tag{10}$$

where

$$\begin{cases} f(u(t,r,\theta), v(t,r,\theta)) = \left(a_1 - b_1 u(t,r,\theta) - \frac{c_1 v(t,r,\theta)}{d_1 u(t,r,\theta) + d_2 v(t,r,\theta) + k_1}\right) u(t,r,\theta), \\ g(u(t,r,\theta), v(t,r,\theta)) = \left(a_2 - \frac{c_2 v(t,r,\theta)}{u(t,r,\theta) + k_2}\right) v(t,r,\theta), \end{cases} \tag{11}$$

A steady state (u_e, v_e) of (10) is a solution of the following system

$$\begin{cases} D_1\Delta_{r\theta}u_e(t,r,\theta) + f(u_e(t,r,\theta), v_e(t,r,\theta)) = 0 \\ D_2\Delta_{r\theta}v_e(t,r,\theta) + g(u_e(t,r,\theta), v_e(t,r,\theta)) = 0 \end{cases} \tag{12}$$

Let us denote the non-negative cone by

$$\mathbb{R}^2_+ = \{(u, v) \in \mathbb{R}^2, u_0 \geq 0, v_0 \geq 0\}$$

and the positive cone by

$$int\mathbb{R}^2_+ = \{(u, v) \in \mathbb{R}^2, u_0 > 0, v_0 > 0\}.$$

The trivial steady states (belonging to the boundary of int $\mathbb{R}^2_+$, i.e. at which one or more of populations has zero density or is extinct) are in the following forms:

$$E_0 = (0, 0), E_1 = \left(\frac{a_1}{b_1}, 0\right), E_2 = \left(0, \frac{a_2 k_2}{c_2}\right). \tag{13}$$

and the homogeneous steady state is given by $E^* = (u^*, v^*)$, where

$$u^* = \frac{-B + \sqrt{B^2 + 4AC}}{2A}, \tag{14}$$

$$v^* = \frac{a_2}{c_2}(u^* + k_2), \tag{15}$$

and

$$B = c_1 a_2 + b_1 c_2 k_1 + b_1 d_2 k_2 a_2 - a_1 d_1 c_2 - a_1 d_2 a_2,$$

$$A = b_1 d_2 a_2 + d_1 b_1 c_2,$$

$$C = k_1 a_1 c_2 + a_1 a_2 d_2 k_2 - c_1 a_2 k_2,$$

We will investigate the asymptotic behavior of orbits starting in the positive cone.

Proposition 1 ([1])
Let Θ be the set defined by

$$\Theta = \left\{ (u, v) \in \mathbb{R}_+^2,\ 0 \le u \le \frac{a_1}{b_1},\ 0 \le v \le \frac{a_2}{b_1 c_2}(a_1 + b_1 k_2) \right\}$$

(i) Θ is a positively invariant region for the flow associated to equation (10).
(ii) All solutions of (10) *initiating in Θ are ultimately bounded with respect to $\mathbb{R}_+^2$ and eventually enter the attracting set Θ.*

To study the existence of Turing instability one needs to prove the stability of spatially independent homogeneous steady state.

Proposition 2 (local stability without diffusion [1])

- *If $0 < u^* < \theta_1$ or $\theta_2 < u^* < \frac{a_1}{b_1}$, then $E^* = (u^*, v^*)$ is asymptotically stable.*
- *If $(a_2^2 d_2 + a_2 d_1 c_2 + k_1 b_1 c_2 < a_1 d_1 c_2)$ and $\theta_1 < u^* < \theta_2$, then $E^* = (u^*, v^*)$ is unstable for system* (16).
- *If $a_1 d_1 < k_1 b_1$, then the positive equilibrium $E^* = (u^*, v^*)$ is locally asymptotically stable.*

The proofs of Propositions 1 and 2 are given in [1].

4 Model with Three Species

In this section, we consider the following reaction-diffusion model [4, 5, 21, 23]

$$\begin{cases} \frac{\partial U(T,x,y)}{\partial T} = D_1 \Delta U(T,x,y) + (a_0 - b_0 U(T,x,y) - \frac{v_0 V(T,x,y)}{U(T,x,y)+d_0}) U(T,x,y), \\ \frac{\partial V(T,x,y)}{\partial T} = D_2 \Delta V(T,x,y) + (-a_1 + \frac{v_1 U(T,x,y)}{U(T,x,y)+d_0} - \frac{v_2 W(T,x,y)}{V(T,x,y)+d_2}) V(T,x,y), \\ \frac{\partial W(T,x,y)}{\partial T} = D_3 \Delta W(T,x,y) + (c_3 - \frac{v_3 W(T,x,y)}{V(T,x,y)+d_3}) W(T,x,y), \\ \frac{\partial U}{\partial n} = \frac{\partial V}{\partial n} = \frac{\partial W}{\partial n} = 0, \\ U(0,x,y) = U_0(x,y) \ge 0,\ V(0,x,y) = V_0(x,y) \ge 0,\ W(0,x,y) = W_0(x,y) \ge 0, \end{cases} \tag{16}$$

$U(T, x, y)$ the density of prey specie, $V(T, x, y)$ the density of intermediate predator specie and $W(T, x, y)$ the density of top-predator specie, at time T and position (x, y), defined on a circular domain (or disc domain) with radius R (i.e. $\Omega = \{(x, y) \in \mathbf{R}^2 / x^2 + y^2 < R^2\}$. Δ is the Laplacian operator. $\frac{\partial U}{\partial \eta}$, $\frac{\partial V}{\partial \eta}$ and $\frac{\partial W}{\partial \eta}$ are respectively the normal derivatives of U, V and W on $\partial\Omega$. The three species are assumed to diffuse at rates D_i (i $= 1, 2, 3$). a_0, b_0, v_0, d_0, a_1, v_1, v_2, d_2, c_3, v_3 and d_3 are assumed to be positive parameters and are defined as follows: a_0 is the growth rate of the prey U, b_0 measures the mortality due to competition between individuals of the species U, v_0 is the maximum extent that the rate of reduction by individual U can reach, d_0 measures the protection whose prey U and intermediate predator V benefit through the environment, a_1 represents the mortality rate V in the absence of U, v_1 is the maximum value that the rate of reduction by the individual U can reach, v_2 is the maximum value that the rate of reduction by the individual V can reach, v_3 is the maximum value that the rate of reduction by the individual W can reach, d_2 is the value of V for which the rate of elimination by individual V becomes $\frac{v_2}{2}$, c_3 described the growth rate of W, assuming that there are the same number of males and females. d_3 represents the residual loss caused by high scarcity of prey V of the species W.

The initial data $U_0(x, y)$, $V_0(x, y)$ and $W_0(x, y)$ are non-negative continuous functions on Ω. The vector η is an outward unit normal vector to the smooth boundary $\partial\Omega$. The homogeneous Neumann boundary condition signifies that the system is self contained and there is no population flux across the boundary $\partial\Omega$.

Following the same algebraic computations as done in Sect. 3, firstly, we write x and y in polar coordinates as follow $x = r\cos\theta$ and $y = r\sin\theta$. By applying the polar coordinate transformation, we find $\Gamma = \{(r, \theta) : 0 < r < R,\ 0 \le \theta < 2\pi\}$. R is the radius of the disk Γ, with $r = \sqrt{x^2 + y^2}$ and $\theta = \tan^{-1}(\frac{y}{x})$.

Without loss of generalities we denote also

$$u(t, x, y) = u(t, r\cos(\theta), r\sin(\theta)) = u(t, r, \theta),$$

$$v(t, x, y) = v(t, r\cos(\theta), r\sin(\theta)) = v(t, r, \theta)$$

and

$$w(t, x, y) = w(t, r\cos(\theta), r\sin(\theta)) = w(t, r, \theta)$$

are the densities of prey, predators and top predators respectively in polar coordinates.

Therefore the Laplacian operator in polar coordinates is given by:

$$\Delta_{r\theta} u = \frac{\partial^2 u}{\partial r^2} + \frac{1}{r}\frac{\partial u}{\partial r} + \frac{1}{r^2}\frac{\partial^2 u}{\partial \theta^2}. \tag{17}$$

To simplify system (16) we introduce some transformations of variables:

$$U = \frac{a_0}{b_0}u, \; V = \frac{a_0^2}{b_0 v_0}v, \; W = \frac{a_0^3}{b_0 v_0 v_2}w, \; T = \frac{t}{a_0}, \; r = \frac{r'}{a_0}, \; \theta = \theta',$$

and
$a = \frac{b_0 d_0}{a_0}$, $b = \frac{a_1}{a_0}$, $c = \frac{v_1}{a_0}$, $d = \frac{d_2 v_0 b_0}{a_0^2}$, $p = \frac{c_3 a_0^2}{v_0 b_0 v_2}$, $q = \frac{v_3}{v_2}$, $s = \frac{d_3 v_0 b_0}{a_0^2}$, $\delta_1 = a_0 D_1$, $\delta_2 = a_0 D_2$, $\delta_3 = a_0 D_3$.
Then the spatio-temporal system (16) in polar coordinates is written as follows:

$$\begin{cases} \frac{\partial u(t,r,\theta)}{\partial t} = \delta_1 \Delta_{r\theta} u(t,r,\theta) + f(u(t,r,\theta), v(t,r,\theta), w(t,r,\theta)), & \forall (r,\theta) \in \Gamma, \; t > 0 \\ \frac{\partial v(t,r,\theta)}{\partial t} = \delta_2 \Delta_{r\theta} v(t,r,\theta) + g(u(t,r,\theta), v(t,r,\theta), w(t,r,\theta)), & \forall (r,\theta) \in \Gamma, \; t > 0 \\ \frac{\partial w(t,r,\theta)}{\partial t} = \delta_3 \Delta_{r\theta} w(t,r,\theta) + h(u(t,r,\theta), v(t,r,\theta), w(t,r,\theta)), & \forall (r,\theta) \in \Gamma, \; t > 0 \\ \frac{\partial u(t,r,\theta)}{\partial n} = \frac{\partial v(t,r,\theta)}{\partial n} = \frac{\partial w(t,r,\theta)}{\partial n} = 0, \; \forall (r,\theta) \in \partial\Gamma \\ u(0,r,\theta) = u_0(r,\theta) \geq 0, \; v(0,r,\theta) = v_0(r,\theta) \geq 0, \; w(0,r,\theta) = w_0(r,\theta) \geq 0. \end{cases} \tag{18}$$

where

$$\begin{cases} f(u(t,r,\theta), v(t,r,\theta), w(t,r,\theta)) = (1 - u(t,r,\theta) - \frac{v(t,r,\theta)}{u(t,r,\theta)+a})u(t,r,\theta), \\ g(u(t,r,\theta), v(t,r,\theta), w(t,r,\theta)) = (-b + \frac{cu(t,r,\theta)}{u(t,r,\theta)+a} - \frac{w(t,r,\theta)}{v(t,r,\theta)+d})v(t,r,\theta), \\ h(u(t,r,\theta), v(t,r,\theta), w(t,r,\theta)) = (p - \frac{qw(t,r,\theta)}{v(t,r,\theta)+s})w(t,r,\theta), \end{cases} \tag{19}$$

Without diffusion, system (18) becomes

$$\begin{cases} \frac{\partial u(t,r,\theta)}{\partial t} = (1 - u(t,r,\theta) - \frac{v(t,r,\theta)}{u(t,r,\theta)+a})u(t,r,\theta), \\ \frac{\partial v(t,r,\theta)}{\partial t} = (-b + \frac{cu(t,r,\theta)}{u(t,r,\theta)+a} - \frac{w(t,r,\theta)}{v(t,r,\theta)+d})v(t,r,\theta), \\ \frac{\partial w(t,r,\theta)}{\partial t} = (p - \frac{qw(t,r,\theta)}{v(t,r,\theta)+s})w(t,r,\theta), \end{cases} \tag{20}$$

A steady state (u_e, v_e, w_e) of (20) is an homogeneous steady state of (18) which is a solution of the following system

$$\begin{cases} \delta_1 \Delta_{r\theta} u_e(t,r,\theta) + f(u_e(t,r,\theta), v_e(t,r,\theta), w_e(t,r,\theta)) = 0, \\ \delta_2 \Delta_{r\theta} v_e(t,r,\theta) + g(u_e(t,r,\theta), v_e(t,r,\theta), w_e(t,r,\theta)) = 0, \\ \delta_3 \Delta_{r\theta} w_e(t,r,\theta) + h(u_e(t,r,\theta), v_e(t,r,\theta), w_e(t,r,\theta)) = 0, \end{cases} \tag{21}$$

Steady States and stability

Simple (and tedious) algebraic computations show that problem (18) has a homogeneous steady-state if and only

$$qc > bq + p \quad \text{and} \quad qc - bq - p > a(bq + p). \tag{22}$$

The homogeneous steady-state in the case when $d = s$, is uniquely given by

$$u^* = \frac{a(bq + p)}{qc - bq - p}, \quad v^* = (1 - u^*)(u^* + a) \text{ and } w^* = \frac{p(v^* + s)}{q}. \tag{23}$$

A similar study can be used when $d \neq s$.

The conditions (22) ensure that the system (18) has a positive homogeneous steady state corresponding to constant coexistence of the three species $E^* = (u^*, v^*, w^*)$.

Proposition 3 *Conditions (22) are satisfied, the set defined by*

$$\Theta \equiv [0, 1] \times [0, 1 + a] \times \left[0, \frac{p}{q}(1 + a + s)\right] \tag{24}$$

is positively invariant region, moreover all solutions of (18) initiating in Θ *are ultimately bounded with respect to* $\mathbb{R}^3_+$ *and eventually enter the attracting set* Θ.

By the same in the last section, we need the following result which states the stability of the homogeneous steady state.

Proposition 4 (local stability without diffusion) *If conditions (22) are satisfied and*

$$\frac{a + 1}{qc} > \frac{2a}{qc - bq - p},$$

and

$$b + \frac{dp((1 - u^*)(u^* + a) + s)}{q((1 - u^*)(u^* + a) + d)^2} > \frac{cu^*}{u^* + a} \tag{25}$$

and

$$\frac{p^2((1 - u^*)(u^* + a) + s)^2}{q(u^* + a)} > b + \frac{dp((1 - u^*)(u^* + a) + s)}{q((1 - u^*)(u^* + a) + d)^2}.$$

Then, the homogeneous steady state $E^* = (u^*, v^*, w^*)$ *is locally asymptotically stable.*

The proofs of Propositions 3 and 4 require long and tedious (albeit simple) algebraic computations, they can be found in [2].

5 Pattern Formation and Turing Instability

Pattern formation is a process by which a spatially uniform state loses stability to a non-uniform state : a pattern.

Two basic types of symmetry-breaking bifurcations, which are responsible for the emergence of spatio-temporal patterns are:

- The space-independent Hopf bifurcation breaks the temporal symmetry of a system and gives rise to oscillations that are uniform in space and periodic in time.
- The (stationary) Turing bifurcation breaks spatial symmetry, leading to the formation of patterns that are stationary in time and oscillatory in space.
  In this section, we mainly focus on this last type of bifurcation.

5.1 *Turing Instability for Two Species Model*

In this section, in order to study the diffusion driven instability for system (10), we have to analyze the stability of the homogeneous steady state $E^* = (u^*, v^*)$ which corresponds to co-existence of prey and predator. The Jacobian evaluated at the equilibrium $E^* = (u^*, v^*)$ is

$$M = \begin{pmatrix} f_u & f_v \\ g_u & g_v \end{pmatrix} = \begin{pmatrix} \frac{\partial}{\partial u} f(u^*, v^*) & \frac{\partial}{\partial v} f(u^*, v^*) \\ \frac{\partial}{\partial u} g(u^*, v^*) & \frac{\partial}{\partial v} g(u^*, v^*) \end{pmatrix}$$

$$= \begin{pmatrix} \frac{(a_1 d_1 - k_1 b_1)u^* - 2b_1 d_1 u^{*2} - b_1 d_2 u^* v^*}{d_1 u^* + d_2 v^* + k_1} & -\frac{c_1 u^* (k_1 + d_1 u^*)}{(d_1 u^* + d_2 v^* + k_1)^2} \\ \frac{a_2^2}{c_2} & -a_2 \end{pmatrix}$$

By setting

$$S = \begin{pmatrix} u - u^* \\ v - v^* \end{pmatrix} \varphi(r, \theta) e^{\lambda t + ikr}$$

where $\phi(r, \theta)$ is a eigenfunction of the Laplacian operator on a disc domain with zero flux boundary, i.e.:

$$\begin{cases} \Delta_{r\theta} \phi = -k^2 \phi, \\ \phi_r(R, \theta) = 0 \end{cases}$$

k is the wave number and λ is the perturbation growth rate. Then by linearizing around (u^*, v^*), we have the following equation:

$$\frac{dS}{dt} = MS + D\Delta S \tag{26}$$

where

$$D = \begin{pmatrix} D_1 & 0 \\ 0 & D_2 \end{pmatrix}$$

by substituting S by $\phi e^{\lambda t}$ in Eq. (26) and canceling $e^{\lambda t}$, we get:

$$\lambda\phi = M - Dk^2\phi \tag{27}$$

We obtain the characteristic equation for the growth rate λ as determinant of

$$det(\lambda I_2 - M + k^2 D) = 0 \Leftrightarrow \begin{vmatrix} \lambda - f_u + D_1 k^2 & -f_v \\ -g_u & \lambda - g_v + D_2 k^2 \end{vmatrix} = 0, \tag{28}$$

By computation we have the expression of the characteristic equation $\Theta(k^2)$:

$$\Theta(k^2) = \lambda^2 + R(k^2)\lambda + B(k^2) \tag{29}$$

where

$$R(k^2) = k^2(D_1 + D_2) - tr(M) \tag{30}$$

and

$$B(k^2) = D_1 D_2 k^4 - (D_2 f_u + D_1 g_v)k^2 + \det(M). \tag{31}$$

Therefore, the eigenvalues are the roots of (29) are given by

$$\lambda_\pm(k) = \frac{-R(k^2) \pm \sqrt{(R(k^2))^2 - 4B(k^2)}}{2} \tag{32}$$

Let

$$\theta_{1,2} = \frac{-z_2 \pm \sqrt{z_2^2 - 4z_1 z_3}}{z_1^2}, \tag{33}$$

and

$$z_1 = 2b_1 d_1 c_2 + b_1 d_2 a_2,$$

$$z_2 = a_2^2 d_2 + a_2 d_1 c_2 + k_1 b_1 c_2 - a_1 d_1 c_2,$$

$$z_3 = a_2^2 d_2 k_2 + b_1 d_2 k_2 a_2 + k_1 a_2 c_2.$$

Proposition 5 *If $a_2^2 d_2 + a_2 d_1 c_2 + k_1 b_1 c_2 > a_1 d_1 c_2$ or $0 < u^* < \theta_1$ or $\theta_2 < u^*$, θ_1 and θ_2 are defined in Eq.* (33) *and if $D_2 < (D_2)_c$, then $E^* = (u^*, v^*)$ is asymptotically stable for system* (10). *If $D_2 > (D_2)_c$ then $E^* = (u^*, v^*)$ is unstable for system* (10), *where,*

$$(D_2)_c = \frac{-(2D_1 f_v g_u - D_1 f_u g_v)}{f_u^2}$$

$$+\frac{\sqrt{(2D_1 f_v g_u - D_1 f_u g_v)^2 - D_1^2 f_u^2 g_v^2}}{f_u^2}$$

Now, we study the conditions leading to Turing instability for the two-species model. These conditions are given by:

$$Tr(M) = f_u + g_v < 0 \tag{34}$$

$$\det(M) = f_u g_v - f_v g_u > 0 \tag{35}$$

$$D_2 f_u + D_1 g_v > 0 \tag{36}$$

$$(D_2 f_u + D_1 g_v)^2 - 4D_1 D_2 \det(A) > 0 \tag{37}$$

For a predator-prey model, the necessary condition to have the instability of Turing is that the predator spreads faster than the prey, namely $D_2 > D_1$. Turing instability corresponds to the onset of patterns periodic in space and stationary in time. Mathematically speaking, the case when $Im(\lambda(k)) = 0$ for $k = k_c$ is called Turing instability.

The conditions $R(k^2) > 0$ and $B(k^2) > 0$ are equivalent to the stability criterion $R(k^2 = 0) > 0$ and $B(k^2 = 0) > 0$ for the local dynamic. In particular this means that $R(k^2) > 0$ for all k, ($tr(M) < 0$ and $k^2(D_1 + D_2) > 0$, then $R(k^2) > 0$), therefore the only choice for $Re(\lambda(k)) > 0$ is $B(k^2) < 0$ for some $k \neq 0$. Thus the instability of the homogeneous solution can occur when $B(k^2)$ is zero for some k. It means that the instability occur at the point where the equation $B(k^2) = 0$ has a multiple root. We find that $B(k^2)$ is a quadratic polynomial with respect to k^2. Its extremum is a minimum at some k^2 [17].

$$B'(k^2) = 4D_1 D_2 k^3 - 2(D_2 f_{ull} + D_1 g_v)k = 0 \Longrightarrow k_{min}^2 = \frac{1}{2}\left(\frac{D_2 f_u + D_1 g_v}{D_1 D_2}\right). \tag{38}$$

Equation (29) is defined if

$$D_2 f_u + D_1 g_v > 0. \tag{39}$$

Then,

$$B_{min} = B(k^2_{min}) = det(M) - \frac{(D_2 f_u + D_1 g_v)^2}{4D_1 D_2}. \tag{40}$$

If $det(M) < \frac{(D_2 f_u + D_1 g_v)^2}{4D_1 D_2}$, then there exists $k^2 \neq 0$ such that $B(k^2) < 0$.

The bifurcation for which $B_{min} = 0$ that is $det(M) = \frac{(D_2 f_u + D_1 g_v)^2}{4D_1 D_2}$ occurs for a critical value $(D_2)_T$ of the diffusion coefficient D_2, which is a solution of the equation:

$$f_u^2 D_2^2 + 2(2D_1 f_v g_u - D_1 f_u g_v) D_2 + D_1^2 g_v^2 = 0 \tag{41}$$

Then the critical value k_c of the wave number k associated with the critical value $(D_2)_T$ is given by

$$k^2_{min} = \frac{1}{2}\left(\frac{(D_2)_T f_u - D_1 a_2}{D_1 (D_2)_T}\right)$$

and the wavelength w_T associated also with the critical value $(D_2)_T$ is given by

$$w_T = \frac{2\pi}{k_T} = 2\pi \sqrt{\frac{2D_1 (D_2)_T}{(D_2)_T f_u - D_1 a_2}}$$

Then, the resolution of Eq. (31) gives us the region of wavenumbers of unstable modes

$$k_1^2 = \frac{D_2 f_u + D_1 g_v - \sqrt{(D_2 f_u + D_1 g_v)^2 - 4D_1 D_2 \det(M)}}{2D_1 D_2}$$

$$k_2^2 = \frac{D_2 f_u + D_1 g_v + \sqrt{(D_2 f_u + D_1 g_v)^2 - 4D_1 D_2 \det(M)}}{2D_1 D_2}$$

5.2 *Turing Instability for Three Species Model*

Let us now analyze this symmetry breaking bifurcation for system (18). We know that Turing instability occurs from a finite number of wave vectors producing stable spatial patterns depending essentially on the initial condition. Let

$$W = \begin{pmatrix} u - u^* \\ \\ v - v^* \\ \\ w - w^* \end{pmatrix} \varphi(r, \theta) e^{\lambda t + ikr} \tag{42}$$

where k is the wave number and $\varphi(r, \theta)$ is an eigenfunction of the Laplacian operator on a disc domain with zero flux on the boundary, i.e.:

$$\begin{cases} \Delta_{r\theta}\varphi = -k^2\varphi, \\ \varphi_r(R,\theta) = 0 \end{cases}$$

Then, by linearizing around (u^*, v^*, w^*), we have the following equation:

$$\frac{dW}{dt} = D\Delta W + L_E(E^*)W. \tag{43}$$

where $E = (u, v, w)^T$ and

$$L(E) = \begin{pmatrix} f(u,v,w) \\ g(u,v,w) \\ h(u,v,w) \end{pmatrix} = \begin{pmatrix} (1-u-\frac{v}{u+a})u \\ (-b+\frac{cu}{u+a}-\frac{w}{v+d})v \\ (p-\frac{qw}{v+s})w \end{pmatrix}$$

Then, problem (20) can be written as: Consider now the system with diffusion (18) and let us substitute W by $\varphi e^{\lambda t}$ in Eq. (43) and canceling $e^{\lambda t}$, we get:

$$\lambda\varphi = L_E(E^*) - Dk^2\varphi. \tag{44}$$

We obtain the characteristic equation for the growth rate λ as the determinant of

$$det(\lambda I_3 - L_E(E^*) + K^2 D) = 0 \Longleftrightarrow$$

$$det\begin{pmatrix} \lambda - a_{11} + \delta_1 k^2 & -a_{12} & -a_{13} \\ -a_{21} & \lambda - a_{22} + \delta_2 k^2 & -a_{23} \\ -a_{31} & -a_{32} & \lambda - a_{33} + \delta_3 k^2 \end{pmatrix} = 0. \tag{45}$$

The characteristic polynomial from (45) is

$$H(k^2) = \lambda^3 + \Phi_1(k^2)\lambda^2 + \Phi_2(k^2)\lambda + \Phi_3(k^2) = 0, \tag{46}$$

with

$$\Phi_1(k^2) = k^2(\delta_1 + \delta_2 + \delta_3) + B_1,$$

$$\Phi_2(k^2) = k^4(\delta_1\delta_2 + \delta_1\delta_3 + \delta_2\delta_3)$$

$$-k^2(\delta_1(a_{22} + a_{33}) + \delta_2(a_{11} + a_{33}) + \delta_3(a_{11} + a_{22})) + B_2,$$

$$\Phi_3(k^2) = k^6\delta_1\delta_2\delta_3 - k^4(\delta_1\delta_2 a_{33} + \delta_1\delta_3 a_{22} + \delta_2\delta_3 a_{11})$$

$$+k^2(\delta_3(a_{11}a_{22} - a_{12}a_{21}) + \delta_2 a_{11}a_{33}) + B_3.$$

For the stability of the equilibrium point, according to the Routh–Hurwitz criteria, $Re(\lambda) < 0$ if

$$\Phi_1(k^2) > 0, \tag{47}$$

$$\Phi_2(k^2) > 0, \tag{48}$$

$$\Phi_1(k^2)\Phi_2(k^2) - \Phi_3(k^2) > 0. \tag{49}$$

The Turing instability requires that the stable homogeneous steady state becomes unstable due to the interaction and diffusion of species.

Under the conditions of Turing:

$$Re(\lambda(k^2 = 0)) < 0, \;\; Re(\lambda(k^2 > 0)) > 0, \;\; \text{for a} \;\;\; k^2 > 0 \tag{50}$$

We have the following Theorem.

Proposition 6 *If one of the following conditions holds:*

$$\Phi_1(k^2) < 0,$$

$$\Phi_2(k^2) < 0,$$

$$\Phi_1(k^2)\Phi_2(k^2) - \Phi_3(k^2) < 0$$

then, the homogeneous steady state $E^* = (u^*, v^*, w^*)$ of system (18) drives instability.

Proof For $k^2 \neq 0$ we have $\Phi_1(k^2) = -(a_{11} + a_{22} + a_{33}) + k^2(\delta_1 + \delta_2 + \delta_3)$. If $a_{11} + a_{22} + a_{33} < 0$, then $\Phi_1(k^2) > 0$ and instability of Turing does not occur. Thereafter, we suppose in Eq. (48) $\rho = k^2 > 0$, to get:

$$\Phi_2(\rho) = \rho^2 p_1 - \rho p_2 + p_3, \tag{51}$$

where

$$p_1 = \delta_1\delta_2 + \delta_1\delta_3 + \delta_2\delta_3,$$

$$p_2 = \delta_1 a_{22} + \delta_1 a_{33} + \delta_2 a_{11} + \delta_2 a_{33} + \delta_3 a_{11} + \delta_3 a_{22},$$

$$p_3 = a_{11}a_{22} + a_{11}a_{33} + a_{22}a_{33} - a_{12}a_{11} - a_{23}a_{23},$$

a necessary condition for $E^* = (u^*, v^*, w^*)$ of (18) becomes unstable is that

$$\Phi_2(\rho) = \rho^2 p_1 - \rho p_2 + p_3 < 0. \tag{52}$$

For the instability, we need that $p_2 > 0$ and $p_2^2 - 4p_1p_3 > 0$ for some ρ. The equation $p_1\rho^2 - p_2\rho + p_3$ has two positive roots given by:

$$\rho_1 = \frac{p_2 - \sqrt{p_2^2 - 4p_1p_3}}{2p_1} \quad \text{and} \quad \rho_2 = \frac{p_2 - \sqrt{p_2^2 + 4p_1p_3}}{2p_1}. \tag{53}$$

The constant positive steady state $E^* = (u^*, v^*, w^*)$ of (18) is unstable and so (18) experiences Turing instability provided that $\rho_1 < \rho < \rho_2$.

The expressions $\Phi_3(k^2)$ and $\Phi_1(k^2)\Phi_2(k^2) - \Phi_3(k^2)$ are a cubic function of k^2 of the form

$$\Phi_3(k^2) = q_1(k^2)^3 + q_2(k^2)^2 + q_3k^2 + q_4, \tag{54}$$

$$q_1 = \delta_1\delta_2\delta_3,$$

$$q_2 = -(\delta_1\delta_2a_{33} + \delta_1\delta_3a_{22} + \delta_2\delta_3a_{11}),$$

$$q_3 = \delta_1a_{22}h_w + \delta_2a_{11}a_{33} + \delta_3a_{11}a_{22} - \delta_1a_{23}a_{32} - \delta_3a_{22}a_{21}$$

$$= \delta_1(a_{22}a_{33} - a_{23}a_{32}) + \delta_2a_{11}a_{33} + \delta_3(a_{11}a_{22} - a_{12}a_{21}),$$

$$q_4 = \Phi_3(0) = a_{12}a_{21}a_{33} + a_{11}a_{23}a_{32} - a_{11}a_{22}a_{33},$$

with $q_1 = det(D) \geq 0$ and $q_4 = -det(L_E(E^*)) > 0$.

If Φ_3 has a minimum, one finds by simple computation that

$$\frac{d\Phi_3}{d(k^2)} = 3q_1(k^2)^2 + 2q_2(k^2) + q_3 = 0 \tag{55}$$

and $\frac{d^2\Phi_3}{d^2(k^2)} > 0$, this minimum is reached for the solution of (55) at

$$k_{inf}^2 = \frac{-q_2 + \sqrt{q_2^2 - 3q_1q_3}}{3q_1}. \tag{56}$$

If $a_{11} > 0, a_{22} > 0$ and $a_{33} > 0$ then $q_2 < 0$.

If $a_{22}a_{33} < a_{23}a_{32}$, $a_{11}a_{33} < 0$, $a_{11}a_{22} < a_{12}a_{21}$ or $a_{22}a_{33} < 0$, $a_{11}a_{33} < 0$ and $a_{11}a_{22} < 0$ then, $q_3 < 0$.

To verify condition (49) let us denote

$$\Psi(k^2) = \Phi_1(k^2)\Phi_2(k^2) - \Phi_3(k^2) = r_1(k^2)^3 + r_2(k^2)^2 + r_3k^2 + r_4, \tag{57}$$

where

$$
\begin{aligned}
r_1 &= 2\delta_1\delta_2\delta_3 + \delta_1^2\delta_3 + \delta_1^2\delta_2 + \delta_1\delta_2^2 + \delta_1\delta_3^2 + \delta_3\delta_2^2 + \delta_2\delta_3^2 \\
&= (\delta_2 + \delta_3)(\delta_1^2 + \delta_2\delta_3 + \delta_1\delta_2 + \delta_1\delta_3),
\end{aligned}
$$

$$
\begin{aligned}
r_2 = &-(\delta_1^2 a_{22} + \delta_1^2 a_{33} + \delta_2^2 a_{11} + \delta_2^2 a_{33} + \delta_3^2 a_{11} + \delta_3^2 a_{22} + 2\delta_1\delta_2 a_{11} + 2\delta_1\delta_2 a_{33} \\
&+2\delta_1\delta_3 a_{11} + 2\delta_1\delta_3 a_{22} + 2\delta_1\delta_2 a_{22} + 2\delta_1\delta_3 a_{33}, +2\delta_2\delta_3 a_{11} + 2\delta_2\delta_3 a_{22} + 2\delta_2\delta_3 a_{33}) \\
= &-a_{11}(\delta_3 + \delta_2)(2\delta_1 + \delta_2 + \delta_3) - a_{22}(\delta_3 + \delta_1)(\delta_1 + 2\delta_2 + \delta_3) \\
&-a_{33}(\delta_1 + \delta_2)(\delta_1 + \delta_2 + 2\delta_3),
\end{aligned}
$$

$$
\begin{aligned}
r_3 = &\ \delta_1 a_{22}^2 + \delta_1 a_{33}^2 + \delta_2 a_{11}^2 + \delta_2 a_{33}^2 + \delta_3 a_{11}^2 + \delta_3 a_{22}^2 + 2\delta_1 a_{11} a_{22} \\
&+2\delta_1 a_{11} a_{33} + 2\delta_1 a_{22} a_{33} - \delta_1 f_v g_u - \delta_1 f_w h_u + 2\delta_2 f_u g_v \\
&+2\delta_2 a_{11} a_{33} + 2\delta_2 a_{22} a_{33} - \delta_2 a_{12} a_{21} - \delta_2 a_{23} a_{32} + 2\delta_3 a_{11} a_{22} \\
&+2\delta_3 a_{11} a_{33} + 2\delta_1 a_{22} a_{33} - \delta_3 a_{23} a_{32} \\
= &\ \delta_1 a_{22}^2 + \delta_1 a_{33}^2 + \delta_2 a_{11}^2 + \delta_2 a_{33}^2 + \delta_3 a_{11}^2 + \delta_3 a_{22}^2 + 2(\delta_1 + \delta_2 \\
&+\delta_3)(a_{11} a_{22} + a_{11} a_{33} + 2a_{33} a_{22}) - \delta_1 a_{12} a_{21} \\
&-\delta_2(a_{12} a_{21} + a_{23} a_{32}) - \delta_3 a_{23} a_{32},
\end{aligned}
$$

$$
\begin{aligned}
r_4 &= \Psi(0) \\
&= -(a_{11}^2 a_{22} + a_{11}^2 a_{33} + 2a_{11} a_{22} a_{33} + a_{11} a_{33}^2 + a_{11} a_{22}^2 \\
&\quad +a_{22}^2 a_{33} + a_{22} a_{33}^2) + a_{12} a_{21} a_{22} + a_{22} a_{23} a_{32}.
\end{aligned}
$$

$r_4 > 0$ if

$$
a_{11}^2 a_{22} + a_{11}^2 a_{33} + 2a_{11} a_{22} a_{33} + a_{11} a_{33}^2 + a_{11} a_{22}^2
$$

$$
+a_{22}^2 a_{33} + a_{22} a_{33}^2 < a_{12} a_{12} a_{22} + a_{22} a_{23} a_{32}.
$$

If Ψ has a minimum, by simple algebraic computation we get

$$
\frac{d\Psi}{d(k^2)} = 3r_1(k^2)^2 + 2r_2(k^2) + r_3 = 0 \tag{58}
$$

and $\frac{d^2\Psi}{d^2(k^2)} > 0$, this minimum is reached for the solution of (58) at

$$
k_{1inf}^2 = k_{inf}^2 = \frac{-r_2 + \sqrt{r_2^2 - 3r_1 r_3}}{3r_1} \tag{59}
$$

$r_2 < 0$ if $a_{11} > 0$, $a_{22} > 0$ and $a_{33} > 0$.

$r_3 < 0$ if $a_{12}a_{21} > 0$, $(a_{12}a_{21} + a_{23}a_{32}) > 0$, $a_{23}a_{32}) > 0$ and $\delta_1 a_{22}^2 + \delta_1 a_{33}^2 + \delta_2 a_{11}^2 + \delta_2 a_{33}^2 + \delta_3 a_{11}^2 + \delta_3 a_{22}^2 + 2(\delta_1 + \delta_2 + \delta_3)(a_{11}a_{22} + a_{11}a_{33} + 2a_{33}a_{22}) < \delta_1 a_{12}a_{21} + \delta_2(a_{12}a_{21} + a_{23}a_{32}) + \delta_3 a_{23}a_{32}$.

By using the conditions for the existence of the homogeneous steady state of the system without diffusion to be stable ($\Phi_1(0) > 0$, $\Phi_2(0) > 0$, $\Phi_3(0) > 0\Phi_1(0)$ $\Phi_2(0) - \Phi_3(0) > 0$) and the necessary condition for the homogeneous steady state of the system with diffusion to be instable that is to say, at least one of the following conditions, ($\Phi_1(k^2) < 0$, $\Phi_2(k^2) < 0$, $\Phi_3(k^2) < 0$, $\Phi_1(k^2)\Phi_2(k^2) - \Phi_3(k^2) < 0$) is satisfied for a certain $k^2 \neq 0$, we can prove the following proposition which gives a necessary condition (not sufficient) for the instability for the homogeneous steady state of the reaction-diffusion system with three species.

Let

$$\Phi_3(k_{inf}^2) = \frac{2q_2^3 - 9q_1q_2q_3 + 27q_1^2q_4 - 2(q_2^2 - 3q_1q_3)^{\frac{3}{2}}}{27q_1^3}$$

$$\Psi(k_{1inf}^2) = \frac{2r_2^3 - 9r_1r_2r_3 + 27r_1^2r_4 - 2(r_2^2 - 3r_1r_3)^{\frac{3}{2}}}{27r_1^3}$$

Therefore, in the following assumptions:
$(H_0) : q_2 < 0$
$(H_1) : q_3 < 0$
$(H_2) : q_2^2 - 3q_1q_3 > 0$
$(H_3) : r_2 < 0, r_3 < 0$ and $q_2^2 - 3q_1q_3 > 0$
$(H_4) : r_2^2 - 3r_1r_3 > 0$
$(H_5) : 2q_2^3 - 9q_1q_2q_3 + 27q_1^2q_4 - 2(q_2^2 - 3q_1q_3)^{\frac{3}{2}} < 0$
$(H_6) : 2r_2^3 - 9r_1r_2r_3 + 27r_1^2r_4 - 2(r_2^2 - 3r_1r_3)^{\frac{3}{2}} < 0$
and using

Lemma 1 *(i)- If (H_0) or (H_1) and (H_2) are verified, then k_{inf}^2 is a positive real.*
(ii)- If (H_0), (H_2) and (H_3) (Resp (H_4)) are verified, then k_{inf}^2 is a positive real (Resp k_{1inf}^2 is a positive real).
(iii)- If (H_5) (Resp (H_6)), then $\Phi_3(k_{inf}^2) < 0$ (Resp $\Psi(k_{1inf}^2) < 0$).

we can easily prove the final result:

Proposition 7 *Suppose*
1—[(H_0) or (H_1) and (H_2)] or [(H_0), (H_2) and (H_3)] or [(H_0), (H_2) and (H_4)].
2—(H_5) or (H_6).
If conditions 1 and 2 are satisfied, then we have emergence of Turing instability for system (18).

5.3 Numerical Simulations

In this subsection, we perform numerical simulations to illustrate the theoretical results given in the previous sections. In Figs. 1 and 2, Patterns formation are shown for systems (10) and (18).

Initial conditions for system (10) have been chosen as

$$u(0, r, \theta) = u^*((rcos\theta)^2 + (rsin\theta)^2) < 400 \tag{60}$$

$$v(0, r, \theta) = v^*((rcos\theta)^2 + (rsin\theta)^2) < 400 \tag{61}$$

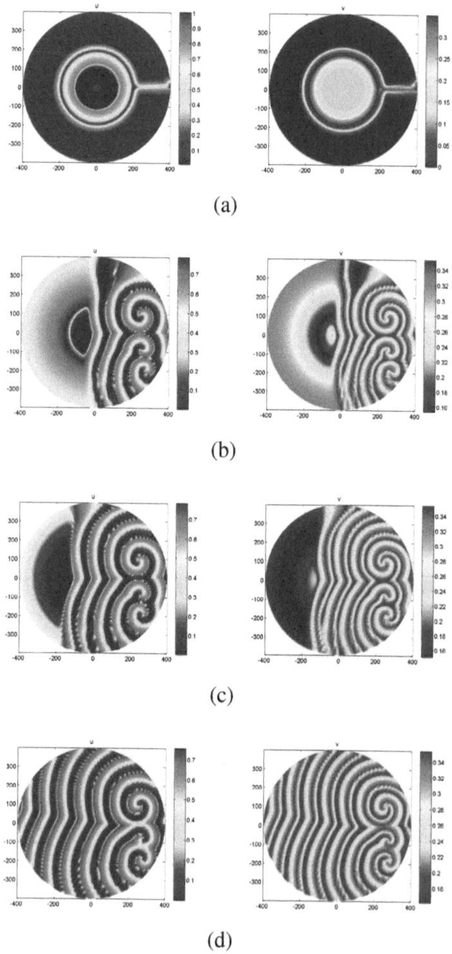

Fig. 1 Spatial distribution of species for system (10) with $D_1 = D_2 = 1, a_1 = 1, a_2 = 0.02,\ b_1 = 1,\ k_1 = 0.2,\ k_2 = 0.1,\ d_1 = 0.9,\ d_2 = 0.1,\ c_1 = 1.1, c_2 = 0.02$ and time varying **a** for $t = 100$, **b** for $t = 2800$, **c** for $t = 3500$, **d** for $t = 6000$. The left figures are spatial evolutions of the prey and the right are for predator

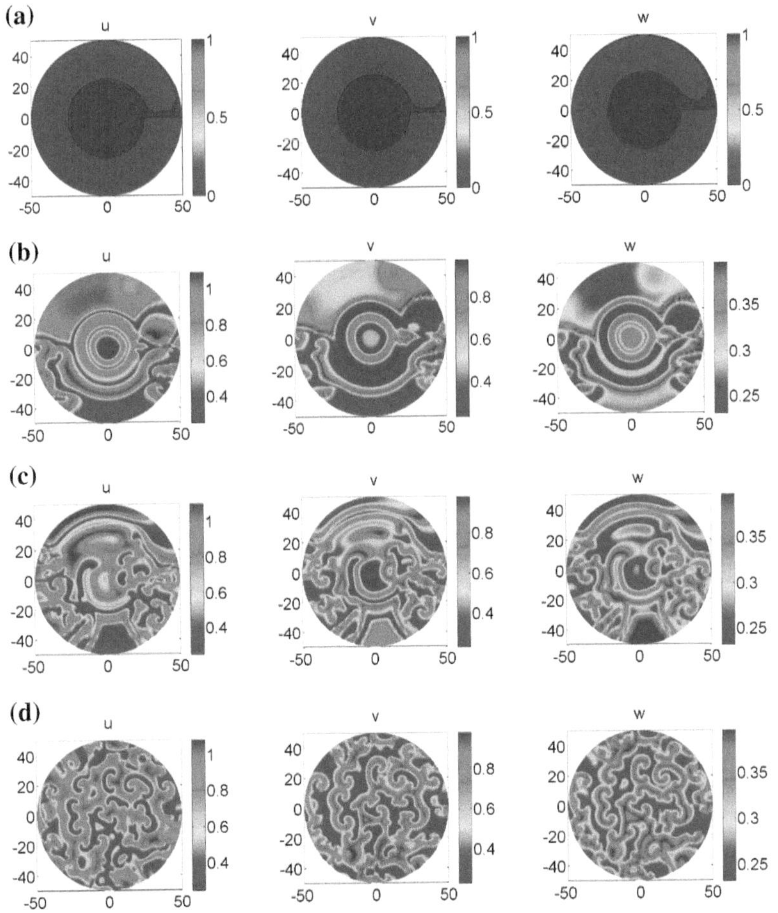

Fig. 2 Spatial distribution of prey (first column), predator (second column) and top predator (third column) for system (18). Spatial patterns are obtained with diffusivity coefficients $\delta_1 = 0.02$, $\delta_2 = 0.01$ and $\delta_3 = 0.05$, $a_0 = 0.5$, $a_1 = 0.4$, $b_0 = 0.36$, $c_3 = 0.2$, $d_0 = 0.3$, $d_2 = 0.4$, $d_3 = 0.4$, $v_0 = 0.4$, $v_1 = 0.8$, $v_2 = 0.4$, $v_3 = 0.6$ at different time levels: for t = 0 (a), t = 1000 (b), t = 2000 (c), t = 20000 (d)

Initial conditions for system (18) have been chosen as,

$$u(0, r, \theta) = u^*((rcos\theta)^2 + (rsin\theta)^2) < 50,$$

$$v(0, r, \theta) = v^*((rcos\theta)^2 + (rsin\theta)^2) < 50,$$

$$w(0, r, \theta) = w^*((rcos\theta)^2 + (rsin\theta)^2) < 50.$$

References

1. Abid, W., Yafia, R., Aziz-Alaoui, M.A., Bouhafa, H., Abichou, A.: Diffusion Driven Instability and Hopf Bifurcation in Spatial Predator-Prey Model on a Circular Domain, Evolution Equations and Control Theory, Volume 4, Number 2, pp. 115–129 (2015)
2. Abid, W., Yafia, R., Aziz-Alaoui, M.A., Bouhafa, H., Abichou, A.: Instability and pattern formation in three-species food chain model via holling type II functional response on a circular domain. Int. J. Bifurc. Chaos **25**, 1550092 (2015). doi:10.1142/S0218127415500923 (25 pages)
3. Araujo, S.B.L., De Aguiar, M.A.M.: Pattern formatiom, outbreaks, and synchronization in food chain with two and three species. Phys. Rev. E **75**, 14 (2007) Article ID 061908
4. Aziz-Alaoui, M.A.: Study of a Leslie-Gower-type tritrophic population model. Sol. Fractals **14**, 1275–1293 (2002)
5. Aziz Alaoui, M.A., Daher Okiye, M.: Boundedness and global stability for a predator-prey model with modified Leslie-Gower and Holling type II shemes. Appl. Math. Lett. **16**, 1069–1075 (2003)
6. Beddington, J.R.: Mutual interference between parasites or predators and its effect on searching efficiency. J. Anim. Ecol. **44**, 331–340 (1975)
7. Camara, B.I., Aziz-Alaoui, M.A.: Complexity in a prey predator model. ARIMA **9**, 109–122 (2008)
8. Camara, B.I., Aziz-Alaoui, M.A.: Dynamics of a predator-prey model with diffusion. Dyn. Contin. Discret. Impul. Syst. Ser. A : Math. Anal. **15**, 897–906 (2008)
9. Camara, B.I., Aziz-Alaoui, M.A.: Turing and Hopf patterns formation in a predator-prey model with Leslie-Gower-type functional response. Dyn. Contin. Discret. Impuls. Syst. Ser. B **16**(4), 479–488 (2009)
10. Chen, F.: On a nonlinear non-autonomous predator-prey model with diffusion and distributed delay. J. Comput. Appl. Math. **180**, 33–49 (2005)
11. Daher Okiye, M., Aziz Alaoui, M.A.: On the dynamics of a predator-prey model with the Holling-Tanner functional, Editor V. Capasso. In: Proceedings of the ESMTB Conference, pp. 270–278 (2002)
12. DeAngelis, D.L., Goldstein, R.A., O'Neill, R.V.: A model for trophic interaction. Ecology **56**, 881–892 (1975)
13. Fisher, R.A.: The advance of advantageous genes. Ann. Eugen. **7**, 335–369 (1937)
14. Hsu, S.B.: Constructing Lyapunov functions for mathematical models in population biology. Taiwan. J. Math. **9**(2), 151–173 (2005)
15. Kolmogorov, A.N., Petrovsky, I.G., Piskunov, N.S.: Etude de l'équation de la diffusion avec croissance de la quantité de matiére et son application á un probléme biologique, Bulletin Université d'Etat á Moscou (Bjul. Moskowskogo Gos. Univ.), Serie internationale A 1, pp. 1–26 (1937)
16. Lotka, A.J.: Elements of Physical Biology. Williams and Wilkins, Baltimore (1925)
17. Murray, J.D.: Mathematical Biology: II. Spatial Models and Biomedical Applications. Springer, Berlin (2003)
18. Nindjin, A.F., Aziz-Alaoui, M.A., Cadivel, M.: Analysis of a predator-prey model with modified Leslie-Gower and Holling-Type II schemes with time delay. Nonlinear Anal. Real World Appl. **7**(5), 1104–1118 (2006)
19. Nindjin, A.F., Aziz-Alaoui, M.A.: Persistence and global stability in a delayed Leslie-Gower type three species food chain. J. Math. Anal. Appl. **340**(1), 340–357 (2008)
20. Volterra, V.: Variations and fluctuations of the number of individuals in animal species living together. In: Chapman, R.N. (ed.) Animal Ecology, pp. 409–448. McGraw Hill, New York (1931)
21. Yafia, R., El Adnani, F., Talibi, H.: Alaoui, Stability of limit cycle in a predator-prey model with modified Leslie-Gower and Holling-type II schemes with time delay. Appl. Math. Sci. **1**, 119–131 (2007)

22. Yafia, R., El Adnani, F., Talibi Alaoui, H.: Limit cycle and numerical similations for small and large delays in a predator-prey model with modified Leslie-Gower and Holling-type II schemes. Nonlinear Anal. Real World Appl. **9**, 2055–2067 (2008)
23. Yafia, R., Aziz, M.A.: Alaoui. Existence of periodic travelling waves solutions in predator prey model with diffusion. Appl. Math. Model. **37**(6), 3635–3644 (2013)
24. Yu, S.B.: Global asymptotic stability of a predator-prey model with modified Leslie-Gower and Holling-type II schemes. Discret. Dyn. Nat. Soc. Article ID 208167 (2012)

Pattern Formation in a Prey-Predator Model with Nonlocal Interaction Terms

Malay Banerjee, Moitri Sen and Vitaly Volpert

Abstract We study a spatio-temporal prey-predator model with nonlocal interaction terms. Nonlocal interactions are considered for prey and predator species to describe the nonlocal intra-specific competition for limited resources. We show that the region of pattern formation increases with the increase of the range of nonlocal interaction. Numerical continuation technique is used to determine the existence of multiple stationary patterns.

Keywords Nonlocal interaction · Pattern formation · Stationary pattern

2010 Mathematics Subject Classification 92Bxx · 35B36

1 Introduction

Spatio-temporal pattern formation in population dynamics continues to attract much attention. Stationary pattern formation for the interacting populations was first studied by Levin and Segel [19] in the case of a planktonic system on the basis of the seminal work of Turing [24] on chemical morphogenesis. Importance of spatio-temporal models to describe the stabilization and long term existence of certain species was first indicated by Gause [12]. Effect of the distribution of species over their habitat in terms of long time stable coexistence of the interacting population was first studied by

---

M. Banerjee (✉)
Department of Mathematics and Statistics, IIT Kanpur, Kanpur, India
e-mail: malayb@iitk.ac.in

M. Sen
SERC, IISc Bangalore, Bangalore, India
e-mail: moitri300784@gmail.com

V. Volpert
Institut Camille Jordan, UMR 5208 CNRS, University Lyon 1,
69622 Villeurbanne, France
e-mail: volpert@math.univ-lyon1.fr

J.M. Cushing et al. (eds.), *Applied Analysis in Biological and Physical Sciences*,
Springer Proceedings in Mathematics & Statistics 186,
DOI 10.1007/978-81-322-3640-5_2

Luckinbill [16, 17]. There are very few experimental evidences on the spatial pattern formation for the interacting populations since it is quite difficult to get data on the long timescale. Some recent literature provides an observation of vegetation pattern formation and planktonic blooms [10, 14, 22]. However, appropriate evidences for stationary, periodic and aperiodic pattern formation in the interacting populations are still lacking.

Mathematical analysis and relevant numerical simulations of spatio-temporal models can explain stationary and oscillatory patterns of population distribution over their habitats. Classical spatio-temporal models of interacting populations are constructed under the assumption that reproduction, predation and competition are based on the local interaction of individuals. In this context, the reaction term describes the growth rate of the population and the diffusion term takes care of random movement of the individuals within their habitats. However a long range interaction of the individuals in the population is not taken into account. Spatio-temporal models with self-diffusion terms also ignore the influence of the population density of one species towards the rate of diffusion of another one. In reality displacement of predators is mostly influenced by the prey location [20]. As a result, generation of spatial patterns depends mostly on the reaction kinetics. This limitation manifests itself in the absence of Turing pattern in the Gause type prey-predator models with prey-dependent functional response and linear death rate of predator population although, such models can produce non-Turing patterns [21]. It is important to note that the non-Turing patterns indicate a continuous change in the population density which does not converge to any stationary state/distribution.

Conventional spatio-temporal models in population dynamics for a single species population growth do not generate spatial patterns. Genieys et al. [13] established that a spatio-temporal model of single species population growth can induce such patterns in the presence of nonlocal consumption of resources. It describes intra-specific competition and it is appropriate for modelling of the emergence of biological species. Various models with nonlocal and global consumption of resources are developed in order to describe Darwin's theory of evolution [11].

The main objective of the present work is to study a spatio-temporal model with nonlocal consumption of resources in the context of predator-prey interaction with Michaelis–Menten type ratio-dependent functional response [3] and to describe the consumption of prey by their specialist predator. Nonlocal interaction terms are introduced for both the prey and predator population with the assumption that the intra-specific competition takes place between the individuals located at two different locations. In Sect. 2 the formulation of basic model is presented. Linear stability results for the nonlocal model are obtained in Sect. 3 and some numerical simulation results are given in Sect. 4. The main results of this work are summarized in Sect. 5 along with a brief discussion on the possible future developments of this study.

2 Basic Model

Spatio-temporal models of prey-predator interaction can produce stationary as well as non-stationary patterns. The concerned models are constructed by adding self-diffusion terms and/or cross-diffusion terms to the temporal models [18, 20]. Local and global bifurcations for the temporal model have significant impact on the resulting spatio-temporal pattern formation [5–7]. Replacing the local interaction terms involved with the reaction kinetics by nonlocal interactions [25], we modify complex nonlinear dynamics and pattern formation in the models of interacting populations.

2.1 *Temporal Model*

In this work we consider the Michaelis–Menten type ratio-dependent model for prey-predator interaction and governed by the following coupled nonlinear ordinary differential equations,

$$\frac{du}{dt} = u(1-u) - \frac{\alpha uv}{u+v} \equiv f(u,v), \tag{1}$$

$$\frac{dv}{dt} = \frac{\beta uv}{u+v} - \gamma v - \delta v^2 \equiv g(u,v), \tag{2}$$

and subjected to non-negative initial conditions $u(0) \geq 0,\ v(0) \geq 0$. Here u and v stand for the population densities of prey and predator species respectively. All the parameters involved in the model (1)–(2) are dimensionless positive numbers. Parameters α, β, γ and δ can be interpreted as dimensionless and rescaled grazing rate of prey by their specialist predators, growth rate of predators, death rate of predators and intra-specific competition rate among the predators respectively (see [5] for details). This model can be considered as an extension of the classical Bazykin type prey-predator model [9] with ratio-dependent functional response. Above model exhibits a wide variety of dynamic behavior and undergoes several types of local and global bifurcations which are related to the stable/oscillatory coexistence of the prey and the predator as well as to the extinction of both populations [5]. The model (1)–(2) admits at most two interior equilibrium points $E_j^*(u_j^*, v_j^*)$ with the coordinates

$$u_j^* = \frac{\alpha(\gamma - 2\beta + \delta) + 2\beta + (-1)^j\sqrt{\alpha^2(\gamma-\delta)^2 + 4\alpha^2\beta\delta + 4\alpha^3\delta(\gamma-\beta)}}{2}$$

$$v_j^* = \frac{u_j^*(1-u_j^*)}{u_j^* + \alpha - 1},\quad j = 1, 2.$$

The first components of E_j^* are the positive real roots of the quadratic equation

$$(\alpha\delta + \beta)u^2 + (2\alpha\beta - 2\beta - \alpha\delta - \alpha\gamma)u + \alpha^2\beta + \alpha\gamma + \beta - \alpha^2\gamma - 2\alpha\beta = 0.$$

Here we assume that the parametric restriction $\alpha^2\beta + \alpha\gamma + \beta - \alpha^2\gamma - 2\alpha\beta < 0$ is satisfied, and hence we get only one feasible coexisting equilibrium point E_*^2.

The temporal model (1)–(2) exhibits a wide variety of local (transcritical, saddle-node, Hopf, Bautin) and global bifurcations (homoclinic, saddle-node bifurcation of limt cycles) (see [5] for further details). Here we consider β, γ, δ as fixed parameters and α as bifurcation parameter. We choose α as a bifurcation parameter since it determines the strength of coupling between two equations. Throughout this work the values of three parameters are fixed, $\beta = 1$, $\gamma = 0.6$ and $\delta = 0.1$. With this choice of parameter values, we find only one feasible interior equilibrium point E_2^* whenever $\alpha < 2.5$. The conditions for local asymptotic stability of E_2^* are given by [18] the inequalities

$$a_{11} + a_{22} < 0, \quad a_{11}a_{22} - a_{12}a_{21} > 0, \tag{3}$$

where

$$a_{11} = \left.\frac{\partial f}{\partial u}\right|_{E_2^*}, \quad a_{12} = \left.\frac{\partial f}{\partial v}\right|_{E_2^*}, \quad a_{21} = \left.\frac{\partial g}{\partial u}\right|_{E_2^*}, \quad a_{22} = \left.\frac{\partial g}{\partial u}\right|_{E_2^*}. \tag{4}$$

The coexisting equilibrium point E_2^* is stable for $\alpha < 2.01 \equiv \alpha_H$ and it is unstable for $\alpha > \alpha_H$. It loses stability through a supercritical Hopf bifurcation. The stable limit cycle generated through the Hopf bifurcation disappears due to a homoclinic bifurcation [5]. Conditions for the Hopf bifurcation are as follows [15]:

$$[a_{11} + a_{22}]_{\alpha=\alpha_H} = 0, \; [a_{11}a_{22} - a_{12}a_{21}]_{\alpha=\alpha_H} > 0, \; \left.\frac{d}{d\alpha}(a_{11} + a_{22})\right|_{\alpha=\alpha_H} \neq 0. \tag{5}$$

2.2 Spatio-Temporal Model

Incorporating the self-diffusion terms into the temporal model described above, we get the spatio-temporal model of predator-prey interaction:

$$\frac{\partial u}{\partial t} = u(1-u) - \frac{\alpha uv}{u+v} + \frac{\partial^2 u}{\partial x^2}, \tag{6}$$

$$\frac{\partial v}{\partial t} = \frac{\beta uv}{u+v} - \gamma v - \delta v^2 + d\frac{\partial^2 u}{\partial x^2}, \tag{7}$$

for $(t, x) \in \mathbb{R}_+ \times \Omega$, where $\Omega \subset \mathbb{R}$ and subjected to the initial conditions

$$u(0, x) = u_0(x) \geq 0, \quad v(0, x) = v_0(x) \geq 0, \quad x \in \Omega, \tag{8}$$

and periodic boundary conditions. The positive parameter d stands for the ratio of diffusivity of the prey and the predator species.

The components of the interior equilibrium point E_2^* correspond to homogeneous steady-state for the system (6)–(7). Instability of the homogeneous steady-state $u(t, x) = u_2^*$, $v(t, x) = v_2^*$, whenever the condition (3) is satisfied, is known as Turing instability [18, 20]. Turing instability conditions are given by

$$a_{11} + a_{22} < 0, \; a_{11}a_{22} - a_{12}a_{21} > 0, \; da_{11} + a_{22} > 2\sqrt{d(a_{11}a_{22} - a_{12}a_{21})}, \tag{9}$$

and the equation of the Turing bifurcation curve is

$$da_{11} + a_{22} = 2\sqrt{d(a_{11}a_{22} - a_{12}a_{21})}. \tag{10}$$

For the chosen parameter values, as mentioned in the Sect. 2.2, the spatio-temporal model (6)–(7) exhibits stationary as well as chaotic pattern for different values of α and d. Thorough numerical simulations reveal stationary and chaotic patterns within and outside the Turing domain. These results are summarized in Fig. 1. For numerical simulations, we have considered a small perturbation of the homogeneous steady-state within a narrow interval around the middle point of the spatial domain $[0, L]$. Throughout this work we will consider only this type of initial conditions. We also

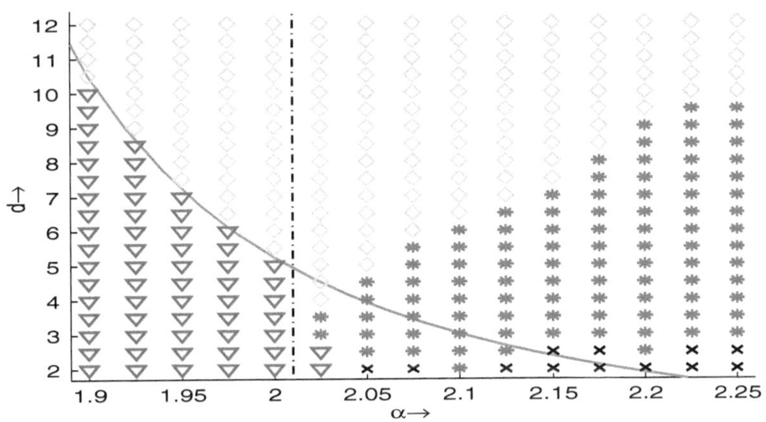

Fig. 1 Stationary and chaotic patterns are observed from numerical simulation of the model (6)–(7) with $\beta = 1$, $\gamma = 0.6$, $\delta = 0.1$ and different values of α and d. Two curves correspond to Turing bifurcation curve (*magenta*) and Hopf-bifurcation curve (*black*) respectively. The symbols represent: $\nabla \to$ homogeneous stead-state, $\diamond \to$ stationary pattern, $* \to$ spatio-temporal chaos, **x** $\to$ extinction

note that the results are independent of the choice of Δt and Δx in the considered range providing a good accuracy of the simulations.

2.3 Spatio-Temporal Model with Nonlocal Interaction Terms

The spatio-temporal model of prey-predator interaction with nonlocal intra-species competition terms are governed by the following two coupled integro-differential equations:

$$\frac{\partial u(t,x)}{\partial t} = u(t,x)(1 - w_1(t,x)) - \frac{\alpha u(t,x)v(t,x)}{u(t,x)+v(t,x)} + \frac{\partial^2 u(t,x)}{\partial x^2}, \quad (11)$$

$$\frac{\partial v(t,x)}{\partial t} = \frac{\beta u(t,x)v(t,x)}{u(t,x)+v(t,x)} - (\gamma + \delta w_2(t,x))v(t,x) + d\frac{\partial^2 v(t,x)}{\partial x^2}, \quad (12)$$

where

$$w_1(t,x) = \int_{-\infty}^{\infty} \phi_1(x-y)u(t,y)dy, \;\; w_2(t,x) = \int_{-\infty}^{\infty} \phi_2(x-y)v(t,y)dy.$$

The above model is subjected to the positive initial condition and periodic boundary condition over the same domain as described in the Sect. 2.2. To be specific, here we consider $x \in [0, L]$.

The model with nonlocal interaction term implies that the intra-specific competition is not limited to the individuals located at the same spatial point x only. Rather the prey located at space point x will compete for resources with other preys located in some area around this space point. Hence the intra-species competition rate among the prey is determined by the integral $\int_{-\infty}^{\infty} \phi_1(x-y)u(t,y)dy$ and the function $\phi_1(x-y)$ measures the efficiency of the competition of prey located at the space point y with the prey located at the space point x. The function ϕ_1 depends on the difference $x - y$ since it is assumed that the efficiency of competition depends on the distance between the individuals. Hence the intra-specific competition among the prey is described by the expression $u(t,x)\int_{-\infty}^{\infty} \phi_1(x-y)u(t,y)dy$, and a similar assumption is true for the intra-specific competition among the predators.

The interaction kernels ϕ_1 and ϕ_2 are supposed to have bounded supports and satisfy the conditions $\int_{-\infty}^{\infty} \phi_1(z)dz = \int_{-\infty}^{\infty} \phi_2(z)dz = 1$. In this work we consider them as piece-wise constant functions:

$$\phi_1(z) = \phi_2(z) \equiv \phi(z) = \begin{cases} \frac{1}{2M}, & |z| < M \\ 0, & |z| \geq M. \end{cases} \quad (13)$$

3 Linear Stability Analysis of Nonlocal Model

In this work we are interested to study Turing pattern formation for the nonlocal model (11)–(12) around the unique coexisting homogeneous steady state only. Accordingly, we carry out linear stability analysis of the homogeneous steady state (u_2^*, v_2^*). Linearizing the system (11)–(12) around the homogeneous steady state $u(t,x) = u_2^*$, $v(t,x) = v_2^*$, we get the following two linear equations:

$$\lambda u_1(x) = (a_{11} + u_2^*)u_1(x) - u_2^* \int_{-\infty}^{\infty} \phi(x-y)u_1(y)dy + a_{12}v_1(x) + u_1''(x),$$

$$\lambda v_1(x) = a_{21}u_1(x) + (a_{22} + \delta v_2^*)v_1(x) - \delta v_2^* \int_{-\infty}^{\infty} \phi(x-y)v_1(y)dy + dv_1''(x).$$

Here u_1 and v_1 are perturbations around the homogeneous steady-state of the variables u and v respectively. Taking Fourier transform of these equations, we get

$$\lambda\overline{u}_1(\xi) = (a_{11} + u_2^*)\overline{u}_1(\xi) - u_2^*\overline{\phi}(\xi)\overline{u}_1(\xi) + a_{12}\overline{v}_1(\xi) - \xi^2\overline{u}_1(\xi), \tag{14}$$

$$\lambda\overline{v}_1(\xi) = a_{21}\overline{u}_1(\xi) + (a_{22} + \delta v_2^*)\overline{v}_1(\xi) - \delta v_2^*\overline{\phi}(\xi)\overline{u}_1(\xi) - d\xi^2\overline{v}_1(\xi), \tag{15}$$

where $\overline{u}_1(\xi), \overline{v}_1(\xi)$ and $\overline{\phi}(\xi)$ are Fourier transforms of the functions $u_1(x), v_1(x)$ and $\phi(x)$, respectively. Stability conditions of the homogeneous steady state are given by the inequalities

$$a_{11} + a_{22} + u_2^* + \delta v_2^* - (1+d)\xi^2 - (u_2^* + \delta v_2^*)\overline{\phi}(\xi) < 0, \tag{16}$$

$$[-(a_{11} + u_2^*) + u_2^*\overline{\phi}(\xi) + \xi^2][-(a_{22} + \delta v_2^*) + \delta v_2^*\overline{\phi}(\xi) + d\xi^2] - a_{12}a_{21} > 0. \tag{17}$$

Like in the case of Turing instability, we are interested to find the instability condition by reversing the inequality (17). Using (13) we get $\overline{\phi}(\xi) = \frac{1}{2M}\int_{-M}^{M}\cos(\xi y)dy = \frac{\sin(\xi M)}{(\xi M)}$, as ϕ is an even function. Hence the condition for instability is given by $D(\xi, M) < 0$ where

$$D(\xi, M) \equiv d\xi^4 + \left[(du_2^* + \delta v_2^*)\frac{\sin(\xi M)}{\xi M} - (da_{11} + a_{22} + du_2^* + \delta v_2^*)\right]\xi^2$$
$$+ \left[-(a_{11} + u_2^*) + u_2^*\frac{\sin(\xi M)}{\xi M}\right]\left[-(a_{22} + \delta v_2^*) + \delta v_2^*\frac{\sin(\xi M)}{\xi M}\right] - a_{12}a_{21}. \tag{18}$$

To determine the stability boundary, we need to find the values of ξ and M such that only one eigenvalue will cross the origin and other eigenvalues have negative real parts. The stability boundary in $\xi\, M$-parameter space is determined by the solutions of the equations

$$D(\xi, M) = 0, \quad \frac{\partial D(\xi, M)}{\partial \xi} = 0, \quad \frac{\partial D(\xi, M)}{\partial M} = 0. \tag{19}$$

According to this definition of the stability boundary, in the stability region the stationary solution is stable for any values of ξ and M. In the instability region, there exists at least one pair of these parameters for which the solution is unstable. Considering eigenvalue problem (14)–(15) over a bounded interval, we find that ξ takes a sequence of real values determined by the length of the interval. If we consider all real values of ξ, then we get the essential spectrum and the interval is not a priori fixed.

Differentiating $D(\xi, M)$ with respect to M and then equating to zero, we get

$$\left[u_2^*\left\{-(a_{22}+\delta v_2^*)+\delta v_2^*\frac{\sin(\xi M)}{\xi M}+d\xi^2\right\}\right.$$
$$\left.+\,\delta v_2^*\left\{-(a_{11}+u_2^*)+u_2^*\frac{\sin(\xi M)}{\xi M}+\xi^2\right\}\right]\left[\frac{\cos(\xi M)}{M}-\frac{\sin(\xi M)}{\xi M^2}\right]=0,$$

and hence the required condition reduces to

$$\tan(\xi M)=\xi M. \tag{20}$$

Writing $z=\xi M$ in the above expression, we get

$$\tan z = z. \tag{21}$$

We will denote the roots of this equation by z_j satisfying $0<z_1<z_2<\cdots$ and we define $\mu_j=\frac{\sin z_j}{z_j}$. Finally, solving the following two equations

$$D(\xi, M)=0,\quad \frac{\partial D(\xi, M)}{\partial \xi}=0,$$

we get,

$$\xi_j^2=\frac{(da_{11}+a_{22})+(1-\mu_j)(du_2^*+\delta v_2^*)}{2d},\quad j=1,2,3,\ldots, \tag{22}$$

and

$$M_j=\frac{z_j}{\xi_j}=\frac{z_j\sqrt{2d}}{\sqrt{(da_{11}+a_{22})+(1-\mu_j)(du_2^*+\delta v_2^*)}},\quad j=1,2,3,\ldots. \tag{23}$$

Existence of admissible ξ_j and M_j depends on the parameter values. They determine the extension of the Turing domain for the model with nonlocal interaction when compared with its local counterpart. We explain this in more detail in Sect. 4 with a numerical example.

4 Numerical Simulation Results

The numerical simulation results presented in this section are obtained for the following model:

$$\frac{\partial u(t,x)}{\partial t} = u(t,x)\left(1 - \frac{1}{2M}\int_{-M}^{M} u(t,y)dy\right) - \frac{\alpha u(t,x)v(t,x)}{u(t,x)+v(t,x)} + \frac{\partial^2 u(t,x)}{\partial x^2}, \tag{24}$$

$$\frac{\partial v(t,x)}{\partial t} = \frac{u(t,x)v(t,x)}{u(t,x)+v(t,x)} - \left(0.6 + \frac{0.1}{2M}\int_{-M}^{M} v(t,y)dy\right)v(t,x) + d\frac{\partial^2 v(t,x)}{\partial x^2}, \tag{25}$$

subjected to initial conditions (as described at Sect. 2.3) and periodic boundary conditions. Numerical simulations are carried out using the standard Euler method for the temporal part, Trapezoidal rule for the integration and a three point finite difference scheme for the diffusion terms with $\Delta t = 0.01$ and $\Delta x = 1$. Accuracy of the simulations is controlled by decreasing space and time discretization. In order to understand the impact of nonlocal interaction on pattern formation, we carry out numerical simulations for the values of parameters $\alpha \in [1.9, 2.25]$, $d \in [2, 12]$ and for a certain range of M.

We observe that increasing M induces stationary patterns over a larger parametric domain. We present a result of numerical simulations for $M = 5$ in Fig. 2. Comparison of the regions filled with different symbols in parametric domains, in Figs. 1 and 2, clearly indicates a significant change in the resulting pattern in the presence of nonlocal interactions. From Fig. 1 we see the extinction scenario in case of local interaction and $d = 2, \alpha \in (2.125, 2.25)$. However the nonlocal interaction induces

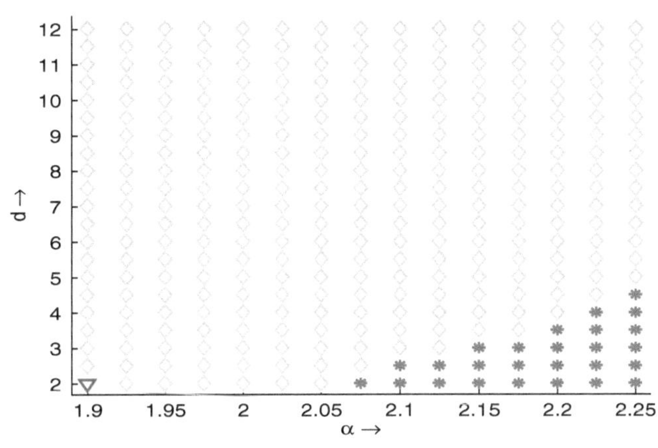

Fig. 2 Stationary and chaotic patterns are observed from numerical simulation of the model (24)–(25) for $M = 5$. Symbols represent: $\nabla \rightarrow$ homogeneous stead-state, $\diamond \rightarrow$ stationary pattern, $* \rightarrow$ spatio-temporal chaos

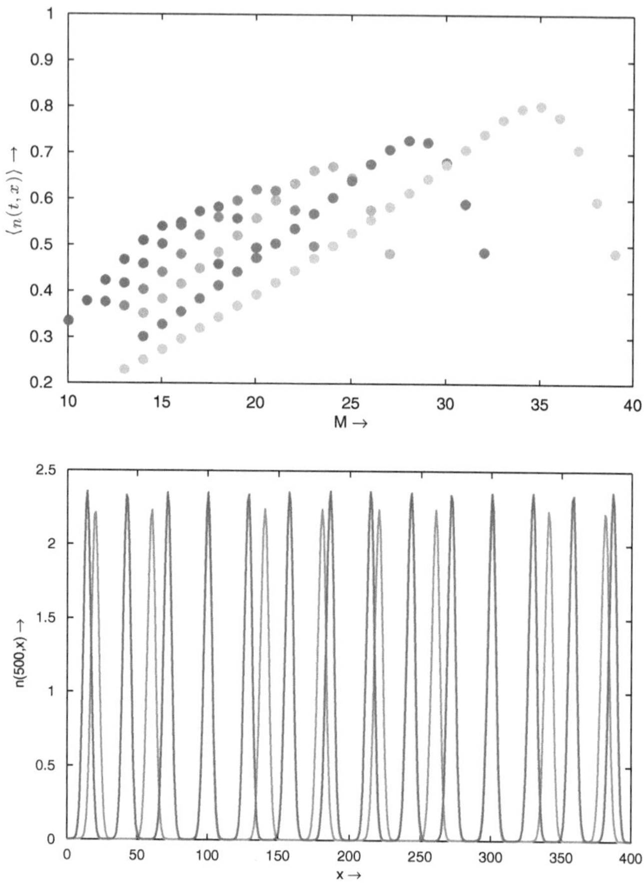

Fig. 3 *Upper panel* Spatial average of stationary distribution of prey population for a range of values of M are plotted here. The curves with different colours represent the branches of solutions obtained through forward and backward continuation method. *Lower panel* Stationary distribution of prey population corresponding to two different branches when $M = 20$. Other parameter values are $\alpha = 2.15$, $\beta = 1$, $\gamma = 0.6$, $\delta = 0.1$ and $d = 2$ (color figure online)

chaotic pattern for the same parametric range and $M = 5$. Further increase of the range of nonlocal interaction induces stationary patterns over a larger domain of the parameter space. In particular we find stationary pattern over the entire range $(\alpha, d) \in [1.9, 2.25] \times [2, 12]$ with $M = 10$ (the figure is not presented here for the sake of brevity).

It is interesting to note that only stationary patterns are observed for $M = 10$ and for all values of $(\alpha, d) \in [1.9, 2.25] \times [2, 12]$. However there exist multiple stationary solutions for these values of α, d and M. Existence of multiple stationary state against a range of values for M is demonstrated in Fig. 3. In the case of multiple stationary solutions, the choice of initial condition determines which one will be observed through numerical simulations. For a particular choice of M, we have performed numerical simulation to obtain the stationary pattern and this pattern is used as the initial condition for the next simulation with another value of M. This continuation technique along a branch of solutions is continued up to the values of M for which transition to another branch of solution is observed. The spatial average of the stationary patterns are plotted against M. The curves presented in Fig. 3 with different colours correspond to the multiple stationary branches. Initially we have obtained the stationary patterns for $M = 10, 15, 20, 25, 30, 35$ and then forward and backward continuation technique is used to obtained six different branches. Number of stationary peaks within a fixed spatial domain varies from one branch to the other (see lower panel of Fig. 3).

Finally we provide here supportive numerical illustration for the extension of Turing instability domain in the presence of nonlocal interaction. Turing instability condition for the model (6)–(7) is not satisfied if we choose $\alpha = 1.925$, $d = 4$ and the homogeneous steady-state $u_{*2} = 0.25$, $v_{*2} = 0.16$ is stable. For this choice of parameter values, first two inequalities of (9) are satisfied and the last inequality does not hold. As a result, the point $(\alpha, d) = (1.925, 4)$ lies outside the Turing domain (see Fig. 1). Our numerical simulation result revealed the appearance of Turing like pattern for the model (11)–(12) with the same choice of parameter values and $M = 5$ (see Fig. 2). Substituting all the relevant parameter values and $M = 5$ in (18) we find $D(\xi, 5) < 0$ for $\xi \in (0.26, 0.64)$. Similar argument holds for other choices of (α, d) and this result explains the extension of Turing domain in the presence of nonlocal interaction.

5 Discussions

In this work we have considered a classical prey-predator model with nonlocal intra-species competitions among the prey and the predator. The nonlocal terms take into account the possibility for the individuals of prey and predator species to hunt for resources and/or favorable habitat in some area around their average location. Hence prey individuals on one location compete with the prey individuals from some other location for the consumption of resources and similar argument is true for the intra-species competition among the predator individuals. This modification of the model influences survival of prey as well as predator. We have considered a preliminary aspect of nonlocal interaction and for mathematical simplicity the ranges of nonlocal interactions for prey and predators are assumed to be same. The length of the support functions of the kernel ϕ involved with the integral terms (equal to $2M$) indicates the area over which prey and predator individuals can rapidly move for favorable resources. Linear stability analysis along with the numerical simulations reveal that long range intra-species competition enhances the stable coexistence of both species.

Stationary patterns indicate formation of some groups of individuals within their habitats, and number of groups is determined by the length of the chosen spatial domain and maintain a correlation with the value $L/2M$. Enhancement in the range of nonlocal interaction for a given number of groups result in the overlapping of foraging areas and it alters the size of groups forming localized niches. This phenomena is responsible for the existence of multiple stationary branches obtained through numerical continuation technique. Over a longer time period and gradual increase of the range of nonlocal interaction is responsible for transition from one stationary branch to the other and hence we can claim that the proposed modelling approach is capable to capture the evolutionary aspects. Reaction-diffusion equations with nonlocal interaction terms are intensively studied [1, 2, 4, 8, 13] but there are still many open questions. Similar to [11, 13] we can expect the existence of periodic travelling waves and more complex spatio-temporal dynamics like modulated traveling wave and wave of chaos. Another important development concerns the interaction of species with global consumption [11] which can show a completely different dynamics.

Finally, we remark that the work presented here is a starting point and several questions are required to be answered in due course of time [25, 26]. First natural question is to consider different ranges of nonlocal interactions for the prey and predator species as these quantities are related with their body sizes and type of the species (invertebrate, vertebrate etc.). Another aspect will be worthy to investigate is the effect of nonlocal consumption of prey by their predators and consequent effect of nonlocal prey-dependent growth rate for the specialist predators. Here for simplicity we have used a step function as the kernels. However we need to validate the outcomes by considering other types of kernel functions [23]. Of course the proposed and other relevant considerations will make the model more complicated and challenging and hence we will continue our work to address them in our forthcoming research works.

Acknowledgements The work of M. Sen is partially supported by Dr. D.S. Kothari postdoctoral fellowship, UGC, India.

References

1. Apreutesei, A., Ducrot, A., Volpert, V.: Competition of species with intra-specific competition. Math. Model. Nat. Phenom. **3**(4), 1–27 (2008)
2. Apreutesei, A., Ducrot, A., Volpert, V.: Travelling waves for integro-differential equations in population dynamics. DCDS B **11**(3), 541–561 (2009)
3. Arditi, R., Ginzburg, L.R.: Coupling in predator-prey dynamics: ratio-dependence. J. Theor. Biol. **139**, 311–326 (1989)
4. Aydogmus, O.: Patterns and transitions to instability in an intraspecific competition model with nonlocal diffusion and interaction. Math. Nodel. Nat. Phenom. **10**(6), 17–29 (2015)
5. Banerjee, M., Abbas, S.: Existence and non-existence of spatial patterns in a ratio-dependent predator-prey model. Ecol. Complex. **21**, 199–214 (2015)
6. Banerjee, M., Banerjee, S.: Turing instabilities and spatio-temporal chaos in ratio-dependent Holling-Tanner model. Math. Biosci. **236**, 64–76 (2012)
7. Banerjee, M., Petrovskii, S.: Self-organised spatial patterns and chaos in a ratio-dependent predator-prey system. Theor. Ecol. **4**, 37–53 (2011)
8. Bayliss, A., Volpert, V.A.: Patterns for competing populations with species specific nonlocal coupling. Math. Nodel. Nat. Phenom. **10**(6), 30–47 (2015)

9. Bazykin, A.D.: Nonlinear Dynamics of Interacting Populations. World Scientific, Singapore (1998)
10. Hillary, R.M., Bees, M.: Plankton lattices and the role of chaos in plankton patchiness. Phy. Rev. E **69**, 031913 (2004)
11. Bessonov, N., Reinberg, N., Volpert, V.: Mathematics of Darwins diagram. Math. Model. Nat. Phenom. **9**(3), 5–25 (2014)
12. Gause, G.F.: The Strugle for Existence. Williams and Wilkins, Baltimore (1935)
13. Genieys, S., Volpert, V., Auger, P.: Pattern and waves for a model in population dynamics with nonlocal consumption of resources. Math. Model. Nat. Phenom. **1**(1), 63–80 (2006)
14. Klausmeier, C.A.: Regular and irregular patterns in semiarid vegetation. Science **284**, 1826–1828 (1999)
15. Kuznetsov, Y.A.: Elements of Applied Bifurcation Theory. Springer, New York (2004)
16. Luckinbill, L.L.: Coexistence in laboratory populations of Paramecium aurelia and its predator Didinium nasutum. Ecology **54**, 1320–1327 (1973)
17. Luckinbill, L.L.: The effects of space and enrichment on a predator-prey system. Ecology **55**, 1142–1147 (1974)
18. Murray, J.D.: Mathematical Biology II. Springer, Heidelberg (2002)
19. Levin, S.A., Segel, L.A.: Hypothesis for origin of planktonic patchiness. Nature **259**, 659–659 (1976)
20. Okubo, A., Levin, S.: Diffusion and Ecological Problems: Modern Perspectives. Springer, Berlin (2001)
21. Petrovskii, S.V., Malchow, H.: A minimal model of pattern formation in a prey-predator system. Math. Comput. Model. **29**, 49–63 (1999)
22. Scarsoglio, S., Laio, F., D'Odorico, P., Ridolfi, L.: Spatial pattern formation induced by Gaussian white noise. Math. Biosci. **229**, 174–184 (2011)
23. Segal, B.L., Volpert, V.A., Bayliss, A.: Pattern formation in a model of competing populations with nonlocal interactions. Phys. D **253**, 12–23 (2013)
24. Turing, A.M.: The chemical basis of morphogenesis. Philos. Trans. R. Soc. Lon. B **237**, 37–72 (1952)
25. Volpert, V.: Elliptic Partial Differential Equations, Volume 2. Reaction-Diffusion Equations. Birkhauser, Basel (2014)
26. Volpert, V.: Pulses and waves for a bistable nonlocal reaction-diffusion equation. Appl. Math. Lett. **44**, 21–25 (2015)

One Dimensional Maps as Population and Evolutionary Dynamic Models

Jim M. Cushing

Abstract I discuss one dimensional maps as discrete time models of population dynamics from an extinction-versus-survival point of view by means of bifurcation theory. I extend this approach to a version of these population models that incorporates the dynamics of a single phenotypic trait subject to Darwinian evolution. This is done by proving a fundamental bifurcation theorem for the resulting two dimensional, discrete time model. This theorem describes the bifurcation that occurs when an extinction equilibrium destabilizes. Examples illustrate the application of the theorem. Included is a short summary of generalizations of this bifurcation theorem to the higher dimensional maps that arise when modeling the evolutionary dynamics of a structured population.

Keywords Discrete time dynamics · Difference equations · Population dynamics · Evolutionary dynamics · Bifurcations · Equilibria · Stability · Allee effects

2010 Mathematics Subject Classification 92D25 · 92D15 · 37G35 · 39A30

1 Introduction

Iterative maps of the form

$$x_{t+1} = f(x_t)\, x_t \tag{1}$$

(often called difference equations) are widely used to model the discrete time (deterministic) dynamics of biological populations. Here x_t is some measure of population density at discrete census times $t = 0, 1, 2, \ldots$ and the expression $f(x)$ describes the per capita (or per unit) contribution to the population at the the next census time. We refer to f as the *population growth rate*. In this context the sequence x_t is

J.M. Cushing (✉)
Department of Mathematics and Program in Applied Mathematics,
University of Arizona, Tucson, AZ 85721, USA
e-mail: cushing@math.arizona.edu

J.M. Cushing et al. (eds.), *Applied Analysis in Biological and Physical Sciences*,
Springer Proceedings in Mathematics & Statistics 186,
DOI 10.1007/978-81-322-3640-5_3

non-negative, and the function f assumes only non-negative values for non-negative values of its argument. An initial condition $x_0 \geq 0$ generates a unique sequence (or trajectory) for $t = 0, 1, 2, \ldots$. The asymptotic properties of the sequence x_t are often of central interest and these depend, of course, on the properties of f. Famous examples include

$$f(x) = b\frac{1}{1+cx}, \quad f(x) = bxe^{-cx} \tag{2}$$

where b, c are positive constants.

The first example in (2) used in (1) gives what historically was called the discrete logistic model (or Pielou's logistic or Beverton-Holt model [23]). For this model, it is well known that the *extinction equilibrium* (fixed point) $x_e = 0$ is globally asymptotically stable for $x_0 \geq 0$ (i.e. it is locally asymptotically stable and attracts all trajectories with $x_0 \geq 0$) if $b < 1$ while the *positive equilibrium* $x_e = c^{-1}(b-1)$ is globally asymptotically stable for $x_0 \geq 0$ when $b > 1$. This is a prototypical example of the fundamental bifurcation that occurs at $b = 1$ where the extinction equilibrium destabilizes and, as a result, a stable positive equilibrium is created. The second example in (2) used in (1) gives the so-called Ricker model. The extinction equilibrium of this model also destabilizes at $b = 1$ with the result that there exists positive equilibrium $x_e = c^{-1}\ln b$ for $b > 1$. The positive equilibrium is (globally) stable for $1 < b < e^2$ but unstable for $b > e^2$. As b increases the Ricker model exhibits a period doubling cascade to chaos, similar to that exhibited by the famous quadratic map given by $f(x) = b(1 - cx)$ (which is often called the logistic map, rather inappropriately from a population dynamic point of view). Thus, both of these basic examples illustrate a fundamental bifurcation: when the extinction equilibrium destabilizes, a stable positive equilibria is created (at least for $b \gtrapprox 1$). As we will see, this is a general phenomenon for population models (1).

The function f is often regarded as describing reproductive processes and, as a result, the map (1) assumes all contributions to the population at time $t+1$ are due to reproductive events (and the survival of offspring until the census at $t+1$). This is appropriate, for example, for so-called semelparous (or monocarpic) populations in which individuals die after reproduction and, consequently, no reproductive individuals at time t are alive at time $t+1$. In this case generations do not overlap. It is often stated that one dimensional maps are only applicable to populations with non-overlapping generations, but this is not true. Suppose $s(x)$ is the fraction of the population x at time t that survives a unit of time. Then, if $b(x)$ is the (per capita) number of offspring (that survive until time $t+1$), we have

$$f(x) = b(x) + s(x). \tag{3}$$

The resulting map (1) allows for overlapping generations. We will refer to $b(x)$ as the birth or fertility rate and $s(x)$ as the survival rate.

If f depends only on the state variable x, as indicated above, then it is only the current population density that determines the population density at the next time census and, as a result, the mathematical model (1) is time autonomous. There are, however, many circumstances under which f also depends explicitly on time t. For example, in a seasonally fluctuating environment model coefficients, such as b and c in the discrete logistic or Ricker models, might be assumed periodic functions of time. In this case, the model equation (1) is periodically forced. Or model parameters might fluctuate randomly, due to random fluctuations in the physical environment (environmental stochasticity) or in individual organism characteristics (demographic stochasticity). In this case, the model equation (1) becomes a stochastic dynamical system.

Another reason that model parameters can change in time is Darwinian evolution, which is a case we will consider here. Suppose v is a quantified, phenotypic trait of an individual that is subject to evolution (i.e. it has a heritable component, it has variability among individuals in the population, and it accounts for differential fitness, e.g. individual differences among vital rates such as fertility and survival). If we assume the per capita contribution to the population made by an individual depends on its trait v, then $f = f(x, v)$ depends on both x and v. It might be the case that this contribution also depends on the traits of other individuals (due, for example, to competition for resources or other interactions among individuals). We can model this situation (frequency dependence) by assuming that f also depends on the mean trait u in the population so that $f = f(x, v, u)$. A canonical way to model Darwinian evolution is to model the dynamics of x_t and the mean trait u_t by means of the equations

$$x_{t+1} = f(x_t, v, u_t)|_{v=u_t} x_t \tag{4a}$$

$$u_{t+1} = u_t + \sigma^2 \partial_v F(x_t, v, u_t)|_{v=u_t}. \tag{4b}$$

The first equation asserts that the population dynamics can be (reasonably well) modeled by assuming the trait v is set equal to the population mean. The second equation (called Lande's or Fisher's or the breeder's equation) prescribes that the change in the mean trait is proportional to the *fitness gradient*, where fitness in this model is denoted by $F(x, v, u)$ [1, 21, 27, 28, 30, 36]. The modeler decides on an appropriate measure of fitness [32], which is often taken to be f or $\ln f$. The constant of proportionality $\sigma^2 \geq 0$ is called the speed of evolution. It is related to the variance of the trait in the population (exactly how depends on the derivation of the mean trait equation), which is assumed constant in time. Thus, if $\sigma^2 = 0$ no evolution occurs (there is no variability) and one has a one-dimensional map for just population dynamics of the form (1). If evolution occurs $\sigma^2 > 0$ then the model is a two dimensional map with state variable $[x_t, u_t]$.

In Sect. 2 we discuss non-evolutionary models of the form (1)–(3) with a focus on the basic question of extinction versus survival from a bifurcation theory point of view. In Sect. 3 we discuss a general class of evolutionary models from the same point of view.

2 One Dimensional Maps as Population Dynamic Models

The linearization principle applied to the extinction equilibrium $x = 0$ of (1)–(3) implies the extinction equilibrium is (locally asymptotically) stable if the *inherent population growth rate* $r_0 \triangleq f(0) \geq 0$ satisfies $r_0 < 1$ and unstable if $r_0 > 1$. In order to expose more explicitly the role of the *inherent* birth and death rates (i.e. the birth and death rates in the absence of any density effects) we write

$$f(x) = b_0\varphi(x) + s_0\sigma(x), \quad \varphi(0) = \sigma(0) = 1$$

where b_0 and s_0 are inherent birth and survival rates. Let Ω be an open interval of real numbers that contains the half line of nonnegative real numbers $\bar{R}_+$ (the closure of the positive real numbers R_+). We assume the following.

A1. *$b_0, s_0 \in \bar{R}_+$ and the functions φ and σ are twice continuously differentiable as maps from Ω to R_+ and $\bar{R}_+$ respectively that satisfy $\varphi(0) = \sigma(0) = 1$ and*

$$\lim_{x\to+\infty} \varphi(x) = 0 \tag{5}$$

$$0 \leq \sup_{x\in\Omega} s_0\sigma(x) < 1. \tag{6}$$

Condition (5) insures that the birth rate drops to 0 as population density x increases without bound. Condition (6) (which implies $0 \leq s_0 < 1$) expresses the fact that some mortality occurs during any time step. Specifically, the fraction $1 - s_0\sigma(x)$ of the population that is lost to mortality is bounded away from 0 uniformly for all $x \geq 0$. We can interpret $s_0\sigma(x)$ as an individual's probability of survival over one time unit.

An introduction of

$$r_0 \triangleq b_0 + s_0$$

into

$$f(x) = (r_0 - s_0)\varphi(x) + s_0\sigma(x) \tag{7}$$

allows easy use of r_0 as a bifurcation parameter in the resulting population model

$$x_{t+1} = ((r_0 - s_0)\varphi(x_t) + s_0\sigma(x_t))\, x_t. \tag{8}$$

Our goal is to study the existence and stability of positive equilibria as they depend on r_0.

The algebraic equation for a positive equilibrium is

$$(r_0 - s_0)\,\varphi(x) + s_0\sigma(x) = 1.$$

Solving this equation for r_0 we obtain

$$r_0 = \gamma(x)$$

where

$$\gamma(x) \stackrel{\circ}{=} \frac{1 - s_0\sigma(x) + s_0\varphi(x)}{\varphi(x)}.$$

By assumption A1, the function $\gamma(x)$ is a twice continuously differentiable on Ω and satisfies

$$\gamma(0) = 1, \quad \gamma(x) > 0 \text{ for } x \in \Omega, \quad \lim_{x\to+\infty} \gamma(x) = +\infty.$$

The graph C of the positive equilibrium pairs $[r_0, x_e]$, $x_e > 0$, is the set of points $[\gamma(x), x]$ obtained from all values of $x > 0$, which is a continuum that contains the point $[1, 0]$ in its closure. We say that the continuum C of equilibrium pairs *bifurcates from the continuum of extinction equilibrium pairs* $[r_0, 0]$ *at* $r_0 = 1$, i.e. at the point $[1, 0]$.

We define the *spectrum* S of C to be the range of the function $\gamma(x)$ for $x > 0$. The spectrum consists of those values of r_0 for which the population model (1) has a positive equilibrium. Under the assumption A1, the spectrum S is infinite and contains 1 in its closure. It is therefore a half line. If we denote the (positive) infimum of $\gamma(x)$ by r_m, then

$$0 < r_m \stackrel{\circ}{=} \inf_{x>0} \gamma(x) \leq 1$$

and the spectrum is

$$\begin{aligned} S &= \{r_0 : 1 < r_0 < +\infty\} \quad \text{if } r_m = 1 \\ S &= \{r_0 : r_m \leq r_0 < +\infty\} \quad \text{if } r_m < 1. \end{aligned} \tag{9}$$

The stability of a positive equilibrium pair $[r_0, x_e] \in C$, as determined by the linearization principle, depends on the quantity

$$\lambda(r_0, x) \stackrel{\circ}{=} \frac{d(f(x)\,x)}{dx} = (r_0 - s_0)\,\varphi(x) + s_0\sigma(x) + \left((r_0 - s_0)\,\varphi'(x) + s_0\sigma'(x)\right)x$$

evaluated at the equilibrium pair $[r_0, x_e]$. Here we use a prime "′" to denote the derivative a function of a single variable. A calculation shows

$$\lambda(r_0, x_e) = 1 - \gamma'(x_e)\,\varphi(x_e)\,x_e \tag{10}$$

from which we can conclude that positive equilibrium pairs $[r_0, x_e]$ near the bifurcation point $[1, 0]$ are (locally asymptotically) stable if $\gamma'(0) > 0$ and are unstable if $\gamma'(0) < 0$. (An equilibrium pair is nonhyperbolic if $\gamma'(x_e) = 1$, in which case the linearization principle is unable to determine stability.) The sign of

$$\gamma'(0) = -\left((1 - s_0)\,\varphi'(0) + s_0\sigma'(0)\right)$$

is the opposite of the sign of the weighted average $(1 - s_0)\,\varphi'(0) + s_0\sigma'(0)$. The derivatives $\varphi'(0)$ and $\sigma'(0)$ are the *sensitivities* of fertility and survival to changes low level population density. If one of these sensitivities is positive, it is called a *component Allee effect* [2, 4]. If a derivative is negative, it implies a *negative feedback effect* is caused by (low level) increased density. This is the most commonly made assumption in population models. Thus we see that if, at low population densities, there are no component Allee effects and at least one negative feedback effect is present, then $\gamma'(0) > 0$ and, in a neighborhood of the bifurcation point, small equilibria are locally asymptotically stable. Moreover, in this case, the bifurcating positive equilibria correspond to r_0 values greater than 1 and the bifurcation is said to be *forward.* On the other hand, if least one sensitivity is positive enough so that $\gamma'(0) < 0$ (or if both sensitivities are positive), then in a neighborhood of the bifurcation point small equilibria are unstable. In this case, the bifurcating positive equilibria correspond to values of r_0 less than 1 and the bifurcation is said to be *backward.* Notice that in this case $r_m < 1$ and the spectrum contains 1 in its interior (see (9)).

We arrive at the conclusion: *if $\gamma'(0) \neq 0$ then in a neighborhood of the bifurcation point a forward bifurcation is stable (meaning that the equilibria on C are stable) and a backward bifurcation is unstable.* See Fig. 1.

This bifurcation scenario occurring in a neighborhood of the bifurcation point $[r_0, x_e] = [1, 0]$ is quite general for population models. Mathematically it is a transcritical bifurcation exhibiting an exchange of stability principle, which is a phenomenon known to occur in quite general settings from nonlinear functional analysis [24, 31]. It has been established for numerous population models of many mathematical types [6]. For this reason, one can refer to the bifurcation described above as a *fundamental bifurcation theorem* for nonlinear population dynamic models.

However, the stability properties of the positive equilibria near the bifurcation point need not persist entirely along the continuum C of equilibrium pairs. It is well known for one dimensional maps that positive equilibria can destabilize and period doublings and routes-to-chaos can occur. These secondary bifurcations are model dependent, being determined by the properties of the density terms $\varphi(x)$ and $\sigma(x)$. They can occur in models with either forward or backward bifurcations at $[r_0, x_e] = [1, 0]$. One thing we can conclude from (10) is that if $\gamma'(x_e) > 0$ then the equilibrium is unstable. This means that the *equilibria along decreasing segments of the continuum C are unstable*, as illustrated in Fig. 1.

Along increasing segments of C, however, stability is uncertain. One fact we can assert from (10) is that increasing segments in a neighborhood of the isolated critical points of $\gamma(x)$ are (locally asymptotically) stable (because $\gamma'(x_e)$ will be small and positive). Isolated critical points at which a generic extrema occurs $(\gamma''(x_e) \neq 0)$

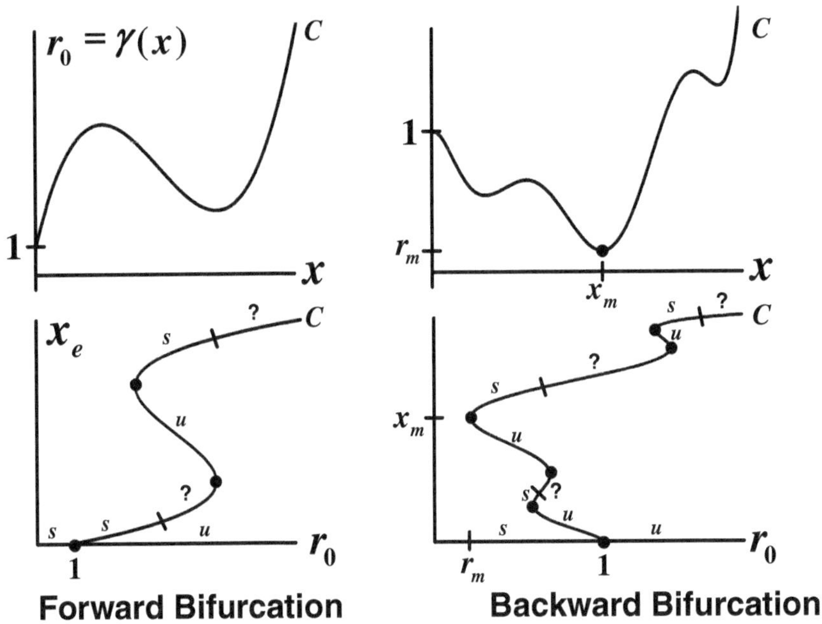

Fig. 1 The graph C of $r_0 = \gamma(x)$ gives the continuum of positive equilibria which bifurcates from the extinction equilibrium $[r_0, x] = [1, 0]$. The second row shows the graphs of equilibria x_e plotted against the inherent population growth rate r_0 obtained by reflecting the graphs above them through the $r_0 = x$ line. On these graphs the letter s indicates a (locally asymptotically) stable positive equilibrium while the letter u indicates an unstable positive equilibrium. The dots in the lower bifurcation diagrams show where blue-sky (*saddle-node*) bifurcations occur. The angled line segments and question marks shown along the curve C indicate that stability is, in general, guaranteed along increasing segments of C in only a neighborhood of a bifurcation point. See Example 1 and Fig. 2

correspond to the "turning or fold points" as seen (and indicated by the solid dots) the lower row of graphs in Fig. 1. These are called *blue-sky bifurcations* (or saddle-node or tangent bifurcations).

Thus, *in the neighborhood of blue-sky bifurcations the lower (decreasing) segment of C will contain unstable equilibria while the upper (increasing) segment of C will have stable equilibria.* See Fig. 1.

We note that a backward bifurcation creates the possibility that positive stable equilibria (or other kinds of attractors) can occur for $r_0 < 1$ when the extinction equilibrium is also stable. Such multiple attractor scenarios, where one attractor is extinction and the other is non-extinction, is called a *strong Allee effect* [4]. A backward bifurcation is not necessary for a strong Allee effect, but it is a common way for them to occur in population models [15].

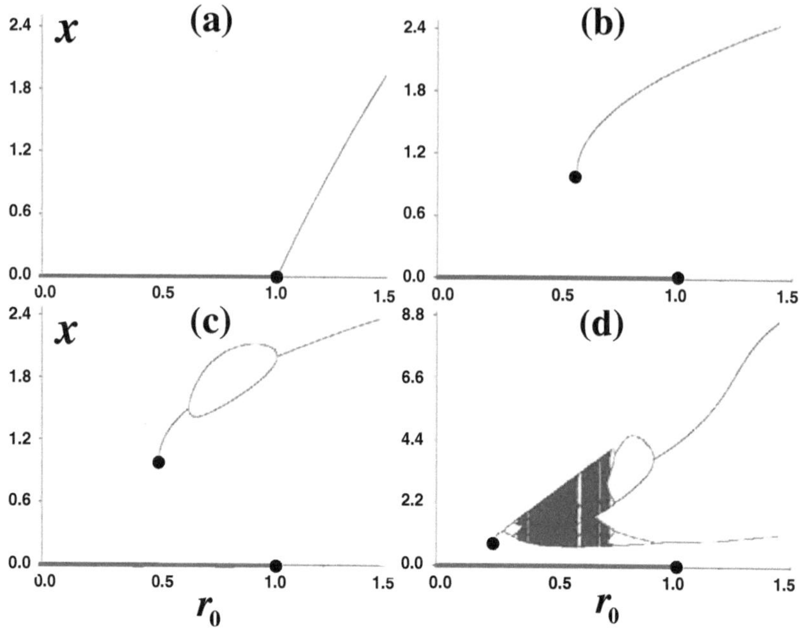

Fig. 2 Shown are bifurcation diagrams for the model Eq. (11) in Example 1. (a) A forward bifurcation of positive equilibria occurs when $a = -5$, $s = 0.5$, and $c = 5$. (b)–(d) Backward bifurcations occur when $a = 1$. The three cases shown in (b)–(d) use $s_0 = 0.5$, 0.4, and 0.1 respectively and all three use $c = 5$. The branches of unstable equilibria that connect the bifurcation point $[r_0, x_e] = [1, 0]$ to the blue-sky bifurcation point (indicated by *solid dots*) are not shown. The blue-sky bifurcation points $[r_m, x_e]$ in Example 1 are $[r_m, a]$ with r_m given by (12), which for these three cases are $r_m = 0.583$, 0.500, and 0.250

Example 1 We consider an example that illustrates both possibilities of forward and backward bifurcations, depending on a parameter value. As seen above, to construct such an example requires a component Allee effect, for at least some parameter values in the model equation. If, in this example, we assume that density dependence is absent in the survival rate and occurs only in fertility rate, then $\sigma(x) \equiv 1$. Thus, to obtain component Allee effects our choice of $\varphi(x)$ must allow for $\varphi'(0) > 0$ for at least some parameter values in addition, of course, to the requirements in A1.

Innumerable functional expressions have been used in the literature to construct difference equation models and, in particular, Allee effects; see for example [4, 20]. For our illustrative purposes here, the specific functional form of $\varphi(x)$ is not so important as that requirements in A1 be satisfied and that it contain a coefficient whose value determines the presence or absence of a component Allee effect. A rationale function that serves these purposes is

$$\varphi(x) = \frac{1 + ca^2}{1 + c(x - a)^2}$$

where $c > 0$ and a is any real number. This expression is that used the discrete Bernoulli equation [33] adapted so as to include the possibility of a component Allee effect. Specifically, since

$$\varphi'(0) = 2a\frac{c}{ca^2+1}$$

we see that a component Allee effect is present when $a > 0$ and is absent when $a < 0$. The resulting difference equation is

$$x_{t+1} = \left[(r_0 - s_0) \frac{1+ca^2}{1+c(x_t-a)^2} + s_0 \right] x_t. \tag{11}$$

To determine the geometry of the bifurcating branch of positive equilibria, we note that

$$\gamma(x) = \frac{1-s_0}{1+ca^2}\left(1+c(x-a)^2\right) + s_0$$

and calculate

$$\gamma'(x) = \frac{2c(1-s_0)}{1+ca^2}(x-a), \quad \gamma''(x) = \frac{2c(1-s_0)}{1+ca^2} > 0.$$

From

$$\gamma'(0) = -2ac\frac{1-s_0}{1+ca^2}$$

we deduce that *a forward bifurcation occurs (i.e.* $\gamma'(0) > 0$*) when* $a < 0$ *and a backward bifurcation occurs (i.e.* $\gamma'(0) < 0$*) if* $a > 0$.

In the case of a backward bifurcation we see from these calculations that $\gamma(x)$ has a minimum at $x_m = a$ and the lower endpoint of the spectrum S is

$$r_m = \frac{1+s_0ca^2}{1+ca^2} \tag{12}$$

at which a blue-sky bifurcation occurs (sometimes called a *tipping point*). There are two positive equilibria for $r_m < r_0 < 1$ and, according to the general principles above (as shown in Fig. 1), the smaller positive equilibria is unstable and the larger equilibrium is stable at least for $r_0 \gtrapprox r_m$. The larger positive equilibrium is not necessarily stable for all values of $r_0 > r_m$ however, a fact demonstrated by the dynamically computed bifurcation diagrams shown in Fig. 2. In those sample diagrams we see the possibility of complicated secondary bifurcations (period doublings, etc.) for $r_0 > r_m$. Specifically, Fig. 2c, d show secondary bifurcations (and apparently chaotic attractors) occurring for values of $r_0 < 1$. In these cases strong Allee effects involving non-equilibrium attractors occur.

3 Evolutionary Dynamics

We begin with a population growth rate in equation (1) given by

$$f(x) = (r_0 - s_0)\varphi(x) + s_0\sigma(x), \quad \varphi(0) = \sigma(0) = 1.$$

Concerning model parameter dependence on a heritable trait, we make the following assumptions. First, the inherent individual fertility and survival rates, and hence r_0 and s_0, depend only on the individual's trait v. A rationale for this is that at low population densities the traits (and hence the characteristics and behavior) of other individuals have negligible effect on the individual's vita rates. Thus, it is only the density terms φ and σ that depend on the mean trait u. Moreover, as is commonly done, we assume that these effects are a function of the difference between v and u and that the effects are maximized (or minimized) when $v = u$, i.e. when the individual is most like other individuals (as represented by the mean trait u) [36]. We incorporate these assumptions by writing

$$f(x, v, u) = (r_0(v) - s_0(v))\varphi(x, v, v-u) + s_0(v)\sigma(x, v, v-u) \tag{13}$$

where, for all values of the arguments v and z in $\varphi(0, v, z)$ and $\sigma(0, v, z)$ we assume

$$\varphi(0, v, z) \equiv \sigma(0, v, z) \equiv 1 \tag{14a}$$
$$\partial_z\varphi(0, v, z)\,|_{z=0} \equiv \partial_z\sigma(0, v, z)\,|_{z=0} \equiv 0 \tag{14b}$$
$$\partial_{zz}\varphi(0, v, z)\,|_{z=0} \neq 0, \quad \partial_{zz}\sigma(0, v, z)\,|_{z=0} \neq 0. \tag{14c}$$

An example is the discrete logistic (Beverton-Holt) or Ricker expressions

$$\varphi = \frac{1}{1+cx}, \quad \varphi = \exp(-cx)$$

in which the coefficient c is under the influence of evolution, that is to say, $c = c(z)$ where the distribution of c values is Gaussian-like (often taken to be the case in evolutionary game theoretic models [36])

$$c(z) = \psi \exp\left(-z^2/w\right)$$

where $\psi > 0$ and $w > 0$ are positive constants. The maximal density effect on fertility is experienced by an individual when its inherited trait v equals the population mean u (i.e. when $z = v - u = 0$). In some models $\psi = \psi(v)$ is assumed a function of the trait v, in which case $c = c(v, z)$. These modeling assumptions on φ satisfy the constraints (14). Similar models can be built using the $c(z) = \psi/\left(1 + wz^2\right)$. In these examples w measures the width of the distribution of c values around $z = 0$.

If, for notational simplicity, we define

$$\varphi(x,v) \mathrel{\hat{=}} \varphi(x,v,0), \quad \sigma(x,v) \mathrel{\hat{=}} \sigma(x,v,0)$$
$$f(x,v) \mathrel{\hat{=}} f(x,v,0)$$

and take fitness to be

$$F(x,v) = \ln f(x,v),$$

then the evolutionary model (15) becomes

$$x_{t+1} = f(x_t,u_t)\,x_t \tag{15a}$$
$$u_{t+1} = u_t + \sigma^2 \frac{\partial_v f(x_t,u_t)}{f(x_t,u_t)}. \tag{15b}$$

Here we have used the notation

$$f(x,u) \mathrel{\hat{=}} f(x,v)|_{v=u}, \quad \partial_v f(x,u) \mathrel{\hat{=}} \partial_v f(x,v)|_{v=u}.$$

More explicitly, in (15)

$$f(x,u) = (r_0(u) - s_0(u))\,\varphi(x,u) + s_0(u)\,\sigma(x,u)$$

$$\partial_v f(x,u) = \left(r_0'(u) - s_0'(u)\right)\varphi(x,u) + (r_0(u) - s_0(u))\,\partial_v\varphi(x,u)$$
$$+ s_0'(u)\,\sigma(x,u) + s_0(u)\,\partial_v\sigma(x,u).$$

The assumption we make on the terms in this model are as follows.

A2. *Assume* $r_0(v)$ and $s_0(v)$ *are twice continuously differentiable functions mapping* R *to* R_+ and $\bar{R}_+$. respectively. *Assume* $\varphi(x,v)$ and $\sigma(x,v)$ *are twice continuously differentiable functions mapping* $\Omega \times R$ *to*R_+ and $\bar{R}_+$ *respectively that satisfy*

$$\varphi(0,v) \equiv \sigma(0,v) \equiv 1 \textit{ for all } v \in R \tag{16}$$

and

$$0 \leq \sup_{x\in\Omega, v\in R} s_0(v)\,\sigma(x,v) < 1. \tag{17}$$

Conditions (16) and (17) imply $0 \leq \sup_{v\in R} s_0(v) < 1$ and

$$\partial_v\varphi(0,v) \equiv \partial_v\sigma(0,v) \equiv 0 \text{ for all } v.$$

The equilibrium equations associated with (15) are

$$x = f(x,u)\,x$$
$$0 = \partial_v f(x,u).$$

Clearly $x = 0$ solves the first equilibrium equation. Thus, a pair $[x, u] = [0, u]$ is an equilibrium if and only if u satisfies

$$\partial_v f(0, u) = r_0'(u) = 0.$$

Definition 1 An **extinction equilibrium** is an equilibrium $[x, u]$ with $x = 0$. A pair $[0, u^*]$ is an extinction equilibrium if and only if u^* is a **critical trait**, i.e. if and only if $r_0'(u^*) = 0$.

We assume throughout that there exists a critical trait u^*. To use

$$r_0^* \overset{\circ}{=} r_0(u^*)$$

as a bifurcation parameter we write

$$r_0(v) = r_0^* \rho(v), \quad \rho(u^*) = 1, \quad \rho'(u^*) = 0$$

where $\rho(v)$ satisfies the same conditions in A2 as does $r_0(v)$. Note that $[0, u^*]$ is an equilibrium for all values of the bifurcation parameter r_0^*.

To investigate the (local asymptotic) stability of the extinction equilibrium $[0, u^*]$ by means of the linearization principle we consider the Jacobian

$$J(x, u) = \begin{pmatrix} f(x, u) + x\partial_x f(x, u) & x\partial_v f(x, u) \\ \partial_{xv} \ln f(x, u) & 1 + \sigma^2 \partial_{vv} \ln f(x, u) \end{pmatrix} \tag{18}$$

of (15). Evaluated at an extinction equilibrium $[x, u] = [0, u^*]$ this Jacobian becomes

$$J(0, u^*) = \begin{pmatrix} r_0^* & 0 \\ \eta / r_0^* & 1 + \sigma^2 \rho''(u^*) \end{pmatrix} \tag{19}$$

where

$$\eta \overset{\circ}{=} \left(-\partial_x \varphi(0, u^*) + \partial_x \sigma(0, u^*)\right) s_0'(u^*) + (1 - s^*)\, \partial_{xv} \varphi(0, u^*) + s_0^* \partial_{xv} \sigma(0, u^*).$$

The eigenvalues appear along the diagonal.

Theorem 1 *Assume A2 and that u^* is a critical trait.*

(a) Suppose

$$\left|1 + \sigma^2 \rho''(u^*)\right| < 1. \tag{20}$$

Then the extinction equilibrium $[0, u^*]$ of (15) is (locally asymptotically) stable if $r_0^* < 1$ and is unstable if $r_0^* > 1$.

(b) Suppose, on the other hand, that

$$\left|1+\sigma^2\rho''\left(u^*\right)\right|>1. \tag{21}$$

Then the extinction equilibrium is unstable.

Note that (20) holds if $\rho''\left(u^*\right)<0$ and σ^2 is small, i.e. ρ has a generic maximum at u^* and evolution is not rapid. Condition (21) holds if $\rho''\left(u^*\right)>0$, i.e. ρ has a generic minimum at u^*.

Definition 2 A **positive equilibrium** $[x_e, u_e]$ of (15) is an equilibrium with $x_e>0$.

The equilibrium equations satisfied by positive equilibria are

$$1=f\left(x,u\right) \tag{22a}$$
$$0=\partial_v f\left(x,u\right) \tag{22b}$$

which we can re-write as

$$g(x,u,r_0^*)=0 \tag{23a}$$
$$h\left(x,u,r_0^*\right)=0 \tag{23b}$$

where

$$\begin{aligned} g\left(x,u,r_0^*\right) &\circeq \left(r_0^*\rho\left(u\right)-s_0\left(u\right)\right)\varphi\left(x,u\right)+s_0\left(u\right)\sigma\left(x,u\right)-1 \\ h\left(x,u,r_0^*\right) &\circeq \left(r_0^*\rho'\left(u\right)-s_0'\left(u\right)\right)\varphi\left(x,u\right)+s_0'\left(u\right)\sigma\left(x,u\right) \\ &\quad+\left(r_0^*\rho\left(u\right)-s_0\left(u\right)\right)\partial_v\varphi\left(x,u\right)+s_0\left(u\right)\partial_v\sigma\left(x,u\right). \end{aligned}$$

Note that the equations (23) are satisfied by $[x,u]=[0,u^*]$ and $r_0^*=1$. To use the implicit function theorem to solve equations (23) for x and u as functions of r_0^* near this solution, we need $\Delta\left(0,u^*,1\right)\neq 0$ where

$$\Delta\left(x,u,r_0^*\right)=\det\begin{pmatrix}\partial_x g\left(x,u,r_0^*\right) & \partial_u g\left(x,u,r_0^*\right) \\ \partial_x h\left(x,u,r_0^*\right) & \partial_u h\left(x,u,r_0^*\right)\end{pmatrix}$$

and

$$\begin{aligned} \partial_x g\left(x,u,r_0^*\right) &= \left(r_0^*\rho\left(u\right)-s_0\left(u\right)\right)\partial_x\varphi\left(x,u\right)+s_0\left(u\right)\partial_x\sigma\left(x,u\right) \\ \partial_u g\left(x,u,r_0^*\right) &= \left(r_0^*\rho'\left(u\right)-s_0'\left(u\right)\right)\varphi\left(x,u\right)+\left(r_0^*\rho\left(u\right)-s_0\left(u\right)\right)\partial_v\varphi\left(x,u\right) \\ &\quad+s_0'\left(u\right)\sigma\left(x,u\right)+s_0\left(u\right)\partial_v\sigma\left(x,u\right) \end{aligned}$$

$$\begin{aligned}
\partial_x h\left(x,u,r_0^*\right) &= \left(r_0^*\rho'(u) - s_0'(u)\right)\partial_x\varphi(x,u) + s_0'(u)\,\partial_x\sigma(x,u) \\
&\quad + \left(r_0^*\rho(u) - s_0(u)\right)\partial_{xv}\varphi(x,u) + s_0(u)\,\partial_{xv}\sigma(x,u) \\
\partial_u h\left(x,u,r_0^*\right) &= \left(r_0^*\rho''(u) - s_0''(u)\right)\varphi(x,u) + \left(r_0^*\rho'(u) - s_0'(u)\right)\partial_v\varphi(x,u) \\
&\quad + s_0''(u)\,\sigma(x,u) + s_0'(u)\,\partial_v\sigma(x,u) \\
&\quad + \left(r_0^*\rho'(u) - s_0'(u)\right)\partial_v\varphi(x,u) + \left(r_0^*\rho(u) - s_0(u)\right)\partial_{vv}\varphi(x,u) \\
&\quad + s_0'(u)\,\partial_v\sigma(x,u) + s_0(u)\,\partial_{vv}\sigma(x,u).
\end{aligned}$$

For notational purposes we let an asterisk denote evaluation at $\left([x,u],r_0^*\right) = ([0,u^*],1)$, i.e.

$$\partial_x^*\varphi \mathrel{\hat{=}} \partial_x\varphi\left(0,u^*\right), \quad r_0^* = r_0\left(u^*\right), \quad \text{etc.}$$

A calculation shows

$$\Delta\left(0,u^*,1\right) = \det\begin{pmatrix} -\kappa & 0 \\ \eta & \rho''(u^*) \end{pmatrix}$$

where

$$\begin{aligned}
\kappa &\mathrel{\hat{=}} -\left[\left(1-s_0^*\right)\partial_x^*\varphi + s_0^*\partial_x^*\sigma\right] \\
\eta &\mathrel{\hat{=}} \left(-\partial_x^*\varphi + \partial_x^*\sigma\right)s_0'\left(u^*\right) + \left(1-s_0^*\right)\partial_{xv}^*\varphi + s_0^*\partial_{xv}^*\sigma.
\end{aligned}$$

It follows that $\Delta(0,u^*,1) \neq 0$ if and only if $\kappa \neq 0$ and $\rho''(u^*) \neq 0$. Under these conditions the Implicit Function Theorem implies the existence of (unique, smooth) solutions $[x,u] = \left[x_e\left(r_0^*\right), u_e\left(r_0^*\right)\right]$ for r_0^* near 1 with $[x_e(1), u_e(1)] = [0,u^*]$. The latter equality means that this branch of solutions bifurcates from the branch of extinction equilibria at $r_0^* = 1$.

To determine whether these solutions are feasible as equilibria of the population model, we need to determine whether $x = x_e(r_0)$ is positive or not. We also want to determine when a positive equilibrium is (locally asymptotically) stable. From the identities

$$\begin{aligned}
g\left(x_e\left(r_0^*\right), u_e\left(r_0^*\right), r_0^*\right) &= 0 \\
g\left(x_e\left(r_0^*\right), u_e\left(r_0^*\right), r_0^*\right) &= 0
\end{aligned}$$

valid for r_0^* near 1 we find, by differentiation with respect to r_0^* followed by an evaluation at $r_0^* = 1$, that

$$\begin{aligned}
x_e'(1)\,\partial_x^* g + u_e'(1)\,\partial_u^* g + \partial_{r_0^*}^* g &= 0 \\
x_e'(1)\,\partial_x^* h + u_e'(1)\,\partial_u^* h + \partial_{r_0^*}^* h &= 0
\end{aligned}$$

or

$$\begin{aligned}
\kappa x_e'(1) + 1 &= 0 \\
\eta x_e'(1) + \rho''\left(u^*\right)u_e'(1) &= 0
\end{aligned}$$

and hence

$$x_e'(1) = \frac{1}{\kappa}, \quad u_e'(1) = \frac{-\kappa}{\eta \rho''(u^*)}.$$

The first equation implies $x_e(r_0^*)$ is positive for $r_0^* \gtrapprox 1$ if $\kappa > 0$ and for $r_0^* \lessapprox 1$ if $\kappa < 0$. As a result we conclude that a bifurcation of positive equilibrium pairs from the extinction equilibrium $[0, u^*]$ occurs at $r_0^* = 1$ and is forward if $\kappa > 0$ or backward if $\kappa < 0$.

We can determine the stability of the positive equilibria, by the linearization principle, from the eigenvalues of the Jacobian (18) evaluated at $(x, u) = (x_e^*(r_0^*), u_e^*(r_0^*))$, namely

$$\begin{pmatrix} f(x(r_0^*), u(r_0^*)) + x(r_0^*)\,\partial_x f(x(r_0^*), u(r_0^*)) & x(r_0^*)\,\partial_v f(x(r_0^*), u(r_0^*)) \\ \partial_{xv} \ln f(x(r_0^*), u(r_0^*)) & 1 + \sigma^2 \partial_{vv} \ln f(x(r_0^*), u(r_0^*)) \end{pmatrix}$$

which, because $(x(r_0^*), u(r_0^*))$ solves the equilibrium equations (22), simplifies to

$$\begin{pmatrix} 1 + x(r_0^*)\,\partial_x f(x(r_0^*), u(r_0^*)) & 0 \\ \partial_{xv} f(x(r_0^*), u(r_0^*)) & 1 + \sigma^2 \partial_{vv} f(x(r_0^*), u(r_0^*)) \end{pmatrix} \tag{24}$$

The eigenvalues, which appear along the diagonal, are

$$\lambda_1 = 1 - (r_0^* - 1) + O\left((r_0^* - 1)^2\right)$$
$$\lambda_2 = 1 + \sigma^2 \rho''(u^*) + O(r_0^* - 1).$$

For r_0^* near 1 the positive equilibrium $[x_e^*(r_0^*), u_e^*(r_0^*)]$ is stable if both $|\lambda_i| < 1$ and is unstable if at least one $|\lambda_i| > 1$.

We summarize these results in the theorem below. In that theorem we make use of the following definitions. If $[x_e, u_e]$ is an equilibrium for a value of r_0^*, then we call $(r_0^*, [x_e, u_e])$ an *equilibrium pair*. If the equilibrium $[x_e, u_e]$ is positive, stable or unstable, then we say the equilibrium pair $(r_0^*, [x_e, u_e])$ is respectively positive, stable or unstable.

Theorem 2 *Assume A2 and that u^* is a critical trait such that $\rho''(u^*) \neq 0$. Assume $\kappa \neq 0$. Then a continuum $\mathcal{C}$ of positive equilibrium pairs $(r_0^*, [x_e, u_e])$ of the evolutionary model (15) bifurcates from the extinction equilibrium* $[1, (0, u^*)]$. *Near the bifurcation point these positive equilibria are approximately*

$$x_e = \frac{1}{\kappa}(r_0^* - 1) + O\left((r_0^* - 1)^2\right) \tag{25a}$$

$$u_e = u^* + \frac{-\kappa}{\eta \rho''(u^*)}(r_0^* - 1) + O\left((r_0^* - 1)^2\right) \tag{25b}$$

The bifurcation is forward if $\kappa > 0$ and backward if $\kappa < 0$.

(a) *If* (20) *holds, then the stability of the bifurcation (meaning the stability of the equilibria* $[x_e, u_e]$ *on the continuum* C *) is determined by the direction of bifurcation. Specifically, a forward bifurcation is stable and a backward bifurcation is unstable.*

(b) *If* (21) *holds, then both forward and backward bifurcations are unstable.*

Note that $\rho''(u^*) > 0$ implies (21) holds. In this case, Theorems 1 and 2 imply that in a neighborhood of the bifurcation point all equilibria – extinction and positive – are unstable. An explanation for this is roughly as follows. The trait dynamic equation (15b) is based on the assumption that the mean trait u_t moves up the fitness gradient, which near the extinction equilibrium is approximately $\ln \rho(v)$. The inequality $\rho''(u^*) > 0$ implies $\ln \rho(v)$ has a local minimum at the critical trait $v = u^*$ and consequently the trait component u_t of orbits in a neighborhood of the extinction equilibrium increases until the orbit leaves the neighborhood. Thus, when $\rho''(u^*) \neq 0$ a necessary condition for stability is $\rho''(u^*) < 0$, which implies $\rho(v)$ has a local maximum at the critical trait $v = u^*$. In this case (20) is equivalent to

$$\sigma^2 < \frac{-2}{\rho''(u^*)}.$$

That is to say, a forward bifurcation will be stable if the speed of evolution σ^2 is not too fast.

Corollary 1 *Assume A2 and that* u^* *is a critical trait such that* $\rho''(u^*) < 0$. *Assume* $\kappa \neq 0$. *If the speed of evolution* σ^2 *is not too fast, then the direction of bifurcation of the positive equilibrium pairs guaranteed by Theorem 2 determines their stability: a forward bifurcation is stable and a backward bifurcation is unstable.*

In A2 we assumed for simplicity that the domain of trait values v is the whole real lie R. Since Theorem 2 and Corollary 1 concern bifurcation phenomena in a neighborhood of an extinction equilibrium $[1, (0, u^*)]$, these results remain valid if R is replaced in A2 by an open set of trait values, so long as the critical trait u^* lies in the set.

Example 2 As mentioned in Sect. 1 the fertility density term

$$\varphi(x) = \frac{1}{1 + c_1 x}, \quad c_1 > 0 \tag{26}$$

is used in the classic discrete logistic (Beverton-Holt) equation for a population with non-overlapping generations ($\sigma(x) \equiv 0$). It expresses a negative feedback of population density on the fertility rate (since $\varphi(x)$ is a *decreasing* function of x). For a population with over-lapping generations the survival rate $\sigma(x)$ would not be identically 0. In our example here, we take $\sigma(x)$ to be an *increasing* function of x, i.e. to have a component Allee effect. The biological rationale for this is that we wish to model a trade-off between density effects on the fertility and survival rates: increased population density suppresses an individual's fertility rate, but enhances an individual's survival probability (through, for example, group defence). Trade-off's of this sort play a fundamental role in the study of life history strategies and usually constitute the driving mechanisms that determine evolutionary dynamics and outcomes [32].

An example mathematical function $\sigma(x)$ that satisfies A1 and that is increasing in x is

$$\sigma(x) = \frac{1 + c_2 s_m x}{1 + c_2 s_0 x}, \quad 0 < s_0 < s_m < 1, \; c_2 > 0 \tag{27}$$

As a function of population density x, the survival rate $s_0\sigma(x)$ increases monotonically from s_0 to s_m. The coefficients $c_i > 0$ in the density terms (26) and (27) measure the strength of the density effects on these vital rates.

We assume that a heritable trait v, in addition to determining an individual's inherent net reproductive rate $r_0^*\rho(v)$, determines an individual's sensitivities to density increases, i.e. we assume $c_1 = c_1(v) > 0$ and $c_2 = c_2(v) > 0$ are functions of v. In this example, we assume inherent survival s_0 is trait independent. We have then the fitness function

$$f(x, v) = \left(r_0^*\rho(v) - s_0\right)\frac{1}{1 + c_1(v)x} + s_0\frac{1 + c_2(v)s_m x}{1 + c_2(v)s_0 x} \tag{28}$$

in the model equations (15).

We assume there exists a trait $v = u^*$ at which the inherent growth rate $\rho(v)$ attains a maximum (and does so with $\rho''(u^*) < 0$) and the the numerical scale for the trait is chosen so that $u^* = 0$. On the other hand, we assume the density coefficients c_i are decreasing functions of the trait v. This means that an increase in the trait v results in weaker density effects on fertility and survival.

For the evolutionary model (15)–(28) we calculate the quantity κ, whose sign determines the properties of the bifurcation at $r_0^* = 1$ (according to Theorem 2), to be

$$\kappa = (1 - s_0)c_1(0) - s_0(s_m - s_0)c_2(0).$$

Suppose the coefficient $c_2(0)$ is small compared to $c_1(0)$ so that $\kappa > 0$. Then by Theorem 2 the bifurcation of positive equilibrium pairs is forward and stable. In other words, *if at the critical trait $v = 0$ survival is less sensitive to density effects than is fertility, then the bifurcation of positive equilibrium pairs is forward and there exist (locally asymptotically) stable survival equilibria for $r_0^* \gtrapprox 1$*. On the other hand, if the reverse is true, i.e. *if at the critical trait $v = 0$ survival is more sensitive to density effects than is fertility, then the bifurcation of positive equilibrium pairs is backward and unstable*. In this latter case, there is a potential for a strong Allee effect, which is to say, a potential for the existence of a stable positive equilibrium when $r_0^* < 1$ (as discussed in Sect. 2). Evidence that this can indeed occur in this example appears in Fig. 3.

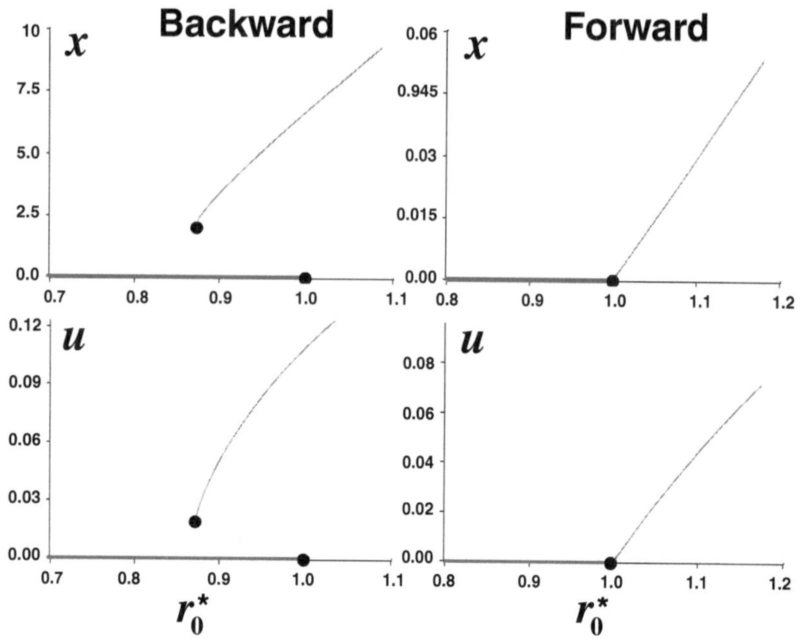

Fig. 3 Bifurcation diagrams plotting the x and u components of (dynamically computed) positive equilibria of the evolutionary model (15) with fitness function (28) and trait dependencies (29). Here $s_0 = 0.3$, $s_m = 0.6$ and $\kappa = 0.7c_1 - 0.09c_2$. The speed of evolution is $\sigma^2 = 0.1$. When $c_1 = 5$, $c_2 = 0.1$ and $\kappa = 3.491 > 0$ a forward, stable bifurcation occurs, as seen in the right column of plots. When $c_1 = 0.1$, $c_2 = 5$ and $\kappa = -0.380 < 0$ a backward, unstable bifurcation occurs, as seen in the left column of plots. In this case a strong Allee effect occurs as can be seen by the existence of positive equilibrium pairs for an interval of r_0^* values less than 1

The bifurcation diagrams shown in Fig. 3 illustrate these results for the evolutionary model (15)–(28) with

$$\rho(v) = e^{-v^2}, \quad c_1(v) = c_1 e^{-v}, \quad c_2(v) = c_2 e^{-v}. \tag{29}$$

In this case, the only critical point of $\rho(v)$ is $v = u^* = 0$ at which inherent fertility has a maximum (note that $\rho''(0) = -2 < 0$). The only extinction equilibrium pairs are $\left(r_0^*, [0, 0]\right)$, which exist for all values of $r_0^* > 0$. The right column of plots in Fig. 1 shows an example when $\kappa > 0$ and the bifurcation from the extinction equilibrium $\left(r_0^*, [0, 0]\right)$ at $r_0^* = 1$ is forward and stable. The left column of plots in Fig. 1 shows an example when $\kappa < 0$ and the bifurcation is backward and unstable. In this case, one can observe the occurrence of a strong Allee effect, i.e. the existence of positive equilibria for an interval of r_0^* values less than 1. Although the intent here is only to illustrate the mathematical results in Theorem 2 by use of the example (15)–(28), we point out some of the resulting biological implications. The biological features implied by (29) are that fertility is optimized at a unique heritable trait $v = 0$ and that an increase in v lessens both the negative density effects on fertility and the positive density effects on survival. The positive equilibrium pairs shown in the bifurcation diagrams of Fig. 3 have a positive mean trait component $u_e > 0$. Thus, evolution does not select to maximize inherent fertility, but selects for a lesser inherent fertility rate and for lower effects of population density on both fertility and survival.

4 Concluding Remarks

We focussed in this paper on one dimensional maps as models of population dynamics. Many of the results given in Sects. 2 and 3 have been extended in several directions. The bifurcation of a global continuum of positive equilibrium as graphically portrayed in Fig. 1 for one dimensional maps has been proved for higher dimensional systems of difference equations of the form

$$x_{t+1} = P(x_t)\, x_t \tag{30}$$

where $x_t \in R^m_+$ and P is an $m \times m$ matrix valued function [6, 8]. This kind of matrix equation arises in structured population dynamics where the vector x_t is the demographic distribution of individuals into specified categories (chronological age, physiological size, weight, etc.) [3]. When the projection matrix P is primitive (non-negative, irreducible and a strictly dominant positive eigenvalue), the direction of bifurcation determines the stability of the bifurcating continuum of positive equilibria (as in Fig. 1). This is a general phenomenon in bifurcation theory [24], as the global extent of the continuum [31].

However, when a non-negative and irreducible projection matrix P is not primitive (its dominant positive eigenvalue is not *strictly* dominant), then it is no longer true in general that the direction of bifurcation determines the stability of the bifurcating positive equilibria. The bifurcation in this case is of higher codimension and is complicated by the possibility of branches of periodic cycles (and even other more complicated invariant sets) that bifurcate simultaneously with positive equilibria [7, 9, 12, 18, 19, 22, 25, 26, 35]. Several lower dimensional species cases have been thoroughly analyzed, but for higher dimensional models a complete theory is lacking. Imprimitive projection matrices do arise in applications [16], and more analysis towards a general theory that determines the properties of the bifurcation at the point where the extinction equilibrium destabilizes would be of interest.

The fundamental bifurcation theorem for matrix equations (30) can be stated using the inherent population growth rate r_0 (the dominant eigenvalue of $P(0)$) as the bifurcation parameter, as is done in Sect. 2 for the $m = 1$ dimensional case. However, it is worth pointing out that for higher dimensional cases it is often mathematically more tractable to use a different bifurcation parameter. As seen in Sect. 2, for the one dimensional case the direction of bifurcation and, consequently, the stability of the bifurcating continuum of positive equilibria is determined by the sign of $r_0 - 1$. Notice that the sign of $r_0 - 1$ is the same as the sign of $R_0 - 1$ where

$$R_0 \doteq b_0 (1 - s_0)^{-1}.$$

This quantity, when written as

$$R_0 = b_0 + b_0 s_0 + b_0 s_0^2 + \cdots = \sum_{i=0}^{\infty} b_0 s_0^i,$$

is seen to be the expected number of newborns per newborn per life time and is referred to as the *inherent net reproduction number*. In higher dimensions, R_0 is defined as follows. The projection matrix is additively decomposed into a fertility matrix F, which accounts for all newborns at the next census, and a transition/survival matrix T, which accounts for survivors at the next census (who might or might not change their classification categories):

$$P(x) = F(x) + T(x)$$

(a direct analog of the decomposition in Sect. 2). Here both F and T are non-negative matrices. The inherent net reproductive number is defined to be the spectral radius of

$$F(0)(I - T(0))^{-1}.$$

Here an additional constraint on the transition/survival matrix T is that it columns sums are less than or equal to 1 (the number of survivors cannot exceed the original number of individuals) with at least one sum strictly less than 1 (there is some mortality in the population). Or more generally, it is assume that the spectral radius of $T(0)$ is less than 1, which (as the generalization of $s_0 < 1$) means the expected life span on an individual is finite. It is known that the signs of $R_0 - 1$ and $r_0 - 1$ (here r_0 is the spectral radius of the projection matrix $F + T$) are the same [5, 11, 29]. Thus, either quantity R_0 or r_0 determine the local stability or instability of the extinction equilibrium and, as a result, the bifurcation point $r_0 = R_0 = 1$ in the fundamental bifurcation theorem. One analytic advantage of this result is that frequently a formula (expressed in terms of the entries of the projection matrix) is available for R_0 while, in higher dimensions, this not the case for r_0 [6].

Backward bifurcations and their role in the creation of strong Allee effects for matrix models (30) are investigated in [15] (for an application see [14]). The stability of the "upper" branch of positive equilibria near blue-sky bifurcation bifurcations (cf. Fig. 1) for higher dimensional matrix equations remains an open question, however.

In Sect. 3 the model tracks the dynamics of only one phenotypic trait v. If a vector of phenotypic traits v and their population means are included in the model, then the evolutionary model takes the form

$$\begin{aligned} x_{t+1} &= P(x_t, v, u_t)|_{v=u_t} x_t \\ u_{t+1} &= u_t + C \nabla_v F(x_t, v, u_t)|_{v=u_t} \end{aligned}$$

where F is a measure of fitness and C is a variance-covariance matrix for the variability of the traits [36]. For example, fitness can be taken to be the logarithm of the spectral radius of the dominant eigenvalue of $P(x, v, u)$, as a generalization of $\ln f(x, v, u)$ used in Sect. 3. Some generalizations of Theorem 2 for the evolutionary matrix model have been established. For a single trait v in a structured population model (of arbitrary size m) with primitive projection matrix see [10] and for multiple traits see [17]. For some applications of imprimitive evolutionary models see [13, 34].

Acknowledgements The author gratefully acknowledges the support of the National Science Foundation (Mathematical Biology Program in the Division of Mathematical Sciences and the Population & Community Ecology Program in the Division of Environmental Biology) under NSF grant DMS 0917435.

References

1. Abrams, P.A.: Modelling the adaptive dynamics of traits involved in inter- and intraspecific interactions: An assessment of three methods. Ecol. Lett. **4**, 166–175 (2001)
2. Allee, W.C.: Animal Aggregations, a Study in General Sociology. University of Chicago Press, Chicago (1931)
3. Caswell, H.: Matrix Population Models: Construction, Analysis and Interpretation, 2nd edn. Sinauer Associates, Inc., Publishers, Sunderland, Massachusetts (2001)
4. Courchamp, F., Berec, L., Gascoigne, J.: Allee Effects in Ecology and Conservation. Oxford University Press, Oxford, Great Britain (2008)
5. Cushing, J.M., Yicang, Z.: The net reproductive value and stability in matrix population models. Nat. Resour. Model. **8**, 297–333 (1994)
6. Cushing, J.M.: An Introduction to Structured Population Dynamics, Conference Series in Applied Mathematics, vol. 71. SIAM, Philadelphia (1998)
7. Cushing, J.M.: Nonlinear semelparous Leslie models. Math. Biosci. Eng. **3**(1), 17–36 (2006)
8. Cushing, J.M.: Matrix Models and Population Dynamics, a chapter in Mathematical Biology. IAS/Park City Mathematics Series. American Mathematical Society, Providence (2009)
9. Cushing, J.M.: Three stage semelparous Leslie models. J. Math. Biol. **59**, 75–104 (2009)
10. Cushing, J.M.: A bifurcation theorem for Darwinian matrix models. Nonlinear Stud. **17**(1), 1–13 (2010)
11. Cushing, J.M.: On the relationship between r and R_0 and its role in the bifurcation of equilibria of Darwinian matrix models. J. Biol. Dyn. **5**, 277–297 (2011)
12. Cushing, J.M., Henson, S.M.: Stable bifurcations in nonlinear semelparous Leslie models. J. Biol. Dyn. **6**, 80–102 (2012)
13. Cushing, J.M., Stump, Simon Maccracken: Darwinian dynamics of a juvenile-adult model. Math. Biosci. Eng. **10**(4), 1017–1044 (2013)
14. Cushing, J.M., Henson, S.M., Hayward, J.L.: An evolutionary game theoretic model of cannibalism. Nat. Resour. Model. **28**(4), 497–521 (2015). doi:10.1111/nrm.12079
15. Cushing, J.M.: Backward bifurcations and strong Allee effects in matrix models for the dynamics of structured populations. J. Biol. Dyn. **8**, 57–73 (2014)
16. Cushing, J.M.: On The Fundamental Bifurcation Theorem for Semelparous Leslie Models, Chapter in Mathematics of Planet Earth: Dynamics, Games and Science. CIM Mathematical Sciences Series. Springer, Berlin (2015)
17. J. M. Cushing, F. Martins, A. A. Pinto and Amy Veprauska, A bifurcation theorem for evolutionary matrix models with multiple traits, (submitted for publication)

18. Davydova, N.V., Diekmann, O., van Gils, S.A.: Year class coexistence or competitive exclusion for strict biennials? J. Math. Biol. **46**, 95–131 (2003)
19. Davydova, N.V., Diekmann, O., van Gils, S.A.: On circulant populations. I. The algebra of semelparity. Linear Algebra Appl. **398**, 185–243 (2005)
20. Dennis, B.: Allee effects: population growth, critical density, and the chance of extinction. Nat. Resour. Model. **3**, 481–538 (1989)
21. Dercole, F., Rinaldi, S.: Analysis of Evolutionary Processes: The Adaptive Dynamics Approach and Its Applications. Princeton University Press, Princeton, New Jersey (2008)
22. Diekmann, O., Davydova, N.V., van Gils, S.: On a boom and bust year class cycle. J. Differ. Equ. Appl. **11**, 327–335 (2005)
23. Elaydi, S.N.: An Introduction to Difference Equations, 3rd edn. Springer-Verlag, New York (2005)
24. Keilhöler, H.: Bifurcation Theory: An Introduction with Applications to PDEs. Applied mathematical sciences, vol. 156. Springer, New York (2004)
25. Kon, R.: Nonexistence of synchronous orbits and class coexistence in matrix population models. SIAM J. Appl. Math. **66**(2), 616–626 (2005)
26. Kon, R., Iwasa, Y.: Single-class orbits in nonlinear Leslie matrix models for semelparous populations. J. Math. Biol. **55**, 781–802 (2007)
27. Lande, R.: Natural selection and random genetic drift in phenotypic evolution. Evolution **30**, 314–334 (1976)
28. Lande, R.: A quantitative genetic theory of life history evolution. Ecology **63**, 607–615 (1982)
29. Li, C.-K., Schneider, H.: Applications of Perron-Frobenius theory to population dynamics. J. Math. Biol. **44**, 450–462 (2002)
30. Lush, J.: Animal Breeding Plans. Iowa State College Press, Ames, Iowa, USA (1937)
31. Rabinowitz, P.H.: Some global results for nonlinear eigenvalue problems. J. Funct. Anal. **7**(3), 487–513 (1971)
32. Roff, D.A.: The Evolution of Life Histories: Theory and Analysis. Chapman and Hall, New York (1992)
33. Thieme, H.R.: Mathematics in Population Biology. Princeton University Press, Princeton, New Jersey, USA (2003)
34. Veprauskas, A., Cushing, J.M.: Evolutionary dynamics of a multi-trait semelparous model. Discret. Contin. Dyn. Syst. Ser. B **21**(2), 655–676 (2015)
35. Veprauskas, A., Cushing, J.M.: A juvenile-adult population model: climate change, cannibalism, reproductive synchrony, and strong Allee effects. Journal of Biological Dynamics (2016). doi:10.1080/17513758.2015.1131853
36. Vincent, T.L., Brown, J.S.: Evolutionary Game Theory. Natural Selection, and Darwinian Dynamics. Cambridge University Press, Cambridge (2005)

An SIR Model with Nonlinear Incidence Rate and Holling Type III Treatment Rate

Preeti Dubey, Balram Dubey and Uma S. Dubey

Abstract We propose a mathematical model with nonlinear incidence rate and treatment rate to study the dynamics of susceptible-infected-recovered population. We consider nonlinear incidence rate as Crowley-Martin type and nonlinear treatment rate as Holling type III (saturated treatment function). The global stability analysis of disease-free equilibrium point and endemic equilibrium point has been investigated using Lasalles' invariance principle and Lyapunov function. A threshold value has been found to ensure the extinction or persistence of infection. The non-existence of periodic solutions have been shown using Dulac's criterion. Numerical simulations are performed to validate these analytical findings.

Keywords Crowley-Martin type incidence rate · Holling type III treatment rate · Periodic solution · Dulac's criterion · Persistence

2010 Mathematics Subject Classification 34D20 · 92B05

1 Introduction

In the field of epidemiology, interventions (e.g. treatment, vaccination, quarantine etc.) play an important role in controlling the disease spread. The diseases for which treatment is available like flu, tuberculosis, measles [1, 2]; treatment is an useful tool to eradicate them. Several researchers [3–7] have studied the effect of treatment using

P. Dubey (✉) · B. Dubey
Department of Mathematics, BITS Pilani-Pilani Campus,
Pilani, Rajasthan, India
e-mail: preeti.dubey@pilani.bits-pilani.ac.in

B. Dubey
e-mail: bdubey@pilani.bits-pilani.ac.in

U.S. Dubey
Department of Biological Sciences, BITS Pilani-Pilani Campus,
Pilani, Rajasthan, India
e-mail: uma@pilani.bits-pilani.ac.in

J.M. Cushing et al. (eds.), *Applied Analysis in Biological and Physical Sciences*,
Springer Proceedings in Mathematics & Statistics 186,
DOI 10.1007/978-81-322-3640-5_4

different type of treatment functions. In classical models treatment rate is considered to be proportional to the number of infectives. This treatment rate is not suitable in case of large number of infectives due to availability of limited resources in a community. To study this effect of limited resources, Wang and Ruan [8] developed the constant removal rate (i.e. recovery per unit time), which is given by:

$$h(I) = \begin{cases} r, & \text{if} \quad I > 0 \\ 0, & \text{if} \quad I = 0 \end{cases}.$$

This removal rate is further improved by taking the following removal rate function [9]:

$$h(I) = \begin{cases} rI, & \text{if} \quad 0 \le I \le I_0 \\ rI_0, & \text{if} \quad I > I_0 \end{cases},$$

where r and I_0 are positive constants. This removal rate shows that when the capacity of treatment is not reached then the removal rate is proportional to the number of infectives otherwise it takes the maximum capacity. Several authors [10, 11] used this removal rate to study the dynamics of their models. Further there was a scope to improve this removal rate. Zhang and Liu [12] introduced the improved treatment rate as a continuous differentiable function which saturates at its maximum value. This removal rate is given by the term $h(I) = \frac{rI}{1+\alpha I}$, where r is positive constant which denotes the cure rate and α is non-negative constant which measures the effect of delay in treatment. The term $\frac{1}{1+\alpha I}$ represents inverse of the effect of delay in treatment. This saturated removal rate is recently studied by Zhou and Fan [13] with little modification. This saturated removal rate also named as Holling type II removal rate and considered by several authors [14–16] to study the dynamics of their models. Dubey et al. [15] proposed an SEIR model with three different types of removal rates: (i) Holling type II removal rate, (as explained above) (ii) Holling type III removal rate, this is given by the term $h(I) = \frac{\beta I^2}{1+\alpha I^2}$, where β is positive constant and α is non-negative constant, $h(I)$ is a continuous differentiable function and approaches to its peak or maximum value when the number of infectives is large, and (iv) Holling type IV removal rate, which is given by $h(I) = \frac{\beta I}{\frac{I^2}{a}+I+b}$, where β and a are positive constants and b is non-negative constant.

In population dynamics, transmission of infection is the process in which susceptibles are getting infected via infected population through the various channels. Transmission plays an important role to study the dynamical behaviour of epidemic models. Recently, several researchers [11, 17–24] have focused on nonlinear type incidence rate whereas in standard models the incidence rate was defined by law of mass action i.e. bilinear incidence rate [3, 14, 25–32]. Different type of nonlinear incidence rates [15, 16, 33–36] (e.g. Holling type II, Beddington-DeAngelis type, etc.) have already been implemented by the authors in their models to study the dynamics of infectious diseases. Considering these facts, we proposed a mathematical model incorporating Crowley-Martin type incidence rate and saturated treatment rate (Holling type III) in SIR model to analyze the cited epidemic situation to control the spread of infection.

In the next section, we present susceptible-infected-recovered model with nonlinear incidence rate and Holling type III treatment rate.

2 The Mathematical Model

We considered compartmental SIR model divided into three compartments; susceptible S, infected I and recovered R compartments respectively. The model is given by following system of differential equations:

$$\begin{cases} \frac{dS}{dt} = A - \delta_0 S - \frac{\alpha SI}{(1+\beta S)(1+\gamma I)}, \\ \frac{dI}{dt} = \frac{\alpha SI}{(1+\beta S)(1+\gamma I)} - \delta_0 I - \delta_1 I - \delta_2 I - \frac{aI^2}{1+bI^2}, \\ \frac{dR}{dt} = \delta_2 I - \delta_0 R + \frac{aI^2}{1+bI^2}, \end{cases} \tag{1}$$

where $S(0) > 0,\ I(0) \geq 0,\ R(0) \geq 0.$

In model equations $\frac{d}{dt}$, represent the rate of change in corresponding compartment. Let A be the recruitment rate of the susceptible and δ_0 be the natural death rate of the population in each class. We assume that the infected individuals die out at the rate δ_1 due to infection. Infected individuals may get recover with auto immunity with the rate δ_2 and join the recovered class. We have also considered the treatment of infected individuals as saturated removal rate. The term $h(I) = \frac{aI^2}{(1+bI^2)}$ represents Holling type III [37, 38] treatment rate (continuously differentiable function), where a and b are non-negative constants and can be understood as treatment given to the infected individuals and limitation to the treatment availability, respectively. Unlike the Holling type II treatment rate this treatment rate grows first very fast and later on increases slowly with increase in number of infection and gets saturated to its maximum level $\frac{a}{b}$ (treatment capacity of community) due to limited availability of resources in the community [15, 16]. The term $\frac{\alpha SI}{(1+\beta S)(1+\gamma I)}$ denotes the monotone nonlinear incidence rate, where α is incidence rate of infection, β and γ are the effects of inhibition due to susceptible individuals and due to infected individuals or γ may also be understood as the crowding effect due to infected individuals. This functional response was introduced by P.H. Crowley and E.K. Martin in 1989 [39, 40] and is known as Crowley-Martin type incidence rate. We notice that other forms of nonlinear incidence rates can be derived from this incidence rate [16]:

(i) If we put $\beta = \gamma = 0$, then αSI which is bilinear incidence rate [14, 25–30].
(ii) If we take $\gamma = 0$, then $\frac{\alpha SI}{(1+\beta S)}$, which is saturated incidence rate with the susceptible individuals [19, 20].

(iii) For $\beta = 0$, we get $\frac{\alpha SI}{(1+\gamma I)}$, which is again saturated incidence rate but with the infected individuals. In such a case, the contact between infective and susceptible individuals may saturate at high infection level due to crowding of infective individuals or due to protection taken by susceptible individuals [11, 23, 24].

Unlike the Beddington-DeAngelis type incidence rate, the Crowley-Martin type incidence rate considers the effect of inhibition among infectives even in case of high density of susceptible populations [47]. This can be seen as follows:
Beddington-DeAngelis type for $S \to \infty$,

$$\lim_{S\to\infty} \frac{\alpha SI}{1+\beta S+\gamma I} = \frac{\alpha}{\beta},$$

and Crowley-Martin type incidence rate for $S \to \infty$,

$$\lim_{S\to\infty} \frac{\alpha SI}{(1+\beta S)(1+\gamma I)} = \frac{\alpha}{\beta(1+\gamma I)}.$$

From the above system (1) we can infer that S and I are free from the effect of R. Thus it is enough to consider the following reduced system for the study:

$$\begin{cases} \frac{dS}{dt} = A - \delta_0 S - \frac{\alpha SI}{(1+\beta S)(1+\gamma I)}, \\ \frac{dI}{dt} = \frac{\alpha SI}{(1+\beta S)(1+\gamma I)} - \delta_3 I - \frac{aI^2}{1+bI^2}, \end{cases} \tag{2}$$

where $\delta_3 = \delta_0 + \delta_1 + \delta_2$ and $S(0) > 0$, $I(0) \geq 0$.

3 Positivity and Boundedness of the System

For system (2), we found that all the solutions initiating in the region defined in Lemma 1 will eventually lie in the same region even after a long time say for $t \to \infty$ or will always stay in the same region. This can be observed as follows:

Let $N = S + I$, then $\dot{N} = \dot{S} + \dot{I} = A - \delta_0 N - (\delta_1 + \delta_2)I - \frac{aI^2}{1+bI^2}$.
Then,

$$N(t) \leq N(0)e^{-\delta_0 t} + \frac{A}{\delta_0}(1 - e^{-\delta_0 t}).$$

Thus,

$$\limsup_{t\to\infty} N(t) \leq \frac{A}{\delta_0}.$$

Furthermore, $\dot{N} < 0$ if $N > \frac{A}{\delta_0}$. This shows that solutions of system (2) point towards Ω the region defined in Lemma 1. Hence Ω is positively invariant and solutions of (2) are bounded. Thus, we can state the following Lemma.

Lemma 1 *The set* $\Omega = \{(S, I) : 0 < S + I \leq \frac{A}{\delta_0}\}$ *is a positively invariant region of system (2).*

The above lemma shows that all solutions of the model are non-negative and bounded. Thus the model is biologically well behaved.

In the next section, we discuss the existence of equilibrium points of system (2).

4 Equilibrium and Stability Analysis

We see that system (2) has only two equilibria: (i) the disease-free equilibrium (DFE) $E_0(\frac{A}{\delta_0}, 0)$, the state when infection dies out i.e. ($I = 0$) and (ii) the endemic equilibrium $E_1(S^*, I^*)$ i.e. the state when infection persists ($I \neq 0$).

We can infer from system (2) that the disease-free equilibrium E_0 always exists and its existence is trivial.

We compute the basic reproduction number using next generation matrix method and describe the stability behaviour of DFE, which is independent of initial status of sub-populations.

4.1 Computation of R_0

Model (2) can be rewritten as $\dot{x} = F(x) - V(x)$, where $x = [I, S]^T$ and $F(x)$ be the rate of appearance of new infections and $V(x)$ be the the rate of transfer of individuals into compartment and out of compartment by all other means. Jacobian of $F(x)$ at E_0 is

$$F = \begin{bmatrix} \frac{\alpha A}{(\delta_0 + A\beta)} & 0 \\ 0 & 0 \end{bmatrix},$$

and Inverse of Jacobian of $V(x)$ at E_0 is

$$V^{-1} = \begin{bmatrix} \frac{1}{\delta_3} & 0 \\ \frac{\alpha A}{(\delta_0 + A\beta)\delta_3\delta_0} & \frac{1}{\delta_0} \end{bmatrix}.$$

Then $\rho(FV^{-1})$ gives the spectral radius (largest eigenvalue) of the next generation matrix (FV^{-1}) [41]. The spectral radius gives the basic reproduction number, thus

$$R_0 = \rho(FV^{-1}) = \frac{\alpha A}{(\delta_0 + A\beta)\delta_3},$$

where R_0 is basic reproduction number.

Theorem 1 *(i) The disease-free equilibrium* $E_0(\frac{A}{\delta_0}, 0)$ *is locally asymptotically stable if* $R_0 < 1$ *and is a saddle point with stable manifold locally in the*

S-direction and unstable manifold locally in the I-direction if $R_0 > 1$.

(ii) The disease-free equilibrium $E_0(\frac{A}{\delta_0}, 0)$ is globally asymptotically stable if $R_0 \leq 1$.

Proof (i) We find the general variational matrix and then compute the variational matrices corresponding to each equilibrium point. The variational matrix corresponding to DFE $E_0(\frac{A}{\delta_0}, 0)$ is given by

$$J_{E_0} = \begin{bmatrix} -\delta_0 & -\frac{\alpha A}{(\delta_0+A\beta)} \\ 0 & \frac{\alpha A}{(\delta_0+A\beta)} - \delta_3 \end{bmatrix}.$$

The above matrix is upper-triangular matrix and has two eigenvalues: $e_1 = -\delta_0$ and $e_2 = \frac{\alpha A}{(\delta_0+A\beta)} - \delta_3$. We note that $e_1 < 0$ and $e_2 < 0$ if $R_0 < 1$. Again $e_2 > 0$ if $R_0 > 1$. Hence the first part of the theorem follows.

(ii) To show the global stability of DFE, we use Lasalle's invariance principle [42]. Let us define the positive definite function

$$L = \frac{1}{1+\beta S_0}\left(S - S_0 - S_0 ln\frac{S}{S_0}\right) + I, \quad \text{where} \quad S_0 = \frac{A}{\delta_0}.$$

Differentiating L along the solutions of (2) and simplifying, we get

$$\dot{L}(t) = -\left[\frac{\delta_0(S-S_0)^2}{S(1+\beta S_0)} + \left(\frac{a}{1+bI^2} + \frac{\delta_3\gamma}{1+\gamma I}\right)I^2\right] + \frac{\delta_3 I}{(1+\gamma I)}[R_0 - 1],$$

$$\dot{L}(t) < 0 \text{ if } R_0 \leq 1 \text{ and } \forall S, I > 0, \dot{L}(t) = 0 \text{ iff } S = S_0 = \frac{A}{\delta_0} \text{ and } I = I_0 = 0.$$

Then let M be the largest invariant set in the set $E = \{(S, I)|\dot{L}(t) = 0\}$ for each element of M, we have $I = 0$. Thus $M = \{E_0\}$ is the singleton set. Thus from Lasalle's invariance principle disease-free equilibrium is globally asymptotically stable.

4.2 Analysis at $R_0 = 1$

In this section, we state and prove the following theorem which characterizes the behavior of the DFE at $R_0 = 1$.

Theorem 2 *The disease-free equilibrium changes its stability from stable to unstable at $R_0 = 1$ and system (2) exhibits transcritical bifurcation with bifurcation parameter $\alpha = \alpha^* = \frac{\delta_3(\delta_0+A\beta)}{A}$.*

Proof Linearization matrix of system (2) at E_0 and bifurcation parameter $\alpha = \alpha^* = \frac{\delta_3(\delta_0+A\beta)}{A}$ is given by

$$J = \begin{bmatrix} -\delta_0 & -\frac{\alpha^* A}{(\delta_0+A\beta)} \\ 0 & \frac{\alpha^* A}{(\delta_0+A\beta)} - \delta_3 \end{bmatrix}.$$

The matrix J has a simple zero eigenvalue at $R_0 = 1$ and other eigenvalue of the matrix has negative real part. At this stage linearization techniques fail to conclude the behaviour of system (2). Centre Manifold Theory is used to study the behaviour of non-hyperbolic equilibrium. Then from Theorem 1 of Castillo-Chavez and Song [43], the bifurcation constants a_1 and b_1 are given by

$$a_1 = \sum_{k,i,j=1}^{2} w_k u_i u_j \left(\frac{\partial^2 f_k}{\partial x_i \partial x_j} \right)_{E_0},$$

and

$$b_1 = \sum_{k,i=1}^{2} w_k u_i \left(\frac{\partial^2 f_k}{\partial x_i \partial \alpha^*} \right)_{E_0},$$

where $u = [\frac{-\alpha^* A}{\delta_0(\delta_0+A\beta)}, 1]^T$ and $w = [0, 1]$ are right eigenvector and left eigenvector of the matrix J corresponding to zero eigenvalue, respectively. Nonzero partial derivatives associated with the system at E_0 and $\alpha = \alpha^*$ are

$$\frac{\partial^2 f_2}{\partial x_1 \partial x_2} = \frac{\alpha^* {\delta_0}^2}{(\delta_0 + A\beta)^2}, \quad \frac{\partial^2 f_2}{\partial x_2^2} = -2\left(a + \frac{\alpha^* \gamma A}{(\delta_0 + A\beta)}\right), \quad \frac{\partial^2 f_2}{\partial x_2 \partial \alpha^*} = \frac{A}{(\delta_0 + A\beta)}.$$

Hence,

$$a_1 = -\frac{{\alpha^*}^2 \delta_0 A}{(\delta_0 + A\beta)^3} - 2\left(a + \frac{\alpha^* \gamma A}{(\delta_0 + A\beta)}\right) < 0,$$

$$b_1 = \frac{A}{(\delta_0 + A\beta)} > 0.$$

This shows that at $R_0 = 1$, DFE changes stability from stable to unstable and positive equilibrium exists when R_0 crosses the threshold value i.e. 'one'. This emphasizes that the system exhibits transcritical bifurcation at $R_0 = 1$.

4.3 *Existence of Endemic Equilibrium $E_1(S^*, I^*)$*

Now we show the existence of endemic equilibrium $E_1(S^*, I^*)$ using isocline method under certain threshold value or conditions. Let us assume that

$$f(S, I) = A - \delta_0 S - \frac{\alpha SI}{(1+\beta S)(1+\gamma I)} = 0, \tag{3}$$

$$g(S, I) = \frac{\alpha S}{(1+\beta S)(1+\gamma I)} - \delta_3 - \frac{aI}{1+bI^2} = 0. \tag{4}$$

From first isocline (3), we observe the following:
(i) when

$$I = 0, \quad \text{then} \quad S = \frac{A}{\delta_0} = S_0.$$

(ii)

$$\frac{dS}{dI} = -\frac{\partial f/\partial I}{\partial f/\partial S},$$

where

$$\frac{\partial f}{\partial I} = -\frac{\alpha S}{(1+\beta S)(1+\gamma I)^2},$$

$$\frac{\partial f}{\partial S} = -\delta_0 - \frac{\alpha I}{(1+\gamma I)(1+\beta S)^2}.$$

This implies that

$$\frac{dS}{dI} = -\frac{\alpha S/(1+\beta S)(1+\gamma I)^2}{\delta_0 + \frac{\alpha I}{(1+\gamma I)(1+\beta S)^2}} < 0.$$

Hence first isocline (3) is decreasing function of I.

From second isocline (4), we have the following observations:

(i) when

$$I = 0, \quad \text{then} \quad S = \frac{\delta_3}{\alpha - \delta_3 \beta} = S_1(say).$$

$$S_1 > 0 \quad \text{if} \quad \alpha > \delta_3 \beta. \tag{5}$$

(ii)

$$\frac{dS}{dI} = -\frac{\partial g/\partial I}{\partial g/\partial S},$$

where

$$\frac{\partial g}{\partial I} = -\frac{\alpha\gamma S}{(1+\beta S)(1+\gamma I)^2} - \frac{a(1-bI^2)}{(1+bI^2)^2},$$

$$\frac{\partial g}{\partial S} = \frac{\alpha I}{(1+\gamma I)(1+\beta S)^2},$$

This implies

$$\frac{dS}{dI} = \frac{\frac{\alpha\gamma S}{(1+\beta S)(1+\gamma I)^2} + \frac{a(1-bI^2)}{(1+bI^2)^2}}{\frac{\alpha I}{(1+\gamma I)(1+\beta S)^2}},$$

It can be noted from the above expression that the denominator is always positive and the numerator is positive if

$$1 - bI^2 > 0 \quad \text{i.e.} \quad bI^2 < 1,$$

After substituting the maximum value of I (i.e. $\frac{A}{\delta_0}$), we get the inequality $bA^2 < \delta_0^2$. Thus, $\frac{dS}{dI}$ is positive if $bA^2 < \delta_0^2$ and $g(S, I)$ is increasing function of I. This implies that the two isoclines (3) and (4) intersects at a unique point $E^*(S^*, I^*)$ if $S_0 > S_1$ i.e. if $R_0 = \frac{\alpha A}{\delta_3(\delta_0+A\beta)} > 1$. Thus we can state the existence and uniqueness of the endemic equilibrium in the following Lemma.

Lemma 2 *The endemic equilibrium $E_1(S^*, I^*)$ exists if the following inequalities hold true:*

$$bA^2 < \delta_0^2, \tag{6}$$

$$R_0 = \frac{\alpha A}{\delta_3(\delta_0 + A\beta)} > 1. \tag{7}$$

Remark 1 It may be noted that if condition (7) holds, then condition (5) is satisfied by default.

Remark 2 If condition (6) fails, then $\frac{dS}{dI}$ for isocline (4) may be positive or negative depending upon the values of parameters. In such a case there may exist more than one endemic equilibrium.

The graphical representation of existence of endemic equilibrium, using the set of parameters given in Table 1, is shown in Fig. 1.

Table 1 Parameter values and units

Parameters	Value (Unit)
Recruitment rate (A)	$2\ (person\ (d)^{-1})$
Natural Death rate of each sub-population (δ_0)	$0.05\ (d)^{-1}$
Disease induced death rate of infected (δ_1)	$0.001\ (d)^{-1}$
Recovery rate of infected due to auto immunity (δ_2)	$0.002\ (d)^{-1}$
Treatment rate (a)	$0.02\ (person)^{-1}\ (d)^{-1}$
Limitation rate in treatment availability (b)	$0.0004\ (person)^{-1}\ (d)^{-1}$
Transmission rate (α)	$0.004\ (person)^{-1}\ (d)^{-1}$
Inhibition rate due to susceptible (β)	$0.004\ (person)^{-1}\ (d)^{-1}$
Inhibition rate due to infected (γ)	$0.002\ (person)^{-1}\ (d)^{-1}$

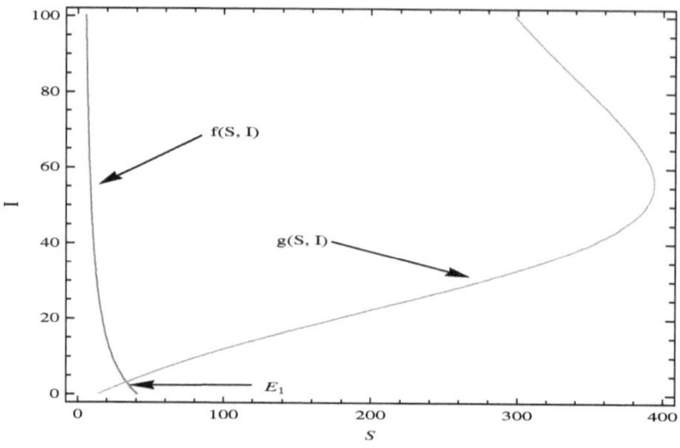

Fig. 1 Plot of two isoclines showing existence of endemic equilibrium (E_1)

The next theorem shows the uniform persistence of system (2). Biologically persistence implies that the sub-populations exist always and will not lead to extinction if initially they are present.

Theorem 3 *Assume that Lemma 1 holds. Let the following inequality is satisfied:* $\max\left\{\frac{\alpha A}{(\delta_0+\beta A)(\delta_0+\gamma A)}, \frac{aA}{\delta_0^2+bA^2}\right\} < 1$. *Then system* (2) *is uniformly persistent.*

Proof In order to define permanence (uniformly persistence) of the system, we assume that $S(0) > 0$ and $I(0) > 0$. Then we say that system (2) is uniformly persistence [45, 46] if there exists positive constants M_1 and M_2 s.t.

$$M_1 \le \liminf_{t\to\infty} S(t) \le \limsup_{t\to\infty} S(t) \le M_2,$$

$$M_1 \le \liminf_{t\to\infty} I(t) \le \limsup_{t\to\infty} I(t) \le M_2.$$

From Lemma 1, it follows that

$$\limsup_{t\to\infty} S(t) \le \frac{A}{\delta_0},$$

$$\limsup_{t\to\infty} I(t) \le \frac{A}{\delta_0}.$$

$\Rightarrow$ For any $\varepsilon > 0$, $\exists$ a $T > 0$ such that

$$S(t) < \frac{A}{\delta_0} + \varepsilon = S_m(say),$$

$$I(t) < \frac{A}{\delta_0} + \varepsilon = S_m, \quad \forall t \geq T.$$

From the first equation of model (1.2), we have

$$\frac{dS}{dt} \geq A - \delta_0 S - \frac{\alpha S_m^2}{(1+\beta S_m)(1+\gamma S_m)},$$

This implies that

$$\liminf_{t\to\infty} S(t) \geq \frac{1}{\delta_0}\left(A - \frac{\alpha S_m^2}{(1+\beta S_m)(1+\gamma S_m)}\right),$$

which is true for every sufficiently small $\varepsilon > 0$. Hence for large t, it follows that

$$\liminf_{t\to\infty} S(t) \geq \frac{A}{\delta_0}\left(1 - \frac{\alpha A}{(\delta_0+\beta A)(\delta_0+\gamma A)}\right) = S_a(say)$$

and $S_a > 0$ if $\frac{\alpha A}{(\delta_0+\beta A)(\delta_0+\gamma A)} < 1$, or $R_0 < \frac{(\delta_0+A\gamma)}{\delta_3}$.

Again from model (2), we have

$$\frac{d}{dt}(S+I) \geq A - \delta_m(S+I) - \frac{aA^2}{\delta_0^2 + bA^2},$$

where $\delta_m = max\{\delta_0, \delta_3\}$.

$$\Rightarrow \liminf_{t\to\infty}(S(t)+I(t)) \geq \frac{A}{\delta_m}\left(1 - \frac{aA}{\delta_0^2 + bA^2}\right) = I_a(say),$$

We note that $I_a > 0$ if

$$\frac{aA}{\delta_0^2 + bA^2} < 1.$$

Hence the theorem follows.

Further, we discuss the local and global stability of the endemic equilibrium point $E_1(S^*, I^*)$. We state and prove the following results:

Theorem 4 *The endemic equilibrium $E_1(S^*, I^*)$ is locally asymptotically stable iff the following inequalities hold true:*

$$\frac{\alpha S^*}{(1+\beta S^*)(1+\gamma I^*)^2} < L_1, \tag{8}$$

$$\frac{\delta_0 \alpha S^*}{(1+\beta S^*)(1+\gamma I^*)^2} < L_2, \tag{9}$$

where

$$L_1 = \delta_0 + \delta_3 + \frac{2aI^*}{(1+bI^{*2})^2} + \frac{\alpha I^*}{(1+\gamma I^*)(1+\beta S^*)^2},$$
$$L_2 = \left(\delta_3 + \frac{2aI^*}{(1+bI^{*2})^2}\right)\left(\delta_0 + \frac{\alpha I^*}{(1+\gamma I^*)(1+\beta S^*)^2}\right).$$

Proof The variational matrix corresponding to endemic equilibrium $E_1(S^*, I^*)$ is given as follows:

$$J_{E_1} = \begin{bmatrix} -\delta_0 - \frac{\alpha I^*}{(1+\gamma I^*)(1+\beta S^*)^2} & -\frac{\alpha S^*}{(1+\beta S^*)(1+\gamma I^*)^2} \\ \frac{\alpha I^*}{(1+\gamma I^*)(1+\beta S^*)^2} & \frac{\alpha S^*}{(1+\beta S^*)(1+\gamma I^*)^2} - \delta_3 - \frac{2aI^*}{(1+bI^{*2})^2} \end{bmatrix}.$$

The characteristic polynomial of the above matrix is given by the following equation

$$\lambda^2 + a_1\lambda + a_2 = 0, \tag{10}$$

where,

$$a_1 = \delta_0 + \frac{\alpha I^*}{(1+\gamma I^*)(1+\beta S^*)^2} - \frac{\alpha S^*}{(1+\beta S^*)(1+\gamma I^*)^2} + \delta_3 + \frac{2aI^*}{(1+bI^{*2})^2},$$
$$a_2 = \left(\delta_3 + \frac{2aI^*}{(1+bI^{*2})^2}\right)\left(\delta_0 + \frac{\alpha I^*}{(1+\gamma I^*)(1+\beta S^*)^2}\right) - \frac{\delta_0\alpha S^*}{(1+\beta S^*)(1+\gamma I^*)^2}.$$

Using the Routh-Hurwitz criteria, it follows that eigenvalues of the above variational matrix have negative real parts iff $a_1 > 0$ and $a_2 > 0$. This implies that the endemic equilibrium $E_1(S^*, I^*)$ is locally asymptotically stable iff inequalities (8) and (9) hold true.

Remark 3 It may be noted that conditions (8) and (9) hold if

$$\frac{\alpha S^*}{(1+\beta S^*)(1+\gamma I^*)^2} < \delta_3 + \frac{2aI^*}{(1+bI^{*2})^2}.$$

In the following theorem, we show that the endemic equilibrium $E_1(S^*, I^*)$ is globally asymptotically stable.

Theorem 5 *Let the following inequality holds in the region* Ω*:*

$$\frac{\alpha^2\gamma S^* I^*}{(1+\gamma I^*)(1+\beta S^*)^2} < X_1X_2, \tag{11}$$

where

$$X_1 = \delta_0 + \frac{\alpha I^* \delta_0^2}{(1+\beta S^*)(\delta_0+\beta A)(\delta_0+\gamma A)},$$

$$X_2 = \frac{\alpha\gamma\delta_0^2 S^*}{(\delta_0+\beta A)(1+\gamma I^*)(\delta_0+\gamma A)} + \frac{a{\delta_0}^2}{({\delta_0}^2+bA^2)(1+bI^{*2})} - \frac{aI^*\sqrt{b}}{2(1+bI^{*2})^2}.$$

Then the positive equilibrium $E_1(S^*, I^*)$ is globally asymptotically stable with respect to all solutions in the interior of the positive quadrant Ω.

Proof We take a positive definite scalar function V as follows:

$$V(S,I) = \frac{1}{2}(S-S^*)^2 + \frac{1}{2}k_1\left(I - I^* - I^* ln\frac{I}{I^*}\right).$$

Differentiating V w.r.t. time t along the solutions of model (2), we get

$$\frac{dV}{dt} = -a_{11}(S-S^*)^2 + a_{12}(S-S^*)(I-I^*) - a_{22}(I-I^*)^2$$

where

$$a_{11} = \delta_0 + \frac{\alpha I^*}{PP^*L} > 0,$$

$$a_{12} = -\frac{\alpha S}{PL} + \frac{\alpha\gamma S^* I^*}{P^*LL^*} + \frac{\alpha k_1}{PP^*L},$$

$$a_{22} = k_1\left(\frac{\alpha\gamma S^*}{P^*LL^*} + \frac{a(1-bII^*)}{(1+bI^{*2})(1+bI^2)}\right),$$

$$P = 1+\beta S, \quad P^* = 1+\beta S^*, \quad L = 1+\gamma I, \quad L^* = 1+\gamma I^*.$$

Sufficient conditions for $\dot{V}$ to be negative definite are $a_{11} > 0$ and $a_{12}^2 < 4a_{11}a_{22}$. The second condition for $\dot{V}$ to be negative definite leads to the inequality (11) for $k_1 = \frac{\gamma S^* I^*}{1+\gamma I^*}$. Hence the theorem follows.

In the following theorem, we show the non-existence of limit cycle under certain condition.

Theorem 6 *If $bA^2 < {\delta_0}^2$, then model (2) does not have any periodic solution in the interior of the positive quadrant of the SI-plane.*

Proof We define a real valued function in the interior of positive quadrant of the SI-plane as follows:

$$H(S,I) = \frac{(1+\beta S)(1+\gamma I)}{SI} > 0.$$

Let us consider,

$$h_1(S, I) = A - \delta_0 S - \frac{\alpha S I}{(1 + \beta S)(1 + \gamma I)},$$

$$h_2(S, I) = \frac{\alpha S I}{(1 + \beta S)(1 + \gamma I)} - \delta_3 I - \frac{a I^2}{1 + b I^2}.$$

Then we have,

$$div(Hh_1, Hh_2) = \frac{\partial}{\partial S}(Hh_1) + \frac{\partial}{\partial I}(Hh_2)$$

$$= -\frac{A(1 + \gamma I)}{I S^2} - \frac{\delta_0 \beta (1 + \gamma I)}{I} - \frac{\delta_3 \gamma (1 + \beta S)}{S} - \frac{a(1 - bI^2 + 2\gamma I)(1 + \beta S)}{S(1 + bI^2)^2}.$$

We can see that the above expression is not zero and this will not change sign in the positive quadrant of the SI-plane if the inequality $bA^2 < \delta_0{}^2$ holds. Then by Dulac's criterion [44], it is apparent that model (2) does not have any periodic solution in the interior of the positive quadrant of the SI-plane.

5 Numerical Simulations

In this section, we present simulation results for model (2) using Mathematica and MatLab 7.10. Mathematica has been used for calculation of symbolic mathematical expressions while Matlab is used to plot the figures.

We chose the dataset of parameters as given in Table 1 for model (2). For this set of parameters, the basic reproduction number R_0 is 2.6025 > 1 and other conditions for the existence of endemic equilibrium are satisfied. Endemic equilibrium point $E_1(S^*, I^*)$ is given by $S^* = 33.0457$ and $I^* = 3.1467$. The phase portrait of susceptible population and infected population (Fig. 2) shows that the trajectories ini-

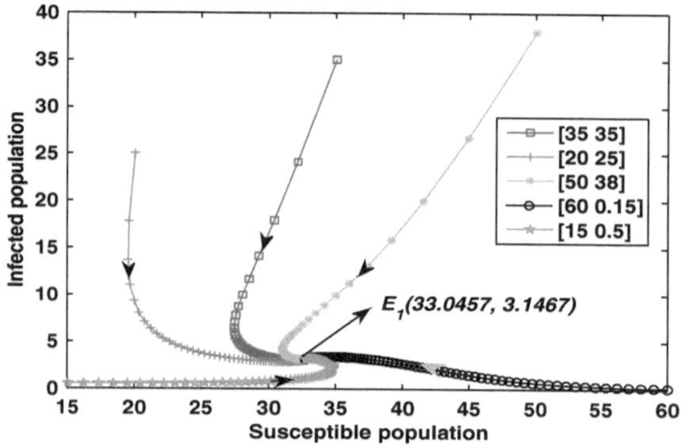

Fig. 2 Phase portrait of endemic equilibrium point E_1

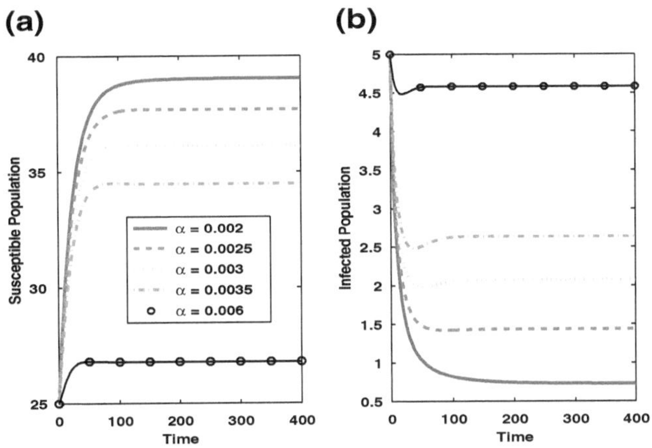

Fig. 3 Effect of incidence rate (α) on S and I

tiating from different initial points (initial values are given in the legend) approach to the unique equilibrium point $E_1(33.0457, 3.1467)$. This is evident from Fig. 2 that the endemic equilibrium point is globally asymptotically stable for this dataset. Thus the stability of the endemic equilibrium point is independent of initial status of susceptibles and infectives.

In Fig. 3a, b, we plotted the effect of variation of incidence rate α on susceptible S and infected population I for the values of parameters given in Table 1. For higher values of α the trajectory corresponding to susceptible population settles down at low level while trajectory for infected population first decreases and then attains its steady state at high level of infection. The initial decrease in infection is due to treatment available in the community. The number of infectives decreases with decrease in incidence of infection which can be controlled by treatment.

We have considered Crowley-Martin type nonlinear incidence rate so the effect of the constant β (involved in the incidence expression) i.e. measure of inhibition with respect to susceptible is plotted in Fig. 4a, b. When β is low the trajectory corresponding to susceptible population settles at a lower level and the trajectory corresponding to infected population settles at high level of infection. This shows that the number of susceptible can be increased and the number of infectives can be decreased by increasing the value of β i.e. by increasing the density of preventive measures taken by susceptible individuals.

The effect of treatment given to the community is shown in Fig. 5a–d using different treatment rates. It may be noted here that the legend for Fig. 5a, c are same. Further, the legend for Fig. 5b, d is same. Figure 5a shows the effect of treatment on the infected and Fig. 5b shows the effect of limitation to the treatment resources on infected population with Holling type III treatment rate while the same has been shown in Fig. 5c, d using Holling type II treatment rate for comparison purposes. In absence of treatment ($a = 0$) the infection increases very rapidly and settles to its

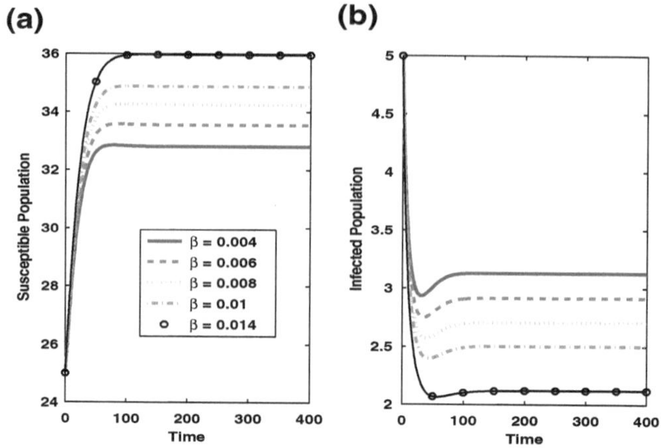

Fig. 4 Effect of β on S and I

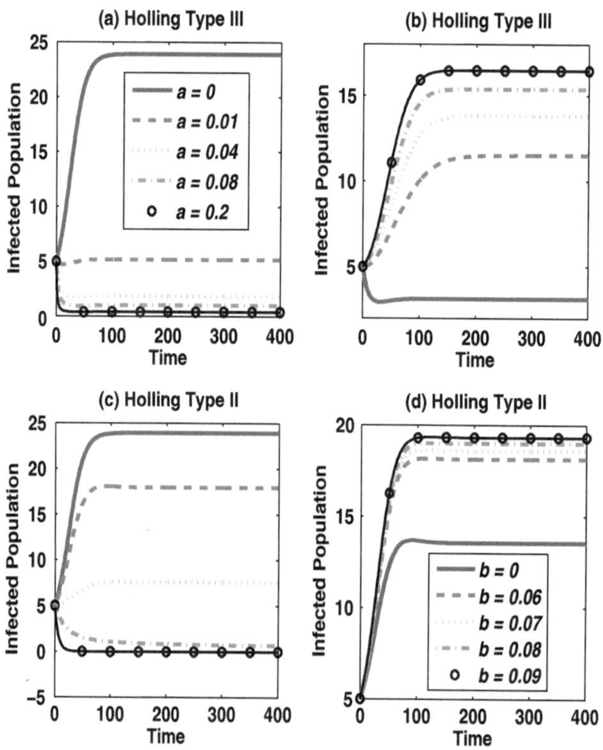

Fig. 5 Effect of a and b on I with HTIII and HTII treatment rates

steady state (Fig. 5a), on the contrary, when there is no restriction to availability of treatment ($b = 0$) the infection decreases sharply and get settled to its steady state (Fig. 5b). Figure 5a, c show that the number of infected individuals can be decreased faster in the case of Holling type III treatment rate in comparison to that of Holling type II treatment rate. When there is low availability of treatment, infection is high. When the ample quantity of treatment is available in the community the infection almost dies out. Infection gets increase with increase in limit to the availability of treatment.

6 Conclusion

In this paper, we addressed the intervention to control the infection and a monotone nonlinear incidence rate to get the better insight of spread of infection among the populations. We found that model has two equilibria: disease-free equilibrium E_0 and endemic equilibrium E_1. It has been shown that the infection persists along with the low availability of treatment when basic reproduction number is greater than one. The local and global stability of each equilibria have been studied and found that persistence or eradication of infection is independent of initial status of the sub-populations and system is uniformly persistence under the condition stated in Theorem 3. This is also evident from numerical simulations that the infection increases with increase in incidence but settles at a lower level due to availability of treatment. Further infection will decrease with the increase in measure of inhibition taken by susceptibles. It is also found that the eradication of infection is possible only when the treatment given to the population managed according to the availability of resources. It has also been observed that the equilibrium point changes its stability from stable to unstable at $R_0 = 1$ i.e. model exhibits transcritical bifurcation at $R_0 = 1$. Non-existence of periodic solution under the condition defined in Theorem 6 ensures that the infection will not reoccur in future under mentioned condition.

Acknowledgements One of the author (PD) gratefully acknowledges the support received from UGC-BSR, New Delhi, India, Grant No. F.4-1/2006(BSR)/7-203/2009(BSR). Authors are also grateful to the anonymous referees for their critical review and suggestion which improved the quality of the paper.

References

1. Earn, D.J.D., Dushoff, J., Levin, S.A.: Ecology and evolution of the flu. Trends Ecol. Evol. **17**, 334–340 (2002)
2. Rohani, P., Keeling, M.J., Grenfell, B.T.: The interplay between determinism and stochasticity in childhood diseases. Am. Nat. **159**, 469–481 (2002)
3. Hethcote, H.W.: The mathematics of infectious diseases. SIAM Rev. **42**, 599–653 (2000)
4. Ma, Z., Zhou, Y., Wang, W., Jin, Z.: Mathematical Modelling and Research of Epidemic Dynamical Systems. Science Press, Beijing (2004)

5. Sun, C., Yang, W.: Global results for an SIRS model with vaccination and isolation. Nonlinear Anal. RWA **11**, 4223–4237 (2010)
6. Qiu, Z., Feng, Z.: Transmission dynamics of an influenza model with vaccination and antiviral treatment. Bull. Math. Biol. **72**, 1–33 (2010)
7. Moghadas, S.M., Alexander, M.E.: Bifurcations of an epidemic model with non-linear incidence and infection-dependent removal rate. Math. Med. Bio. **23**, 231–254 (2006)
8. Wang, W., Ruan, S.: Bifurcation in an epidemic model with constant removal rate of the infectives. J. Math. Anal. Appl. **291**, 775–793 (2004)
9. Wang, W.: Backward bifurcation of an epidemic model with treatment. Math. Biosci. **201**, 58–71 (2006)
10. Hu, Z., Liu, S., Wang, H.: Backward bifurcation of an epidemic model with standard incidence rate and treatment rate. Nonlinear Anal. RWA **9**(5), 2302–2312 (2008)
11. Li, X., Li, W., Ghosh, M.: Stability and bifurcation of an SIR epidemic model with nonlinear incidence and treatment. Appl. Math. Comput. **210**, 141–150 (2009)
12. Zhang, X., Liu, X.: Backward bifurcation of an epidemic model with saturated treatment function. J. Math. Anal. Appl. **348**, 433–443 (2008)
13. Zhou, L., Fan, M.: Dynamics of an SIR epidemic model with limited medical resources revisited. Nonlinear Anal. RWA **13**, 312–324 (2012)
14. Zhang, Z., Suo, S.: Qualitative analysis of a SIR epidemic model with saturated treatment rate. J. Appl. Math. Comput. **34**, 177–194 (2010)
15. Dubey, B., Patra, A., Srivastava, P.K., Dubey, U.S.: Modeling and analysis of an SEIR model with different types of nonlinear treatment rates. J. Biol. Sys. **21** (2013)
16. Dubey, B., Dubey, P., Dubey, U.S.: Dynamics of an SIR model with nonlinear incidence and treatment rate. Appl. Appl. Math. **10**(2), 718–737 (2015)
17. Capasso, V., Serio, G.: A generalization of the Kermack C. Mckendrick deterministic epidemic model. Math. Biosci. **42**, 43–75 (1978)
18. Liu, W.M., Hethcote, H.W., Levin, S.A.: Dynamical behavior of epidemiological models with nonlinear incidence rates. J. Math. Biol. **25**, 359–380 (1987)
19. Zhang, J.Z., Jin, Z., Liu, Q.X., Zhang, Z.Y.: Analysis of a delayed SIR model with nonlinear incidence rate. Discrete. Dyn. Nat. Soc. **2008**, Article ID 66153 (2008)
20. Gao, S., Chen, L., Nieto, J.J., Torres, A.: Analysis of a delayed epidemic model with pulse vaccination and saturation incidence. Vaccine **24**(35), 6037–6045 (2006)
21. Mukhopadhyay, B., Bhattacharyya, R.: Analysis of a spatially extended nonlinear SEIS epidemic model with distinct incidence for exposed and infectives. Nonlinear Anal. RWA **9**, 585–598 (2008)
22. Korobeinikov, A., Maini, P.K.: Nonlinear incidence and stability of infectious disease models. Math. Med. Biol. **22**, 113–128 (2005)
23. Alexander, M.E., Moghadas, S.M.: Periodicity in an epidemic model with a generalized nonlinear incidence. Math. Biosci. **189**, 75–96 (2004)
24. Xu, R., Ma, Z.: Stability of a delayed SIRS epidemic model with a nonlinear incidence rate. Chaos Soliton Fract. **41**, 2319–2325 (2009)
25. Anderson, R. M., May, R.M.: Infectious Diseases of Humans: Dynamics and Control. Oxford University Press, Oxford (1992)
26. Bailey, N.T.J.: The Mathematical Theory of Infectious Diseases and its Applications. Griffin (1975)
27. Brauer, F., Castillo-Chavez, C.: Mathematical Models in Population Biology and Epidemiology. Texts in Applied Mathematics, Springer, New York (2001)
28. Kermack, W.O., McKendrick, A.G.: A Contribution to the mathematical theory of epidemics. Proc. R. Soc. Lond. **A115**, 700–721 (1927)
29. McKendrick, A.G.: Applications of mathematics to medical problems. Proc. Edin. Math. Soc. **44**, 98–130 (1925)
30. Shulgin, B., Stone, L., Agur, Z.: Pulse vaccination strategy in the SIR epidemic model. Bull. Math. Biol. **60**, 1123–1148 (1998)

31. Ghosh, M., Chandra, P., Sinha, P., Shukla, J.B.: Modelling the spread of carrier-dependent infectious diseases with environmental effect. Appl. Math. Comput. **152**, 385–402 (2004)
32. Shukla, J.B., Singh, V., Misra, A.K.: Modeling the spread of an infectious disease with bacteria and carriers in the environment. Nonlinear Anal. RWA. **12**, 2541–2551 (2011)
33. DeAngelis, D.L., Goldstein, R.A., O'Neill, R.V.: A model for tropic interaction. Ecology. **56**, 881–892 (1975)
34. Beddington, J.R.: Mutual interference between parasites or predators and its effect on searching efficiency. J. Anim. Ecol. **44**, 331–340 (1975)
35. Kaddar, A.: Stability analysis in a delayed SIR epidemic model with a saturated incidence rate. Nonlinear Anal. Model. Control **15**, 299–306 (2010)
36. Elaiw, A.M., Azoz, S.A.: Global properties of a class of HIV infection models with Beddington-DeAngelis functional response. Math. Method Appl. Sci. **36**, 383–394 (2013)
37. Lamontagne, Y., Coutu, C., Rousseau, C.: Bifurcation analysis of a predatorprey system with generalised Holling type III functional response. J. Dyn. Differ. Equ. **20**(3), 535–571 (2008)
38. Hethcote, H.W., van den Driessche, P.: Some epidemiological models with nonlinear incidence. J. Math. Biol. **29**(3), 271–287 (1991)
39. Crowley, P.H., Martin, E.K.: Functional responses and interference within and between year classes of a dragonfly population. J. N. Am. Benthol. Soc. **8**(3), 211–221 (1989)
40. Shi, X., Zhou, X., Song, X.: Analysis of a stage-structured predator-prey model with Crowley-Martin function. J. Appl. Math. Comput. **36**(1–2), 459–472 (2011)
41. den Driessche, P.V., Watmough, J.: Reproduction numbers and sub-threshold endemic equilibria for compartmental models of disease transmission. Math. Biosci. **180**(1), 29–48 (2002)
42. La Salle, J.P.: The Stability of Dynamical Systems. SIAM (1976)
43. Castillo-Chavez, C., Song, B.: Dynamical models of tuberculosis and their applications. Math. Biosci. Eng. **1**, 361–404 (2004)
44. Sastry, S.: Analysis, Stability and Control. Springer, New York (1999)
45. Sarwardi, S., Haque, M., Mandal, P.K.: Persistence and global stability of Bazykin predatorprey model with Beddington De Angelis response function. Commun. Nonlinear Sci. Numer. Simul. **19**(1), 189–209 (2014)
46. Wang, W., Mulone, G., Salemi, F., Salone, V.: Permanence and stability of a stage-structured predatorprey model. J. Math. Anal. Appl. **262**(2), 499–528 (2001)
47. Edwin, A.: Modeling and analysis of a two prey-one predator system with harvesting, Holling Type II and ratio-dependent responses. Doctoral dissertation, Makerere University (2010)

Dynamic Complexities in a Pest Control Model with Birth Pulses

Anju Goel and Sunita Gakkhar

Abstract In this paper, an impulsive system of differential equations is proposed to model a pest control system. The stage-structured system consists of immature and mature pest population. Birth pulses occur at regular intervals to release immature pest. The pest is controlled by spraying chemical pesticides affecting both immature and mature pest. The stroboscopic map of the impulsive system is analyzed for the stability of pest-free and non-trivial period-1 solution. Numerical simulations with MATLAB reveal the complex dynamical behavior. Period doubling cascade, chaos and period halving bifurcations are observed above the threshold level.

Keywords Stage-structure · Birth-pulse · Ricker function · Bifurcation · Chaos

1 Introduction

Integrated pest management (IPM) is a long-term control strategy which is a combination of biological, cultural and chemical tactics to reduce pest population in such a way that economic loss to the grower is minimized with least impact on the environment [9]. Many authors have investigated different population models concerning the impulsive pest control. In terms of the mathematical treatment, such models are described by the system of impulsive differential equations [6]. These equations describe the phenomena of steep/ instantaneous changes. Many mathematical models are developed to predict success of an IPM plan. These models incorporate the dynamics of pest when chemical control/biological control is applied continuously [2, 8] or impulsively [7, 9, 12].

Researchers have mostly considered that the mature pest reproduce throughout the year, but quite frequently births are seasonal or occur in regular pulses. The

A. Goel (✉) · S. Gakkhar
Department of Mathematics, Indian Institute of Technology Roorkee (IITR), Roorkee, Uttarakhand 247667, India
e-mail: goelanju23@gmail.com

S. Gakkhar
e-mail: sungkfma@iitr.ernet.in

J.M. Cushing et al. (eds.), *Applied Analysis in Biological and Physical Sciences*, Springer Proceedings in Mathematics & Statistics 186,
DOI 10.1007/978-81-322-3640-5_5

term of continuous reproduction of mature pest is replaced with birth pulse [1]. An impulsive system with birth pulses has been discussed by [3, 4, 10, 11].

In this paper, an impulsive stage-structured system with birth pulses is proposed to study the effects of spraying pesticides to control the pest population. The objective of this paper is to investigate the dynamics of impulsive stage-structured system subjected to time-dependent impulsive control strategy with birth pulses. The threshold conditions for the stability of pest free solution as well as period-1 solution is obtained. Numerical simulation has been carried out to study the analytical results and to explore the complex dynamics of the system.

2 Mathematical Model

Consider the dynamics of stage-structured single species pest. Let total pest density $N(t)$ be divided into two classes: immature class with density x_1 and mature class with density x_2 at time t. The mortality rate of immature and mature pest is d_r. The maturation rate $a_r > 0$ determines the mean length of the juvenile period. Let mature pest population release immature pest continuously and the birth function $B_f(N)$ is assumed to be of Ricker type as:

$$B_f(N(t)) = be^{-N(t)}; \quad N(t) = x_1(t) + x_2(t). \tag{1}$$

Accordingly, the dynamics of system is governed by the following mathematical model:

$$\begin{aligned} \frac{dx_1(t)}{dt} &= B_f(N(t))x_2(t) - (d_r + a_r)x_1(t) \\ \frac{dx_2(t)}{dt} &= a_r x_1(t) - d_r x_2(t) \end{aligned} \tag{2}$$

Now, it is assumed that the mature population reproduce in pulses periodically at an interval T instead of continuous births as in (2). Let $x(t^+)$ and $y(t^+)$ be the quantities of biomass of immature and mature pest just after birth pulse at t. The following system of differential equations for $t \neq mT$ with impulsive conditions incorporating birth pulses at $t = mT$ is given as:

$$\begin{aligned} \frac{dx_1(t)}{dt} &= -d_r x_1(t) - a_r x_1(t) \\ \frac{dx_2(t)}{dt} &= a_r x_1(t) - d_r x_2(t) \qquad t \neq mT \\ x_1(mT^+) &= x_1(mT) + B_f(N(mT))x_2(mT) \\ x_2(mT^+) &= x_2(mT) \qquad t = mT, m = 1, 2, 3, ... \end{aligned} \tag{3}$$

Assuming chemical spray is also periodic of period T, but it is applied at a time different then the birth. Accordingly, l_r is introduced and chemical spray is applied between $(m-1)^{th}$ and m^{th} pulse at fixed time $t=(m+l_r-1)T$, $0<l_r<1$. Further, the fraction of immature and mature pest which die due to pesticide spray instantaneously are $\beta_1, \beta_2 (0<\beta_1, \beta_2<1)$ respectively. Accordingly the system (3) is subjected to the following additional impulsive condition at $t=(m+l_r-1)T$:

$$x_1(t^+)=(1-\beta_1)x_1(t), \qquad x_2(t^+)=(1-\beta_2)x_2(t), \qquad t=(m+l_r-1)T \tag{4}$$

The stage-structured impulsive pest control model with pesticide spray and birth pulse is defined on set $\Re^2_+=\{(x_1,x_2)\in\Re^2 \mid x_1\geq 0, x_2\geq 0\}$. The complete model is now written as:

$$\begin{aligned}
\frac{dx_1(t)}{dt} &= -(d_r+a_r)x_1(t) \\
\frac{dx_2(t)}{dt} &= a_r x_1(t)-d_r x_2(t) && t\neq(m+l_r-1)T, t\neq mT \\
x_1(t^+) &= (1-\beta_1)x_1(t) \\
x_2(t^+) &= (1-\beta_2)x_2(t) && t=(m+l_r-1)T \\
x_1(t^+) &= x_1(t)+B_f(N(t))x_2(t) \\
x_2(t^+) &= x_2(t) && t=mT
\end{aligned} \tag{5}$$

$$x_1(0)=x_{1_0}>0, x_2(0)=x_{2_0}>0$$

Here, x_{1_0} and x_{2_0} are initial densities of immature and mature pest respectively. All model parameters are positive constants.

3 Model Analysis

Let $x_1=x_{1m-1}$ and $x_2=x_{2m-1}$ be the initial densities of immature and mature pest respectively at $t=(m-1)T$. The analytical solution of the differential equations of system (5) between the pulses can be written as:

$$\begin{aligned}
x_1(t) &= x_{1m-1}e^{-(a_r+d_r)(t-(m-1)T)} && (m-1)T\leq t<(m+l_r-1)T \\
x_2(t) &= e^{-d_r(t-(m-1)T)}[x_{2m-1}+x_{1m-1}(1-e^{-a_r(t-(m-1)T)})]
\end{aligned} \tag{6}$$

$$\begin{aligned}
x_1(t) &= (1-\beta_1)x_{1m-1}e^{-(a_r+d_r)(t-(m-1)T)} \qquad (m+l_r-1)T\leq t<mT \\
x_2(t) &= [\{(\beta_2-\beta_1)e^{-a_r l_r T}-(1-\beta_1)e^{-a_r T}+(1-\beta_2)\}x_{1m-1}+(1-\beta_2)x_{2m-1}] \\
&\quad\times e^{-d_r(t-(m-1)T)}
\end{aligned}$$

Applying impulsive conditions and using (6), gives stroboscopic map of system (5) after each successive birth pulse:

$$\begin{aligned} x_{1m} &= (1-\beta_1)x_{1m-1}e^{-(a_r+d_r)T} + be^{-(\beta_2-\beta_1)x_{1m-1}e^{-(d_r+a_rl_r)T}-(1-\beta_2)(x_{2m-1}+x_{1m-1})e^{-d_rT}} \\ &\quad \times [(\beta_2-\beta_1)x_{1m-1}e^{-a_rl_rT} + (1-\beta_2)(x_{2m-1}+x_{1m-1}) - (1-\beta_1)x_{1m-1}e^{-a_rT}]e^{-d_rT} \\ x_{2m} &= [(\beta_2-\beta_1)x_{1m-1}e^{-a_rl_rT} + (1-\beta_2)(x_{2m-1}+x_{1m-1}) - (1-\beta_1)x_{1m-1}e^{-a_rT}]e^{-d_rT} \end{aligned} \tag{7}$$

The dynamical behavior of the system (5) will be given by the dynamical behavior of the system (7) coupled with system (6).

Let R_0 be the intrinsic net reproductive number denoting average number of offspring that an individual produces over the period of its lifetime. It can be computed as:

$$R_0 = \frac{be^{-d_rT}((\beta_2-\beta_1)e^{-a_rl_rT} + (1-\beta_2) - (1-\beta_1)e^{-a_rT})}{(1-(1-\beta_1)e^{-d_rT-a_rT})(1-(1-\beta_2)e^{-d_rT})} = \frac{b}{b_0} \tag{8}$$

For the system (7), the following fixed points are obtained:

(i) A unique pest-free fixed point $E_0 = (0, 0)$ exists without any parametric restriction.
(ii) The non-trivial interior fixed point $E^* = (x_1^*, x_2^*)$ is obtained as:

$$\begin{aligned} x_1^* &= \frac{(1-(1-\beta_2)e^{-d_rT})\log(R_0)}{e^{-d_rT}\{(\beta_2-\beta_1)e^{-a_rl_rT} + (1-\beta_2) - (1-\beta_2)(1-\beta_1)e^{-(d_r+a_r)T}\}} \\ x_2^* &= \frac{((\beta_2-\beta_1)e^{-a_rl_rT} + (1-\beta_2) - (1-\beta_1)e^{-a_rT})\log(R_0)}{(\beta_2-\beta_1)e^{-a_rl_rT} + (1-\beta_2) - (1-\beta_2)(1-\beta_1)e^{-(d_r+a_r)T}} \end{aligned}$$

The interior fixed point exists if $R_0 > 1$. It does not exist for $R_0 < 1$. In other words, the interior fixed point is feasible if birth rate of pest is more than a critical value b_0 which depends upon all model parameters. It may be noted that $E^* = E_0$ if $R_0 = 1$.

3.1 Stability Analysis of the Fixed Point

The linearized system about any arbitrary fixed point $X = (x_1, x_2)$ can be written as:

$$X_m = AX_{m-1} \tag{9}$$

The linearized coefficients of the matrix $A = (a_{ij})_{2\times 2}$ are computed as:

$$
\begin{aligned}
a_{11} &= (1-\beta_1)e^{-(d_r+a_r)T} + be^{-Dx_1-(1-\beta_2)x_2e^{-d_rT}}[D-(1-\beta_1)e^{-(d_r+a_r)T} \\
&\quad - D\{Dx_1+(1-\beta_2)x_2e^{-d_rT}-(1-\beta_1)x_1e^{-(d_r+a_r)T}\}] \\
a_{12} &= b(1-\beta_2)e^{-d_rT-Dx_1-(1-\beta_2)x_2e^{-d_rT}}[1-\{Dx_1+(1-\beta_2)x_2e^{-d_rT} \\
&\quad -(1-\beta_1)x_1e^{-(d_r+a_r)T}\}] \\
a_{21} &= D-(1-\beta_1)e^{-(d_r+a_r)T} \\
a_{22} &= (1-\beta_2)e^{-d_rT} \\
D &= (\beta_2-\beta_1)e^{-(d_r+a_rl_r)T}+(1-\beta_2)e^{-d_rT}
\end{aligned} \tag{10}
$$

Let characteristic equation in terms of trace *Tr* and determinant *Det* be written as:

$$\lambda^2 - Tr\lambda + Det = 0$$

The fixed point $X = (x_1, x_2)$ is stable when the magnitude of eigenvalues of A are less than unity. For this, Jury conditions are:

$$1 - Tr + Det > 0 \tag{11}$$

$$1 + Tr + Det > 0 \tag{12}$$

$$1 - Det > 0. \tag{13}$$

If inequality (11) is violated, then one of the eigenvalues of A is larger than 1. If inequality (12) is violated, then one of the eigenvalues of A is less than -1. Finally, If inequality (13) is violated, then A has a complex-conjugate pair of eigenvalues lying outside the unit circle [5].

Theorem 1 *The pest-free fixed point* $E_0 = (0, 0)$ *is locally asymptotically stable if*

$$R_0 < 1. \tag{14}$$

Proof Using (10) coefficients of the linearized matrix $A = (a_{ij})_{2\times2}$ are evaluated about the pest-free fixed point (0, 0) as:

$$
\begin{aligned}
a_{11} &= (1-\beta_1)e^{-(d_r+a_r)T} + b\{(1-\beta_2)+(\beta_2-\beta_1)e^{-a_rlT}-(1-\beta_1)e^{-a_rT}\}e^{-d_rT} \\
a_{12} &= b(1-\beta_2)e^{-d_rT} \\
a_{21} &= (\beta_2-\beta_1)e^{-(d_r+a_rl)T}+(1-\beta_2)e^{-d_rT}-(1-\beta_1)e^{-(d_r+a_r)T} \\
a_{22} &= (1-\beta_2)e^{-d_rT}
\end{aligned}
$$

Accordingly, the trace *Tr* and determinant *Det* are computed as:

$$
\begin{aligned}
Tr &= [(1-\beta_1)e^{-a_rT} + b\{(1-\beta_2)+(\beta_2-\beta_1)e^{-a_rlT}-(1-\beta_1)e^{-a_rT}\}]e^{-d_rT} \\
&\quad + (1-\beta_2)e^{-d_rT} \\
Det &= (1-\beta_1)(1-\beta_2)e^{-(2d_r+a_r)T}
\end{aligned}
$$

In the following it is observed that conditions (12) and (13) are always satisfied:

$$\begin{aligned}1 + Tr + Det &= 1 + (1-\beta_1)e^{-(d_r+a_r)T} + be^{-d_rT}[\{(1-\beta_2) + (\beta_2-\beta_1)e^{-a_rlT} \\ &\quad - (1-\beta_1)e^{-a_rT}\} + (1-\beta_2) + (1-\beta_2)(1-\beta_1)e^{-(d_r+a_r)T})] > 0 \\ 1 - Det &= 1 - (1-\beta_1)(1-\beta_2)e^{-(2d_r+a_r)T} > 0\end{aligned}$$

The expression $1 - Tr + Det$ simplifies to:

$$\begin{aligned}1 - Tr + Det &= 1 - (1-\beta_1)e^{-(d_r+a_r)T} - b[\{(1-\beta_2) + (\beta_2-\beta_1)e^{-a_rlT} \\ &\quad - (1-\beta_1)e^{-a_rT}\} + (1-\beta_2) - (1-\beta_1)(1-\beta_2)e^{-(d_r+a_r)T}]e^{-d_rT} \\ &= (1-(1-\beta_1)e^{-d_rT-a_rT})(1-(1-\beta_2)e^{-d_rT}) - b\{(1-\beta_2) \\ &\quad + (\beta_2-\beta_1)e^{-a_rlT} - (1-\beta_1)e^{-a_rT}\}e^{-d_rT}\end{aligned}$$

Accordingly, the condition (11) gives

$$\begin{aligned}(1-(1-\beta_1)e^{-d_rT-a_rT})(1-(1-\beta_2)e^{-d_rT}) > b\{(1-\beta_2) + (\beta_2-\beta_1)e^{-a_rlT} \\ - (1-\beta_1)e^{-a_rT}\}e^{-d_rT}\end{aligned}$$

i.e.

$$b < \frac{(1-(1-\beta_1)e^{-d_rT-a_rT})(1-(1-\beta_2)e^{-d_rT})}{e^{-d_rT}((\beta_2-\beta_1)e^{-a_rl_rT} + (1-\beta_2) - (1-\beta_1)e^{-a_rT})} = b_0 \qquad (15)$$

Using (8) and (15), the stability condition (14) is obtained. □

Accordingly, the fixed point (0, 0) is locally stable for $b \in (0, b_0)$. The trajectories in the neighborhood of (0, 0) tend to origin and the pest will extinct. Thus, the pest eradication is possible when $R_0 < 1$. The existence of non-trivial fixed point is overruled in this case. When $R_0 > 1$, the fixed point (0, 0) is unstable. The pest-free fixed point becomes non-hyperbolic at $R_0 = 1$ and it collides with the interior fixed point. Further analysis in this case is carried out later in the next section.

Theorem 2 *The non-trivial fixed point $E^* = (x_1^*, x_2^*)$ is locally asymptotically stable provided*

$$b < b_0 e^{\frac{2((\beta_2-\beta_1)e^{-a_rl_rT}+(1-\beta_2)-(1-\beta_2)(1-\beta_1)e^{-(d_r+a_r)T})(1+(1-\beta_1)(1-\beta_2)e^{-(2d_r+a_r)T})}{(1-(1-\beta_1)e^{-(d_r+a_r)T})(1-(1-\beta_2)e^{-d_rT})((\beta_2-\beta_1)e^{-a_rl_rT}+(1-\beta_2)+(1-\beta_2)(1-\beta_1)e^{-(d_r+a_r)T})}} (= b_c) \qquad (16)$$

Proof Using (10) coefficient of linearized matrix A are computed around $E^* = (x_1^*, x_2^*)$ as:

$$
\begin{aligned}
a_{11} &= (1-\beta_1)e^{-(d_r+a_r)T} + b_0 e^{-d_rT}[(\beta_2-\beta_1)e^{-a_rl_rT} + (1-\beta_2)\\
&\quad - (1-\beta_1)e^{-a_rT} - \{(\beta_2-\beta_1)e^{-a_rl_rT} + (1-\beta_2)\}x_2^*]\\
a_{12} &= b_0(1-\beta_2)[1-x_2^*]e^{-d_rT}\\
a_{21} &= (\beta_2-\beta_1)e^{-(d_r+a_rl_r)T} + (1-\beta_2)e^{-d_rT} - (1-\beta_1)e^{-(d_r+a_r)T}\\
a_{22} &= (1-\beta_2)e^{-d_rT}
\end{aligned}
$$

Accordingly, the trace Tr and determinant Det are computed as:

$$
\begin{aligned}
Tr &= (1-\beta_1)e^{-(d_r+a_r)T} + b_0 e^{-d_rT}[(\beta_2-\beta_1)e^{-a_rl_rT} + (1-\beta_2)\\
&\quad - (1-\beta_1)e^{-a_rT} - \{(\beta_2-\beta_1)e^{-a_rl_rT} + (1-\beta_2)\}x_2^*] + (1-\beta_2)e^{-d_rT}\\
Det &= (1-\beta_1)(1-\beta_2)e^{-(2d_r+a_r)T}[1-b_0x_2^*]
\end{aligned}
$$

It is observed that conditions (11) and (13) are always satisfied:

$$
\begin{aligned}
1-Tr+Det &= 1-(1-\beta_1)e^{-(d_r+a_r)T} - b_0[(\beta_2-\beta_1)e^{-a_rl_rT} + (1-\beta_2)\\
&\quad - (1-\beta_1)e^{-a_rT} - \{(\beta_2-\beta_1)e^{-a_rl_rT} + (1-\beta_2)\}x_2^*]e^{-d_rT}\\
&\quad - (1-\beta_2)e^{-d_rT} + (1-\beta_1)(1-\beta_2)e^{-(2d_r+a_r)T}[1-b_0x_2^*]\\
&= (1-(1-\beta_1)e^{-(d_r+a_r)T})(1-(1-\beta_2)e^{-d_rT}) - b_0[(\beta_2-\beta_1)e^{-a_rl_rT}\\
&\quad + (1-\beta_2) - (1-\beta_1)e^{-a_rT}](1-x_2^*)e^{-d_rT}\\
&= (1-(1-\beta_1)e^{-(d_r+a_r)T})(1-(1-\beta_2)e^{-d_rT})x_2^* > 0
\end{aligned}
$$

$$
1-Det = 1-(1-\beta_1)(1-\beta_2)e^{-(d_r+a_r)T}[1-b_0x_2^*] > 0
$$

The expression $1+Tr+Det$ simplifies to:

$$
\begin{aligned}
1+Tr+Det &= 1+(1-\beta_1)e^{-(d_r+a_r)T} + b_0[(\beta_2-\beta_1)e^{-(d_r+a_rl_r)T} + (1-\beta_2)e^{-d_rT}\\
&\quad - (1-\beta_1)e^{-(d_r+a_r)T} - \{(\beta_2-\beta_1)e^{-(d_r+a_rl_r)T} + (1-\beta_2)e^{-d_rT}\}x_2^*]\\
&\quad + (1-\beta_2)e^{-d_rT} + (1-\beta_1)(1-\beta_2)e^{-(2d_r+a_r)T}[1-b_0x_2^*]\\
&= (1+(1-\beta_1)e^{-d_rT-a_rT})(1+(1-\beta_2)e^{-d_rT}) + b_0[(\beta_2-\beta_1)e^{-a_rl_rT}\\
&\quad + (1-\beta_2) - (1-\beta_1)e^{-a_rT} - \{(\beta_2-\beta_1)e^{-a_rl_rT} + (1-\beta_2)\\
&\quad + (1-\beta_1)(1-\beta_2)e^{-(d_r+a_r)T}\}x_2^*]e^{-d_rT}\\
&= (1+(1-\beta_1)e^{-d_rT-a_rT})(1+(1-\beta_2)e^{-d_rT})\\
&\quad + (1-(1-\beta_1)e^{-d_rT-a_rT})(1-(1-\beta_2)e^{-d_rT})\\
&\quad \times \left[1 - \frac{(\beta_2-\beta_1)e^{-a_rl_rT} + (1-\beta_2) + (1-\beta_1)(1-\beta_2)e^{-(d_r+a_r)T}}{(\beta_2-\beta_1)e^{-a_rl_rT} + (1-\beta_2) - (1-\beta_1)(1-\beta_2)e^{-(d_r+a_r)T}}\log(R_0)\right]
\end{aligned}
$$

Accordingly, condition (12) gives:

$$
\begin{aligned}
R_0 &< exp(CF^{-1}) \qquad (17)\\
C &= 2((\beta_2 - \beta_1)e^{-a_r l_r T} + (1-\beta_2) - (1-\beta_2)(1-\beta_1)e^{-(d_r+a_r)T})\\
&\quad \times (1 + (1-\beta_1)(1-\beta_2)e^{-(2d_r+a_r)T})\\
F &= (1-(1-\beta_1)e^{-(d_r+a_r)T})(1-(1-\beta_2)e^{-d_r T})\\
&\quad \times ((\beta_2-\beta_1)e^{-a_r l_r T} + (1-\beta_2) + (1-\beta_2)(1-\beta_1)e^{-(d_r+a_r)T})
\end{aligned}
$$

The stability condition (16) is obtained from (17). Hence, if $b < b_c$ then E^* is locally asymptotically stable. □

Accordingly, when $b_0 < b < b_c$, the fixed point E^* exists and is locally asymptotically stable. The trajectories of system (7) tend to asymptotically stable period-1 solution $(x_{1e}(t), x_{2e}(t))$:

$$
\begin{aligned}
x_{1e}(t) &= (1-\beta_1)x_1^* e^{-(a_r+d_r)(t-(m-1)T)}\\
x_{2e}(t) &= [(\beta_2-\beta_1)x_1^* e^{-(d_r+a_r)l_r T} + (1-\beta_2)(x_2^* + x_1^*)e^{-d_r l_r T}]e^{-d_r(t-(m+l_r-1)T)}\\
&\quad - (1-\beta_1)x_1^* e^{-(a_r+d_r)(t-(m-1)T)}
\end{aligned}
$$

Further increasing parameter value $b > b_c$, the fixed point E^* losses its stability and the system may exhibit complex dynamics. The complex dynamical behavior will be shown in later section. The interior fixed point becomes non-hyperbolic at $b = b_c$ and one of the eigenvalues becomes -1. So, there is a possibility of flip bifurcation.

4 Transcritical Bifurcation

The pest-extinction point $E_0 = (0,0)$ becomes non-hyperbolic at $R_0 = 1$ (i.e. $b = b_0$) as one of the eigenvalues is 1. It is observed that at $b = b_0$, $E_0 = (0,0)$ and $E^* = (x_1^*, x_2^*)$ exchange stability. The center manifold theorem and Lemma 1 is used to characterize the nature of bifurcation point $b = b_0$.

Lemma 1 *Let $x \to f(x,b), x \in \Re, b \in \Re$ be a one-parameter family of one-dimensional map having a non-hyperbolic fixed point with an eigenvalue 1 then the map undergoes transcritical bifurcation at $(x,b) = (0,0)$ if*

$$
\frac{\partial f(0,0)}{\partial b} = 0, \qquad \frac{\partial^2 f(0,0)}{\partial x \partial b} \neq 0, \qquad \frac{\partial^2 f(0,0)}{\partial^2 x} \neq 0
$$

Theorem 3 *The system (7) undergoes transcritical bifurcation at $b = b_0$.*

Proof Let $\Upsilon = e^{-d_r T}((\beta_2 - \beta_1)e^{-a_r l_r T} + (1 - \beta_2) - (1 - \beta)e^{-a_r T})$. Consider the map

$$\begin{pmatrix} x_1 \\ x_2 \end{pmatrix} \rightarrow \begin{pmatrix} be^{-((\beta_2-\beta_1)x_1 e^{-a_r l r_r T}+(1-\beta_2)(x_2+x_1))e^{-d_r T}}[\Upsilon x_1 + (1-\beta_2)e^{-d_r T}x_2] \\ +(1-\beta_1)x_1 e^{-(a_r+d_r)T} \\ \Upsilon x_1 + (1-\beta_2)e^{-dT}x_2 \end{pmatrix} \quad (18)$$

Let $x_1 = u, x_2 = v, b = b_1 + b_0, b_0 = \Upsilon^{-1}(1 - (1-\beta_1)e^{-d_r T - a_r T})(1 - (1-\beta_2) e^{-d_r T})$. The fixed point E_0 of the map is transformed to (u, v) and the map (18) becomes: $\begin{pmatrix} u \\ v \end{pmatrix} \rightarrow \begin{pmatrix} (b_1 + b_0)\Theta(\Upsilon u + (1-\beta_2)ve^{-d_r T}) + (1-\beta_1)ue^{-(a_r+d_r)T} \\ \Upsilon u + (1-\beta_2)ve^{-d_r T} \end{pmatrix}$
where $\Theta = exp\,(-(\beta_2 - \beta_1)ue^{-(d_r+a_r l)T} - (1-\beta_2)(v+u)e^{-d_r T}))$.

It can be rewritten as:

$$\begin{pmatrix} u \\ v \end{pmatrix} \rightarrow M \begin{pmatrix} u \\ v \end{pmatrix} + \begin{pmatrix} c_{11}u^2 + c_{12}uv + c_{13}v^2 + c_{14}b_1 u + c_{15}b_1 v \\ 0 \end{pmatrix} \quad (19)$$

The coefficients of the matrix $M = m_{ij_{2\times 2}}$ and coefficients $c_{1j}, j = 1, 2, 3, 4, 5$ are obtained as:

$$\begin{aligned}
m_{11} &= (1-\beta_1)e^{-(d_r+a_r)T} + b_0\Upsilon, \\
m_{21} &= \Upsilon, \\
m_{12} &= b_0(1-\beta_2)e^{-d_r T}, \\
m_{22} &= (1-\beta_2)e^{-d_r T} \\
c_{11} &= [-(\beta_2-\beta_1)e^{-a_r lT}(e^{-a_r l_r T} + 2(1-\beta_2) + (1-\beta_1)e^{-a_r T}) - (1-\beta_2)^2 \\
&\quad + (1-\beta_2)(1-\beta_1)e^{-a_r T}]b_0 e^{-2d_r T} \\
c_{12} &= (-2(\beta_2-\beta_1)e^{-a_r l_r T} - 2(1-\beta_2) + (1-\beta_1)e^{-a_r T})b_0(1-\beta_2)e^{-2dT} \\
c_{13} &= -b_0(1-\beta_2)^2 e^{-2d_r T} \\
c_{14} &= \Upsilon \\
c_{15} &= (1-\beta_2)e^{-dT}
\end{aligned}$$

The eigenvalues of M are 1 and $(1-\beta_2)(1-\beta_1)e^{-(a_r+2d_r)T}$. The corresponding eigenvectors are $\{J_{11}, 1\}^T$ and $\{J_{12}, 1\}^T$ where $J_{11} = \Upsilon^{-1}(1 - (1-\beta_2)e^{-d_r T})$ and $J_{12} = \Upsilon^{-1}e^{-d_r T}(1-\beta_2)((1-\beta_1)e^{-(d_r+a^r)T} - 1)$.
Consider the transformation $\begin{pmatrix} u \\ v \end{pmatrix} = J \begin{pmatrix} \bar{x} \\ \bar{y} \end{pmatrix}$ with $J = \begin{pmatrix} J_{11} & J_{12} \\ 1 & 1 \end{pmatrix}$. Now, the map (19) can be written as:

$$\begin{pmatrix} \bar{x} \\ \bar{y} \end{pmatrix} \rightarrow \begin{pmatrix} 1 & 0 \\ 0 & (1-\beta_2)(1-\beta_1)e^{-(a_r+2d_r)T} \end{pmatrix} \begin{pmatrix} \bar{x} \\ \bar{y} \end{pmatrix} + \begin{pmatrix} f_1(\bar{x}, \bar{y}, b_1) \\ f_2(\bar{x}, \bar{y}, b_1) \end{pmatrix} \quad (20)$$

where

$$f_1(\bar{x},\bar{y},b_1) = d_1 b_1\bar{x} + d_2 b_1\bar{y} + d_3\bar{x}\bar{y} + d_4\bar{x}^2 + d_5\bar{y}^2$$
$$f_2(\bar{x},\bar{y},b_1) = -f_1(\bar{x},\bar{y},b_1)$$

$$\begin{aligned}
d_1 &= (1-(1-\beta_2)(1-\beta_1)e^{-(a_r+2d_r)T})\Upsilon^{-1}\\
d_2 &= (1-\beta_2)(1-\beta_1)e^{-(a_r+d_r)T}(1-(1-\beta_2)(1-\beta_1)e^{-(a_r+2d_r)T})\Upsilon^{-1}e^{-d_rT} + c_{12}\Upsilon^{-1}\\
d_3 &= 2c_{13} + 2c_{11}\Upsilon^{-2}e^{-d_rT}(1-\beta_2)\{(1-\beta_1)e^{-(d_r+a_r)T}-1\}(1-(1-\beta_2)e^{-d_rT})\Upsilon^{-1}\\
&\times[1-(1-\beta_2)e^{-d_rT}\{1+\{(1-\beta_1)e^{-(d_r+a_r)T}-1\}\}][1-(1-\beta_2)(1-\beta_1)e^{-(a_r+2d_r)T}]\\
d_4 &= [(c_{11}+c_{12})\Upsilon^{-1}(1-(1-\beta_2)e^{-d_rT})+c_{13}]\\
&\times[1-(1-\beta_2)(1-\beta_1)e^{-(a_r+2d_r)T}]\Upsilon^{-1}\\
d_5 &= [(c_{11}+c_{12})(1-\beta_2)((1-\beta_1)e^{-(d_r+a_r)T}-1)\Upsilon^{-1}e^{-d_rT}+c_{13}]\\
&\times[1-(1-\beta_2)(1-\beta_1)e^{-(a_r+2d_r)T}]\Upsilon^{-1}
\end{aligned}$$

Now the center manifold theorem is used to determine the nature of bifurcation of the fixed point $(0,0)$ at $b_1 = 0$. The center manifold for the map (20) can be represented as:

$$w^c(0) = \{(\bar{x},\bar{y},b_1)\in\Re^3 | \bar{y} = f(\bar{x},b_1), f(0,0) = 0, Df(0,0) = 0\}$$

Let $\bar{y} = f(\bar{x},b_1) = B_0b_1 + B_1b_1\bar{x} + B_2\bar{x}^2 + O(|b_1|^2 + |b_1\bar{x}^2| + |\bar{x}|^3)$. The coefficients in $\bar{y}$ can be computed as:

$$B_0 = 0,\quad B_1 = \frac{-d_1}{(1-(1-\beta_2)(1-\beta_1)e^{-(a_r+2d_r)T})},$$
$$B_2 = \frac{d_4}{(1-\beta_2)(1-\beta_1)e^{-(a_r+2d_r)T}-1}$$

The map restricted to the center manifold is given by:

$$\bar{f}:\bar{x}\to\bar{x}+f_1(\bar{x},\bar{y},b_1) = \bar{x}+d_1b_1\bar{x}+d_2b_1\bar{y}+d_3\bar{x}\bar{y}+d_4\bar{x}^2+d_5\bar{y}^2$$
$$=\bar{x}+d_1b_1\bar{x}+\frac{d_3d_4\bar{x}^3}{(1-\beta_2)(1-\beta_1)e^{-(a_r+2d_r)T}-1}+d_4\bar{x}^2+O(|b_1|^2+|b_1\bar{x}^2|+|\bar{x}|^4)$$

From the Lemma 1, it can be observed that

$$\frac{\partial\bar{f}(0,0)}{\partial b_1} = 0,\qquad \frac{\partial^2\bar{f}(0,0)}{\partial x\partial b_1} = d_1 \neq 0,\qquad \frac{\partial^2\bar{f}(0,0)}{\partial^2 x} = 2d_4 \neq 0$$

Note that, all conditions of Lemma 1 are satisfied at $(\bar{x},b_1) = (0,0)$. Further, E^* becomes E_0 as $b = b_0$. Hence the system (7) undergoes to transcritical bifurcation between $E_0 = (0,0)$ and $E^* = (x^*,y^*)$ at $b = b_0$.

5 Numerical Simulation

In this section, numerical analysis of the system is performed based on the analytical results. The critical parameters for investigation are identified as T, l_r, β_1, β_2, a_r and d_r that are involved in R_0 and affects the dynamics of the system. Consider the following parameter set

$$a_r = 0.4, d_r = 0.2, l_r = 0.5, \beta_1 = 0.4, \beta_2 = 0.8 \tag{21}$$

Considering $T = 1.0$ and data set (21), the constant b_0 is computed as $b_0 = 5.4674$. For $b = 5$, the basic reproduction number is obtained as ($R_0 = 0.9145 < 1$). According to Theorem 1, the pest-free fixed point is locally asymptotically stable and coexistence is not possible. Taking $b = 11$, the pest-free fixed point becomes unstable. The non-trivial fixed point $E^* = (1.5106, 0.1898)$ is stable as $b_0 < b < 101.7607$ (see Theorem 2). Further, transcritical bifurcation occurs at $b = 5.4674$.

Figure 1 shows variation of equilibrium level of pest population versus l_r for different killing rates of immature and mature pest $\beta_1, \beta_2 \in (0, 1)$. It clearly shows that the lowest equilibrium level of pest is possible when $l_r = 1$. It also follows from Fig. 1 that the number of the immature and mature pest is dependent on the killing rate and pesticide spray timing. From biological point of view, the aim is to reduce the pest to the low level not to eliminate it. If pest is removed early in the season, higher amounts of pesticides are needed to reduce the pest. Since pesticides have negative impact on the environment this is not desirable. Therefore, the best timing of pesticide spraying is before the end of season that is when half of the season has been passed or before the end of season. Further, it can be easily seen that equilibrium density of mature pest is a decreasing function with respect to pesticide spraying time parameter l_r:

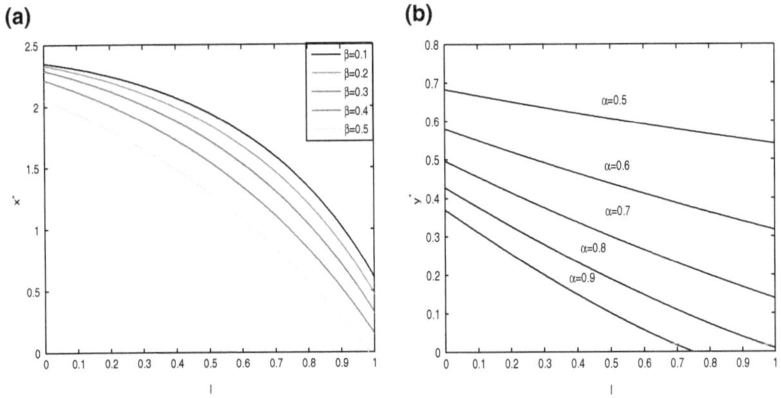

Fig. 1 Equilibrium density versus l_r for data set (21)

$$\frac{dx_2^*}{dl_r} = -\frac{(a_rT(1-\beta_1)(\beta_2-\beta_1)e^{-a_rl_rT-a_r}(1-(1-\beta_2)e^{-d_rT}))\log(R_0)}{((\beta_2-\beta_1)e^{-a_rl_rT}+(1-\beta_2)-(1-\beta_2)(1-\beta_1)e^{-(d_r+a_r)T})^2}$$
$$-\frac{(a_rT(\beta_2-\beta_1)e^{-a_rl_rT})}{((\beta 2-\beta_1)e^{-a_rl_rT}+(1-\beta_2)-(1-\beta_2)(1-\beta_1)e^{-(d_r+a_r)T})} < 0$$

Similarly, it can be proved that equilibrium density of immature pest is monotonic decreasing function with respect to parameter l_r for the data set (21).

The expression of R_0, which includes killing rate of immature and mature pest, clearly shows the effect of pesticide spray on threshold R_0. The first order derivatives of the threshold R_0 with respect to β and α respectively are found to be negative:

$$\frac{dR_0}{d\beta_1} = -\frac{be^{-d_rT}((1-\beta_2)e^{-(d_r+a_r)T}(1-e^{-a_rl_rT})+e^{-a_rT}-e^{-a_rlT})}{(1-(1-\beta_2)e^{-d_rT})(1-(1-\beta_1)e^{-(d_r+a_r)T})^2} < 0$$
$$\frac{dR_0}{d\beta_2} = -\frac{be^{-d_rT}((1-\beta_1)e^{-d_rT}(e^{-a_rT}-e^{-a_rl_rT})+1-e^{-a_rl_rT})}{(1-(1-\beta_2)e^{-d_rT})^2(1-(1-\beta_1)e^{-(d_r+a_r)T})} < 0$$

It can be easily observed that the larger killing (or poisoning) rate β_1 and β_2 reduces the threshold value R_0. The effects of pesticide spraying on immature and mature pest have strong effect on threshold R_0.

Figure 2a shows the variation of R_0 with impulsive period T. The non-monotonic behavior of R_0 with respect to the impulsive period T is observed. As the values of T increases, R_0 first decreases and attains a peak then it decreases with increase in T. From the point of pest control, it is necessary that the pulse time period should be selected carefully. As, once the threshold value R_0 is less than 1, pest goes to extinction.

To see complexity due to pesticide spraying time and birth rate, two parameter bifurcation diagram is drawn in $l_r - b$ plane see Fig. 2b. In this figure, region of pest

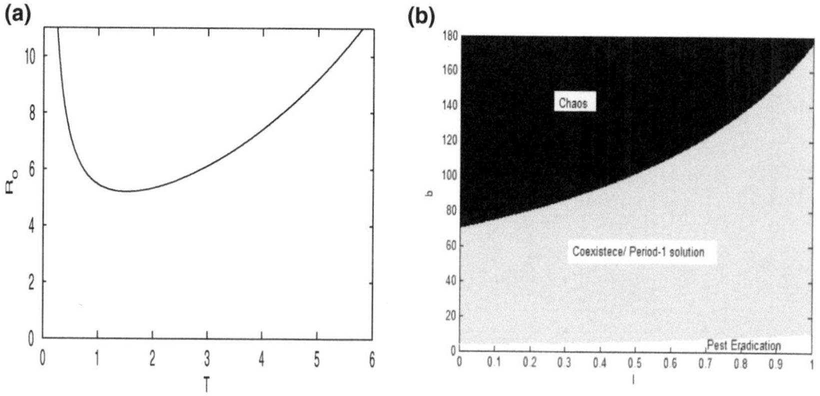

Fig. 2 **a** Contours plot shows non-monotonicity of R_0. **b** Two parameter bifurcation diagram in $l_r - b$ plane with respect to T for data set (21)

extinction is shown by white, while the region of stable period-1 solution is shown by light-grey. For higher birth rates of the pest, the system becomes chaotic which is shown in the black region. At lower birth rates pest will be eradicated for all values of pesticide spray time. It has been observed that pest will go to extinction in a small neighborhood of (0, 0). The periodic doubling leads to chaos in the system for higher birth rate.

Typical bifurcation diagram is drawn for total pest in Fig. 3a with respect to critical parameter b which is involved in R_0 as well as in $R_c = \frac{b}{b_c}$. The bifurcation diagram shows existence of chaos through period-doubling route. The critical value for period-doubling bifurcation parameter is $b_c = 101.7607$ as obtained from Eq. (16) is confirmed from Fig. 3a. The period-1 solution occurs in the range $b \in (5.4674, 101.7607)$. As parameter value of b increases further, successive period-doubling with period-2, period-4, period-8 and period-16 occur in the intervals (101.7607, 296.841), (296.841, 406.783), (406.783, 434.8518) and (434.8518, 441.16) respectively. The cascades of period-doubling is observed in the bifurcation diagram, which is route of chaos in the system. Several periodic windows are visible in the interval (500, 700). A region of Fig. 3a is separately blown up in Fig. 3b in the interval (690, 720) and a periodic window is clearly visible. Thus, the pest coexist in periodic solution/ chaotic attractor in range of values of b beyond b_c.

The Fig. 4 shows bifurcation diagrams with respect to pesticide spray time $l_r \in (0, 1)$ with $b = 450$. The system (7) depicts very complex dynamical behavior if pesticide is sprayed just after birth pulse. As parameter l_r increases, chaotic behavior is followed by period-halving bifurcation. It can be observed that immature and mature pest decreases as the pesticide spray timing l_r increases. From an ecological point of view, it can be observed that pesticide spray time may reduce complexity with increasing l_r and may stabilize the system.

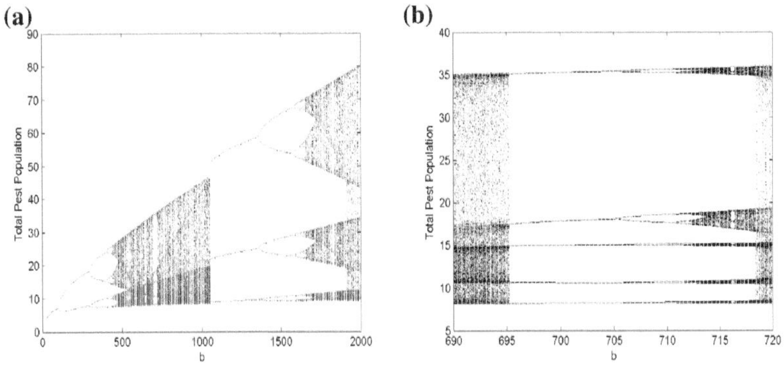

Fig. 3 **a** Bifurcation diagram with respect to parameter b. **b** Blown up of the bifurcation diagram for data set (21)

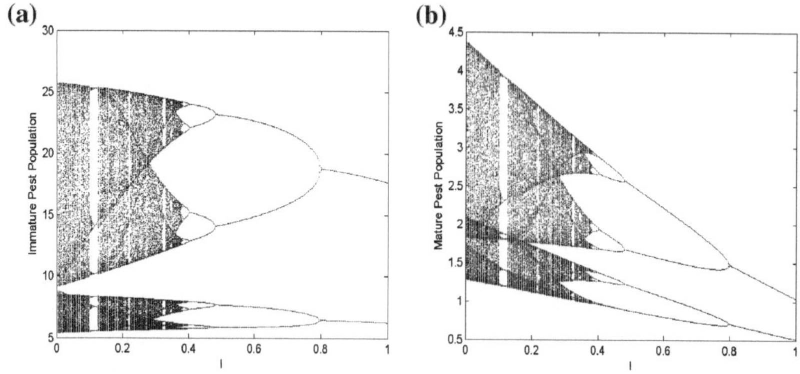

Fig. 4 Bifurcation diagrams **a** for immature pest **b** for mature pest with respect to time of spraying pesticide l_r for data set (21)

6 Discussion

In this paper, a stage-structured pest model with birth pulse and impulsive spraying pesticide at fixed time is considered. By using the stroboscopic map, the complete expression for periodic solution with period-1 is obtained. Also the threshold conditions for the stability of two fixed points is obtained. The effects of pesticide spraying timing on the number of the pest is considered. The results show that the best time of pesticide spraying is at the end of the season, that is before and near the time of birth. The system with Ricker type birth function shows a very complex dynamical behavior. As the parameter of pesticide spray timing increases, there exist period-halving bifurcations followed by chaotic behavior. It is observed that the number of mature pest is a decreasing function with respect to pesticide spray. Also there exist a cascade of period-halving bifurcations from chaos to cycles.

Acknowledgements The first author expresses thanks to *Ministry of Human Resources and Development (MHRD), India* for providing financial support without which this research effort would not be possible.

References

1. Caughley, G.: Analysis of Vertebrate Populations. Wiley, New York (1977)
2. Chatterjee, S., Isaia, M., Venturino, E.: Effects of spiders predational delays in intensive. Nonlinear Anal. **10**, 3045–3058 (2009)
3. Hui, J., Chen, L.S.: Dynamic complexities in ratio-dependent predator-prey ecosystem models with birth pulse and pesticide pulse. Int. J. Bifurc. Chaos **14**(8), 2893–2903 (2004)
4. Jiao J.J., Cai S.H., Chen L.S.: Analysis of a stage-structured predator prey system with birth pulse and impulsive harvesting at different moments. Nonlinear Anal. RWA **12**, 2232–2244 (2011)

5. Jury, E.I.: Inners and Stability of Dynamic Systems. Wiley, New York (1974)
6. Lakshmikantham, V., Bainov, D.D., Simeonov, P.S.: Theory of Impulsive Differential Equations. World Scientific, Singapore (1989)
7. Nundloll, S., Mailleret, L., Grognard, F.: Two models of interfering predators in impulsive biological control. J. Biol. Dyn. **4**(1), 102–114 (2010)
8. Rafikov, M., Balthazar, J.M., Von Bremen, H.F.: Mathematical modeling and control of population systems: application in biological pest control. Appl. Math. Comput. **200**, 557–573 (2008)
9. Tang, S., Cheke, R.A.: Models for integrated pest control and their biological implications. Math. Biosci. **215**(1), 115–125 (2008)
10. Tang, S., Chen, L.: Density-dependent birth rate, birth pulses and their population dynamic consequences. J. Math. Biol. **44**(2), 185–199 (2002)
11. Tang, S., Chen, L.: The effect of seasonal harvesting on stage-structured population models. J. Math. Biol. **48**, 357–374 (2004)
12. Tang, S., Xiao, Y., Chen, L., Cheke, R.A.: Integrated pest management models and their dynamical behaviour. Bull. Math. Biol. **67**, 115–135 (2005)

Dynamical Behavior of a Modified Leslie–Gower Prey–Predator Model with Michaelis–Menten Type Prey-Harvesting

R.P. Gupta and Peeyush Chandra

Abstract In this article we discuss the dynamical behavior of a modified Leslie–Gower prey–predator model in presence of nonlinear harvesting in prey under the assumption that the protection provided by environment to prey and predator is different. The objective of this work is to find the parametric conditions so that extinction of the species can be prevented in presence of continuous harvesting of the prey population. We analyze the effect of harvesting on the proposed model by considering the harvesting as a bifurcation and control parameter. The existence and stability of equilibrium points are discussed and singular optimal control has been derived through Pontryagin's Maximum Principle. This study provides important tools for investigations pertaining to controllability of the system. Numerical simulations using MATLAB are carried out as supporting evidences of our analytical findings.

Keywords Stability · Hopf-bifurcation · Bionomic equilibria · Singular control

2010 Mathematics Subject Classification 34D20 · 37G10

1 Introduction

In population dynamics one of the fundamental interactions which influences the dynamics of species is predation. Hence prey–predator models are receiving a considerable attention by researchers in recent years. More realistic models have been developed keeping in view the laboratory experiments and observations. Also

R.P. Gupta (✉)
Department of Mathematics, VNIT, Nagpur 440 010, India
e-mail: rravipratap.86@gmail.com

P. Chandra
Department of Mathematics and Statistics, IIT Kanpur,
Kanpur 208 016, India
e-mail: peeyush@iitk.ac.in

J.M. Cushing et al. (eds.), *Applied Analysis in Biological and Physical Sciences*,
Springer Proceedings in Mathematics & Statistics 186,
DOI 10.1007/978-81-322-3640-5_6

harvesting of resources is important issue from the ecological and the economic point of view. Depending upon the types of species, harvesting of all or one of the species has been considered in the literature [3–5, 9]. There are several types of harvesting functions available in the literature. One of the most popular one is proportional harvesting [3] which is given by

$$H_1(x, E) = qEx. \tag{1}$$

where, x is population density (stock abundance), q is the catchability coefficient and E is the effort applied to harvest individuals.

It may be noted that this harvesting function (1) has limited applicability as it accounts for unbounded linear increase of $H_1(x, E)$ with x for fixed E and unbounded linear increase of $H_1(x, E)$ with E for fixed x. In view of this Michaelis–Menten type harvesting has been suggested which is given by [3]

$$H_2(x, E) = \frac{qEx}{m_1E + m_2x}. \tag{2}$$

Here m_1, m_2 are suitable positive constants. It may be noted that the nonlinear harvesting function (2) exhibits saturation effects with respect to both the stock abundance and the effort-level.

The Leslie–Gower prey–predator model assumes that interacting species grow according to the logistic law and that the environmental carrying capacity for the predator is not a constant but proportional to the population size of the prey. However, due to the rarity of the prey, the predator can switch over to other food, but its growth is still limited by the fact that its favorite prey is not available in abundance. Therefore in modified Leslie–Gower functional response a positive constant is added into the function of the carrying capacity for the predator. With these assumptions Aziz-Alaoui and Daher Okiye [1] proposed a two-dimensional system for a prey–predator which incorporates a modified version of Leslie–Gower and Holling-type II functional response.

Gupta and Chandra [7] extended this model for prey–predator system in presence of nonlinear prey-harvesting under the assumption that the protection provided by the environment to prey and predator is same. They studied the permanence, stability and bifurcation (saddle-node bifurcation, transcritical, Hopf-Andronov and Bogdanov-Takens) of this model. They also observed that whenever there are two interior equilibria, the one lying on the left is always a saddle and the other can be either an attractor or a repeller surrounded by a limit cycle. The present study is an extension of the model studied by Gupta and Chandra [7]. Here it is assumed that the protection provided by the environment to prey and predator is different.

2 Mathematical Model

In this section we briefly describe the model formulation which is being studied in the current article.

2.1 Model with Prey Harvesting

We consider here the modified Leslie–Gower prey–predator model in presence of nonlinear prey-harvesting under the assumption that the protection provided by the environment to prey and predator is different. The model is given by

$$\begin{cases} \frac{dx}{dt} = rx\left(1-\frac{x}{k}\right) - \frac{axy}{n_1+x} - \frac{qEx}{m_1E+m_2x}, \\ \frac{dy}{dt} = sy\left(1-\frac{by}{n_2+x}\right), \end{cases} \tag{3}$$

subject to positive initial conditions $x(0) > 0,\ y(0) > 0$. Here, $x(t)$ and $y(t)$ are the prey and predator population densities respectively. r and k are intrinsic growth rate and environmental carrying capacity for the prey species respectively. a is the maximum value which per capital reduction rate of prey can attain, n_1 and n_2 measures the extent to which environment provides protection to prey and predator respectively and s is the intrinsic growth for the predator species. b for predator has similar meaning to a. For biological considerations all the parameters are assumed to be positive.

Gupta and Chandra [7] have considered a particular case of above model where they assumed that the environment provides similar protection to both prey and predator (i.e. $n_1 = n_2 = n$). In the following we discuss the positivity, boundedness and permanence of solutions of the model system (3).

2.2 Positivity of Solution

Integrating Eq. (3) we get

$$x(t) = x(0)\exp\left[\int_0^t \left(r\left(1-\frac{x(\tau)}{k}\right) - \frac{ay(\tau)}{n_1+x(\tau)} - \frac{qE}{m_1E+m_2x(\tau)}\right)d\tau\right],$$

and

$$y(t) = y(0)\exp\left[\int_0^t s\left(1-\frac{by(\tau)}{n_2+x(\tau)}\right)d\tau\right],$$

showing that $x(t) \geq 0$ and $y(t) \geq 0$ whenever $x(0) > 0$ and $y(0) > 0$. Hence all solutions remain within the first quadrant of the xy-plane starting from an interior point of it. Further we can easily establish that solution trajectories starting from $(x_0, 0)$ with $x_0 > 0$, remain within the positive x-axis at all future time and similar result holds for trajectories starting from a point on the positive y-axis. Hence, $R^2_{+0} = \{(x, y) : x, y \geq 0\}$ is an invariant set.

2.3 *Boundedness of Solution*

Consider, $(x(t), y(t))$ be an arbitrary positive solution of the system (3) subject to a positive initial condition. Using the positivity of variables x, y and the first equation of the system (3), we can write,

$$\frac{dx}{dt} = rx\left(1 - \frac{x}{k}\right) - \frac{axy}{n_1 + x} - \frac{qEx}{m_1E + m_2x} \leq rx\left(1 - \frac{x}{k}\right), \tag{4}$$

From Lemma 1 of [7], we have

$$x(t) \leq \max\{x(0), k\} \equiv M_1 \ \text{ for all } \ t \geq 0.$$

Further, from the second equation of the system (3), we have

$$\frac{dy}{dt} = sy\left(1 - \frac{by}{n_2 + x}\right) \leq sy\left(1 - \frac{by}{n_2 + M_1}\right), \tag{5}$$

Again from Lemma 1 of [7], we have

$$y(t) \leq \max\left\{y(0), \frac{n_2 + M_1}{b}\right\} \equiv M_2 \ \text{ for all } \ t \geq 0.$$

This completes the proof of the boundedness of solutions and hence the system under consideration is dissipative.

2.3.1 Permanence

Here we prove the permanence result for the system (3). Biologically it ensures the long term co-existence of both the species.

Proposition 1 *The system (3) is permanent if,* $\frac{a(n_2+k)}{bn_1} + \frac{q}{m_1} < r$.

Proof From the above result it is clear that $0 < x(t) < k$ and so $y(t) \leq \frac{n_2+k}{b}$ for sufficiently large t. Therefore from first equation of system (3), we can write

$$\frac{dx}{dt} = rx\left(1 - \frac{x}{k}\right) - \frac{axy}{n_1 + x} - \frac{qEx}{m_1E + m_2x} \geq x\left(r - \frac{rx}{k} - \frac{a(n_2 + k)}{bn_1} - \frac{q}{m_1}\right).$$

From Lemma 1 of [7] if, $\omega_1 = 1 - \frac{a(n_2+k)}{brn_1} - \frac{q}{m_1 r} > 0$, then

$$\liminf_{t\to+\infty} x(t) \geq k\omega_1,$$

i.e. for each $\epsilon > 0$ there exists $t_1(\epsilon) > 0$ such that $x(t) \geq k\omega_1 - \epsilon$, for all $t > t_1$.

Further, since $x(t)$ is the solution of a differential equation so it is continuous on the interval $[0, t_1]$ and hence $x(t) \geq m_3$ on $t \in [0, t_1]$ for some $m_3 > 0$. Therefore, if we choose $0 < \epsilon < k\omega_1$ and $m_4 = \min\{k\omega_1 - \epsilon, m_3\}$, then $x(t) \geq m_4 > 0$ for all t and hence from predator equation of system (3), we can write

$$\frac{dy}{dt} \geq sy\left(1 - \frac{by}{n_2 + m_4}\right),$$

which on using Lemma 1 of [7] gives the following result

$$\liminf_{t\to+\infty} y(t) \geq \frac{n_2 + m_4}{b} \equiv \omega_2.$$

Also from inequalities (4) and (5), together with the Lemma 1 of [7], we have

$$\limsup_{t\to\infty} x(t) \leq k \text{ and } \limsup_{t\to\infty} y(t) \leq \frac{n_2 + M_1}{b}.$$

Thus,

$$\min\left\{\liminf_{t\to+\infty} x(t),\ \liminf_{t\to+\infty} y(t)\right\} \geq \min(k\omega_1, \omega_2)$$

and

$$\max\left\{\limsup_{t\to+\infty} x(t),\ \limsup_{t\to+\infty} y(t)\right\} \leq \max\left(k, \frac{n_2 + M_1}{b}\right).$$

Hence result follows from [10].

3 Existence and Stability of Equilibria

The system (3) has following trivial equilibrium points

(i) Origin $S_0(0, 0)$,
(ii) Prey-extinction equilibrium is $S_1(0, \frac{n_2}{b})$,
(iii) Predator-free equilibrium points are $S_L(x_L, 0)$ and $S_H(x_H, 0)$ where x_L and x_H are positive roots the following quadratic equation

$$rm_2x^2 + (rm_1E - rkm_2)x + qEk - rkm_1E = 0 \tag{6}$$

The interior equilibrium points are the points of intersection of the following two non-trivial nullclines,

$$y = -\frac{[rm_2x^2 + (rm_1E - rkm_2)x + qEk - rkm_1E](n_1 + x)}{ka(m_1E + m_2x)}, \tag{7}$$

$$y = \frac{n_2 + x}{b}, \tag{8}$$

in the interior of the first quadrant. A portion of the first nullcline (7) lies in R_+^2 for $x \in [x_L, x_H]$ and it is a continuous smooth curve joining the points $S_L(x_L, 0)$ and $S_H(x_H, 0)$. For the first nullcline $y \to \infty$ when $x \to -\frac{m_1E}{m_2}$. The second nullcline (8) is a straight line which intersects the x-axis at $(-n_2, 0)$ and y-axis at $\left(0, \frac{n_2}{b}\right)$.

Using the Eqs. (7) and (8) we see that the first component of interior equilibrium is a positive root of following cubic equation.

$$\begin{aligned}&rbm_2x^3 + (-rbkm_2 + akm_2 + rbm_1E + rbn_1m_2)x^2\\&+(-rbkn_1m_2 - rbkm_1E + akn_2m_2 + akm_1E + qEkb + rbn_1m_1E)x\\&+akn_2m_1E + Ekbn_1(q - rm_1) = 0\end{aligned} \tag{9}$$

Also, if $q > rm_1$ it can be ensured from Descartes' rule of signs, the Eq. (9) has at least one negative real root.

It is difficult to derive analytical conditions to determine the number of interior equilibrium points, however the possible number of feasible interior equilibrium points can be explained from the relative positions and shapes of the non-trivial nullclines as presented in Fig. 1. It can be observe from the Eq. (7) that the nontrivial prey nullcline intersects x-axis at $(x_l, 0)$, $(x_h, 0)$ and $(-n_1, 0)$, where x_l and x_h are real roots of the quadratic equation (6). Since the nontrivial predator nullcline (8) is a straight line therefore we can easily observe that there will be at most two positive interior equilibrium points. From this diagram one can see that whenever the point $(-n_2, 0)$ lies between the points $(x_l, 0)$ and $(x_h, 0)$, where x_l and x_h are such that $x_l < 0$ and $x_h > 0$, then the system (3) has exactly one equilibrium point. We denote interior equilibrium point by $S_*(x_*, y_*)$.

One can notice from the Fig. 1 that whenever $(x_l, 0)$ lies between the points $(-n_2, 0)$ and $(x_h, 0)$ such that $x_l < x_h$, then the system (3) can have either zero, one or two interior equilibrium points. Therefore, the dynamics of the system (3) is similar to that of the model studied in [7]. Hence we omit this case here.

Theorem 1 *(i) The origin $S_0(0, 0)$ is a saddle point if $q > rm_1$ and unstable if $q < rm_1$.*

(ii) The axial equilibrium point $S_L(x_L, 0)$ is always unstable.

(iii) The equilibrium point $S_H(x_H, 0)$ is always a saddle point.

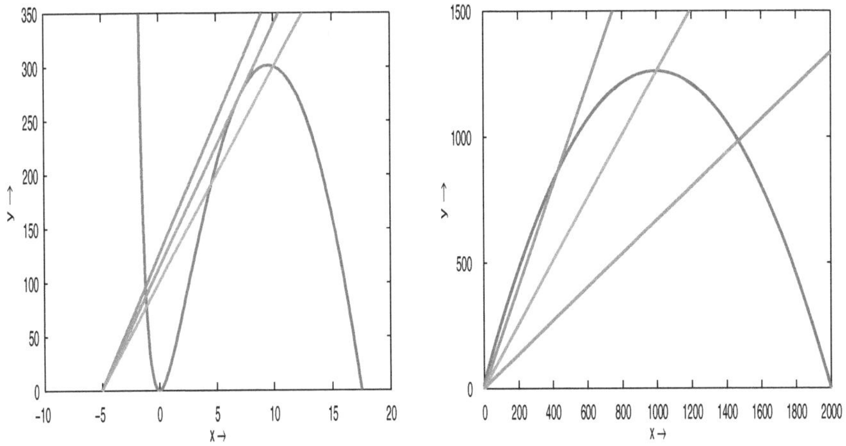

Fig. 1 The two nullclines (prey and predator) are drawn to get possible number of interior equilibrium points

(iv) The axial equilibrium point $S_1(0, \frac{n_2}{b})$ *is stable if* $\frac{a n_2}{b n_1} + \frac{q}{m_1} > r$ *and a saddle point if* $\frac{a n_2}{b n_1} + \frac{q}{m_1} < r$.

(v) System (3) undergoes a transcritical bifurcation around $S_1(0, \frac{n_2}{b})$ *if* $\frac{a n_2}{b n_1} + \frac{q}{m_1} = r$.

Proof Proof of these results can be done as in [7]. Therefore we omit it here.

3.1 Stability and Hopf-Bifurcation of Interior Equilibrium

Theorem 2 *(a) The equilibrium point* $S_*(x_*, y_*)$ *is locally asymptotically stable if,* $\frac{s}{x_*} + \frac{r}{k} > \frac{a(n_2+x_*)}{b(n_1+x_*)^2} + \frac{qEm_2}{(m_1E+m_2x_*)^2}$ *and* $\frac{r}{k} + \frac{a}{b(n_1+x_*)} > \frac{a(n_2+x_*)}{b(n_1+x_*)^2} + \frac{qEm_2}{(m_1E+m_2x_*)^2}$

(b) The system undergoes a Hopf-bifurcation with respect to bifurcation parameter s around the equilibrium point $S_*(x_*, y_*)$ *if,* $\frac{s}{x_*} + \frac{r}{k} = \frac{a(n_2+x_*)}{b(n_1+x_*)^2} + \frac{qEm_2}{(m_1E+m_2x_*)^2}$ *and* $\frac{r}{k} + \frac{a}{b(n_1+x_*)} > \frac{a(n_2+x_*)}{b(n_1+x_*)^2} + \frac{qEm_2}{(m_1E+m_2x_*)^2}$.

Proof These results can be proved by using linearization techniques and Hopf-bifurcation theorem as in [7].

3.2 Numerical Simulation

For the parameter values $r = 0.5, k = 2000, a = 0.2, n_1 = 10, q = 0.03, E = 1, m_1 = 0.2, m_2 = 0.1, b = 0.5, n_2 = 1$, we observe that a unique interior equilibrium point $S_*(x_*, y_*) = (430, 862)$. A small amplitude stable periodic solution appears around it for $s^{[H]} = 0.2761145338$. The trivial equilibrium points $S_0(0, 0)$, $S_1(0, \frac{n_2}{b}) = (0, 2)$ and $S_H(x_H, 0) = (1999.4, 0)$ are all unstable or saddle. For $s > s^{[H]}$ the unique interior equilibrium point $S_*(x_*, y_*) = (430, 862)$ is stable and for $s < s^{[H]}$ the size of periodic solution increases. These results are shown in Fig. 2. The time series diagram for the above set of parameter value is also provided in Fig. 3 to verify the dynamics of the system. The local maximum and minimum of x and y

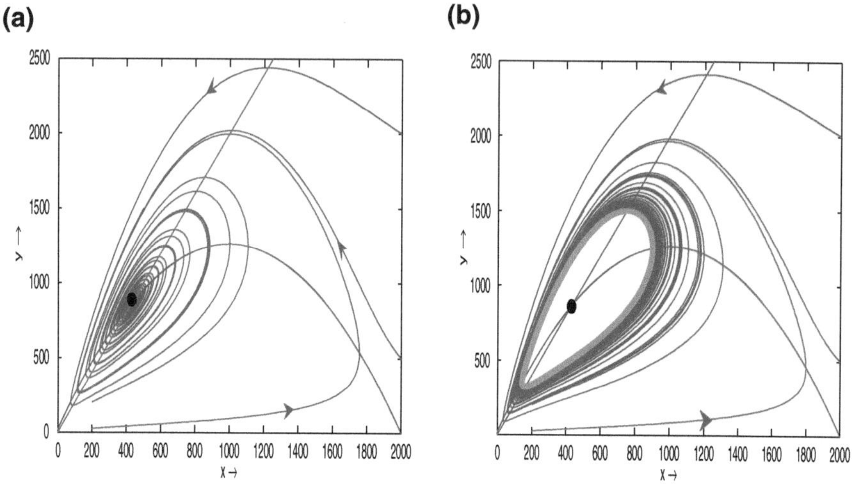

Fig. 2 **a** The unique interior equilibrium point $S_*(x_*, y_*)$ is stable. **b** A stable periodic solution bifurcates from the interior equilibrium through Hopf-bifurcation for $s^{[H]} = 0.27$

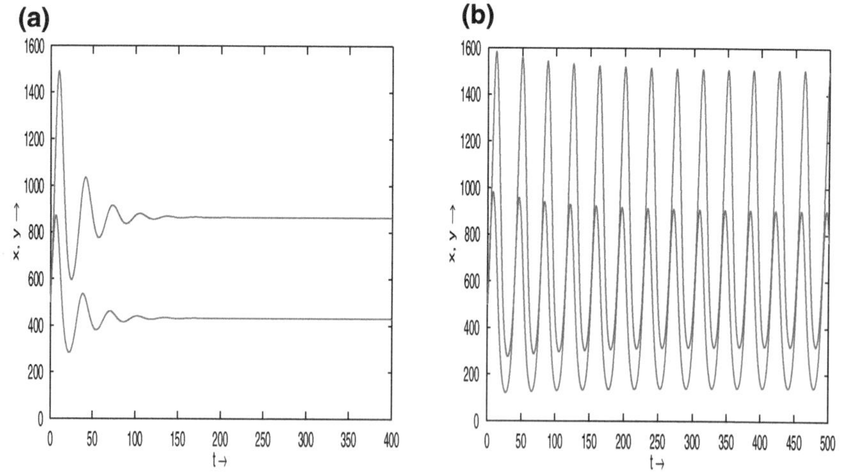

Fig. 3 **a** Time series diagram for $S_*(x_*, y_*)$ is stable. **b** Time series diagram for periodic solution through Hopf-bifurcation for $s^{[H]} = 0.27$

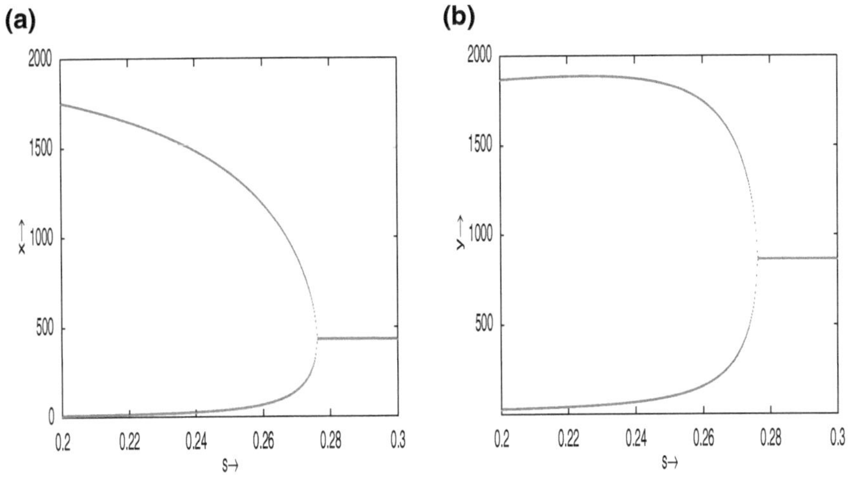

Fig. 4 **a** Bifurcation diagram for prey population with respect to growth rate of predator. **b** Bifurcation diagram for predator population with respect to growth rate of predator

are plotted with respect to a range of values of s in Fig. 4 to ensure the existence of Hopf-bifurcation.

4 Bionomic Equilibria

The bionomic equilibrium is the combination of biological and economic equilibrium point, i.e. the bionomic equilibrium is the intersection of biological equilibrium curve and the zero profit line. The Net Economic Revenue is given by:

Net Economic Revenue (N.E.R.) = Total Revenue (T.R.) - Total Cost (T.C.),

where $T.R. = pqE\,x/(m_1E + m_2x)$ and $T.C. = CE$ with C being the constant harvesting cost per unit effort and p being the constant price per unit biomass of prey species.

Thus the net profit at any time [5] is given by

$$P(x, E) = \left(\frac{pq\,x}{m_1E + m_2x} - C\right)E. \tag{10}$$

Note that if the harvesting cost is greater than the revenue for prey species (i.e. $C > \frac{pqx}{m_1E+m_2x}$), then harvesting in prey species is not profitable and so it is not of interest. Hence to continue the harvesting we consider here the cost must be less than the revenue for prey species (i.e. $C < \frac{pqx}{m_1E+m_2x}$).

The bionomic equilibrium $(x_\infty, y_\infty, E_\infty)$ is given by the positive solutions of $\frac{dx}{dt} = \frac{dy}{dt} = P = 0$, i.e.,

$$r\left(1 - \frac{x}{k}\right) - \frac{ay}{n_1 + x} - \frac{qE}{m_1 E + m_2 x} = 0, \tag{11}$$

$$by = n_2 + x, \tag{12}$$

$$\frac{pqx}{m_1 E + m_2 x} - C = 0. \tag{13}$$

From Eqs. (11), (12) and (13), we get the bionomic equilibrium points as $(x_\infty, y_\infty, E_\infty)$ where x_∞ is positive root of following quadratic equation

$$\begin{aligned} &rbm_1 px^2 + (akm_1 p + rbn_1 m_1 p + qkbp - rbkm_1 p - kbCm_2)x \\ &+ qkbn_1 p + akn_2 m_1 p - rbkn_1 m_1 p - kbn_1 Cm_2 = 0, \end{aligned} \tag{14}$$

with $y_\infty = \frac{n_2 + x_\infty}{b}$, $E_\infty = \frac{pq - Cm_2}{Cm_1} x_\infty$ and $pq > Cm_2$.

5 Singular Optimal Control

In this section, our objective is to maximize the current value of continuous time stream of revenues which is given by

$$J(x, E) = \int_0^\infty e^{-\delta t} P(x, E) dt, \tag{15}$$

where δ denotes the continuous annual discount rate which is fixed by harvesting agencies. We shall maximize (15) together with the steady state equations (11) and (12) with the help of Pontryagin's Maximum Principle [11]. The control variable E is subjected to the constraint $0 \le E \le E_{max}$, where E_{max} is a feasible upper limit for the harvesting effort.

Therefore, the optimal control problems over an infinite time horizon is given by

$$\max_{0 \le E(t) \le E_{max}} \int_0^\infty e^{-\delta t} P(x, E) dt, \tag{16}$$

subject to the system (3) and $x(0) = x_0, \ y(0) = y_0$.

The associated Hamiltonian function is given by

$$H(x, y, E, t) = \left(\frac{pqx}{m_1 E + m_2 x} - C\right) E e^{-\delta t} + \lambda_1 \left(rx\left(1 - \frac{x}{k}\right) - \frac{axy}{n_1 + x} - \frac{qEx}{m_1 E + m_2 x}\right) + \lambda_2 sy \left(1 - \frac{by}{n_2 + x}\right), \tag{17}$$

where $\lambda_i = \lambda_i(t),\ i = 1, 2$ are adjoint variables.

Differentiating the Hamiltonian H with respect to the control variable E we get

$$\frac{\partial H}{\partial E} = \left(\frac{pqx^2 m_2}{(m_1 E + m_2 x)^2} - C\right) e^{-\delta t} + \frac{\lambda_1 q x^2 m_2}{(m_1 E + m_2 x)^2}, \tag{18}$$

$$\text{and} \quad \frac{\partial^2 H}{\partial E^2} = -\frac{2m_1}{m_1 E + m_2 x}\left(\frac{\partial H}{\partial E} + Ce^{-\delta t}\right). \tag{19}$$

Notice that Hamiltonian is nonlinear in control variable and from Eqs. (18) and (19) we can not claim that the Hamiltonian is strictly monotone with respect to the control variable E [14]. Therefore we can not assure here that the optimal control strategy involves bang-bang controls.

Now, we find the singular path and hence singular solutions in the following. The considered control problem admits a singular solution on the control set $[0, E_{max}]$, if $\frac{\partial H}{\partial E} = 0$ which gives

$$\lambda_1 e^{\delta t} = p - \frac{C(m_1 E + m_2 x)^2}{q m_2 x^2}, \tag{20}$$

where $\lambda_1 e^{\delta t}$ is the usual shadow price [6].

In order to find the path of a singular control, Pontryagins Maximum Principle [11] is utilized and the adjoint variables must satisfy the adjoint equations given by

$$\frac{d\lambda_1}{dt} = -\frac{\partial H}{\partial x} \quad \text{and} \quad \frac{d\lambda_2}{dt} = -\frac{\partial H}{\partial y}. \tag{21}$$

Since, we are looking for singular optimal equilibrium solution, so we use steady state equations (11) and (12), hence x, y and E can be taken as constant [8]. Thus Eq. (21) along with steady state equations (11) and (12) give

$$\frac{d\lambda_1}{dt} = -\frac{pqm_1 E^2 e^{-\delta t}}{(m_1 E + m_2 x)^2} - \lambda_1\left(-\frac{rx}{k} + \frac{ax(n_2 + x)}{b\,(n_1 + x)^2} + \frac{qE^2 m_1}{(m_1 E + m_2 x)^2}\right) - \frac{\lambda_2 s}{b}, \tag{22}$$

$$\frac{d\lambda_2}{dt} = \frac{\lambda_1 a x}{n_1 + x} + \lambda_2 s. \tag{23}$$

Due to the presence of the term $e^{-\delta t}$ no steady state is possible for the above system. Hence we consider the following transformation [13].

$$\lambda_i(t) = \mu_i(t)e^{-\delta t}, \quad i = 1, 2,$$

where μ_i represents the present value of the adjoint variable λ_i.

Using Eq. (20), the Eq. (23) can be written in terms of μ_2 as follows:

$$\frac{d\mu_2}{dt} - (s + \delta)\,\mu_2 = -P_1(x). \tag{24}$$

where $P_1(x) = \frac{\alpha x}{(n_1+x)}\left(\frac{C(m_1E+m_2x)^2}{qm_2x^2} - p\right)$.

The shadow prices $\mu_i = \lambda_i(t)e^{\delta t}, i = 1, 2$ should remain constant over time in singular equilibrium to satisfy the transversality conditions at ∞ (i.e. $\lim_{t\to\infty} \lambda_i(t) = 0$, for $i = 1, 2$). Thus the solution of Eq. (24) is given by

$$\mu_2(t) = \frac{P_1(x)}{s + \delta}.$$

Using the above value of μ_2 the Eq. (22) can be written in the terms of μ_1 as follows:

$$\frac{d\mu_1}{dt} - (Q_1(x) + \delta)\mu_1 = -Q_2(x), \tag{25}$$

where, $Q_1(x) = \left(\frac{rx}{k} - \frac{ax(n_2+x)}{b(n_1+x)^2} - \frac{qE^2m_1}{(m_1E+m_2x)^2}\right)$ and $Q_2(x) = \frac{pqm_1E^2}{(m_1E+m_2x)^2} + \frac{s}{b}\frac{P_1(x)}{(s+\delta)}$.

Solution of the Eq. (25) satisfying the transversality condition at ∞ is given by

$$\mu_1(t) = \frac{Q_2(x)}{Q_1(x) + \delta}. \tag{26}$$

From Eqs. (20) and (26) we get

$$\frac{C(m_1E + m_2x)^2}{qm_2x^2} + \left(\frac{Q_2}{Q_1 + \delta}\right) = p, \tag{27}$$

Equation (27) gives the desired singular path.

6 Discussion

In this paper, we have considered a modified Leslie–Gower predator-prey model with Michaelis–Menten type prey harvesting. This model is applicable for those prey–predator systems where the predator is more capable of switching from its favorite food (the prey) to other food options and it can survive more easily when the prey is lacking severely.

The positivity, boundedness and permanence of the solutions of the proposed system are discussed. Biologically the permanence ensures the long term co-existence of the two species. We have discussed the stability and Hopf bifurcation of a unique coexisting steady-state. We choose the growth rate of predator as Hopf bifurcation parameter to observe the existence of periodic solutions near positive equilibrium. Using Hopf bifurcation theorem, we have established the existence of Hopf bifurcation and have shown our conclusions by numerical simulations. The bifurcation diagrams are provided which indicate that how the period of periodic solutions changes with respect to the growth rate of predator.

We have obtained the conditions for the existence of bionomic equilibrium point of the exploited system. The problem of singular optimal control has been discussed by using Pontryagins Maximum Principle. If the unharvested system is persistent [1], then a sufficiently small harvesting rate will not change drastically the qualitative behavior of the system, but the region of coexistence shrinks as the harvesting rate increases. This provides a theoretical support for safe harvesting in biological resource management in terms of co-existence of species.

References

1. Aziz-Alaoui, M.A., Okiye, M.D.: Boundedness and global stability for a predator-prey model with modified Leslie-Gower and holling-type II schemes. Appl. Math. Lett. **16**(7), 1069–1075 (2003)
2. Birkhoff, G., Rota, G.C.: Ordinary Differential Equations. Wiley, Boston (1982)
3. Clark, C.W.: Mathematical Bioeconomics: The Optimal Management of Renewable Resources. Wiley, New York (1976)
4. Clark, C.W.: Mathematical models in the economics of renewable resources. SIAM Rev. **21**, 81–99 (1979)
5. Das, T., Mukherjee, R.N., Chaudhari, K.S.: Bioeconomic harvesting of a prey-predator fishery. J. Biol. Dyn. **3**, 447–462 (2009)
6. Dubey, B., Chandra, P., Sinha, P.: A model for fishery resource with reserve area. Nonlinear Anal. Real World Appl. **4**, 625–637 (2003)
7. Gupta, R.P., Chandra, P.: Bifurcation analysis of modified Leslie-Gower predator-prey model with Michaelis-Menten type prey harvesting. J. Math. Anal. Appl. **398**, 278–295 (2013)
8. Khamis, S.A., Tchuenche, J.M., Lukka, M., Heilio, M.: Dynamics of fisheries with prey reserve and harvesting. Int. J. Comput. Math. **88**(8), 1776–1802 (2011)
9. Krishna, S.V., Srinivasu, P.D.N., Kaymackcalan, B.: Conservation of an ecosystem through optimal taxation. Bull. Math. Biol. **60**, 569–584 (1998)
10. Pal, P.J., Sarwardi, S., Saha, T., Mandal, P.K.: Mean square stability in a modified Leslie-Gower and holling-typeII predator-prey model. J. Appl. Math. Inform. **29**(3–4), 781–802 (2011)
11. Pontryagin, L.S., Boltyonskii, V.S., Gamkrelidze, R.V., Mishchenko, E.F.: The Mathematical Theory of Optimal Processes. Wiley, New York (1962)
12. Song, X., Chen, L.: Optimal harvesting and stability for a two-species competitive system with stage structure. Math. Biosci. **170**, 173–186 (2001)
13. Srinivasu, P.D.N.: Bioeconomics of a renewable resource in presence of a predator. Nonlinear Anal. Real World Appl. **2**, 497–506 (2001)
14. Srinivasu, P.D.N., Prasad, B.S.R.V.: Time optimal control of an additional food provided predator-prey system with applications to pest management and biological conservation. J. Math. Biol. **60**, 591–613 (2010). doi:10.1007/s00285-009-0279-2

A Special Class of Lotka–Volterra Models of Bacteria-Virus Infection Networks

Daniel A. Korytowski and Hal L. Smith

Abstract We show that the classical Volterra Lyapunov function can be used to obtain useful results for a Lotka–Volterra model of bacteria-virus infection networks. In particular, if a positive equilibrium exists, then it is stable and all positive trajectories are bounded and persistent in the sense that the limit inferior of each component is positive.

Keywords Virus · Bacteria · Infection network · Lotka–Volterra system

2010 Mathematics Subject Classification 92D40 · 93D30

1 Introduction

Motivated by the work of Jover et al. on virus-bacteria infection networks in aquatic environments [3] modeled by Lotka–Volterra systems, we showed in [4] that balanced bacteria-virus communities with either one-to-one or nested infection networks are persistent and that they can be assembled, one species at a time, through a sequence of persistent bacteria-virus subcommunities provided certain trade-offs between bacterial growth rate and defence against infection and between virus efficiency at infection and their host range are satisfied. Our modeling was based on a chemostat-like model involving a single limiting nutrient. In [5], we showed how these same results could be obtained using a slight modification of the Lotka–Volterra model employed in [3]. We also exploited the Lotka–Volterra framework and Volterra's famous Lyapunov function to obtain strong results on the global behavior of positive solutions of our model, in some cases obtaining global convergence to the positive equilibrium. Here,

D.A. Korytowski · H.L. Smith (✉)
School of Mathematical and Statistical Sciences, Arizona State University, Tempe, AZ, USA
e-mail: halsmith@asu.edu

D.A. Korytowski
e-mail: daniel.korytowski@asu.edu

J.M. Cushing et al. (eds.), *Applied Analysis in Biological and Physical Sciences*, Springer Proceedings in Mathematics & Statistics 186,
DOI 10.1007/978-81-322-3640-5_7

we show that the Volterra Lyapunov function can be employed to obtain useful results in the case of general infection networks under suitable conditions.

Let $B_i,\ 1 \le i \le n$, denote bacteria strain i density and $V_j,\ 1 \le j \le m$, denote phage strain j density. The specific growth rate of B_i is denoted by r_i and we assume without loss of generality that

$$r_1 > r_2 > \cdots > r_n > 0. \tag{1}$$

We assume that the effect of density dependence and intra-strain competition on growth rate of bacteria is independent of bacteria strain as in [3–5]. Decay rates of phage strains are denoted by μ_j. The infection network, which phage strain infects which bacteria strain is encoded by the non-negative $n \times m$ matrix Φ:

$$\Phi_{ij} = \text{attack rate of virus j on host i.}$$

The non-negative matrix β, of the same size as Φ, captures the "burst size", or number of progeny virus released by each lysed bacteria:

$$\beta_{ij} = \text{burst size of virus j from host i}$$

We tacitly assume that matrices Φ and β have the same set of zero entries. The equations of our model are the following.

$$\begin{aligned} B_i' &= B_i\left(r_i - \sum_{k=1}^{n} B_k\right) - B_i \sum_{j=1}^{m} \Phi_{ij} V_j,\ 1 \le i \le n \\ V_j' &= V_j\left(\sum_{k=1}^{n} \beta_{kj}\Phi_{kj} B_k - \mu_j\right),\ 1 \le j \le m. \end{aligned} \tag{2}$$

A special feature of system (2), also assumed in [5], is that intra-specific and inter-specific competition among bacteria is assumed to be identical for all bacteria.

A solution of (2) is said to be positive if all components of it are positive for all t; the form of (2) ensures that a solution is positive if and only if it has all positive components for some particular value of t.

We assume that an equilibrium $E^* = (B^*, V^*)$ exists with all components positive. It satisfies

$$\begin{aligned} r &= 1_n^T B^* 1_n + \Phi V^* \\ \mu &= [\beta \cdot \Phi]^T B^* \end{aligned} \tag{3}$$

where $r = (r_1, \ldots, r_n)^T$, 1_n is the n-vector with all entries one, and $\mu = (\mu_1, \ldots, \mu_m)^T$. $[\beta \cdot \Phi]$ denotes the entry-wise product of β and Φ.

Generically, no such equilibrium will exist if $n \ne m$. If $n = m$, then the second equation has a positive solution B^* if μ belongs to the interior of the polyhedral cone spanned by the columns of $[\beta \cdot \Phi]$. The existence of such a solution is independent of the length of the column vectors of the matrix and the length of μ, depending only on their respective directions. In view of (1), B^* must not be so large that $r_n - 1_n^T B^* 1_n < 0$ since then the first equation will not have a positive solution V^*.

If E^* exists, then the system may be rewritten as:

$$B_i' = B_i \left(\sum_{k=1}^{n} (B_k^* - B_k) + \sum_{j=1}^{m} \Phi_{ij}(V_j^* - V_j) \right), \quad 1 \le i \le n \tag{4}$$

$$V_j' = V_j \sum_{k=1}^{n} \beta_{kj} \Phi_{kj}(B_k - B_k^*), \ 1 \le j \le m.$$

Let $U(x, x^*) = x - x^* - x^* \log x/x^*, \ x, x^* > 0$, be the familiar Volterra function and let

$$W(B, V) = \sum_{i=1}^{n} c_i U(B_i, B_i^*) + \sum_{j=1}^{m} d_j U(V_j, V_j^*) \tag{5}$$

where $c_1, \ldots, c_n$ and $d_1, \ldots, d_m$ are to be determined.

Then the derivative of W along solutions of (4), $\dot{W}$, is given by

$$\begin{aligned}
\dot{W} &= -\sum_{i=1}^{n} c_i(B_i - B_i^*) \cdot \sum_{k=1}^{n} (B_k - B_k^*) + \sum_{i=1}^{n} c_i(B_i - B_i^*) \sum_{j=1}^{m} \Phi_{ij}(V_j^* - V_j) \\
&\quad + \sum_{j=1}^{m} d_j(V_j - V_j^*) \sum_{i=1}^{n} \beta_{ij}\Phi_{ij}(B_i - B_i^*) \\
&= -\sum_{i=1}^{n} c_i(B_i - B_i^*) \cdot \sum_{k=1}^{n} (B_k - B_k^*) \\
&\quad + \sum_{j=1}^{m} (V_j - V_j^*) \sum_{i=1}^{n} (d_j\beta_{ij} - c_i)\Phi_{ij}(B_i - B_i^*)
\end{aligned}$$

If

$$0 = (d_j \beta_{ij} - c_i)\Phi_{ij}, \ c_i = 1, \ 1 \le i \le n, \ 1 \le j \le m, \tag{6}$$

then we have

$$\dot{W} = -\left(\sum_{i=1}^{n} (B_i - B_i^*) \right)^2.$$

Evidently, (6) holds if, for example, $d_j \beta_{ij} = 1$, a very restrictive assumption. If we require that β_{ij}, the burst size for virus j is independent of the host i that it infects,

$\beta_{ij} = \beta_j$, then (6) holds with $d_j = \beta_j^{-1}$. Clearly, the latter is a strong assumption but one that has been made for strategic reasons in [3–5].

We proceed, assuming that (6) is satisfied. It follows that E^* is a stable equilibrium (see Theorem X.1.1 in [1]). Furthermore, a weak form of persistence holds since positive solutions satisfy $W(B(t), V(t)) \leq W(B(0), V(0))$, $t \geq 0$ and therefore $0 < \liminf_{t\to\infty} x(t) \leq \limsup_{t\to\infty} x(t) < \infty$ for each component $x = B_i, V_j$.

By LaSalle's invariance principle, the asymptotic dynamics (on the omega limit set) of a positive solution takes place on the set

$$\left\{(B, V) : \dot{W} = 0\} = \{(B, V) : \sum_{i=1}^{n}(B_i - B_i^*) = 0\right\}$$

and therefore is governed by the system

$$B_i' = -B_i\left(\sum_{j=1}^{m}\Phi_{ij}(V_j - V_j^*)\right),\ 1 \leq i \leq n \tag{7}$$

$$V_j' = V_j\sum_{k=1}^{n}\beta_{kj}\Phi_{kj}(B_k - B_k^*),\ 1 \leq j \leq m.$$

Observe that if $\beta_{kj}\Phi_{kj}$ is independent of k for some j, then $V_j' = 0$ on the limit set since $\sum_{i=1}^{n}(B_i - B_i^*) = 0$.

System (7) is conservative if $\beta_{ij} = \beta_j,\ \forall i, j$. Indeed,

$$\sum_{i=1}^{n}\frac{B_i'}{B_i}(B_i - B_i^*) = -\sum_{i=1}^{n}\sum_{j=1}^{m}\Phi_{ij}(V_j - V_j^*)(B_i - B_i^*) = -\sum_{j=1}^{m}\frac{V_j'}{\beta_j V_j}(V_j - B_j^*).$$

Adding the last term to the first and integrating gives that

$$\sum_{i=1}^{n}U(B_i(t), B_i^*) + \sum_{j=1}^{m}\frac{1}{\beta_j}U(V_j(t), V_j^*) = \text{constant}.$$

We summarize our results as follows.

Theorem 1 *Let the system (2) have a positive equilibrium $E^* = (B^*, V^*)$ and suppose that (6) holds. Then E^* is locally stable. Moreover, every positive solution of (2) is weakly persistent: there exists $m, M > 0$, which depend on the solution, such that $m \leq B_i(t), V_j(t) \leq M, \forall i, j, t \geq 0$. Trajectories on the omega limit set of a positive* *solution belong to the hyperplane* $\sum_{i=1}^{n} B_i = \sum_{i=1}^{n} B_i^*$ *and satisfy (7). If* E^* *is the unique positive equilibrium, then*

$$\lim_{t\to\infty} \frac{1}{t} \int_0^t (B(s), V(s))ds = E^*$$

holds for every positive solution.

The final assertion follows from Theorem 5.2.3 in [2].

2 Two Virus and Two Bacteria

We start by assuming that $\beta_{ij} = \beta_j$ and define

$$e_{ij} = \frac{\beta_j \Phi_{ij}}{\mu_j} \tag{8}$$

the efficiency of virus j at exploiting bacteria i. Our assumption for this model is that virus one specializes on infecting bacteria one while virus two specializes on infecting bacteria two. Thus virus one should have a lower virus efficiency for infecting bacteria two than for bacteria one and it should be superior in infecting bacteria one than virus two is. Symmetric considerations hold for virus two. This leads to the following inequalities.

$$e_{11} > e_{21}, \quad e_{11} > e_{12}, \quad e_{22} > e_{21}, \quad e_{22} > e_{12}. \tag{9}$$

Note that they imply $e_{12}/e_{22} < 1 < e_{11}/e_{21}$. (9) imply that the second equation of (3) has a unique positive solution given by

$$\begin{aligned} B_1^* &= \frac{e_{22} - e_{21}}{e_{11}e_{22} - e_{12}e_{21}} \\ B_2^* &= \frac{e_{11} - e_{12}}{e_{11}e_{22} - e_{12}e_{21}} \end{aligned} \tag{10}$$

A necessary condition for the first equation of (3) to have a positive solution is that

$$r_i > B_1^* + B_2^*, \ i = 1, 2. \tag{11}$$

If, in addition to (11), the following holds

$$\frac{e_{12}}{e_{22}} < \frac{r_1 - B_1^* - B_2^*}{r_2 - B_1^* - B_2^*} < \frac{e_{11}}{e_{21}}, \tag{12}$$

then the virus components of the equilibrium are positive:

$$V_1^* = \frac{\beta_1}{\mu_1} \frac{e_{22}(r_1 - B_1^* - B_2^*) - e_{12}(r_2 - B_1^* - B_2^*)}{e_{11}e_{22} - e_{12}e_{21}} \tag{13}$$
$$V_2^* = \frac{\beta_2}{\mu_2} \frac{e_{11}(r_2 - B_1^* - B_2^*) - e_{21}(r_1 - B_1^* - B_2^*)}{e_{11}e_{22} - e_{12}e_{21}}$$

In summary, a unique positive equilibrium exists if (9), (11), and (12) hold. Theorem 1 implies that the positive equilibrium is stable and all positive solutions have time-averages equal to it.

3 Conclusion

In [4] we showed persistence for a balanced bacteria-virus community with either one-to-one or nested infection networks in a chemostat-based model. In [5] we established the same result for a Lotka–Volterra model with the same two special network structures. In addition, we introduced the Volterra-type Lyapunov function and used it to obtain global qualitative features of the dynamics. In this paper we use a similar Lyapunov function to extend our earlier results to more general infection networks than one-to-one and nested networks, for example, allowing each virus to infect each host, under suitable conditions. We find that if the burst size for a virus is independent of which host it infects, also assumed in [3–5], then the coexistence equilibrium is stable and weak persistence of the host-virus community holds. In addition, the time averages of host and virus densities are asymptotic to their appropriate equilibrium

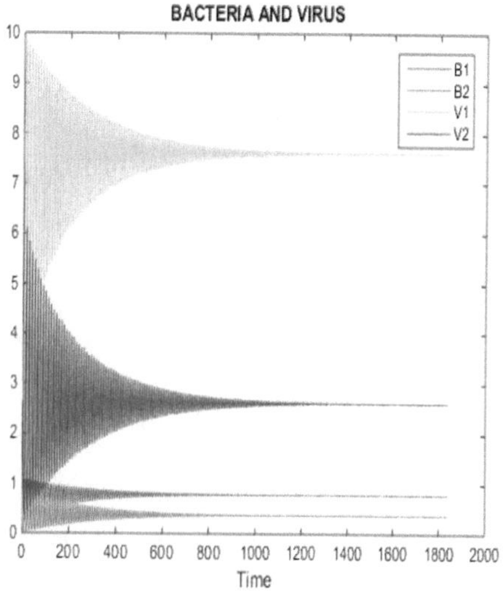

Fig. 1 Parameters: $r_1 = 6, r_2 = 5, \Phi_{11} = 0.5, \Phi_{12} = 0.4, \Phi_{21} = 0.3, \Phi_{22} = 0.6, \beta_1 = 4, \beta_2 = 5, \mu_1 = 2, \mu_2 = \frac{8}{3}$

values. Figure 1 depicts a representative example of system dynamics. Parameters specified in the figure are chosen to satisfy conditions (9), (11), and (12), and are not intended to be biologically realistic. Solutions are highly oscillatory, and seem to be aperiodic. It is an open question as to whether our conclusions hold without the restrictive conditions (6).

Acknowledgements This work was supported by a grant from the Simons Foundation (#355819, hal smith)

References

1. Hale, J.: Ordinary Differential Equations. Robert E. Krieger Publishing Co., Malabar (1980)
2. Hofbauer, J., Sigmund, K.: Evolutionary Games. Cambridge University Press, Cambridge (1998)
3. Jover, L.F., Cortez, M.H., Weitz, J.S.: Mechanisms of multi-strain coexistence in host phage systems with nested infection networks. J. Theor. Biol. **332**, 65–77 (2013)
4. Korytowski, D.A., Smith, H.L.: How nested and monogamous infection networks in host-phage communities come to be. Theor. Ecol. **8**(1), 111–120 (2015)
5. Korytowski, D.A., Smith, H.L.: Persistence in Phage-Bacteria Communities with Nested and One-to-One Infection Networks, Discrete and Continuous Dynamical Systems-B (to appear)

Plant Disease Propagation in a Striped Periodic Medium

Arnaud Ducrot and Hiroshi Matano

Abstract This work deals with the existence and non-existence of travelling wave solutions for a two-dimensional reaction-diffusion equation coupled with ordinary differential equations. The system models the spatial spread of a fungal disease over a field of crops whose spatial configuration exhibits a periodic stripe pattern, such as a vineyard. The standard comparison principle does not hold for this system. We establish a sharp criterion for the existence of (directional) travelling waves in terms of what we call the epidemic threshold $\mathscr{R}_0$, which is independent of the direction of the travelling wave. We then study the minimal speed c_θ^* of travelling waves for each direction $\theta \in [0, 2\pi)$ and prove its monotone dependence on θ. Our analysis is based on the fixed point argument, a variational characterization of principal eigenvalues and Harnack type inequalities for elliptic and parabolic problems.

Keywords Reaction-diffusion equations · Travelling waves · Periodic striped medium · Minimal wave speed · Epidemic model

2010 Mathematics Subject Classification 35J60 · 35K57 · 35C07 · 92D30

1 Introduction

In this paper we study the existence and non-existence of travelling wave solutions for a system of equations modelling the spatial spread of a fungal disease over a field of crops. The problem we shall consider consists of a reaction-diffusion equation coupled with ordinary differential equations and is a simplified version

---

A. Ducrot
IMB, UMR CNRS 5251, University of Bordeaux, 33076 Bordeaux, France
e-mail: arnaud.ducrot@u-bordeaux.fr

H. Matano (✉)
Graduate School of Mathematical Sciences, University of Tokyo, Komaba, Tokyo 153-8914, Japan
e-mail: matano@ms.u-tokyo.ac.jp

J.M. Cushing et al. (eds.), *Applied Analysis in Biological and Physical Sciences*, Springer Proceedings in Mathematics & Statistics 186,
DOI 10.1007/978-81-322-3640-5_8

of the epidemic model proposed by Burie et al. [8] to study the spatial spread of "powdery mildew" epidemics over vineyards.

Fungal diseases of crops are roughly caused by the germination of spores of a fungus and the development of infected lesions on the leafs of plants. During the infectious (or sporulating) period, the infected lesions produce new spores that disperse through the cropping system due to environmental turbulence which leads to new infections by falling down on healthy vegetable tissue. In the context of fungal diseases, infected plants do not recover and remain infected until death. The normalized model we shall consider in this work takes into account the main feature of such an epidemic cycle as well as the spatial structure of the cropping system. It is posed for $t > 0$ and $(x, y) \in \mathbb{R}^2$ and reads as follows:

$$\begin{cases} (\partial_t - D\Delta)\, S(t,x,y) = -S(t,x,y) + rI(t,x,y), \\ \partial_t H(t,x,y) = -\alpha(x)H(t,x,y)S(t,x,y), \\ \partial_t I(t,x,y) = \alpha(x)H(t,x,y)S(t,x,y) - \beta I(t,x,y). \end{cases} \tag{1}$$

Here $H(t,x,y)$ and $I(t,x,y)$ respectively denote the density of healthy and infectious plant tissue at time $t > 0$ and spatial location $(x, y) \in \mathbb{R}^2$ while $S(t,x,y)$ denotes the density of spores. The infected tissue remains infectious for a period of average length $\frac{1}{\beta}$ during which spores are produced by this infected tissue at a rate $r > 0$.

The contamination rate is denoted by $\alpha(x)$. It depends on the spatial location in order to take into account the varying germination probability due to the spatial structure of the field. Indeed this probability is larger in regions with high density of plants, so that the function $\alpha(x)$ reflects the spatial distribution of the plants over the field. Here we assume that this function depends only on one variable x, meaning that the cropping system exhibits a spatial configuration consisting of parallel rows. We furthermore assume the following periodicity condition

$$\alpha(x + L) \equiv \alpha(x), \tag{2}$$

for some given period $L > 0$, meaning that the rows are periodically arrayed. Such a spatial heterogeneity is particularly well suited for the modelling of vineyards, but it also applies to various other types of cropping systems.

As mentioned above, System (1) is a simplified version of the model studied by Burie et al. in [8], where the authors consider a dual-range dispersal process for the spores — short and long ranges — which is formulated by a system of two reaction-diffusion equations coupled with ODEs. This system with dual range dispersal has been further investigated by Mammeri et al. in [26] by including spatial heterogeneities. Numerical experiments performed in the aforementioned work indicate that the epidemic propagates in the form of a pulsating travelling wave, in particular when the environment exhibits a striped periodic pattern. Let us also mention the earlier work of Zawolek and Zadoks [38], in which the authors have used reaction-diffusion equations with time delay to model and investigate the spatial spread of fungal diseases of plants. We also refer to [30] and the references therein for models of plant diseases using ODEs and delay differential equations, and to the monograph of Okubo and Levin [29] and that of Shigesada and Kawazaki [32] for more details about passive and stratified diffusion processes in biology.

Travelling wave solutions for System (1) have been studied in [9, 10] in the case of a spatially homogeneous environment, namely when α is a constant function. The aim of the present paper is to study the existence and non-existence of travelling waves for System (1) posed in a periodic striped environment. Since the medium is spatially periodic, one has to deal with the so-called pulsating travelling waves, which generalize the usual notion of travelling waves in homogeneous media. Roughly speaking, a pulsating travelling wave is a solution whose profile and speed fluctuate as its front passes through varying environments, yet it keeps a certain coherent structure during the entire course of evolution. See Definition 2.1 below for the precise notion of such solutions in the context of System (1). We also refer the reader to [33, 35, 36] for earlier works related to this notion.

Front propagations in spatially and/or temporally heterogenous media have received much attention in the past two decades. We refer the reader to [37] for a review of relatively early works on this topic. We also refer to [1, 3, 4, 6, 7, 22, 24, 25, 34] and the references cited therein for detailed mathematical analysis of pulsating travelling waves in scalar equations and order preserving evolution problems. More general notions of propagating solutions can be found in Berestycki and Hamel [2] for general spatially heterogeneous media, and in Huang and Shen [18] and Shen [31] for spatio-temporally heterogeneous media. Let us finally mention some works on front propagation in a striped periodic medium. Kinezaki et al. [21] studies biological invasions in such a medium from the ecological viewpoint. The work of Liang and Matano [23] studies the problem of finding a periodic coefficient that maximizes the speed of spreading fronts. These works are for scalar KPP type equations, but their problems are posed on a striped periodic medium similar to our System (1).

In the present paper, we shall first prove (see Theorem 1 below) that the quantity $\mathscr{R}_0$ defined in (8) below gives a sharp threshold for the existence of (pulsating) travelling wave solutions of (1). More precisely, we shall show that no travelling wave exists in any direction if $\mathscr{R}_0 < 1$, while, if $\mathscr{R}_0 > 1$, travelling waves exists in every direction. Furthermore, for each $\theta \in [0, 2\pi)$, and for any $c \in [c_\theta^*, \infty)$, there exists a travelling wave solution in the direction θ with speed c. The minimal wave speed c_θ^* will be characterized in a closed form (10). The minimality of (10) will be proved by deriving suitable Harnack like inequalities for some elliptic and parabolic problems coupled with ODEs. Since our analysis does not make use of comparison arguments, we expect that such a methodology can be extended to other problems.

Next we study properties of the directional minimal wave speed c_θ^* (see Theorem 2). We shall first prove that c_θ^* has a certain monotonicity property with respect to the angle θ. This implies, in particular, that the epidemic spreads faster along the rows than in the transverse directions. Next we shall derive asymptotic formulas for c_θ^* in the homogenization limit as $L \to 0$ and also in the limit as $L \to \infty$ (see Theorem 3). The former corresponds to the fast diffusion ($D >> 1$) approximation in the original variables, while the latter corresponds to the slow diffusion limit ($D << 1$). We refer to Hamel et al. [16] and Hamel et al. [17] for some similar results for the scalar Fisher-KPP equation.

Remark 1.1 The reason why we pay much attention to the minimal wave speed c^*_θ is because we suspect that it is closely related to the so-called *spreading speed*, namely the speed of the expanding front arising in solutions with compactly supported initial data. In epidemiology, it is important to estimate how fast the epidemic spreads starting from localized initial data, and this is precisely what the notion of spreading speed is about. In the case of scalar KPP type equations or more general order-preserving evolution problems with periodic spatial inhomogeneity, the relation between the minimal wave speed and the spreading speed is well understood: the spreading speed is given by the Wulff shape associated with the minimal wave speed, as defined in (14); see [4, 34]. We believe that the same relation holds for the epidemic model (1), though at present, very little is known rigorously about the spreading properties of epidemic-diffusion models in general.

The one-dimensional travelling wave profiles for System (1) are illustrated in Fig. 1 above for a homogeneous medium (left) and for a periodic medium (right). Since no influx of population is considered in System (1), the infection dies out after the propagation of the wave of infection. Hence the pulsating travelling waves we shall investigate in the present paper involve travelling pulses of infections. Such propagating profiles are commonly observed in the Kermack and McKendick model. We refer to Hosono and Ilyas [15] for the study of such a system in a homogeneous medium, in which the existence of travelling pulses have been proved. We also refer to Ducrot and Giletti in [11] where a Kermack and McKendick like system of equations in a periodic environment is considered. Note that System (1), as well as the problems studied in the aforementioned papers, admits a continuum of stationary states. Therefore one needs to prescribe the state of the healthy tissue component H ahead of the epidemic front. To that aim we fix a periodic function $v^+ \equiv v^+(x)$ that

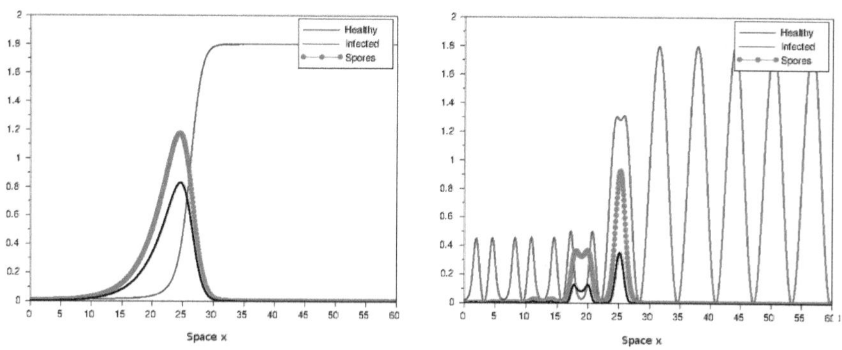

Fig. 1 Typical one-dimensional front propagation: in a homogeneous medium (*left*) and in a periodic medium (*right*)

represents the spatial structure of the cropping system before the arrival of epidemic. Recalling (2), the function v^+ is assumed to satisfy the same periodicity condition as the function α, that is,

$$v^+(x+L) \equiv v^+(x).$$

Hence we shall look for travelling wave solutions for (1) connecting the prescribed stationary state $(S, H, I) = \big(0, v^+, 0\big)$ at $t = -\infty$ and the stationary state $(S, H, I) = \big(0, v^-, 0\big)$ at $t = +\infty$. Here $v^- \equiv v^-(x)$ is an unknown $L-$periodic function that should be found together with the travelling wave profile and the wave speed. This function v^- represent the surviving healthy vegetable tissue behind the epidemic wave.

This paper is organized as follows. In Sect. 2, we state the main results of the present paper. Section 3 deals with preliminary results including the formulation of the problem that determines the travelling wave profile and the study of the principal eigenvalue of some elliptic problem with periodic boundary conditions. These preliminary results will allow us to characterize the minimal wave speed in a closed form. In Sect. 4 we provide a key estimate for the solutions by using suitable Harnack like inequalities. This estimate will be used in Sect. 5 to prove the criticality of the minimal wave speed. Section 5 is devoted to the proof of our non-existence and existence results. Finally Sect. 6 investigates several qualitative properties of this minimal wave speed.

2 Main Results

In this section we shall state the main results of this paper. Before doing so, let us observe that (1) can be reduced to the case $D = 1$ by replacing the period L by $\frac{L}{\sqrt{D}}$. Thus, in the sequel we shall assume, without loss of generality, that

$$D = 1. \tag{3}$$

With this simple change of variable, the fast (resp. slow) diffusion approximation $D \to \infty$ (resp. $D \to 0$) in the original variables corresponds to small (resp. large) period limit as $L \to 0$ (resp. $L \to \infty$).

From now on we set $\mathbb{T}_L = \mathbb{R}/(L\mathbb{Z})$, the one-dimensional torus with period $L > 0$ and, in addition to (3), we assume that the parameters arising in (1) satisfy the following set of conditions:

Assumption 2.1 We assume that $r > 0$ and $\beta > 0$ are given positive parameters while the functions α and v^+ belong to $L^\infty_+(\mathbb{T}_L)$ and satisfy

$$\kappa(x) := \alpha(x)v^+(x) \in L^\infty_+(\mathbb{T}_L) \setminus \{0\}. \tag{4}$$

The condition $r > 0$ implies that infected plants produce spores. The condition $\kappa(x) \not\equiv 0$ implies that some healthy plants exist in the area where infection is possible (i.e. where $\alpha(x) \neq 0$) before the epidemic arrives. Clearly these conditions are necessary for the propagation of epidemic. The condition $\beta > 0$ implies that the population of infectious plants decays with positive rate, which is also a natural assumption from the epidemiological point of view.

Next let us recall the definition of pulsating travelling waves in a given direction θ. We state it in the context of System (1) in a periodic striped medium (Fig. 2); see [23] for a similar definition.

Definition 2.1 (*Directional travelling wave*) An entire solution $(S, H, I)(t, x, y)$ of system (1) defined on $\mathbb{R} \times \mathbb{R}^2$ is called a (pulsating) travelling wave solution of (1) in the direction $\theta \in [0, 2\pi)$ with speed $c > 0$ if it is written in the form

$$(S, H, I)(t, x, y) = (u, v, w)(x \cos\theta + y \sin\theta - ct, x), \tag{5}$$

where $(u, v, w) \equiv (u, v, w)(z, x) : \mathbb{R} \times \mathbb{T}_L \to \mathbb{R}^3$ is a bounded function satisfying the following conditions:

(i) The function (u, v, w) is positive and $L-$periodic with respect to x. Here positivity is understood in the following sense:

$$u > 0, \ \ v \geq 0 \text{ and } w \geq 0.$$

(ii) The function (u, v, w) possesses the following asymptotics at $z = \pm\infty$:

$$\lim_{z\to\pm\infty} (u, v, w)(z, x) = \big(0, \ v^{\pm}(x), \ 0\big) \text{ in } L^{\infty}(\mathbb{T}_L),$$

where v^+ is a prescribed $\mathbb{T}_L-$periodic function while v^- is a priori an unknown $\mathbb{T}_L-$periodic function.

As mentioned in the Introduction, $v^+ \equiv v^+(x)$ in the above definition is a prescribed nonnegative $L-$periodic function that describes the density of the vegetable tissue before the epidemic, while $v^- \equiv v^-(x)$ is an unknown $L-$periodic function that represents the density of the healthy tissue after the epidemic.

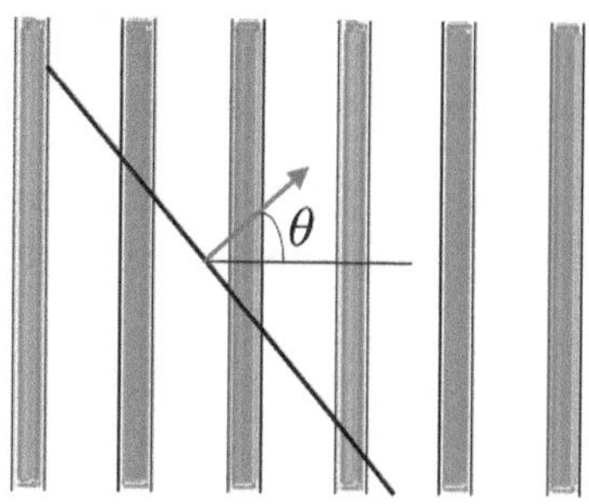

Fig. 2 A travelling wave in the direction θ. The *diagonal line* shows schematically the position of the front. Note that, in general, the actual front is not flat; it undulates periodically along this line

Next, for each $\theta \in [0, 2\pi)$, we define a function $\mu^\theta : \mathbb{R} \times (-\beta, \infty) \to \mathbb{R}$ as follows, where κ is a function defined in (4) in Assumption 2.1.

$$\mu^\theta(\lambda, \nu) = \max_{g \in \mathscr{A}} \left\{ \int_{\mathbb{T}_L} \left[-g'(x)^2 + \frac{r\kappa(x) g^2(x)}{\beta + \nu} \right] dx + \lambda^2 \cos^2\theta \left(\mathscr{J}(g) - 1 \right) \right\},$$

where

$$\mathscr{A} := \left\{ g \in C^1(\mathbb{T}_L) : \; g > 0, \; \int_{\mathbb{T}_L} g^2(x) dx = 1 \right\}, \tag{6}$$

while $\mathscr{J} : \mathscr{A} \to \mathbb{R}$ denotes the functional defined by

$$\mathscr{J}(g) = \frac{L^2}{\int_{\mathbb{T}_L} \frac{dx}{g^2(x)}}, \quad \forall g \in \mathscr{A}. \tag{7}$$

As we shall explain later, $\mu^\theta(\lambda, \nu)$ is the principal eigenvalue of a certain differential operator (see (31)). Now we introduce the following quantity:

$$\mathscr{R}_0 := \mu^\theta(0, 0) = \max_{g \in \mathscr{A}} \left\{ \int_{\mathbb{T}_L} \left[-g'(x)^2 + \frac{r\kappa(x) g^2(x)}{\beta} \right] dx \right\}. \tag{8}$$

As we explain in Remark 2.1 below, this quantity is closely related to the so-called basic reproduction number of the epidemic system (1). We shall call $\mathscr{R}_0$ the *epidemic threshold*, as it plays the role of a threshold for the existence of travelling wave solutions for (1). Note that $\mathscr{R}_0$ does not depend on the angle θ.

Remark 2.1 In epidemiology, "the basic reproduction number" of an infection roughly means the number of cases of infection that one case generates on average during its infectious period. Let us explain how the above quantity $\mathscr{R}_0$ is related to this notion. When the medium is spatially homogeneous, namely $\nu^+(x) = \nu^+$ and $\alpha(x) = \alpha$, the above quantity $\mathscr{R}_0$ becomes

$$\mathscr{R}_0 = \frac{r\kappa}{\beta},$$

while System (1) reduces to the following ODE system:

$$\dot{S} = -S + rI, \quad \dot{H} = -\alpha HS, \quad \dot{I} = \alpha HS - \beta I.$$

To compute the basic reproduction number, we replace the last equation by $\dot{I} = -\beta I$ and plug $I(t) = e^{-\beta t} I(0)$ into the first equation. Then the total amount of infection caused by the (infinitesimally small) initial data $I(0)$ is given by

$$\alpha H(0) \int_0^\infty S(t) dt = \frac{r}{\beta} \alpha I(0) H(0) = \frac{r\kappa}{\beta} I(0).$$

Hence the basic reproduction number for the homogeneous problem is $r\kappa/\beta$, which coincides with the above-mentioned value of $\mathscr{R}_0$. Next we consider the original system with $D = 1$. Note first that $\mathscr{R}_0$ in (8) coincides with the principal eigenvalue μ of the following eigenvalue problem on the one-dimensional torus $\mathbb{T}_L := \mathbb{R}/(L\mathbb{Z})$:

$$\varphi'' + \frac{r\kappa(x)}{\beta}\varphi = \mu\varphi, \quad x \in \mathbb{T}_L.$$

As for the basic reproduction number, assuming that the initial data of infection $I(0, x) << 1$ is independent of y, and making similar calculations as above, we see that the total amount of infection caused by $I(0, x)$ is given by

$$\frac{r\kappa(x)}{\beta}(1 - \Delta_1)^{-1} I(0, x), \quad x \in \mathbb{T}_L,$$

where Δ_1 denotes the operator d^2/dx^2 on the one-dimensional torus $\mathbb{T}_L$. Thus the basic reproduction number may be defined by the largest eigenvalue of the operator $I_0 \mapsto \frac{r\kappa(x)}{\beta}(1 - \Delta_1)^{-1} I_0$. Let $\hat{\mu}$ denote this largest eigenvalue. Then

$$\psi'' + \hat{\mu}^{-1}\frac{r\kappa(x)}{\beta}\psi = \psi, \quad x \in \mathbb{T}_L \quad \left(\psi := (1 - \Delta_1)^{-1} I_0\right).$$

It follows that $\hat{\mu} > 1$ (resp. < 1) if and only if $\mu > 1$ (resp. < 1). For the convenience of later calculations we define $\mathscr{R}_0$ by μ, not $\hat{\mu}$. As we shall see later, the value $\mathscr{R}_0 = 1$ gives a sharp threshold between the existence and non-existence of travelling waves in any direction.

Our first main result is the following:

Theorem 1 (Travelling wave solutions) *Let Assumption 2.1 be satisfied. Then the following statements hold true.*

(i) Assume that $\mathscr{R}_0 \le 1$, then System (1) *does not admit any travelling wave solution in any given direction $\theta \in [0, 2\pi)$.*
(ii) Assume that $\mathscr{R}_0 > 1$. Then for each direction $\theta \in [0, 2\pi)$ there exists a constant $c_\theta^* > 0$ such that:

(1) for any $c \in \left[c_\theta^*, \infty\right)$, System (1) *possesses a travelling wave solution in the direction θ;*
(2) for any $c \in \left(0, c_\theta^*\right)$, System (1) *does not admit any travelling wave in the direction θ.*

Remark 2.2 When $\mathscr{R}_0 > 1$, we will henceforth refer to the quantity c_θ^*, for each $\theta \in [0, 2\pi)$, as the minimal wave speed of System (1) in the direction θ.

When $\mathscr{R}_0 > 1$, the minimal wave speed c_θ^* in the above theorem is characterized in a close form as follows.

Here we only give a sketch. Detailed arguments will be given in Proposition 3.1 below. Fix $\theta \in [0, 2\pi)$. Then for each $\lambda \in \mathbb{R}$, the fixed point equation

$$\Lambda^\theta(\lambda, \nu) := \lambda^2 + \mu^\theta(\lambda, \nu) - 1 = \nu, \tag{9}$$

has a unique solution $\nu^\theta(\lambda) > 0$ with the following properties:

$$\nu^\theta(0) > 0 \text{ and } \lim_{\lambda\to\infty} \frac{\nu^\theta(\lambda)}{\lambda} = \infty.$$

Using this notation, the minimal wave speed c^*_θ is defined by

$$\begin{aligned} c^*_\theta = \min_{\lambda>0} \frac{\nu^\theta(\lambda)}{\lambda} &= \inf\{c > 0 : \ \exists \lambda > 0 \ \ \lambda c - \Lambda^\theta(\lambda, c\lambda) = 0\}, \\ &= \sup\{c > 0 : \ \lambda c - \Lambda^\theta(\lambda, c\lambda) \le 0, \ \forall \lambda \ge 0\}. \end{aligned} \tag{10}$$

Remark 2.3 Note that the epidemic threshold $\mathscr{R}_0$ given in (8) depends upon the heterogeneity functions α and ν^+ only through their product $\kappa := \alpha\nu^+$. The same is true of $\mu^\theta(\lambda, \nu)$ for any λ, ν; hence, by (9), (10), the minimal wave speed c^*_θ depends on α, ν^+ only through κ. Hereafter we shall write $\mathscr{R}_0(\kappa)$ and $c^*_\theta(\kappa)$, whenever we want to emphasize their dependence on the heterogeneity function κ.

Remark 2.4 Since $\mu^\theta(\lambda, \nu)$ depends on the angle θ only through $\cos^2\theta$, c^*_θ can be extended to all $\theta \in \mathbb{R}$ as a π-periodic function.

The next theorem is concerned with the minimal wave speed c^*_θ:

Theorem 2 (The minimal speed) *Let $\mathscr{R}_0$ the epidemic threshold defined in* (8).

(i) Assume that $\mathscr{R}_0 > 1$. Then the function $\theta \mapsto c^*_\theta$ satisfies

$$c^*_{\theta+\pi} = c^*_\theta \quad \textit{and} \quad c^*_{\pi-\theta} = c^*_\theta$$

for all $\theta \in [0, 2\pi)$ (or all $\theta \in \mathbb{R}$), and it is increasing on the interval $[0, \frac{\pi}{2}]$.

(ii) Let $\left(\widehat{\alpha}, \widehat{\nu^+}\right)$ and $\widehat{\kappa}$ denote the periodic Schwarz rearrangement of the pair $\left(\alpha, \nu^+\right)$ and κ, respectively, over $\mathbb{T}_L$ as defined in Definition 6.1. Then

$$\mathscr{R}_0(\kappa) \le \mathscr{R}_0\left(\widehat{\kappa}\right) \quad \textit{and} \quad \mathscr{R}_0(\kappa) \le \mathscr{R}_0\left(\widehat{\alpha}\widehat{\nu^+}\right).$$

Furthermore, if $\mathscr{R}_0(\kappa) > 1$ then for all $\theta \in [0, 2\pi)$, there hold

$$c^*_\theta(\kappa) \le c^*_\theta\left(\widehat{\kappa}\right) \quad \textit{and} \quad c^*_\theta(\kappa) \le c^*_\theta\left(\widehat{\alpha}\widehat{\nu^+}\right).$$

Our next result deals with the asymptotic properties of the epidemic threshold and the minimal wave speed for fast diffusion $D{>>}1$ (resp. slow diffusion $0 < D{<<}1$).

As mentioned at the beginning of Sect. 2, this problem is equivalent to the case where $D = 1$ and $L << 1$ (resp. $D = 1$ and $L >> 1$). Hence we consider two given functions α and v^+ in $L^\infty_+(\mathbb{T}_1)$ such that $\kappa(x) := \alpha(x)v^+(x) \in L^\infty_+(\mathbb{T}_1)\backslash\{0\}$ and we define, for each $L > 0$, the following $L-$periodic functions

$$\alpha_L(x) = \alpha\left(\frac{x}{L}\right), \quad v_L^+(x) = v^+\left(\frac{x}{L}\right) \text{ and } \kappa_L(x) = \alpha_L(x)v_L^+(x), \quad x \in \mathbb{T}_L.$$

Then we have the following theorem:

Theorem 3 (Slow/fast diffusion limit) *The function $L \mapsto \mathscr{R}_0(\kappa_L)$ is increasing and satisfies*

$$\lim_{L\to 0} \mathscr{R}_0(\kappa_L) = \mathscr{R}_{0,0} := \frac{r}{\beta}\int_{\mathbb{T}_1} \kappa(x)dx,$$

and

$$\lim_{L\to\infty} \mathscr{R}_0(\kappa_L) = \mathscr{R}_{0,\infty} := \frac{r\|\kappa\|_\infty}{\beta}.$$

Moreover the following properties hold true:

(i) (**Fast diffusion limit**) *If $\mathscr{R}_{0,0} > 1$, then $\mathscr{R}_0(\kappa_L) > 1$ for all $L > 0$. Let $c^*_{\theta,L}$ denote, for each $\theta \in [0, 2\pi)$ and $L > 0$, the minimal wave speed in the direction θ for System* (1) *with (α, v^+) replaced by (α_L, v_L^+). Then for each $\theta \in [0, 2\pi)$, the function $L \to c^*_{\theta,L}$ is increasing and has the following limit:*

$$\lim_{L\to 0} c^*_{\theta,L} = \overline{c^*}, \quad \text{with } \frac{1}{\overline{c^*}^2} = \sup_{x>0}\left\{\frac{1}{x^2}\left(1 - \frac{r\int_0^1 \kappa(s)ds}{\beta + x} + x\right)\right\}. \tag{11}$$

(ii) (**Slow diffusion limit**) *If $\mathscr{R}_{0,\infty} > 1$ then there exists $L_0 > 0$ such that, for all $L > L_0$, the minimal wave speed $c^*_{\theta,L}$ is well defined in each direction θ, and it converges as $L \to \infty$ to a limit $c^*_{\theta,\infty}$ that is characterized as follows*

$$c^*_{\theta,\infty} = \inf\left\{c > 0 : \exists\lambda > 0 \ \lambda^2\sin^2\theta - c\lambda - 1 + \frac{F\left(\cos\theta\,\lambda\sqrt{\beta + c\lambda}\right)}{\beta + c\lambda} < 0\right\},$$

where F is a function as defined below in (12).

The function $F : \mathbb{R} \to \mathbb{R}$ in the above theorem is an even function defined by

$$F(x) = \begin{cases} r\|\kappa\|_\infty & \text{if } |x| < j(r\|\kappa\|_\infty), \\ j^{-1}(|x|) & \text{if } |x| \geq j(r\|\kappa\|_\infty), \end{cases} \tag{12}$$

where $j : [r\|\kappa\|_\infty, \infty) \to [j(r\|\kappa\|_\infty), \infty)$ is an nondecreasing function defined by

$$j(k) = \int_0^1 \sqrt{k - r\kappa(x)}\,dx.$$

Now, as we mentioned in Remark 1.1, we suspect that the directional wave speed c_θ^* is closely related to the "spreading speed", namely the speed of expanding front of a solution starting from compactly supported initial data for S, I. In the context of System (1), this is interpreted as the speed of propagation of the epidemic that is triggered by a localized infection. If we denote by w_θ^* the spreading speed in the direction θ, then, by the analogy of known results on scalar KPP type equations, we suspect that the spreading speed is given in the form

$$w_\theta^* = \min_{|\phi|<\frac{\pi}{2}} \frac{c_{\theta+\phi}^*}{\cos\phi}. \tag{13}$$

Once the spreading speed is computed, then asymptotic shape of the spreading front is roughly given by the following form in the polar coordinates

$$r = tw_\theta^* + \varepsilon(t) \quad \text{with} \quad \lim_{t\to\infty} \frac{\varepsilon(t)}{t} = 0.$$

Equivalently, the spreading speed and the shape can be expressed by using the Wulff shape associated with c_θ^*, which is defined by

$$\mathscr{W} = \bigcap_{\theta\in[0,2\pi)} \left\{(x, y) \in \mathbb{R}^2 : \; x\cos\theta + y\sin\theta \le c_\theta^*\right\}. \tag{14}$$

In fact, the boundary of $\mathscr{W}$ is given by $r = w_\theta^*$ with w_θ^* as in (13). The validity of the formula (13) for scalar KPP type equations in periodic media, or even in more general problems, is well established; see, for instance, [4, 12, 13, 34]. However, in all these previous works, the proof of the validity of the formula (13) relies heavily on the comparison principle (or the maximum principle), which, unfortunately, does not hold in the epidemic system (1). Therefore, at present, it is just a speculation that the spreading speed is given by (13), or, equivalently, by the Wulff shape (14). We nonetheless believe that this is true.

In the special case where the function κ ia a step function of the form

$$\kappa(x) = \begin{cases} m & \text{if } x \in (0, l), \\ 0 & \text{if } x \in (l, 1), \end{cases}$$

for some $m > 0$ and $l \in (0, 1)$, one can compute the limit wave speed $c_{\theta,\infty}^*$ explicitly. Consequently, the spreading speed and the Wulff shape associated with $c_{\theta,\infty}^*$ can be computed explicitly. Figure 3 below shows this limit Wulff shape for different values of l (with $m = 2$, $r = 1$ and $\beta = 1$ fixed).

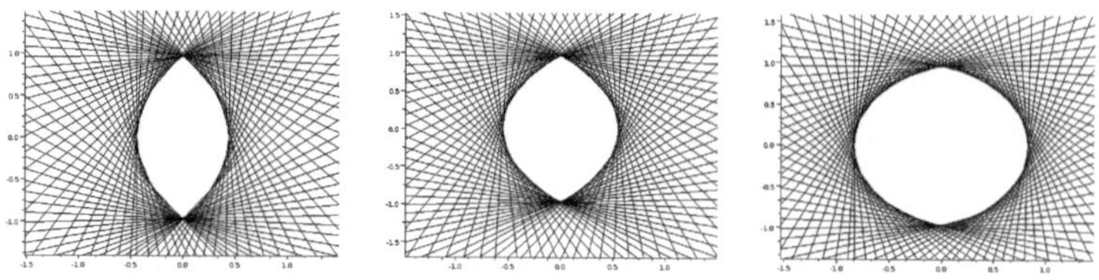

Fig. 3 The Wulff shapes in the slow diffusion limit for varying parameters: (from *left* to *right*) $l = 0.2, 0.4$ and 0.8

3 Preliminaries

3.1 Travelling Wave Formulation and Remarks

In this subsection we write down the equations satisfied by the profile function of the directional travelling wave solution of (1) and reformulate the equations in a suitable form.

Consider a pulsating travelling wave (S, H, I) in the direction $\theta \in [0, 2\pi)$ with speed $c > 0$, and let (u, v, w) denote its profile function. Then it satisfies the following elliptic problemcoupled with ODE

$$(\Delta_\theta + c\partial_z - 1)\, u(z, x) + rw(z, x) = 0, \quad z \in \mathbb{R}, \quad x \in \mathbb{T}_L, \tag{15}$$

$$c\partial_z v(z, x) = \alpha(x)u(z, x)v(z, x), \quad z \in \mathbb{R}, \quad x \in \mathbb{T}_L, \tag{16}$$

$$c\partial_z w(z, x) + \alpha(x)u(z, x)v(z, x) - \beta w(z, x) = 0 \quad z \in \mathbb{R}, \quad x \in \mathbb{T}_L, \tag{17}$$

where Δ_θ denotes the differential operator

$$\Delta_\theta = \partial_x^2 + 2\cos\theta\ \partial_x\partial_z + \partial_z^2. \tag{18}$$

This system is supplemented with the following limit behaviour

$$\lim_{z \to \pm\infty} \big(u,\ v,\ w\big)\,(z, x) = \big(0,\ v^{\pm}(x),\ 0\big)\,. \tag{19}$$

Next we shall derive an alternative integral reformulation of System (15)–(19). To that aim let us prove the following lemma.

Lemma 3.1 *Let* (u, v, w) *be a positive solution of* (15)–(19) *with a speed* $c > 0$ *then one has*

$$\sup_{x\in\mathbb{T}_L}\int_{\mathbb{R}} w(z,x)dz < \infty \textit{ and } \sup_{x\in\mathbb{T}_L}\int_{\mathbb{R}} u(z,x)dz < \infty.$$

Proof Let $M > 0$ be given. Integrating (16) with respect to $z \in (-M, M)$ yields for almost every $x \in \mathbb{T}_L$

$$c\,[v(M,x) - v(-M,x)] = \int_{-M}^{M} \alpha(x)u(z,x)v(z,x)dz.$$

Passing to the limit $M \to \infty$ yields using (19)

$$\sup_{x\in\mathbb{T}_L}\int_{\mathbb{R}} \alpha(x)u(z,x)v(z,x)dz < \infty.$$

Next we infer from (17) that

$$\sup_{x\in\mathbb{T}_L}\int_{\mathbb{R}} w(z,x)dz < \infty. \tag{20}$$

We now split the argument into two cases: $\theta \neq 0, \pi$ and $\theta \in \{0, \pi\}$. For the latter values of the parameter θ, the operator Δ_θ defined in (18) is not uniformly elliptic.

First case: $\theta \neq 0, \pi$.

Now observe that because of elliptic regularity (recall that $\theta \neq 0, \pi$) the limit behaviour in (19) implies that

$$\lim_{z\to\pm\infty} u(z,x) = 0 \text{ in } C^1(\mathbb{T}_L). \tag{21}$$

Now for each $K > 0$ we set

$$\overline{u}^K(x) = \int_{-K}^{K} u(z,x)dz \text{ and } w^K(x) = \int_{-K}^{K} w(z,x)dz.$$

Then integrating the u−equation with respect to $z \in (-K, K)$ yields for any $x \in \mathbb{T}_L$:

$$\begin{aligned}&\left[\frac{d^2}{dx^2} + 2\cos\theta\frac{d}{dx} - 1\right]\overline{u}^K(x)\\&+\left[(\partial_z u(K,x) - \partial_z u(-K,x)) + c(u(K,x) - u(-K,x))\right]\\&+rw^K(x) = 0.\end{aligned}$$

Because of (20) the family $\{w^K\}_{K>0}$ is bounded so that, using (21), we infer from the above equation that the family $\{\overline{u}^K\}_{K>0}$ is bounded in $H^2(\mathbb{T}_L)$ and Lemma 3.1 follows.

Second Case: $\theta \in \{0, \pi\}$.

To study that case we come back to the parabolic equation by introducing the functions

$$(S, I)(t, x) = (u, w)(\cos\theta x - ct, x).$$

Observe that (20) rewrites as

$$\sup_{x\in\mathbb{R}} \int_{\mathbb{R}} w(t, x)dt < \infty.$$

Next note that the function S becomes a solution of

$$\partial_t S = \partial_x^2 S - S + rI, \ (t, x) \in \mathbb{R} \times \mathbb{R}.$$

Moreover due to parabolic regularity, the limit behaviour in (19) rewrites as

$$\lim_{t\to\pm\infty} S(t, x) = 0 \text{ locally uniformly for } x \in \mathbb{R}.$$

Using similar arguments as in the proof of the non-degenerate case $\theta \neq 0, \pi$, namely integrating in time over some interval $(-K, K)$ for $K > 0$, we easily conclude that for each compact set $[-K, K]$ one has

$$\sup_{x\in[-K,K]} \int_{\mathbb{R}} S(t, x)dt = \frac{1}{c} \sup_{x\in[-K,K]} \int_{\mathbb{R}} u(z, x)dz < \infty,$$

that completes the proof of Lemma 3.1.

Now let $\theta \in [0, 2\pi)$ be given and fixed. Let (u, v, w) be a solution of (15)–(19) for a given wave speed $c > 0$. Then from (16) and (19) one gets

$$v(z, x) = v^+(x) \exp\left(-\frac{\alpha(x)}{c} \int_z^{\infty} u(s, x)ds\right), \ \forall (z, x) \in \mathbb{R} \times \mathbb{T}_L. \tag{22}$$

Remark 3.1 Passing to the limit $z \to -\infty$ in the above equation and using Lemma 3.1, one obtains that there exists $\eta > 0$ such that

$$\eta v^+(x) \leq v^-(x) \leq v(z, x) \leq v^+(x), \ \forall (z, x) \in \mathbb{R} \times \mathbb{T}_L.$$

Moreover we can re-write (17) by using the following integral formulation:

$$w(z,x) = \frac{\alpha(x)}{c}\int_0^\infty u(z+s,x)v(z+s,x)e^{-\frac{\beta}{c}s}ds,\ \forall(z,x)\in\mathbb{R}\times\mathbb{T}_L. \qquad (23)$$

Finally we infer from (15), (22) and (23) that Problem (15)–(19) reduces to the following single scalar equation for u on the cylinder $\mathbb{R}\times\mathbb{T}_L$:

$$(\Delta_\theta + c\partial_z - 1)\,u(z,x) + \frac{r\kappa(x)}{c}\int_0^\infty u(z+s,x)e^{-\frac{\alpha(x)}{c}\int_{z+s}^\infty u(\sigma,x)d\sigma}e^{-\frac{\beta}{c}s}ds = 0, \qquad (24)$$

together with

$$\lim_{z\to\pm\infty} u(z,x) = 0, \text{ uniformly for } x\in\mathbb{T}_L. \qquad (25)$$

It is also easy to check that a solution u of (24)–(25) provides a solution for (15)–(19) via the expressions (22) and (23). Therefore the two problems are equivalent.

Now let $U(z,x) := \int_z^\infty u(z,x)dz$. Then U is a decreasing (with respect to $z\in\mathbb{R}$) solution of the following monotone non-local equation on the cylinder $\mathbb{R}\times\mathbb{T}_L$:

$$(\Delta_\theta + c\partial_z - 1)\,U(z,x) + rv^+(x)\int_0^\infty H\left(\frac{\alpha(x)}{c}U(z+s,x)\right)e^{-\frac{\beta}{c}s}ds = 0, \quad (26)$$

where we have set $H(u) = 1 - e^{-u}$. This monotone reformulation of Problem (15)–(19) will not be used in this work to prove our existence and non-existence results. Instead, we shall directly work with the integro-differential formulation given (24)–(25) to derive our existence result. Our non-existence results will be discussed by using the original formulation given in (15)–(19).

To proceed, let us come back to Problem (24)–(25). By analogy with the Fisher-KPP equation, we expect that solutions of this equation would have an exponential decay as $z\to\infty$, namely at the leading edge of the front. Thus we make the following ansatz for some $\lambda > 0$ and some positive $L-$periodic function φ:

$$u(z,x) \approx e^{-\lambda z}\varphi(x), \quad \text{for } z >> 1 \text{ and } x\in\mathbb{T}_L. \qquad (27)$$

Next observe that $\exp\left(-\frac{\alpha(x)}{c}\int_{z+s}^\infty u(\sigma,x)d\sigma\right) \approx 1$ for $z >> 1$. Therefore, plugging the above ansatz into (24) yields the following equation for φ:

$$\varphi''(x) - 2\lambda\cos\theta\varphi'(x) + \frac{r\kappa(x)}{\beta+\lambda c}\varphi(x) = \left(c\lambda - \lambda^2 + 1\right)\varphi(x), \quad x\in\mathbb{T}_L.$$

This implies that φ is an eigenfunction of the elliptic operator

$$\mathscr{L}_{\lambda,\nu}^{\theta} := \frac{d^2}{dx^2} - 2\lambda\cos\theta\frac{d}{dx} + \frac{r\kappa(x)}{\beta+\nu}, \quad x\in\mathbb{T}_L, \qquad (28)$$

with $\nu = \lambda c$. Since $\varphi > 0$, by the Krein-Rutman theorem φ is the principal eigenfunction of this operator. Denote by $\mu^\theta(\lambda, \nu)$ the principal eigenvalue of $\mathscr{L}^\theta_{\lambda,\nu}$. Then we see from the above equation that

$$\mu^\theta(\lambda, c\lambda) = c\lambda - \lambda^2 + 1. \tag{29}$$

Thus, under the ansatz (27), c can be the speed of a travelling wave in the direction θ only if there exists $\lambda > 0$ satisfying the identity (29). Assuming further that the converse is true (that is, any $c > 0$ that satisfies (29) for some $\lambda > 0$ is the speed of some travelling wave in the direction θ), the problem of finding the minimal wave speed for (1) in the direction θ amounts to finding the smallest $c > 0$ that satisfies (29) for some $\lambda > 0$.

As one can see, the observation here bears much similarity to that for scalar KPP type equations in a striped periodic medium:

$$u_t = \Delta u + b(x) f(u), \quad (x, y) \in \mathbb{R}, t \in \mathbb{R},$$

where $b(x + L) = b(x)$ and f is a KPP type nonlinearity. This problem was studied in [23] and the following relation that is analogous to (29) was established for the speed of travelling waves in the direction θ:

$$\mu^\theta(\lambda) = c\lambda - \lambda^2, \tag{30}$$

where $\mu^\theta(\lambda)$ is the principal eigenvalue of the operator

$$\frac{d^2}{dx^2} - 2\lambda \cos\theta \frac{d}{dx} + b(x), \quad x \in \mathbb{T}_L.$$

In [23], the formula (30) was proved even for the case where b is a measure on $\mathbb{R}$. If b is a bounded function, the same formula also follows directly from [5], in which the eigenfunction characterization of the speed is established in a highly general setting. Note also that the validity of the asymptotics (27) is confirmed in [14] for a large class of scalar KPP type equations in periodic medium.

On the other hand, there are a number of notable differences between the present problem and scalar KPP type equations. For one thing, the characteristic equation (29) is more complicated than (30) as the former μ^θ involves λ in two parts. More important difference is that the rigorous proof of the formula (30) for scalar KPP type equations relies largely on the comparison principle, which we do not have for System (1). Therefore more involved arguments will be needed to determine the minimal wave speed in each direction θ for System (1).

This characteristic equation (29) will be studied in the next subsection. This will allow us to (formally) define the set of admissible wave speeds as the set of speed $c > 0$ such that there exists $\lambda > 0$ satisfying (29). This formal heuristic argument based on the ansatz (27) will turn out to be useful in the rigorous proof of our existence and non-existence results in Sect. 5.

3.2 Analysis of the Eigenvalue Problem

In this subsection we study the characteristic equation (29) and characterize the set of admissible wave speeds. To that aim we write down the principal eigenvalue problem for the operator $\mathscr{L}^\theta_{\lambda,\nu}$ in (28) for each $\theta \in [0, 2\pi)$, $\lambda \in \mathbb{R}$ and $\nu > -\beta$:

$$\begin{cases} \mathscr{L}^\theta_{\lambda,\nu}\varphi := \varphi'' - 2\lambda\cos\theta\varphi' + G(\nu, x)\varphi = \mu^\theta(\lambda,\nu)\varphi, \quad \forall x \in \mathbb{T}_L, \\ \varphi \in \bigcap_{p\geq 1} W^{2,p}(\mathbb{T}_L) \text{ and } \varphi > 0, \end{cases} \tag{31}$$

where the function $G : (-\beta, \infty) \times \mathbb{T}_L \to \mathbb{R}$ is defined by

$$G(\nu, x) = \frac{r\kappa(x)}{\beta + \nu}. \tag{32}$$

Note that the operator $\mathscr{L}^\theta_{\lambda,\nu}$ is not self-adjoint if $\lambda \neq 0$, and because of the periodic boundary conditions, some eigenvalues (not the principal one) can indeed be of complex values. Therefore, the standard variational formulation of eigenvalues do not apply to (31). However, thanks to the formula due to Nadin in [27, 28], we have the following variational representation of the principal eigenvalue of (31):

Lemma 3.2 *For each $\theta \in [0, 2\pi)$, $\lambda \in \mathbb{R}$ and $\nu > -\beta$, one has*

$$\mu^\theta(\lambda,\nu) = \max_{g\in\mathscr{A}} \left\{ \int_{\mathbb{T}_L} \left[-g'(x)^2 + \frac{r\kappa(x)g(x)^2}{\beta+\nu} \right] dx + \lambda^2\cos^2\theta(\mathscr{J}(g) - 1) \right\}, \tag{33}$$

where $\mathscr{A}$ is defined in (6) *while the functional $\mathscr{J}$ is defined in* (7).

From this variational formulation we obtain the following qualitative properties for the principle eigenvalue $\mu^\theta(\lambda, \nu)$.

Lemma 3.3 *The function $\mu : [0, 2\pi) \times \mathbb{R} \times (-\beta, \infty) \to \mathbb{R}$ defined by*

$$\mu(\theta, \lambda, \nu) = \mu^\theta(\lambda, \nu),$$

have the following properties:

(i) *$\mu(\pi - \theta, \lambda, \nu) = \mu(\theta, \lambda, \nu)$ and $\mu(2\pi - \theta, \lambda, \nu) = \mu(\theta, \lambda, \nu)$ for all $\theta \in [0, 2\pi)$ and all $(\lambda, \nu) \in \mathbb{R} \times (-\beta, \infty)$. For each $\lambda \in \mathbb{R}$ and $\nu > -\beta$ the function $\theta \to \mu(\theta, \lambda, \nu)$ is increasing on $[0, \frac{\pi}{2}]$.*
(ii) *For each $\theta \in [0, 2\pi)$ and each $\nu > -\beta$ the map $\lambda \to \mu(\theta, \lambda, \nu)$ is convex on $\mathbb{R}$.*
(iii) *For each $\theta \in [0, 2\pi)$ and each $\lambda \in \mathbb{R}$ the map $\nu \to \mu(\theta, \lambda, \nu)$ is decreasing and convex on $(-\beta, \infty)$.*

Proof The proof of (i) and (iii) directly follows from the variational formula given in Lemma 3.2 while the convexity property stated in (ii) follows from the following bound for the functional $\mathscr{J}$:

$$\mathscr{J}(g) \geq 1 \text{ for each } g \in \mathscr{A}.$$

Indeed from Hölder inequality, we have for each $g \in \mathscr{A}$:

$$L^2 = \left(\int_{\mathbb{T}_L} g(x) \frac{1}{g(x)} dx \right)^2 \leq \int_{\mathbb{T}_L} g(x)^2 dx \times \int_{\mathbb{T}_L} \frac{dx}{g(x)^2} \leq \int_{\mathbb{T}_L} \frac{dx}{g(x)^2}.$$

Recalling (9), the quantity $\Lambda^\theta(\lambda, \nu)$ enjoys the following properties that will allow us to define the set of admissible wave speed as well as the minimal wave speed.

Proposition 3.1 *The function* $(\theta, \lambda, \nu) \in [0, 2\pi) \times \mathbb{R} \times (-\beta, \infty) \mapsto \Lambda^\theta(\lambda, \nu) \in \mathbb{R}$ *satisfies the following properties:*

(i) *For each* $\theta \in [0, 2\pi)$ *the map* $(\lambda, \nu) \mapsto \Lambda^\theta(\lambda, \nu)$ *is convex and*

$$\lim_{\lambda \to \infty} \frac{1}{|\lambda|} \Lambda^\theta(\lambda, \nu) = \infty,$$

where the above convergence holds uniformly for $\nu > -\beta$ *and* $\theta \in [0, 2\pi)$. *For each* $(\lambda, \nu) \in \mathbb{R} \times (-\beta, \infty)$, *the function* $\theta \mapsto \Lambda^\theta(\lambda, \nu)$ *is increasing on* $[0, \frac{\pi}{2}]$.

(ii) *For each* $\lambda \in \mathbb{R}$ *and* $\theta \in [0, 2\pi)$ *the map* $\nu \mapsto \Lambda^\theta(\lambda, \nu)$ *is non-increasing on* $(-\beta, \infty)$.

(iii) *If* $\mathscr{R}_0 > 1$ *then for all* $\theta \in [0, 2\pi)$ *one has*

$$\Lambda^\theta(\lambda, \nu) > 0, \ \forall \lambda \in \mathbb{R}, \ \nu > -\beta.$$

Proof The convexity and increasing properties stated in (i) directly follow from Lemma 3.3. Next observe that one has

$$\Lambda^\theta(\lambda, \nu) \geq \lambda^2 - 1, \ \forall \lambda \in \mathbb{R}, \ \nu > -\beta, \tag{34}$$

that completes the proof of (i). The point (ii) directly follows from Lemma 3.3.

It remains to prove (iii). Recalling that $\mathscr{R}_0 - 1 = \Lambda^\theta(0, 0) > 0$, we shall argue by contradiction by assuming that there exist $\theta \in [0, 2\pi)$, $\lambda \in \mathbb{R}$ and $\nu > -\beta$ such that $\Lambda^\theta(\lambda, \nu) = 0$. This means that there exists $\varphi \in \bigcap_{p \geq 1} W^{2,p}(\mathbb{T}_L)$ with $\varphi > 0$ satisfying

$$\varphi''(x) - 2\lambda \cos\theta \varphi'(x) + \left[G(\nu, x) + \lambda^2 - 1 \right] \varphi(x) = 0, \ x \in \mathbb{T}_L.$$

Moreover according to (34) one has $\lambda^2 - 1 \leq 0$. Now note that the function $\varphi > 0$ satisfies the equation on $\mathbb{T}_L$

$$\varphi''(x) - 2\lambda \cos\theta \varphi'(x) + \left(\lambda^2 - 1\right) \varphi(x) = -f(x), \tag{35}$$

where the function f is defined by

$$f(x) := \frac{r\kappa(x)}{\nu + \beta} \varphi(x) \geq 0 \text{ and } f(x) \not\equiv 0.$$

As a consequence the function $-f$ belongs to the range of the periodic elliptic operator $M := \frac{d^2}{dx^2} - 2\lambda \cos\theta \frac{d}{dx} + \left(\lambda^2 - 1\right)$ acting on $L^2(\mathbb{T}_L)$. Hence Fredholm alternative ensures that $-f$ is orthogonal to the kernel of the adjoint operator M^* on $\mathbb{T}_L$. However this kernel is spanned by the eigenvector φ^* defined by

$$\varphi^*(x) = \begin{cases} 1 \text{if} \lambda \cos\theta = 0 \text{ and } \lambda^2 = 1, \\ \left(1 - e^{L\gamma^-}\right) e^{\gamma^+ x} + \left(e^{L\gamma^+} - 1\right) e^{\gamma^- x} \text{ else,} \end{cases}$$

where we have set

$$\gamma^{\pm} = -\lambda \cos\theta \pm \sqrt{\lambda^2 \cos^2\theta + 1 - \lambda^2}.$$

Once again since $\lambda^2 - 1 \leq 0$, one obtains that $\varphi^* > 0$ so that, since $f \geq 0$ and $f \not\equiv 0$, the solvability condition $\int_{\mathbb{T}_L} f(x)\varphi^*(x)dx = 0$ provides a contradiction. This completes the proof of (iii).

We now come back to the study of the characteristic equation given (29). To that aim we assume that $\mathscr{R}_0 > 1$. Let $\theta \in [0, 2\pi)$ be given. As discussed above, we aim to characterize of the set of admissible wave speed, denoted by $\mathscr{V}_\theta$, defined as follows

$$\mathscr{V}_\theta = \left\{c > 0 : \ \exists \lambda > 0 \ \mu^\theta(\lambda, c\lambda) = c\lambda - \lambda^2 + 1\right\}.$$

Let us observe that, using the above notation, this set can be re-written as

$$\mathscr{V}_\theta = \left\{c > 0 : \ \exists \lambda > 0 \ \Lambda^\theta(\lambda, c\lambda) = c\lambda\right\}. \tag{36}$$

On the one hand, as a consequence of Proposition 3.1(ii)–(iii), one may observe that for each $\lambda \in \mathbb{R}$, for each $\theta \in [0, 2\pi)$ there exists a unique $\nu^\theta(\lambda) > 0$ solution of the fixed point equation

$$\nu^\theta(\lambda) = \Lambda^\theta\left(\lambda, \nu^\theta(\lambda)\right). \tag{37}$$

In addition due to the convexity property stated in Proposition 3.1 (i), for each $\theta \in [0, 2\pi)$ the function $\lambda \mapsto \nu^\theta(\lambda)$ is also continuous and convex on $\mathbb{R}$ and satisfies

$$\lim_{\lambda \to \pm\infty} \frac{\nu^\theta(\lambda)}{|\lambda|} = \infty, \quad \text{uniformly for } \theta \in [0, 2\pi). \tag{38}$$

Now we define for each $\theta \in [0, 2\pi)$ the positive real number c_θ^* as in (10), namely

$$c_\theta^* = \inf_{\lambda>0} \frac{\nu^\theta(\lambda)}{\lambda}. \tag{39}$$

Note that because of the limit behaviour (38) and since $\nu^\theta(0) > 0$ the above infimum is actually a minimum.

On the other hand, due to the reformulation of $\mathscr{V}_\theta$ in (36), one obtains that

$$\mathscr{V}_\theta = \left[c_\theta^*, \infty\right),$$

where c_θ^* is defined in (39).

Moreover one may observe that, due the properties of the function Λ^θ provided in Proposition 3.1, the following definitions for c_θ^* are equivalent to the above one:

$$\begin{aligned} c_\theta^* &= \sup\left\{c > 0 : \ \Lambda^\theta(\lambda, c\lambda) - c\lambda \geq 0, \ \forall \lambda \geq 0\right\} \\ &= \inf\left\{c > 0 : \ \exists \lambda > 0 \ \ \Lambda^\theta(\lambda, c\lambda) - c\lambda < 0\right\}. \end{aligned}$$

4 Harnack Inequality

The aim of this section is to prove an important bound for the travelling wave profiles. This estimate will be crucially used in Sect. 5 to prove the minimality of the wave speed c_θ^* defined in the previous section. We shall more specifically prove the following.

Proposition 4.1 *Let* (u, v, w) *be a travelling wave profile of* (1) *with speed* $c > 0$ *in the direction* $\theta \in [0, 2\pi)$. *Then there exists some constant* $M > 0$ *such that*

$$w(z, x) \leq Mu(z, x), \ \forall (z, x) \in \mathbb{R} \times \mathbb{T}_L.$$

This section is devoted to the proof of this key estimate. For that purpose we shall develop suitable Harnack inequality like arguments for a reaction-diffusion equation coupled with an ordinary differential equation. This section is split into two parts. We shall first deal with the case $\theta \in (0, 2\pi)\backslash\{\pi\}$ and, then we shall focus on the case $\theta = 0, \ \pi$. This splitting follows from the properties of the elliptic operator Δ_θ defined in (18). Indeed, as already mentioned, when $\theta \in (0, 2\pi)\backslash\{\pi\}$ then Δ_θ is uniformly elliptic while when $\theta \in \{0, \pi\}$ it is degenerate.

4.1 *The Case* $\theta \in (0, 2\pi)\backslash\{\pi\}$

Let $\theta \in (0, 2\pi)\backslash\{\pi\}$ be given and fixed. In order to prove Proposition 4.1 we will need a slightly modified version of the usual Harnack inequality for elliptic equations. This modified version reads as follows.

Lemma 4.1 *Let $f \in L^\infty_+(\mathbb{R}^2)$ be given. Let $c > 0$ be a given constant. Let $u \equiv u(z, x)$ be a nonnegative continuous function that is L−periodic with respect to its second argument and that satisfies the equation*

$$(\Delta_\theta + c\partial_z - 1)\, u(z, x) + f(z, x) = 0, \quad z \in \mathbb{R}, \quad x \in \mathbb{R}.$$

Then for each $h > 0$ there exists some constant $C(h) > 0$ such that for any $z \in \mathbb{R}$

$$\max_{D_h(z)} u \leq C(h) \min_{D_h(z)} u + C(h) \int_0^L \sup_{z-2h \leq s \leq z+2h} f(s, x)dx.$$

Here we have set $D_h(z) = [z - h, z + h] \times [0, L]$.

Proof To prove this lemma, let us introduce for each $h > 0$ and $z \in \mathbb{R}$ the rectangle

$$D'_h(z) = (z - 2h, z + 2h) \times (-L, 2L).$$

Let $h > 0$ and $z \in \mathbb{R}$ be given. Consider the function k defined by the resolution of the elliptic problem:

$$\begin{cases} Lk + f = 0 \text{ in } D'_h(z), \\ k = 0 \text{ on } \partial D'_h(z), \end{cases}$$

where the operator L is defined by

$$L = \Delta_\theta + c\partial_z - 1.$$

Now note that, since $u \geq 0$, one has

$$\begin{cases} L(u - k) = 0 \text{ in } D'_h(z), \\ (u - k) \geq 0 \text{ on } \partial D'_h(z). \end{cases}$$

Thus, due to the maximum principle, ones obtains that $u - k \geq 0$. Moreover the classical Harnack inequality for homogeneous elliptic equations ensures there exists some constant $C(h) > 0$ such that

$$\max_{D_h(z)}(u - k) \leq C(h) \min_{D_h(z)}(u - k).$$

This readily yields

$$\max_{D_h(z)} u \leq C(h) \min_{D_h(z)} u + \max_{D_h(z)} k.$$

Let us now consider the one-dimensional function $f^* = f^*(x)$ defined by

$$f^*(x) = \sup_{z-2h \leq s \leq z+2h} f(s, x), \quad x \in \mathbb{R},$$

as well as the positive function $h^* = h^*(x)$ defined by the resolution of the one-dimensional elliptic boundary value problem

$$h^{*''}(x) + f^*(x) = 0 \text{ in } (-L, 2L), \quad h^*(-L) = h^*(2L) = 0.$$

Next let us observe that $f \leq f^*$ in $D'_h(z)$ so that

$$L(k - h^*) \geq 0, \text{ in } D'_h(z) \text{ and } \quad k - h^* \leq 0 \text{ on } \partial D'_h(z).$$

Hence the maximum principle applies and ensures that $k \leq h^*$ in $D'_h(z)$. Thus this yields

$$\max_{D_h(z)} k \leq \max_{D_h(z)} h^*.$$

One the other hand, let us notice that simple computations ensure that

$$h^*(x) \leq (x + L) \int_0^L f^*(x)dx, \quad \forall x \in (-L, 2L).$$

This implies that

$$\max_{D_h(z)} u \leq C(h) \min_{D_h(z)} u + 3L \int_0^L \sup_{z-2h \leq s \leq z+2h} f(s, x)dx,$$

and this completes the proof of the lemma.

Equipped with this lemma we are now able to prove Proposition 4.1 for directions $\theta \neq 0, \pi$.

Proof (Proof of Proposition 4.1 for $\theta \neq 0, \pi$) Fix $\theta \in (0, 2\pi)\setminus\{\pi\}$. Let (u, v, w) be a solution of (15)–(19) for some speed $c > 0$. Now observe that due to Remark 3.1, there exists $\eta > 0$ such that

$$\eta v^+(x) \leq v(z, x) \leq v^+(x), \ \forall z \in \mathbb{R} \times \mathbb{T}_L. \tag{40}$$

In that proof the notation C or $C(h)$ will denote any constant that may depend on h but that is independent of $(z, x) \in \mathbb{R} \times \mathbb{T}_L$.
Let $h > 0$ and $z \in \mathbb{R}$ be given. Then applying Lemma 4.1 to the u−equation (see (15)) ensures that there exists some constant $C = C(h)$ such that for all $z \in \mathbb{R}$

$$\max_{D_z(h)} u \leq C \min_{D_h(z)} u + C \int_0^L \sup_{z-2h \leq s \leq z+2h} w(z, x)dx. \tag{41}$$

Next let us recall that w is given in (23). Hence (40) yields

$$\int_{\mathbb{T}_L} w(z,x)dx \geq \eta \int_{\mathbb{T}_L} \frac{\kappa(x)}{c} \int_0^\infty u(z+s,x)e^{-\frac{\gamma}{c}s}dsdx. \tag{42}$$

Hence, since $\kappa \not\equiv 0$ (see Assumption 2.1), one deduces that there exists some constant $C(h)$ (independent of z) such that for all $z' \in [-3h, -2h]$

$$\int_0^L w(z+z',x')dx' \geq C(h) \min_{(z',x)\in D_h(z)} u(z',x).$$

On the other hand let us consider the function $K(t,z,x)$ defined as the heat kernel associated to the heat operator $\partial_t - \Delta_\theta$ on $\mathbb{R}^2$. Then observe that the function Γ defined by

$$\Gamma(z,x) = \int_0^\infty e^{-t} K(t, z-ct, x)dt,$$

becomes the Green function of the elliptic operator $\Delta_\theta + c\partial_z - 1$. As a consequence using (15), one obtains

$$u(z,x) = r\iint_{\mathbb{R}^2} \Gamma(z-z', x-x')w(z',x')dz'dx'. \tag{43}$$

Thus using the $\mathbb{T}_L$−periodicity with respect to x of the functions u and w, one obtains that for all $k \in (-h,h)$

$$\begin{aligned}\min_{x\in[0,L]} u(z+k,x) &\geq r\int_{\mathbb{R}} \left\{\left[\min_{x\in[L,3L]} \Gamma(z+k-z',x)\right]\int_0^L w(z',x')dx'\right\}dz' \\ &\geq C(h) \inf_{-3h\leq z'\leq -2h} \int_0^L w(z+z',x')dx'.\end{aligned}$$

Here the constant $C(h) > 0$ is given by

$$C(h) = r\int_{-4h}^{-h} \left\{\left[\min_{x\in[L,3L]} \Gamma(z',x)\right]dz'\right\}.$$

We now come back to (43) to get an upper estimate of the last term arising in the right hand side of (41). To that aim let us observe that due to (17) the function $z \mapsto e^{\frac{-\gamma z}{c}} w(z,x)$ is decreasing for each $x \in \mathbb{T}_L$. Now observe that for each $k \in (-h,h)$ and $(z,x) \in \mathbb{R}\times[0,L]$ one has

$$\begin{aligned}u(z+k,x) &\geq r\int_{z-3h}^{z-2h}\int_{-3L}^{-2L} \Gamma(z+k-z',x-x')e^{\frac{\gamma z'}{c}}e^{-\frac{\gamma z'}{c}}w(z',x')dz'dx', \\ &\geq r\int_{-3h}^{-2h} \min_{x\in[2L,4L]} [\Gamma(l,x)]\, e^{\frac{\gamma l}{c}}dl e^{-\frac{2\gamma h}{c}}\overline{w}(z-2h)dz', \\ &\geq C(h)\overline{w}(z-2h).\end{aligned}$$

Here we have set $\overline{w}(z) := \int_0^L w(z,x)dx$. As a consequence one has obtained that for each $h > 0$ there exists some constant $C(h) > 0$ such that

$$C(h)\overline{w}(z-2h) \leq \min_{D_h(z)} u, \ \forall z \in \mathbb{R}. \tag{44}$$

Finally let us observe that since $c\overline{w}'(z) \leq \gamma \overline{w}(z)$, one has

$$\sup_{k\in(-2h,2h)} \overline{w}(z+k) \leq e^{\frac{4h\gamma}{c}} \overline{w}(z-2h), \ \forall z \in \mathbb{R}, \ h > 0. \tag{45}$$

Finally coupling (41) together with (44) and (45) ensures that the following Harnack inequality holds true: For each $h > 0$ there exists $C(h) > 0$ such that for all $z \in \mathbb{R}$, there holds

$$\max_{D_h(z)} u \leq C(h) \min_{D_h(z)} u.$$

To complete the proof of Proposition 4.1, let us observe that from the above estimate, there exists some constant $C > 1$ such that for all $(z,x) \in \mathbb{R} \times \mathbb{T}_L$, one has

$$C^{-1} \int_0^L u(z,x)dx \leq u(z,x) \leq C \int_0^L u(z,x)dx.$$

Next using the expression of w in (23) together with (44) and (45), one obtains that there exists some constant $C > 0$ such that for all $(z,x) \in \mathbb{R} \times \mathbb{T}_L$, it holds that

$$w(z,x) \leq C\overline{w}(z) \leq Cu(z,x).$$

This completes the proof of Proposition 4.1 in the case where $\theta \neq 0, \pi$.

4.2 Proof of Proposition 4.1 in the Case $\theta = 0, \pi$

Here we assume that $\theta = 0, \pi$. As mentioned above, in that case, since Δ_θ is not uniformly elliptic our proof will rely on the parabolic formulation of travelling wave solutions instead of the elliptic wave profile formulation.

Let $(u, v, w) : \mathbb{R} \times \mathbb{T}_L \to \mathbb{R}^3$ be a travelling wave profile in the direction θ and speed $c > 0$. Consider the vector valued function $(S, H, I) \equiv (S, H, I)(t, x)$ defined by

$$(S, H, I)(t, x) = (u, v, w)(\cos\theta x - ct, x). \tag{46}$$

Hence the function (S, H, I) satisfies (1) on $\mathbb{R} \times \mathbb{R}$, that reads as for all $(t, x) \in \mathbb{R} \times \mathbb{R}$:

$$\begin{aligned}
&\partial_t S(t,x) = \partial_x^2 S(t,x) - S(t,x) + rI(t,x),\\
&\partial_t H(t,x) = -\alpha(x)H(t,x)S(t,x),\\
&\partial_t I(t,x) = \alpha(x)H(t,x)S(t,x) - \beta I(t,x).
\end{aligned} \tag{47}$$

Moreover the $\mathbb{T}_L$−periodicity of the profile (u, v, w) reformulates as

$$(S, H, I)(t, x + Lk) = (S, H, I)\left(t - \frac{Lk\cos\theta}{c}, x\right), \ \forall (t,x) \in \mathbb{R}^2, \ \forall k \in \mathbb{Z}. \tag{48}$$

In order to prove Proposition 4.1 in that context, namely for $\theta = 0, \pi$, and similarly as in the above elliptic proof, we need a slightly modified version of the usual one-dimensional parabolic Harnack inequality that reads as follows.

Lemma 4.2 *Let $f \in L_+^\infty(\mathbb{R} \times \mathbb{R})$ be given and let u be a nonnegative function satisfying*

$$(\partial_t - L)\, u(t,x) = f(t,x) \text{ with } L = \partial_x^2 - 1.$$

Then for each $\tau > 0$ and $h > 0$ there exists some constant $C > 1$ such that for all $t \in \mathbb{R}$ and $x \in \mathbb{R}$ one has

$$\max_{y \in [-h,h]} u(t-\tau, x+y) \le C\left[\min_{y\in[-h,h]} u(t, x+y) + \sup_{s \in (-2\tau,-\tau)} \int_{-2h}^{2h} f(t+s, x+y)dy\right].$$

Proof Let $\tau > 0$ and $h > 0$ be given. Up to translation in time and space, we assume for notational simplicity that $t = 0$ and $x = 0$.

Consider the nonnegative function k defined as the solution of the parabolic problem

$$\begin{cases} \partial_t k - Lk = f \text{ in } Q := (-2\tau, 0) \times (-2h, 2h),\\ k(-2\tau, .) = 0 \text{ and } k(t,x) = 0 \text{ on } (-2\tau, 0) \times \{-2h, 2h\}. \end{cases}$$

Then from the parabolic comparison principle one has $k \ge u$ and the function $u - k$ satisfies on Q the homogeneous parabolic equation

$$\partial_t(u-k) - L(u-k) = 0.$$

Now the parabolic Haranck inequality applies and ensures that there exists some constant $C > 1$ such that for all $t \in (-\tau, 0)$

$$\max_{y\in[-h,h]} (u-k)(t-\tau, y) \le C \min_{y \in [-h,h]} (u-k)(t,y).$$

Hence this yields for all $t \in (-\tau, 0)$

$$\max_{y\in[-h,h]} u(t-\tau, y) \le C \min_{y\in[-h,h]} u(t,y) + \max_{y\in[-h,h]} k(t-\tau, y).$$

Now constructing the nonnegative function $w^* \equiv w^*(x)$ defined by

$$-\left(w^*\right)''(x) = \sup_{-2\tau \le t \le -\tau} f(t,x) \text{ and } w^*(-2h) = w^*(2h) = 0,$$

one obtains from the comparison principle that $k(t,x) \le w^*(x)$ for all $t \in (-2\tau, -\tau)$ and $x \in (-2h, 2h)$. Finally the explicit expression for w^* completes the proof of the lemma.

Using the above modified Harnack inequality we are now able to complete the proof of Proposition 4.1 in the case $\theta = 0, \pi$.

Proof (Proof of Proposition 4.1 for $\theta = 0, \pi$) Let $h > 0$ and $\tau > 0$ be given and set $D_h(x) = [x-h, x+h]$.

Now from the one-dimensional parabolic Harnack inequality derived in Lemma 4.2 applied to each $S-$equation, there exists some constant $C = C(h) > 0$ such that for all $t \in \mathbb{R}$ one has

$$\max_{y \in D_h(x)} S(t-\tau, y) \le C\left(\min_{y \in D_h(x)} S(t,y) + \sup_{s \in [t-2\tau, t-\tau]} \int_{x-2h}^{x+2h} I(s,y)dsdy\right).$$

Since the function $t \mapsto e^{\beta s} I(s,x)$ is increasing, one obtains, for some constant C independent of t and x, that

$$\max_{y \in D_h(x)} S(t-\tau, y) \le C\left[\min_{y \in D_h(x)} S(t,y) + \int_{x-2h}^{x+2h} I(t-\tau, y)dsdy\right]. \tag{49}$$

On the other hand, from the parabolic $S-$equation we get that

$$S(t,x) = r\int_{-\infty}^{t}\int_{-\infty}^{\infty} K(t-s, x-y)I(s,y)dyds,$$

where the function $K : (0,\infty) \times \mathbb{R} \to \mathbb{R}$ is defined by

$$K(s,\xi) = \frac{1}{\sqrt{4\pi s}} e^{-\frac{\xi^2}{4s} - s}.$$

Using the above integral reformulation one obtains that for any $k \in [-h, h]$

$$S(t, x+k) \ge r\int_{t-\tau}^{t}\int_{x-2h}^{x+2h} K(t-s, x+k-y)I(s,y)dyds.$$

Since the function $t \mapsto e^{\beta s} I(s,x)$ is increasing, one obtains that

$$
\begin{aligned}
S(t,x+k) \geq & r \int_{t-\tau}^{t} \int_{x-2h}^{x+2h} e^{-\beta s} K(t-s, x+k-y) e^{\beta s} I(s,y) dy ds \\
\geq & r \int_{t-\tau}^{t} \int_{x-2h}^{x+2h} e^{-\beta s} K(t-s, x+k-y) e^{\beta(t-\tau)} I(t-\tau, y) dy ds.
\end{aligned}
$$

Hence there exists some constant $C = C(\tau, h) > 0$ such that for all $(t,x) \in \mathbb{R} \times \mathbb{R}$ one has

$$
\min_{D_h(x)} S(t,.) \geq C \int_{x-2h}^{x+2h} I(t-\tau, y) dy. \tag{50}
$$

Coupling this inequality together with (49) yields for some constant $C > 0$

$$
\max_{y \in D_h(x)} S(t-\tau, y) \leq C \min_{y \in D_h(x)} S(t,y), \; \forall (t,x) \in \mathbb{R} \times \mathbb{R}. \tag{51}
$$

However integrating the $I-$equation in (47) ensures that

$$
I(t,x) = \alpha(x) \int_{-\infty}^{t} H(\sigma, x) S(\sigma, x) e^{\beta(\sigma - t)} d\sigma. \tag{52}
$$

Moreover, recalling Remark 3.1, there exists $\eta > 0$ such that $\eta v^+(x) \leq H(t,x) \leq v^+(x)$ and by using the same arguments as in the proof of the elliptic case above, one obtains that

$$
\max_{y \in D_h(x)} S(t-\tau, y) \leq \min_{y \in D_h(x)} S(t,y), \; \forall (t,x) \in \mathbb{R} \times \mathbb{R}.
$$

Now we use a similar argument as in the proof of the elliptic case above to complete the proof of the proposition. Let $h > 0$ and $\tau > 0$ be given. Then by using (52), one obtains that for some constant $C > 0$ (independent of t and x and that may change from line to line)

$$
\begin{aligned}
\max_{y \in D_h(x)} I(t-2\tau, y) \leq & C \int_{-\infty}^{0} \max_{y \in D_h(x)} S(t-2\tau+l, y) e^{\beta l} dl \\
\leq & C \int_{-\infty}^{0} \min_{y \in D_h(x)} S(t-\tau+l, y) e^{\beta l} dl \\
\leq & C \int_{-\infty}^{0} \int_{x-2h}^{x+2h} S(t-\tau+l, y) e^{\beta l} dl \\
\leq & C \int_{x-2h}^{x+2h} I(t-\tau, y) dy.
\end{aligned}
$$

Using (50), one obtains that for all $h > 0$, $\tau > 0$, there exists some constant $C > 0$ such that for all $(t, x) \in \mathbb{R} \times \mathbb{R}$, one has

$$\sup_{y \in D_h(x)} I(t, y) \leq C \min_{D_h(x)} S(t + 2\tau, .).$$

Now we choose $\tau = \frac{L}{2c}$ and $h = 2L$. Recalling that (see (48))

$$S\left(t + \frac{L}{c}, x\right) \equiv S(t, x - \cos\theta L),$$

one gets, since $h > L$, that

$$\min_{y \in D_h(x)} S(t + 2\tau, y) \leq S(t, x), \ \forall (t, x) \in \mathbb{R} \times \mathbb{R}.$$

Therefore there exists some constant $C > 0$ such that for all $(t, x) \in \mathbb{R} \times \mathbb{R}$

$$I(t, x) \leq CS(t, x).$$

This completes the proof of Proposition 4.1 in this parabolic case $\theta = 0, \pi$.

5 Proof of Theorem 1

This section is concerned with the proof of Theorem 1. As in the statement of this result, this section is split into three parts, the proof of (i), the proof of (ii)–(1) and those of (ii)–(2).

5.1 *Proof of Theorem 1 (i)*

This section is devoted to prove that $\mathscr{R}_0 > 1$ is a necessary condition for the existence of travelling of System (1). To that aim we shall prove the following lemma.

Lemma 5.1 *Let (u, v, w) be a travelling wave profile of (1) in some direction $\theta \in [0, 2\pi)$ and for some wave speed $c > 0$. Define the function $\Omega : \mathbb{T}_L \to \mathbb{R}$ by*

$$\Omega(x) = \frac{1}{c} \int_{-\infty}^{\infty} u(z, x) dz, \quad \forall x \in \mathbb{T}_L.$$

Then it is a solution of the following nonlinear elliptic equation on $\mathbb{T}_L$

$$\Omega''(x) - \Omega(x) + \frac{rv^+(x)}{\beta}\left(1 - e^{-\alpha(x)\Omega(x)}\right) = 0,\ x \in \mathbb{T}_L. \tag{53}$$

Before proving this lemma one may first observe that it allows us to complete the proof of Theorem 1(i). Indeed it is easy to note that the second order equation in (53) has a positive solution if and only if the principal eigenvalue $\sigma(L)$ of the linear operator $L = \frac{d^2}{dx^2} + \frac{r}{\beta}\kappa(x) - 1$ on the torus $\mathbb{T}_L$ is positive. This linear operator corresponds to the linearization of (53) at the trivial solution $\Omega = 0$. However, recalling the definition of $\mathscr{R}_0$ in (8), one has $\sigma(L) = \mu^\theta(0,0) - 1 = \mathscr{R}_0 - 1$ so that this completes the proof of Theorem 1(i). Moreover, one may also notice that when $\mathscr{R}_0 > 1$ then (53) has a unique positive solution Ω and it follows from the above lemma that the following corollary holds true.

Corollary 5.1 *Assume that* $\mathscr{R}_0 > 1$. *Let* (u, v, w) *be a travelling wave profile of* (1) *in some direction* $\theta \in [0, 2\pi)$ *for some wave speed* $c > 0$. *Then, for all* $x \in \mathbb{T}_L$, *it holds that*

$$\int_{\mathbb{R}} u(z,x)dz = c\Omega(x) \text{ and } v^-(x) = v^+(x)\exp\left(-\alpha(x)\Omega(x)\right), \tag{54}$$

where Ω *is the unique positive solution of* (53).

Proof (Proof of Lemma 5.1) Let (u, v, w) be a solution of (15)–(19) for some given and fixed value of $c > 0$. Next due to Lemma 3.1, one knows that there exists some constant $M > 0$ such that

$$\int_{\mathbb{R}} u(z,x)dz \le M \quad \forall x \in \mathbb{T}_L.$$

This allows us to introduce the bounded and $\mathbb{T}_L$−periodic function

$$\Omega(x) := \frac{1}{c}\int_{-\infty}^{\infty} u(z,x)dz, \quad \forall x \in \mathbb{T}_L.$$

Note that since $u > 0$ then $\Omega > 0$. Next integrating (15) with respect to $z \in \mathbb{R}$ yields

$$\Omega''(x) - \Omega(x) + \frac{r}{c}\int_{-\infty}^{\infty} w(s,x)ds = 0, \quad \forall x \in \mathbb{T}_L. \tag{55}$$

Integrating (16) provides

$$\begin{aligned} &v(z,x) = v^+(x)\exp\left(-\frac{\alpha(x)}{c}\int_z^{\infty} u(s,x)ds\right), \ \forall (z,x) \in \mathbb{R} \times \mathbb{T}_L, \\ &c(v^+(x) - v^-(x)) = \alpha(x)\int_{-\infty}^{\infty} u(s,x)v(s,x)ds, \ \forall x \in \mathbb{T}_L. \end{aligned} \tag{56}$$

Letting $z \to -\infty$ in the first equality in (56) yields, for any $x \in \mathbb{T}_L$,

$$v^-(x) = v^+(x) \exp\left(-\alpha(x)\Omega(x)\right). \tag{57}$$

Now integrating (17) over $\mathbb{R}$ and combining it together with the second equality in (56) lead, for any $x \in \mathbb{T}_L$, to

$$\alpha(x) \int_{-\infty}^{\infty} u(s,x)v(s,x)ds = \beta \int_{-\infty}^{\infty} w(s,x)ds = c(v^+(x) - v^-(x)). \tag{58}$$

Finally we infer from (55), (57) and (58), that the function Ω satisfies (53). Since $\Omega > 0$, this completes the proof of Lemma 5.1.

5.2 *Proof of Theorem 1(ii)–(1)*

In this section we investigate the existence of travelling wave solution for (1) when $c \geq c_\theta^*$. This part uses rather classical arguments that we shall sketch for the sake of completeness. Note that instead of using the monotone reformulation (26) for the travelling waves, we shall construct a sub and super solution pair and we shall directly deal with the formulation given in (24). We split this section into two parts. We first investigate the case of super-critical wave speed $c > c_\theta^*$ and then we recover the critical case $c = c_\theta^*$ by using limiting arguments.

5.2.1 Existence of Waves for $c > c_\theta^*$

The aim of this section is to prove the following result:

Proposition 5.1 *Let Assumption 2.1 be satisfied and assume furthermore that* $\mathcal{R}_0 > 1$. *Let* $\theta \in [0, 2\pi)$ *be given and let* $\widehat{c} > c_\theta^*$ *be given. Then there exists a positive solution* (u, v, w) *of system* (15)–(19) *for the given wave speed* $\widehat{c} > 0$.

The proof of this result is split into several steps and will use the equivalent integral formulation, namely system (24)–(25). The first step consists in constructing sub and super solutions that allow us to construct a solution for any $\theta \in (0, 2\pi)\backslash\{0, \pi\}$ when the elliptic operator Δ_θ defined in (18) is non-degenerate. Finally we consider the case $\theta = 0$ and $\theta = \pi$ by coming back to the original evolution problem in order to get some useful local compactness properties for the solutions.

First step: $\theta \neq 0, \pi$

Let us first fix some notation. Since $\widehat{c} > c_\theta^*$ is given and fixed, then, according to Proposition 3.1 and the definition of c_θ^* given in (39), there exist $\lambda > 0$ and $\eta > 0$ such that

(H1) $\Lambda^{\theta}(\lambda, \widehat{c}\lambda) - \widehat{c}\lambda = 0$.
(H2) $\Lambda^{\theta}(\lambda+\varepsilon, \widehat{c}(\lambda+\varepsilon)) - (\lambda+\varepsilon)\widehat{c} < 0 \ \forall \varepsilon \in (0, \eta]$.

Next for each $\widehat{\lambda} > 0$ we denote by $\psi_{\widehat{\lambda}}$ a positive eigenvector associated to the principal eigenvalue problem (31) with $\nu = \widehat{c\lambda}$. Using this notation, straightforward computations allow us to check that the following lemmas hold true.

Lemma 5.2 *The function $(z, x) \mapsto \overline{u}(z, x)$ defined by $\overline{u}(z, x) = e^{-\lambda z}\psi_{\lambda}(x)$, satisfies the linear equation on $\mathbb{R} \times \mathbb{T}_L$:*

$$(\Delta_\theta + \widehat{c}\partial_z - 1)\, u(z, x) + \frac{r\kappa(x)}{\widehat{c}} \int_0^\infty u(z+s, x) e^{-\frac{\beta}{c}s} ds = 0.$$

Lemma 5.3 *For each $\widetilde{k} > 0$ there exists $\beta(\widetilde{k}) > 0$ such that for each $\widehat{\beta} > \beta(\widetilde{k})$ the function $\underline{u} : \mathbb{R} \times \mathbb{T}_L \to \mathbb{R}$ defined by*

$$\underline{u}(z, x) = e^{-\lambda z}\psi_{\lambda}(x) - \widehat{\beta} e^{-(\lambda+\eta)z}\psi_{\lambda+\eta}(x),$$

satisfies the following inequality, on the set $\{(z, x) \in \mathbb{R} \times \mathbb{T}_L : \ \underline{u}(z, x) \geq 0\}$,

$$(\Delta_\theta + \widehat{c}\partial_z - 1)\, u(z, x) + \frac{r\kappa(x)}{\widehat{c}} \int_0^\infty u(z+s, x) \left(1 - \widetilde{k} e^{-\lambda(z+s)}\right)^+ e^{-\frac{\beta}{c}s} ds \geq 0.$$

In the next step we consider a similar problem as (24)–(25) posed on an semi-infinite interval of the form $(-a, \infty)$ for some $a > 0$. Then we prove the existence of solution for this approximated problem by using a fixed point argument and finally we pass to the limit $a \to \infty$ in order to get a solution of (24)–(25).

To that aim we consider for any $a > 0$ the following problem: find a non zero, continuous and positive function u such that $u(., x) \in L^1(-a, \infty)$ for all $x \in \mathbb{T}_L$ and that satisfies, for all $(z, x) \in (-a, \infty) \times \mathbb{T}_L$, the following equation

$$\begin{aligned}&(\Delta_\theta + \widehat{c}\partial_z - 1)\, u(z, x) \\ &+ \frac{r\kappa(x)}{\widehat{c}} \int_0^\infty u(z+s, x) \exp\left(-\frac{\alpha(x)}{\widehat{c}} \int_{z+s}^\infty u(\sigma, x) d\sigma\right) e^{-\frac{\beta}{c}s} ds = 0, \qquad (59)\end{aligned}$$

with the boundary condition

$$u(-a, x) = 0 \ \forall x \in \mathbb{T}_L. \qquad (60)$$

In order to handle this equation, for any $a > 0$, we introduce the Banach space

$$X_a = \{\varphi \in C([-a, \infty) \times \mathbb{T}_L) : \ (z, x) \mapsto e^{\lambda z}\varphi(z, x) \text{ bounded}\},$$

endowed with the weighted norm

$$\|\varphi\|_{X_a} := \sup_{z\geq -a,\, x\in\mathbb{T}_L} e^{\lambda z}|\varphi(z,x)|.$$

To prove the existence of a nontrivial solution of (59)–(60), we shall construct a suitable closed and convex subset of X_a as well as a suitable map on X_a in order to apply a fixed point argument. Using the notation of Lemma 5.3, we now fix

$$\widetilde{k} = \sup_{x\in[0,L]} \frac{\alpha(x)\psi_\lambda(x)}{\lambda\widehat{c}} \text{ and } \beta_0 > \beta(\widetilde{k}). \tag{61}$$

Next we consider the function $\underline{u}$ provided by Lemma 5.3 with $\widehat{\beta} = \beta_0$ and recalling that since $\mathscr{R}_0 > 1$, Corollary 5.1 ensures that the function $\Omega > 0$, the unique solution (53), is well defined. Next we consider the closed and convex subset $\mathscr{C}_{a,\beta_0} \subset X_a$ defined by

$$\mathscr{C}_{a,\beta_0} = \left\{u \in X_a : \ \max(0,\underline{u}) \leq u \leq \overline{u},\ \int_{-a}^{\infty} u(s,x)ds \leq \widehat{c}\Omega(x)\ \forall x \in \mathbb{T}_L\right\},$$

as well as the map $\mathscr{F} : C_{a,\beta_0} \to C\left([-a,\infty)\times\mathbb{T}_L\right)$ defined by

$$\mathscr{F}(u)(z,x) = \int_0^{\infty} \widetilde{u}(z+s,x)\exp\left(-\frac{\alpha(x)}{\widehat{c}}\int_{z+s}^{\infty}\widetilde{u}(\sigma,x)d\sigma\right)e^{-\frac{\beta}{\widehat{c}}s}ds.$$

We also introduce the map $\Phi : \mathscr{C}_{a,\beta_0} \to X_a$ defined by

$$\Phi\left(\widetilde{u}\right) = u,$$

where $u \in X_a$ is defined by the resolution of the equation

$$\begin{cases} \left(\Delta_\theta + \widehat{c}\partial_z - 1\right)u(z,x) + \frac{r\kappa(x)}{\widehat{c}}\mathscr{F}\left(\widetilde{u}\right)(z,x) = 0 \text{ on } (-a,\infty)\times\mathbb{T}_L, \\ u(-a,x) = 0 \ \ \forall x \in \mathbb{T}_L. \end{cases} \tag{62}$$

Next it is easy to check that the map Φ has a fixed point in $\mathscr{C}_{a,\beta_0}$ by applying the Schauder fixed point theorem. For that purpose we prove that

$$\Phi\left(\mathscr{C}_{a,\beta_0}\right) \subset \mathscr{C}_{a,\beta_0} \text{ and } \overline{\Phi\left(\mathscr{C}_{a,\beta_0}\right)} \text{ compact in } X_a.$$

The proof of that claim is rather classical and the details are left to the reader.

Finally to complete the proof of Proposition 5.1 it is sufficient to pass to the limit $a \to \infty$ in the solution $u_a \in \mathscr{C}_{a,\beta_0}$ of (59)–(60) for any $a > 0$. This limiting procedure directly follows from elliptic regularity. One can note that the solution u constructed in that step satisfies

$$\max\left(\underline{u},0\right)\leq u\leq\overline{u}\text{ and }\int_{-\infty}^{\infty}u(s,x)ds\leq\widehat{c}\Omega(x),\ \forall x\in\mathbb{T}_L.$$

Note that the lower estimate above ensures that $u\not\equiv 0$, hence $u>0$ due the uniform ellipticity of Δ_θ. This completes the proof of Proposition 5.1 when $\theta\neq 0,\pi$.

Using this first step, one can use a limiting argument to prove Proposition 5.1 in the case $\theta=0,\pi$.

Second step: $\theta=0,\pi$

Here we consider the case $\theta=0$. Note the case $\theta=\pi$ can be similarly handled and the proof is omitted. In that situation we fix $\widehat{c}>c_0^*$. Therefore there exist $\lambda_0>0$ and $\eta_0>0$ such that

$$\begin{aligned}&\Lambda^0\left(\lambda_0,\widehat{c}\lambda_0\right)=\widehat{c}\lambda_0,\\&\Lambda^0\left(\lambda_0+\varepsilon,\widehat{c}(\lambda_0+\varepsilon)\right)-\widehat{c}\left(\lambda_0+\varepsilon\right)<0,\ \forall\varepsilon\in(0,\eta_0].\end{aligned}$$

Since the map $(\theta,\lambda)\to\Lambda^\theta(\lambda,\widehat{c}\lambda)$ is continuous there exists $\theta_0>0$ small enough such that $\widehat{c}>c_\theta^*$ for each $\theta\in[0,\theta_0)$. Moreover, recalling the properties of the function Λ^θ in Proposition 3.1, for each $\theta\in(0,\theta_0]$, there exist $\lambda_\theta>0$ and a value $\eta\in(0,\eta_0)$ such that $\lambda_\theta\to\lambda_0$ as $\theta\to 0$ and such that for all $\theta\in[0,\theta_0]$ one has

$$\begin{aligned}&\Lambda^\theta\left(\lambda_\theta,\widehat{c}\lambda_\theta\right)=\widehat{c}\lambda_\theta,\\&\Lambda^\theta\left(\lambda_\theta+\eta,\widehat{c}(\lambda_\theta+\eta)\right)-\widehat{c}\left(\lambda_\theta+\eta\right)<0.\end{aligned}$$

Thus from the first step of the proof, for each $\theta\in(0,\theta_0]$ there exists u_θ a non zero and positive solution of (24)–(25) for the wave speed $\widehat{c}$. Moreover, recall that from our construction in the first step, the function u_θ satisfies, for each $\theta\in(0,\theta_0]$, the following estimates on $\mathbb{R}\times\mathbb{T}_L$

$$e^{-\lambda_\theta z}\psi_\theta^{\lambda_\theta}(x)-\widehat{\beta}e^{-(\lambda_\theta+\eta)z}\psi_\theta^{\lambda_\theta+\eta}(x)\leq u_\theta(z,x)\leq e^{-\lambda_\theta z}\psi_\theta^{\lambda_\theta}(x),\tag{63}$$

where ψ_θ^λ denotes the principal eigenvector associated to (31) with the angle θ and $\nu=\widehat{c}\lambda>0$ while $\widehat{\beta}>0$ is a constant large enough independent of $\theta\in(0,\theta_0]$.

Next we introduce the functions

$$\begin{aligned}v_\theta(z,x)&=v^+(x)\exp\left(-\frac{\alpha(x)}{\widehat{c}}\int_z^\infty u_\theta(s,x)ds\right),\\w_\theta(z,x)&=\frac{\alpha(x)}{\widehat{c}}\int_0^\infty u_\theta(z+s,x)v_\theta(z+s,x)e^{-\frac{\beta}{c}s}ds,\end{aligned}\tag{64}$$

and we set for all $t\in\mathbb{R}$, $(x,y)\in\mathbb{R}^2$ and $\theta\in(0,\theta_0]$:

$$\left(S_\theta,H_\theta,I_\theta\right)(t,x,y)=\left(u_\theta,v_\theta,w_\theta\right)\left(x\cos\theta+y\sin\theta-\widehat{c}t,x\right).\tag{65}$$

Recall that the vector valued function $(S_\theta, H_\theta, I_\theta)$ is a travelling wave of Problem (1) with the speed $\widehat{c}$ and the angle θ.

Now let $\{\theta_n\} \subset (0, \theta_0)$ be a given sequence tending to 0 as $n \to \infty$. Due to (65) the sequence $\{I_{\theta_n}\}$ is uniformly bounded. Therefore by using parabolic regularity one obtains that the sequence $\{S_{\theta_n}\}$ is locally bounded in $W^{1,p}_{\text{loc}}(\mathbb{R} \times \mathbb{R}^2)$ for any $p \in (1, \infty)$. As a consequence, the sequence $\{u_{\theta_n}\}$ is also bounded in $W^{1,p}_{\text{loc}}(\mathbb{R} \times \mathbb{T}_L)$. Hence, up to a subsequence, one may assume that u_{θ_n} and S_{θ_n} converge locally uniformly towards some function u and S respectively. Moreover u satisfies (63) with $\theta = 0$ while $S(t, x, y) = u(x - \widehat{c}t, x)$. The lower bound in (63) more particularly ensures that $u \not\equiv 0$ and thus $S \not\equiv 0$ while the upper bound implies that $u(z, x) \to 0$, uniformly for $x \in \mathbb{T}_L$, as $z \to \infty$. Note also that from our construction in the first step, the uniform L^1 bound for the sequence $\{u_{\theta_n}\}$ holds true

$$\int_{-\infty}^{\infty} u_{\theta_n}(\sigma, x)d\sigma \le \widehat{c}\Omega(x), \quad \forall x \in \mathbb{T}_L, \ n \ge 0,$$

that ensures, due to Fatou Lemma, that $u(., x) \in L^1(\mathbb{R})$ for all $x \in \mathbb{T}_L$.

In addition, due to Lebesgue convergence theorem, the sequence of function $\left(v_{\theta_n}, w_{\theta_n}\right)$ defined in (64) with $\theta = \theta_n$ converges in $L^\infty_{\text{loc}}(\mathbb{R} \times \mathbb{T}_L)$ to (v, w) defined by

$$v(z, x) = v^+(x) \exp\left(-\frac{\alpha(x)}{\widehat{c}} \int_z^\infty u(s, x)ds\right),$$

$$w(z, x) = \frac{\alpha(x)}{\widehat{c}} \int_0^\infty u(z + s, x)v(z + s, x)e^{-\frac{\beta}{\widehat{c}}s}ds.$$

Hence, the sequence of functions $\left(H_{\theta_n}, I_{\theta_n}\right)(t, x, y)$ converges to (H, I) for the topology of $L^\infty_{\text{loc}}\left(\mathbb{R} \times \mathbb{R}^2\right)$ with $(H, I)(t, x) = (v, w)(x - \widehat{c}t, x)$. And, the vector valued function (S, H, I) is an entire solution of System (1).

Finally it remains to notice, using the parabolic equation satisfied by S, that the condition $u \not\equiv 0$ or $S \not\equiv 0$, ensures that $S > 0$, that is $u > 0$. This proves that the vector valued function (S, H, I) is positive in the sense of Definition 2.1 and, that (S, H, I) is a travelling wave of (1) with the wave speed $\widehat{c}$ and angle $\theta = 0$. This completes the proof of Proposition 5.1 in the case $\theta = 0, \pi$.

5.2.2 Existence of Waves for $c = c^*_\theta$

The aim of this section is to prove the existence of wave solution for the limit wave speed c^*_θ. More precisely one will show the following result:

Proposition 5.2 *Let Assumption 2.1 be satisfied and assume furthermore that $\mathscr{R}_0 > 1$. Then for each $\theta \in [0, 2\pi)$ there exists a wave solution for the wave speed c^*_θ defined in* (39).

The idea of the proof is to use a limiting procedure $c_n \to c^*_\theta$. Our proof is split into two parts, the case $\theta \in (0, 2\pi)\backslash\{\pi\}$ and the case $\theta = 0$ or $\theta = \pi$.

Proof Let $\theta \in [0, 2\pi)$ be given let $\{c_n\}_{n\geq 0}$ a sequence of real number such that

$$c_n > c^*_\theta, \quad \forall n \geq 0 \text{ and } \lim_{n\to\infty} c_n = c^*_\theta.$$

Recalling the definition of v^- in Corollary 5.1, fix $\alpha^* > 0$ such that

$$\int_{\mathbb{T}_L} v^-(x)dx < \alpha^* < \int_{\mathbb{T}_L} v^+(x)dx. \tag{66}$$

Due to Proposition 5.1 combined together with the translation invariance (with respect to z) of the travelling wave problem, for each $n \geq 0$ one may consider (u_n, v_n, w_n) a positive solution of (15)–(19) with $c = c_n$ and satisfying the normalization condition

$$\int_{\mathbb{T}_L} v_n(0, x)dx = \alpha.$$

Then we get from the first equation in (56) that

$$\alpha^* = \int_{\mathbb{T}_L} v^+(x)\exp\left(-\frac{\alpha(x)}{c_n}\int_0^\infty u_n(s, x)ds\right)dx. \tag{67}$$

We now split our arguments into two parts: $\theta \neq 0, \pi$ and $\theta = 0, \pi$.

Let us start with some fixed angle $\theta \neq 0, \pi$.

Since the operator Δ_θ is uniformly elliptic, from standard elliptic regularity, possibly along a subsequence one may assume that u_n and (v_n, w_n) convergences to some functions u and (v, w) respectively for the strong topology of $C_{\text{loc}}(\mathbb{R} \times \mathbb{T}_L)$ and for the weak star topology of $L^\infty(\mathbb{R} \times \mathbb{T}_L) \times L^\infty(\mathbb{R} \times \mathbb{T}_L)$. Now due to the upper bound for the function u_n, we have

$$u_n(z, x) \leq e^{-\lambda_n z}\psi^{\lambda_n}(x) \;\; \forall (z, x) \in \mathbb{R} \times \mathbb{T}_L,$$

where $\lambda_n > 0$ is the smallest solution of the characteristic equation $\Lambda^\theta(\lambda_n, c_n\lambda_n) = c_n\lambda_n$. Now note that $\lambda_n \to \lambda_0$ as $n \to \infty$ where $\lambda_0 > 0$ is the unique solution of the equation $\Lambda^\theta(\lambda_0, c^*_\theta\lambda_0) = c^*_\theta\lambda_0$. Hence we get from Lebesgues convergence theorem that

$$\int_0^\infty u_n(s, x)ds \to \int_0^\infty u(s, x)ds \text{ uniformly for } x \in \mathbb{T}_L \text{ as } n \to \infty.$$

Therefore we infer from (67) that

$$\alpha^* = \int_{\mathbb{T}_L} v^+(x)\exp\left(-\frac{\alpha(x)}{c}\int_0^\infty u(s, x)ds\right)dx,$$

and, because of the definition of α^* in (66), one concludes that $u \neq 0$. Hence Problem (15)–(17) has a nontrivial solution for the wave speed $c = c_\theta^*$ and the result follows for any angle $\theta \neq 0, \pi$.

For the cases $\theta = 0$ or π, we use the same arguments as above. The lack of local compactness for this elliptic degenerate operators Δ_0 and Δ_π is handled similarly as in the proof of the existence of wave for $\theta = 0$ by using the corresponding parabolic equation. The details are left to the reader and this completes the proof of Proposition 5.2.

5.3 *Proof of Theorem 1(ii)–(2)*

The aim of this section is to prove Theorem 1(ii)–(2), that is that System (1) does not admit any travelling wave solution in any direction θ and for any wave speed $0 < c < c_\theta^*$. Our precise result is given in the following Theorem:

Theorem 4 *Let Assumption 2.1 be satisfied and assume that $\mathscr{R}_0 > 1$. Let us assume that* (15)–(19) *has a positive solution (u, v, w) in some direction $\theta \in [0, 2\pi)$ for some wave speed $c > 0$. Then there exists $\lambda > 0$ such that*

$$c\lambda = \Lambda^\theta(\lambda, c\lambda).$$

Here recalling the definition of c_θ^* in (39), this implies that $c \geq c_\theta^*$ and this completes the proof of Theorem 1(ii)–(2).

In order to prove Theorem 4, once again we split the arguments into two parts: the case $\theta \in (0, 2\pi)\backslash\{\pi\}$ and the case where $\theta = 0$ or π. The key argument in that proof relies on the estimate derived in Proposition 4.1.

Proof (Proof of Theorem 4 for $\theta \neq 0, \pi$) Let $\theta \in (0, 2\pi)\backslash\{\pi\}$ be given and fixed. Let (u, v, w) be a positive solution of (15)–(19).

Now, recalling that the function u satisfies (15), let us observe that due to Proposition 4.1, the function $\frac{\nabla u}{u}$ is bounded on $\mathbb{R} \times \mathbb{T}_L$. Hence consider the real number $\Lambda \in \mathbb{R}$ defined by

$$\Lambda := \liminf_{\substack{z \to \infty \\ x \in \mathbb{T}_L}} \frac{\partial_z u(z, x)}{u(z, x)}. \tag{68}$$

Let us consider a sequence $\{(z_n, x_n)\}_{n \geq 0} \subset \mathbb{R} \times \mathbb{T}_L$ such that

$$\lim_{n \to \infty} z_n = \infty \text{ and } \Lambda = \lim_{n \to \infty} \frac{\partial_z u(z_n, x_n)}{u(z_n, x_n)}.$$

Next consider the sequences of functions $\tilde{u}^n$ and $\tilde{w}^n$ defined as

$$\tilde{u}^n(z, x) = \frac{u(z + z_n, x + x_n)}{u(z_n, x_n)} \text{ and } \tilde{w}^n(z, x) = \frac{w(z + z_n, x + x_n)}{u(z_n, x_n)}.$$

Now note that these sequences of functions are locally bounded (see Proposition 4.1). Hence, due to elliptic regularity, one may assume, possibly along a subsequence, that

$$\tilde{u}^n(z,x) \to \tilde{u}(z,x) \text{ for the topology of } C^1_{loc}(\mathbb{R}\times\mathbb{T}_L),$$
$$\tilde{w}^n(z,x) \to \tilde{w}(z,x) \text{ and } \partial_z\tilde{w}^n(z,x) \to \partial_z\tilde{w}(z,x) \text{ for the } L^\infty_{loc} - \text{ weak } * \text{topology},$$

while $x_n \to x_\infty$ in $\mathbb{T}_L$.
Furthermore these limit functions satisfy the following system of equations

$$\begin{cases} \left[\Delta_\theta + c\partial_z - 1\right]\tilde{u}(z,x) + r\tilde{w}(z,x) = 0, & (z,x)\in\mathbb{R}\times\mathbb{T}_L, \\ \left[c\partial_z - \beta\right]\tilde{w}(z,x) + \kappa(x+x_\infty)\tilde{u}(z,x) = 0, & (z,x)\in\mathbb{R}\times\mathbb{T}_L. \end{cases} \tag{69}$$

Note that the definition of $\tilde{u}^n$ ensures that $\tilde{u}(0,0)=1$. As a consequence one gets that $\tilde{u}>0$. Furthermore the definition of Λ in (68) implies that

$$\partial_z\tilde{u} \geq \Lambda\tilde{u} \text{ and } \partial_z\tilde{u}(0,0) = \Lambda\tilde{u}(0,0). \tag{70}$$

Moreover formula (23) for $\tilde{w}^n$ rewrites as

$$\tilde{w}^n(z,x) = \frac{\alpha(x+x_n)}{c}\int_0^\infty \tilde{u}^n(z+s,x)v(z_n+z+s,x+x_n)e^{-\frac{\beta}{c}s}ds.$$

Hence, recalling that $\kappa=\alpha v^+$, Fatou lemma ensures that, for all $(z,x)\in\mathbb{R}\times\mathbb{T}_L$, one has

$$\kappa(x+x_\infty)\int_0^\infty \tilde{u}(z+s,x)e^{-\frac{\beta}{c}s}ds \leq c\tilde{w}(z,x).$$

On the other hand, since $v(z,x)\leq v^+(x)$, one also obtains the reserve inequality, so that

$$\lim_{n\to\infty}\tilde{w}^n(z,x) = \frac{1}{c}\kappa(x+x_\infty)\int_0^\infty \tilde{u}(z+s,x)e^{-\frac{\beta}{c}s}ds = \tilde{w}(z,x), \tag{71}$$

where the above limit is understood for the weak star topology of $L^\infty(\mathbb{R}\times\mathbb{T}_L)$.

Now observe that the above equation for $\tilde{w}^n$ coupled with the property $\partial_z v\geq 0$ (see (16)) ensures that for all $n\geq 0$

$$\partial_z\tilde{w}^n(z,x) \geq \frac{\alpha(x+x_n)}{c}\int_0^\infty \frac{\partial_z\tilde{u}^n(z+s,x)}{\tilde{u}^n(z+s,x)}\tilde{u}^n(z+s,x)v(z_n+z+s,x+x_n)e^{-\frac{\beta}{c}s}ds.$$

Thus due to the definition of Λ and using (71) one gets

$$\partial_z\tilde{w}(z,x) \geq \Lambda\tilde{w}(z,x). \tag{72}$$

In order to complete the proof of the result let us consider the function $\widehat{u}$ defined by

$$\widehat{u}(z,x)=\Lambda-\frac{\partial_z\tilde{u}(z,x)}{\tilde{u}(z,x)}.$$

Observe that due to (70) it holds

$$\widehat{u}\leq 0 \text{ and } \widehat{u}(0,0)=0.$$

Moreover this function satisfies the following elliptic problem

$$\left(\Delta_\theta+2\begin{pmatrix}1 & \cos\theta\\ \cos\theta & 1\end{pmatrix}\frac{\nabla\tilde{u}}{\tilde{u}}\cdot\nabla+c\partial_z-r\frac{\tilde{w}}{\tilde{u}}\right)\widehat{u}=\frac{r}{\tilde{u}}\left(\partial_z\tilde{w}-\Lambda\tilde{w}\right). \tag{73}$$

Now, using (72), one concludes that

$$\widehat{u}\equiv 0 \text{ and } \partial_z\tilde{w}\equiv\Lambda\tilde{w},$$

that is, recalling that $\tilde{u}>0$,

$$\tilde{u}(z,x)\equiv e^{\Lambda z}\varphi(x) \text{ and } \tilde{w}(z,x)=e^{\Lambda z}w_0(x),$$

for some functions $\varphi>0$ and $0\leq w_0$ with $w_0\not\equiv 0$.

Next, on the one hand, plugging the above expression for $\tilde{u}$ into the integrability condition (71) ensures that

$$c\Lambda<\beta. \tag{74}$$

On the other hand, plugging the above expression for $\tilde{w}$ into the $\tilde{w}-$equation (69) yields

$$(c\Lambda-\beta)\,w_0(x)+\kappa(x+x_\infty)\varphi(x)=0.$$

Finally plugging the above expressions into (69), recalling (74) and that $\varphi>0$, implies that $\lambda:=-\Lambda$ satisfies

$$c\lambda>-\beta \text{ and } c\lambda=\Lambda^\theta(\lambda,c\lambda).$$

This completes the proof of the theorem in the case where $\theta\neq 0,\pi$.

Proof (Proof of Theorem 4 with $\theta=0$ or π) In order to avoid the lack of uniform ellipticity, we shall work with the parabolic problem. For that purpose consider (u,v,w) to be a positive solution of (15)–(19) for some wave speed $c>0$ and let us introduce the function

$$(S,H,I)(t,x)=(u,v,w)(\cos\theta x-ct,x).$$

Note that it is a one-dimensional entire solution of (1). Moreover with these notation, Proposition 4.1 ensures that the function $\frac{I}{S}$ is globally bounded on $\mathbb{R} \times \mathbb{R}$. So that parabolic regularity applies and ensures that the ratios $\frac{\partial_t S}{S}$, $\frac{\partial_x S}{S}$ are all bounded on $\mathbb{R} \times \mathbb{R}$. Similarly as in the proof for the case $\theta \neq 0, \pi$, we define $\Lambda \in \mathbb{R}$ by

$$\Lambda := \limsup_{t \to -\infty} \inf_{x \in [0,L]} \frac{\partial_t S(t,x)}{S(t,x)}.$$

Then using similar arguments as the ones developed above for the elliptic case, one obtains that $\lambda = \frac{\Lambda}{c}$ satisfies the following properties:

$$c\lambda > -\beta \text{ and } c\lambda = \Lambda^\theta(\lambda, c\lambda).$$

This completes the proof of Theorem 4 in the case where $\theta = 0, \pi$.

6 Qualitative Properties of the Minimal Wave Speed

The aim of this section is to derive further qualitative properties of the minimal wave speed and to prove Theorem 2 as well as Theorem 3.

6.1 *Proof of Theorem 2*

In this section we prove Theorem 2. To that aim let us recall that according to Proposition 3.1, for each $(\lambda, \nu) \in \mathbb{R} \times (-\beta, \infty)$, the function $\theta \mapsto \Lambda^\theta(\lambda, \nu)$ is increasing on $\left[0, \frac{\pi}{2}\right]$. Hence it easily follows from (39) that for all $\lambda \in \mathbb{R}$, the function $\theta \mapsto \nu^\theta(\lambda)$ is increasing on that interval so that the function $\theta \mapsto c^*_\theta$ is also increasing on $\left[0, \frac{\pi}{2}\right]$.

Furthermore the symmetries of the minimal wave speed stated in Theorem 2(i) directly follow from the symmetries of the function $\theta \mapsto \mu^\theta(\lambda, \nu)$ stated in Lemma 3.3.

We shall now focus on proving the second part of Theorem 2. Recall that the influence of the periodic Schwarz rearrangement of the heterogeneity of the medium increases the spreading rate of invasion for the scalar Fisher-KPP equation (we refer to Berestycki et al. in [6] and Nadin in [27, 28]). We shall prove that the same property holds true for the epidemic system under consideration, namely System (1). For that purpose let us recall the definition of the periodic Schwarz rearrangement. More details are given by Kawohl in [19, 20].

Definition 6.1 (*Schwarz rearrangement*) Let $f \in L^\infty_+(\mathbb{T}_L)$ be a given function. There exists a unique measurable, non-negative, bounded and $L-$periodic function $\widehat{f}$ such that:

(i) $\widehat{f}$ is symmetric with respect to $\frac{L}{2}$,
(ii) $\widehat{f}$ is nondecreasing on $\left(0, \frac{L}{2}\right)$,
(iii) $\widehat{f}$ and f have the same distribution function, that is that for each $t \in \mathbb{R}$ the sets $\{x \in \mathbb{T}_L : \ f(x) > t\}$ and $\{x \in \mathbb{T}_L : \ \widehat{f}(x) > t\}$ have the same Lebesgue measure.

This function $\widehat{f}$ is called the **periodic Schwarz rearrangement** of the function f.

In order to prove that Schwarz rearrangement increases the epidemic threshold as well as the minimal wave speed when it is defined, let us explicitly write down the dependence of the function κ in the function $\mu^\theta(\lambda, \nu)$ defined in (31). It is now written as $\mu^\theta(\lambda, \nu; \kappa)$.

Recall that the following result has been proved by Nadin [27, 28]: for all $\theta \in [0, 2\pi)$ and all $(\lambda, \nu) \in \mathbb{R} \times (-\beta, \infty)$ one has

$$\mu^\theta(\lambda, \nu; \kappa) \leq \mu^\theta(\lambda, \nu; \widehat{\kappa}) \text{ and } \mu^\theta(\lambda, \nu; \kappa) \leq \mu^\theta(\lambda, \nu; \widehat{\widehat{\alpha}\widehat{\nu^+}}).$$

Note that this result follows from the variational formula (33) together with Polya and Hardy–Littlewood inequalities for rearrangements. Now, using the above inequalities, Theorem 2(ii) follows by recalling definition (8) and (39). This completes the proof of Theorem 2.

6.2 *Influences of the Period*

In order to study the influence of the periodicity of the medium upon the epidemic threshold and upon the minimal wave speed and to prove Theorem 3, we consider two functions α and ν^+ in $L^\infty_+(\mathbb{T}_1)$ such that $\kappa := \alpha \times \nu^+ \not\equiv 0$ and we define for any $L > 0$ the $L-$periodic functions

$$\alpha_L(x) = \alpha\left(\frac{x}{L}\right), \quad \nu^+_L(x) = \nu^+\left(\frac{x}{L}\right), \text{ and } \kappa_L(x) = \kappa\left(\frac{x}{L}\right) \quad x \in \mathbb{T}_L.$$

As above, we explicitly write down the dependence of $\mu^\theta(\lambda, \nu)$ with respect to $L > 0$ by using the notation $\mu^\theta_L(\lambda, \nu)$. Let us also observe that because of the symmetry property stated in Theorem 2, it is sufficient to study the dependence of the period L for the angles $\theta \in \left[0, \frac{\pi}{2}\right]$. The proof of Theorem 3 will follow from the properties of that function $L \mapsto \mu^\theta_L(\lambda, \nu)$. We first collect and recall some needed properties.

Lemma 6.1 *For each $\theta \in \left[0, \frac{\pi}{2}\right]$, $\lambda \in \mathbb{R}$ and $\nu > -\beta$ the following hold true.*

(i) The function $L \mapsto \mu^\theta_L(\lambda, \nu)$ is increasing on $(0, \infty)$.
(ii) One has

$$\lim_{L \to 0^+} \mu^\theta(\lambda, \nu) = \frac{r}{\nu + \beta} \int_{\mathbb{T}_1} \kappa(x)dx,$$

uniformly with respect to $\theta \in \left[0, \frac{\pi}{2}\right]$, locally uniformly for $\lambda \in \mathbb{R}$ and $\nu > -\beta$.

(iii) One has

$$\lim_{L\to\infty} \mu_L^\theta(\lambda,\nu) = -(\lambda\cos\theta)^2 + K_\infty(\lambda\cos\theta,\nu),$$

locally uniformly for $\lambda \geq 0$, $\nu \geq 0$ *and* $\theta \in \left[0, \frac{\pi}{2}\right]$ *where we have set*

$$K_\infty(\lambda,\nu) = \begin{cases} \frac{1}{\nu+\beta} j^{-1}\left(\lambda\sqrt{\nu+\beta}\right) & \text{if } \lambda\sqrt{\nu+\beta} \geq j\left(r\|\kappa\|_\infty\right), \\ \frac{r\|\kappa\|_\infty}{\beta+\nu} & \text{if } \lambda\sqrt{\nu+\beta} \leq j\left(r\|\kappa\|_\infty\right), \end{cases}$$

and $j(k) = \int_0^1 \sqrt{k - r\kappa(x)}dx$ *for any* $k \geq r\|\kappa\|_\infty$.

Parts (i) and (ii) in the above result have been proved by Nadin in [27, 28] while the part (iii) in the above lemma has been proved by Hamel et al. in [17] (see Proposition 3.2 in that paper).

Because of the above lemma, one may first obtain that the epidemic threshold $\mathscr{R}_0 = \mathscr{R}_{0,L} := \mu_L^\theta(0,0)$ is increasing with respect to $L > 0$ and satisfies

$$\lim_{L\to 0} \mathscr{R}_{0,L} = \mathscr{R}_{0,0} := \frac{r}{\beta}\int_{\mathbb{T}_1} \kappa(x)dx \text{ and } \lim_{L\to\infty} \mathscr{R}_{0,L} = \mathscr{R}_{0,\infty} = \frac{r\|\kappa\|_\infty}{\beta}.$$

Next let us define the function $\Lambda_L^\theta(\lambda,\nu)$ by

$$\Lambda_L^\theta(\lambda,\nu) := \lambda^2 - 1 + \mu_L^\theta(\lambda,\nu),$$

and observe that the above function is increasing with respect to L for any $(\lambda,\nu) \in \mathbb{R}\times(-\beta,\infty)$ and any $\theta \in \left[0, \frac{\pi}{2}\right]$. As a consequence if $\mathscr{R}_{0,0} > 1$, then $\mathscr{R}_{0,L} > 1$ for any $L > 0$ and the minimal wave speed $c_{\theta,L}^*$ is well defined for any $L > 0$. Recalling (39), it follows from Lemma 6.1 (i) that $T \mapsto c_{\theta,L}^*$ is increasing with respect to L. Hence it converges as $L \to 0$ and the limit, denoted by $\overline{c^*}$, is given, due to Lemma 6.1(ii) by the following expression

$$\overline{c^*} = \sup\left\{c \geq 0 : \frac{r}{c\lambda+\beta}\int_{\mathbb{T}_1}\kappa(x)dx + \lambda^2 - 1 - c\lambda \geq 0,\ \forall\lambda \geq 0\right\}.$$

This formula is equivalent to the formulation given in Theorem 3(i) and this completes the proof of part (i).

Now in order to prove part (ii) of Theorem 3, we assume that $\mathscr{R}_{0,\infty} > 1$. Then, since the $L \mapsto \mu_L^\theta(0,0)$ is increasing (see Lemma 6.1(i)), there exists $L_0 > 0$ large enough such that $\mathscr{R}_{0,L} > 1$ for all $L \geq L_0$. Hence the minimal wave speed $c_{\theta,L}^*$ is well defined for any $L \geq L_0$ and it is increasing with respect to L. Now to complete the proof of Theorem 3(ii), due to Lemma 6.1(iii), it is sufficient to prove that $c_{\theta,L}^*$ is uniformly bounded with respect to $L \geq L_0$. To that aim, let us observe that for each $L > 0$, for each $\theta \in \left[0, \frac{\pi}{2}\right]$ and each $(\lambda,\nu) \in \mathbb{R}\times(-\beta,\infty)$, one has

$$\Lambda_L^\theta(\lambda, \nu) - c\lambda \leq \lambda^2 - c\lambda - 1 + \frac{r\|\kappa\|_\infty}{\beta + \nu}. \tag{75}$$

Now recalling that $\mathscr{R}_{0,\infty} > 1$, let us defined the quantity $\overline{c} \in (0, \infty)$ by

$$\overline{c} = \inf \left\{ c > 0 : \exists \lambda > 0 \; \lambda^2 - c\lambda - 1 + \frac{r\|\kappa\|_\infty}{\beta + c\lambda} < 0 \right\}.$$

Moreover, recalling the definition of $c^*_{\theta,L}$ in (39), the upper estimate in (75) ensures that for each $L > 0$ sufficiently large one has

$$c^*_{\theta,L} \leq \overline{c}, \; \forall \theta \in \left[0, \frac{\pi}{2}\right].$$

As a consequence of the boundedness of that quantity and due to Lemma 6.1(iii), one obtains that

$$\lim_{L \to \infty} c^*_{\theta,L} = c^*_{\theta,\infty},$$

where this limit wave speed in defined in Theorem 3(ii). This completes the proof of the theorem.

Acknowledgements This work has been partially done while the first author was visiting the University of Tokyo. He would like to thank the second author and the University of Tokyo for their hospitality.

References

1. Berestycki, H., Hamel, F.: Front propagation in periodic excitable media. Comm. Pure Appl. Math. **55**, 949–1032 (2002)
2. Berestycki, H., Hamel, F.: Generalized travelling waves for reaction-diffusion equations. In: Perspectives in Nonlinear Partial Differential Equations, vol. 446, pp. 101–123 (2007). (In honor of H. Brezis, Contemp. Math. Am. Math. Soc.)
3. Berestycki, H., Hamel, F.: Generalized transition waves and their properties. Comm. Pure Appl. Math. **65**, 592–648 (2012)
4. Berestycki, H., Hamel, F., Nadin, G.: Asymptotic spreading in heterogeneous diffusive excitable media. J. Funct. Anal. **255**, 2146–2189 (2008)
5. Berestycki, H., Hamel, F., Nadirashvili, N.: The speed of propagation for KPP type problems. I. Periodic Framew. J. Eur. Math. Soc. **7**, 173–213 (2005)
6. Berestycki, H., Hamel, F., Roques, L.: Analysis of the periodically fragmented environment model I. Influence of periodic heterogeneous environment on species persistence. J. Math. Biol. **51**, 75–113 (2005)
7. Berestycki, H., Hamel, F., Roques, L.: Analysis of the periodically fragmented environment model II. Biological invasions and pulsating traveling fronts. J. Math. Pures Appl. **84**(2005), 1101–1146 (2005)
8. Burie, J.-B., Calonnec, A., Langlais, M.: Mathematical modeling of biological systems. In: Deutsch, A., de la Bravo Parra, R., de Boer, R., Diekmann, O., Jagers, P., Kisdi, E., Kretzschmar, M., Lansky, P., Metz, H. (eds.) Modeling of the Invasion of a Fungal Disease Over a Vineyard, pp. 11–21. Birkhauser, Boston (2007)

9. Burie, J.-B., Calonnec, A., Ducrot, A.: Singular perturbation analysis of travelling waves for a model in phytopathology. Math. Model. Nat. Phen. **1**, 49–63 (2006)
10. Burie, J.-B., Ducrot, A.: Travelling wave solutions for some models in phytopathology. Nonlinear Analysis RWA **10**, 2307–2325 (2009)
11. Ducrot, A., Giletti, T.: Convergence to a pulsating travelling wave for an epidemic reaction-diffusion system with non-diffusive susceptible population. J. Math. Biol. **69**, 533–552 (2014)
12. Freidlin, M.: On wave front propagation in periodic media. In: M. Pinsky (ed.), Stochastic Analysis and Applications. In: Advanced Probability Related Topics, vol. 7, pp. 147–166 (1984)
13. Gärtner, J., Freidlin, M.: On the propagation of concentration waves in periodic and random media. Soviet Math. Dokl. **20**, 1282–1286 (1979)
14. Hamel, F.: Qualitative properties of monostable pulsating fronts: exponential decay and monotonicity. J. Math. Pures Appl. **89**, 355–399 (2008)
15. Hosono, Y., Ilyas, B.: Traveling waves for a simple diffusive epidemic model. Math. Models Methods Appl. Sci. **05**, 935–966 (1995)
16. Hamel, F., Fayard, J., Roques, J.: Spreading speeds in slowly oscillating environments. Bul. Math. Biol. **72**, 1166–1191 (2010)
17. Hamel, F., Nadin, G., Roques, L.: A viscosity solution method for the spreading speed formula in slowly varying media. Indiana Univ. Math. J. **60**, 1229–1247 (2011)
18. Huang, J.H., Shen, W.: Speeds of spread and propagation for KPP models in time almost and space periodic media. SIAM J. Appl. Dyn. Syst. **8**, 790–821 (2009)
19. Kawohl, B.: Rearrangements and Convexity of Level Sets in PDE. Springer, Berlin (1985)
20. Kawohl, B.: On the isoperimetric nature of a rearrangement inequality and its consequences for some variational problems. Arch. Ration. Mech. Anal. **94**, 227–243 (1986)
21. Kinezaki, N., Kawasaki, K., Takasu, F., Shigesada, N.: Modeling biological invasions into periodically fragmented environments. Theo. Pop. Biol. **64**, 291–302 (2003)
22. Liang, X., Lin, X., Matano, H.: A variational problem associated with the minimal speed of travelling waves for spatially periodic reaction-diffusion equations. Trans. Amer. Math. Soc. **362**, 5605–5633 (2010)
23. Liang, X., Matano, H.: Maximizing the spreading speed of KPP fronts in two-dimensional stratified media. Proc. London Math. Soc. **109**, 1137–1174 (2014)
24. Liang, X., Yi, Y., Zhao, X.-Q.: Spreading speeds and traveling waves for periodic evolution systems. J. Diff. Eq. **231**, 57–77 (2006)
25. Liang, X., Zhao, X.-Q.: Spreading speeds and traveling waves for abstract monostable evolution systems. J. Funct. Anal. **259**, 857–903 (2010)
26. Mammeri, Y., Burie, J.-B., Langlais, M., Calonnec, A.: How changes in the dynamic of crop susceptibility and cultural practices can be used to better control the spread of a fungal pathogen at the plot scale? Ecol. Model. **290**, 178–191 (2014)
27. Nadin, G.: Équations de réaction-diffusion et propagation en milieu hétérogène. Ph.D. thesis, Université Pierre et Marie Curie (2008)
28. Nadin, G.: The effect of the Schwarz rearrangement on the periodic principal eigenvalue of a nonsymmetric operator. SIAM J. Math. Anal. **41**, 2388–2406 (2010)
29. Okubo, A., Levin, S.A.: Diffusion and Ecological Problems: Modern Perspectives. Springer, New York (2001)
30. Segarra, J., Seger, M.J., Van den Bosch, F.: Epidemic dynamics and patterns of plant diseases. Phytopathology **91**, 1001–1010 (2001)
31. Shen, W.: Variational principle for spreading speeds and generalized propagating speeds in time almost periodic and space periodic KPP models. Trans. Amer. Math. Soc. **362**, 5125–5168 (2010)
32. Shigesada, N., Kawasaki, K.: Biological Invasions: Theory and Practice. Oxford University Press, Oxford (1997)
33. Shigesada, N., Kawasaki, K., Teramoto, E.: Traveling periodic waves in heterogeneous environments. Theor. Pop. Bio. **30**, 143–160 (1986)

34. Weinberger, H.F.: On spreading speeds and traveling waves for growth and migration in periodic habitat. J. Math. Biol. **45**, 511–548 (2002)
35. Xin, X.: Existence and stability of traveling waves in periodic media governed by a bistable nonlinearity. J. Dyn. Diff. Eq. **3**, 541–573 (1991)
36. Xin, X.: Existence of planar flame fronts in convective-diffusive periodic media. Arch. Ration. Mech. Anal. **121**, 205–233 (1992)
37. Xin, J.X.: Analysis and modeling of front propagation in heterogeneous media. SIAM Rev. **42**, 161–230 (2000)
38. Zawolek, M.W., Zadoks, J.C.: Studies in focus development: an optimum for the dual dispersal of plant pathogens. Phytopathology **82**, 1288–1297 (1992)

A Within-Host Model of Dengue Viral Infection Dynamics

Arti Mishra

Abstract In this paper, a non-linear within-host viral infection model has been proposed and analyzed for primary dengue infection. The model incorporates the dynamics of virus particles, T cells and antibodies during pathogenesis of dengue infection. The basic reproduction number has been computed. The five equilibrium states exist. The existence conditions for various equilibrium points are obtained. The stability analysis of equilibrium states have been discussed. The existence of bi-stability of states E_2 and E_3 is possible under certain parametric conditions. The model shows that the viral load decreases within 7 to 14 days after the onset of symptoms which validates the clinical feature of dengue fever. The model also concludes that the humoral response is more powerful in viral clearance as compared to immune response by CD_8 T cells.

Keywords T cells · Threshold · Stability · Dengue infection · Antibodies

2010 Mathematics Subject Classification One primary · One secondary

1 Introduction

Dengue disease has spread almost all over the world and has become the disease of important international health concern. As there is no commercial dengue vaccine [7], there is need to understand the infection dynamics inside the body. Many mathematical models on dengue infection dynamics at population level are found in literature [5, 6, 14, 17, 18]. However, very few models are available in literature for within-host dengue dynamics [1, 2, 4, 8, 15].

As the virions enter into the body, they infect monocytes, macrophages, dendritic cells and hepatocytes [9, 10, 19]. This stimulates and expands the dengue virus specific CD_4 and CD_8 T cells [19, 20]. The CD_4 T cells obstruct the spread of virions

---

A. Mishra (✉)
Department of Mathematics, Indian Institute of Technology, Roorkee, Uttarakhand, India
e-mail: mishraarti21@gmail.com

J.M. Cushing et al. (eds.), *Applied Analysis in Biological and Physical Sciences*,
Springer Proceedings in Mathematics & Statistics 186,
DOI 10.1007/978-81-322-3640-5_9

through the secretion of antiviral cytokines such as IFN_γ that block viral replication [20]. The CD_8 T cells are also important for viral clearance as they directly kill the infected cells which reduces virus load [16, 19]. The CD_4 T cells are needed to elicit the antibody responses and for the production of both B cells as well as CD_8 T cells [12, 16, 19, 20]. The B cells further produce antibodies to bind the virions [12, 16, 19].

Particularly, Nuraini et al. have incorporated the class of immune cells for primary dengue infection dynamics [15]. These immune cells activate the CTL response which are basically carried by CD_8 T cells. They kill infected cells and are also activated by them. They have not considered the T cell mediated immune response as well as humoral response (by antibodies) explicitly. Ansari and Hesaaraki have considered the Beddington-DeAngelis functional response for target cells and virus particles interaction in the model proposed by Nuraini [1]. They have discussed the existence and stability of various equilibrium states. The dynamics of primary dengue infection at cellular level has also been modeled by Ambika and Gulati [8]. They have considered only the humoral response in the model.

In this paper, a within host nonlinear model for primary dengue infection has been proposed and analyzed to reveal the biological process. To model the virus clearance dynamics in primary dengue infection, both the responses namely, T cell response as well as humoral response should be incorporated. It is known that both the responses play very important role in controlling primary dengue infection dynamics [3, 8]. The paper incorporates the dynamics of T cell mediated response by CD_4 and CD_8 T cells and humoral response by antibodies. In Sect. 2, the mathematical model has been formulated. In Sect. 3, model analysis has been done. In Sect. 4, existence of bi-stability has been discussed. The last section discusses the conclusion of the paper.

2 The Mathematical Model

The within-host nonlinear model assumes the presence of single serotype only. Consider S, I and V be the number of healthy target cells (monocytes, macrophages or dendritic cells, etc.), infected cells and dengue virions respectively. The basic mathematical model having interaction among S, I and V cells was proposed by Nowak and May [13]. In the present model, the dynamics of cellular response by CD_4 T cells (let H) and CD_8 T cells (let Z) as well as the humoral response by antibodies (let A) have been incorporated. Let ω be the constant recruitment of target cells. The virus particles attack target cells and make them infected with the rate β_1. Let β_2 be the decay rate of virus particles due to activation of dengue specific CD_4 T cells. The rate of activation of CD_4 T cells by virus particles be β_3. Let β_4 and β_5 be the rates of activation of CD_8 T cells and CD_4 T cells by infected cells and virus particles respectively. As CD_4 T cells elicit the antibody response, let the rate of activation be c_1. Assume d, d_1, d_2, d_3, d_4 and d_5 be the natural death rates of target cells, infected cells, virus particles, CD_4 T cells, CD_8 T cells and antibodies respectively. The burst rate for virus particles be k and the killing rate of infected cells by CD_8 T cells be

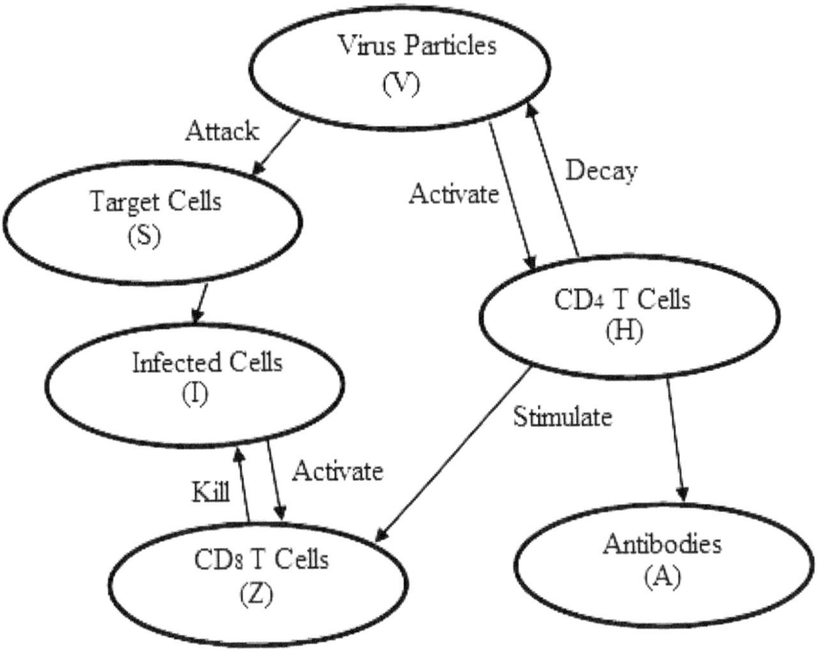

Fig. 1 Schematic diagram

p. Let p_1 be the rate at which antibodies neutralize the virus particles. The Fig. 1 represents the schematic diagram for virus-cell dynamics.

Keeping above in view, the model is formulated below:

$$\frac{dS}{dt} = \omega - \beta_1 SV - dS \tag{1}$$

$$\frac{dI}{dt} = \beta_1 SV - pIZ - d_1 I \tag{2}$$

$$\frac{dV}{dt} = kI - \beta_2 HV - p_1 AV - d_2 V \tag{3}$$

$$\frac{dH}{dt} = \beta_3 HV - d_3 H \tag{4}$$

$$\frac{dZ}{dt} = \beta_4 IZ + \beta_5 HZ - d_4 Z \tag{5}$$

$$\frac{dA}{dt} = c_1 H - d_5 A \tag{6}$$

The model is associated with the following non-negative initial conditions:

$$S(0) \geq 0, I(0) \geq 0, V(0) \geq 0, H(0) \geq 0, Z(0) \geq 0, A(0) \geq 0$$

3 Model Analysis

3.1 Basic Reproduction Number

The basic reproduction number for the model (1)–(6) is defined as the average number of secondary infected target cells generated by a single infected target cell placed in an uninfected target cell population [15]. The basic reproduction number in absence of any immune response is computed by next generation approach. It is given as:

$$R_0 = \frac{\beta_1 k \omega}{d_1 d_2 d} \tag{7}$$

3.2 Equilibrium Points and Stability Analysis

It is observed that the non-linear system of equations (1)–(6) have five equilibrium points. The existence conditions for all equilibrium states are computed in terms of basic reproduction number. For stability analysis of various equilibrium states, the system (1)–(6) is linearized and the Jacobian matrix [J] about the point $E(S, I, V, H, Z, A)$ is given as:

$$J[E] = \begin{bmatrix} -\beta_1 V - d & 0 & -\beta_1 S & 0 & 0 & 0 \\ \beta_1 V & -d_1 - pZ & \beta_1 S & 0 & -pI & 0 \\ 0 & k & -\beta_2 H - d_2 - p_1 A & -\beta_2 V & 0 & -p_1 V \\ 0 & 0 & \beta_3 H & \beta_3 V - d_3 & 0 & 0 \\ 0 & \beta_4 Z & 0 & \beta_5 Z & \beta_4 I + \beta_5 H - d_4 & 0 \\ 0 & 0 & 0 & c_1 & 0 & -d_5 \end{bmatrix}$$

1. The disease-free equilibrium point $E_0 = (\hat{S}, 0, 0, 0, 0, 0, 0)$; $\hat{S} = \frac{\omega}{d}$
   The eigenvalues of jacobian matrix [J] about the disease-free point E_0 are as follows:

$$- \text{d}, -d_3, -d_4, -d_5, \frac{-A \pm B}{2}$$
$$\text{where, A } = d_1 + d_2$$
$$\text{B} = \sqrt{(d_1 - d_2)^2 + 4\beta_1 k \hat{S}}$$

It is observed that all the eigenvalues are having negative real part when $R_0 < 1$. Therefore, the disease-free point (E_0) is locally asymptotically stable.

For the global stability of the disease-free equilibrium state (E_0), the Lyapunov second method of stability [11] has been used. Consider the following positive definite function $L(I, V)$ as:

$$L(I, V) = \frac{k}{dd_1} I + \frac{V}{d}$$

Taking derivative of $L(I, V)$ with respect to t gives,

$$\dot{L}(I, V) = \frac{k}{dd_1}\dot{I} + \frac{\dot{V}}{d}$$

$$\dot{L}(I, V) \leq -V\left(d_2 - \frac{\beta_1 k\omega}{dd_1}\right)$$

$$\text{or } \dot{L}(I, V) \leq -Vd_2\left(1 - \frac{\beta_1 k\omega}{dd_1 d_2}\right) < -Vd_2 \quad \text{for } \frac{\beta_1 k\omega}{dd_1 d_2}(=R_0) < 1$$

$$\text{or } \dot{L}(I, V) < 0 \quad \text{for } R_0 < 1$$

$L(I, V)$ is a Lyapunov function for $R_0 < 1$. Accordingly, the global stability of E_0 can now be stated as following:

Theorem 1 *The locally asymptotically stable state E_0 is also globally asymptotically stable.*

Numerical simulation for disease-free state (E_0) has been performed for the choice of following data from literature [8]:

$$\omega = 10, \beta_1 = 0.001, \beta_2 = 0.001, \beta_3 = 0.001, \beta_4 = 0.001, k = 1, d = 0.05,$$
$$d_1 = 0.5, d_2 = 0.5, d_3 = 0.05, d_4 = 0.02, d_5 = 0.051, p_1 = 0.001, p = 0.001,$$
$$c_1 = 0.01, \beta_5 = 0.015$$

R_0 is found to be less than 1. It is observed from time series analysis (Fig. 2) that, virus particles get cleared within 7–14 days which is in line with the clinical observation of dengue infection.

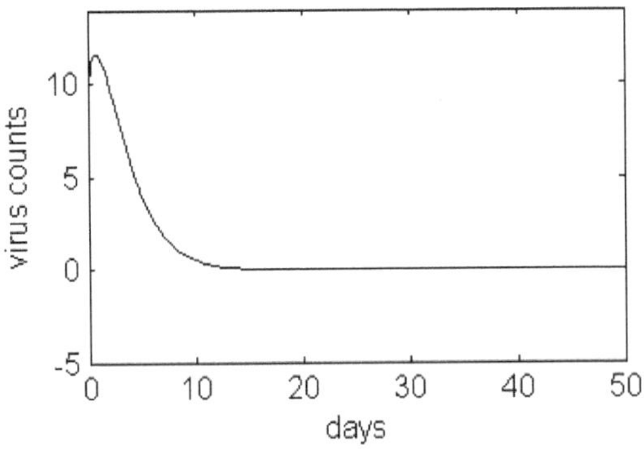

Fig. 2 Time series for the virus particles

2. The ineffective immune response state $E_1 = (\bar{S}, \bar{I}, \bar{V}, 0, 0, 0)$ where

$$\bar{S} = \frac{d_1 d_2}{\beta_1 k}; \quad \bar{I} = \frac{(R_0 - 1) d d_2}{\beta_1 k}; \quad \bar{V} = \frac{(R_0 - 1) d}{\beta_1}$$

Accordingly, the state E_1 exists for

$$R_0 > 1 \tag{8}$$

For the stability of E_1 state, the three eigenvalues of the Jacobian matrix [J] about E_1 are

$$-d_5, -d_4 + \frac{(R_0 - 1) d d_2 \beta_4}{\beta_1 k}, -d_3 + \frac{(R_0 - 1) d \beta_3}{\beta_1} \tag{9}$$

It is noted that the eigenvalues given in (9) will have negative real part when the following condition is satisfied:

$$R_0 < 1 + \min\left(\frac{\beta_1 k d_4}{\beta_4 d d_2}, \frac{\beta_1 d_3}{\beta_3 d}\right) \tag{10}$$

The remaining three eigenvalues are the roots of the following cubic polynomial:

$$\begin{aligned}
\lambda^3 + M_1 \lambda^2 + M_2 \lambda + M_3 &= 0 \ \text{ where,} \\
M_1 &= d + d_1 + d_2 + \beta_1 \bar{V} > 0 \\
M_2 &= d d_1 + d d_2 + \beta_1 d_1 \bar{V} + \beta_1 d_2 \bar{V} > 0 \\
M_3 &= \beta_1 d_1 d_2 \bar{V} > 0
\end{aligned}$$

It is observed that the $M_1 M_2 - M_3 > 0$. Therefore, by Routh–Hurwitz criterion, all the three eigenvalues will have negative real part. Hence, all the six eigenvalues of the system will have negative real part under condition (10).

Theorem 2 *The state E_1 is locally asymptotically stable under condition (10).*

The immune response is ineffective and could not activate T cell response as well as humoral response to a level where infection could be eradicated.

3. The ineffective antibody response state $E_2 = (\tilde{S}, \tilde{I}, \tilde{V}, 0, \tilde{Z}, 0)$ where

$$\tilde{S} = \frac{\omega \beta_4 d_2}{\beta_1 k d_4 + \beta_4 d d_2}; \quad \tilde{I} = \frac{d_4}{\beta_4}; \quad \tilde{V} = \frac{d_4 k}{\beta_4 d_2}; \quad \tilde{Z} = \frac{\beta_4 \omega \beta_1 k - \beta_4 d d_2 d_1 - k \beta_1 d_4 d_1}{p(\beta_1 k d_4 + \beta_4 d d_2)}$$

Accordingly, the state E_2 exists for

$$\beta_4\omega\beta_1 k > \beta_4 d d_2 d_1 + k\beta_1 d_4 d_1$$

or

$$R_0 > 1 + \frac{\beta_1 k d_4}{\beta_4 d d_2} \tag{11}$$

For stability of antibody response ineffective equilibrium point (E_2), the two of the eigenvalues of Jacobian matrix about E_2 for the system are given as:

$$-d_5, -d_3 + \frac{\beta_3 d_4 k}{\beta_4 d_2} \tag{12}$$

It is observed that the eigenvalues given in (12) are having negative real part provided:

$$\frac{\beta_1 k d_4}{\beta_4 d d_2} < \frac{\beta_1 d_3}{\beta_3 d} \tag{13}$$

The remaining eigenvalues are the roots of following fourth degree polynomial:

$$\lambda^4 + A_1\lambda^3 + A_2\lambda^2 + A_3\lambda + A_4 = 0 \quad \text{where}$$
$$A_1 = d + d_1 + d_2 + \beta_1\tilde{V} + p\tilde{Z}$$
$$A_2 = d(d_1 + d_2) + d_1 d_2 - \beta_1 k\tilde{S} + \beta_1\tilde{V}(d_1 + d_2) + p\tilde{Z}(d + d_2 + d_4 + \beta_1\tilde{V})$$
$$A_3 = \frac{\beta_1 k d_4(\beta_4(d_2 + d) + \beta_1 d_2(d_2 + d_4)k)\omega}{\beta_4 d d_2 + \beta_1 k d_4} - \frac{d_1 d_4(\beta_4(d + d_2 + \beta_1 d_2 d_4 k))}{\beta_4}$$
$$A_4 = d_2 d_4 p\tilde{Z}(d + \beta_1\tilde{V})$$

Applying the Routh–Hurwitz criteria to the above fourth degree polynomial, it is observed that $A_1 > 0$, $A_4 > 0$, $A_3 > 0$ and $A_1A_2A_3 > A_3^2 + A_1^2A_4$ provided the conditions (11) and (13) are satisfied.

Theorem 3 *The state E_2, if exists, is locally asymptotically stable when the condition (13) is satisfied.*

This state is obtained due to the presence of infected cells in the body. For humoral response to be activated, there should be sufficient number of virus particles.

4. The ineffective CD_8 T cells response state $E_3 = (\breve{S}, \breve{I}, \breve{V}, \breve{H}, 0, \breve{A})$

$$\breve{S} = \frac{\omega\beta_3}{\beta_1 d_3 + d\beta_3};\ \breve{I} = \frac{\beta_1\omega d_3}{(\beta_1 d_3 + d\beta_3)d_1};\ \breve{V} = \frac{d_3}{\beta_3};$$
$$\breve{H} = \frac{(\omega\beta_1 k\beta_3 - \beta_1 d_1 d_2 d_3 - dd_1 d_2\beta_3)d_5}{d_1(\beta_1\beta_2 d_5 d_3 + \beta_1 p_1 c_1 d_3 + d\beta_2 d_5\beta_3 + dp_1 c_1\beta_3)};$$
$$\breve{A} = \frac{c_1(\omega\beta_1 k\beta_3 - \beta_1 d_1 d_2 d_3 - dd_1 d_2\beta_3)}{d_1(\beta_1\beta_2 d_5 d_3 + \beta_1 p_1 c_1 d_3 + d\beta_2 d_5\beta_3 + dp_1 c_1\beta_3)}$$

The state E_3 exists for,

$$\omega\beta_1 k\beta_3 > \beta_1 d_1 d_2 d_3 + dd_1 d_2\beta_3$$

or

$$R_0 > 1 + \frac{\beta_1 d_3}{\beta_3 d} \tag{14}$$

For the stability of CD_8 T cells response ineffective equilibrium point (E_3), the one of the eigenvalues of Jacobian matrix about E_3 for the system are given as:

$$-d_4 + \beta_5\breve{H} + \beta_4\breve{I} \tag{15}$$

It is observed that the eigenvalue in (15) has negative real part provided:

$$R_0 < W\left(1 + \frac{\beta_1 d_3}{\beta_3 d}\right);\ W = \frac{(\beta_5 d_2 d_5 + \beta_2 d_4 d_5 + c_1 d_4 p_1)k\beta_3 d}{(\beta_5\beta_3 k d_5 + \beta_2\beta_4 d_3 d_5 + \beta_4 c_1 d_3 p_1)d_1 d_2} \tag{16}$$

The rest of the eigenvalues are the roots of following fifth degree polynomial:

$$\lambda^5 + B_1\lambda^4 + B_2\lambda^3 + B_3\lambda^2 + B_4\lambda + B_5 = 0$$
$$B_1 = d + d_1 + d_2 + d_3 + d_5 + \beta_2\breve{H} + \breve{A}p_1 + \beta_1\breve{V}$$
$$B_2 = \beta_2 d_3\breve{H} + (d_2 + \beta_2\breve{H} + p_1\breve{A})(d_5 + d + \beta_1\breve{V} + d_1) + (\omega\beta_1 k\beta_3 - \beta_1 d_1 d_2 d_3 - dd_1 d_2\beta_3)d_1 d_3 + d(d_1 + d_5) + d_1(d_5 + \beta_1\breve{V})$$
$$B_3 = (dd_2 d_5 + d_1 d_2 d_5 + \beta_2(d_1 d_3 + d_1 d_5 + d_3 d_5)\breve{H} + \breve{A}p_1(d_1 d_3 + d_1 d_5 + d_3 d_5 + dd_1 + d_3 + d_5) - (-\omega\beta_1 k\beta_3 + \beta_1 d_1 d_2 d_3 + dd_1 d_2\beta_3)(d + d_3 + d_5) + \beta_3 c_1\breve{H}p_1 + \beta_1(d_3 d_5 + d_2(d_3 + d_5) + \beta_2 d_3\breve{H}$$
$$B_4 = (d_3 d_5(d_1(d_2 + \beta_2\breve{H} + \breve{A}p_1) - (\beta_1 d_1 d_2 d_3 + \beta_3 dd_1 d_2 - \beta_1 k\omega\beta_3)(d_3 d_5 + d(d_3 + d_5)) + \beta_1(d_3 d_5(d_2 + \beta_2\breve{H} + \breve{A}p_1) + d_1(d_3 d_5 + d_2(d_3 + d_5) + \beta_2 d_3\breve{H} + \beta_2 d_5\breve{H} + \breve{A}(d_3 + d_5)p_1) + \beta_3 d_5 k\breve{S}))\breve{V} + d(d_2 d_3 d_5 + \beta_2 d_3 d_5\breve{H} + \breve{A}d_3 d_5 p_1 + d_1(d_2 d_3 + d_2 d_5 + d_3 d_5 + \beta_2 d_3\breve{H} + \beta_2 d_5\breve{H} + \breve{A}d_3 p_1 + \breve{A}d_5 p_1)))$$

$$B_5 = dd_1\breve{H}(\beta_2 d_3 d_5 + \beta_3 c_1 p_1 \breve{V}) + \beta_1 d_1 d_3 d_5(\beta_2 \breve{H} + p_1 \breve{A})$$

For the above fifth degree polynomial, it is observed that $B_1 > 0$, $B_2 > 0$, $B_3 > 0$, $B_4 > 0$, $B_5 > 0$, $B_1 B_2 B_3 > B_3^2 + B_1^2 B_4$ and $(B_1 B_4 - B_5)(B_1 B_2 B_3 - B_3^2 - B_1^2 B_4) > B_5(B_1 B_2 - B_3)^2 + B_1 B_5^2$ for the conditions (14) and (16) to be satisfied. Hence, by Routh–Hurwitz criteria, all the five roots will have negative real part.

Theorem 4 *The state E_3, if exists, is locally asymptotically stable under the condition (16).*

The activation of CD_8 T cells depends not only on CD_4 T cells but also occurs due to the presence of infected cells. There is a need of sufficient amount of infected cells for the CD_8 T cells to be activated.

5. The Endemic equilibrium point $E^* = (S^*, I^*, V^*, H^*, Z^*, A^*)$

$$S^* = \frac{\omega\beta_3}{\beta_1 d_3 + d\beta_3}; \quad I^* = \frac{d_3(\beta_5 d_2 d_5 + d_5 \beta_2 d_4 + p_1 c_1 d_4)}{(\beta_4 d_3 \beta_2 d_5 + \beta_4 d_3 p_1 c_1 + \beta_5 k d_5 \beta_3)}; \quad V^* = \frac{d_3}{\beta_3};$$
$$H^* = \frac{d_5(-\beta_4 d_3 d_2 + \beta_3 k d_4)}{(\beta_4 d_3 \beta_2 d_5 + \beta_4 d_3 p_1 c_1 + \beta_5 k d_5 \beta_3)};$$
$$Z^* = \frac{-d_1 + \beta_1 S^* d_3 - d_1 \beta_3 (1 + \alpha S^*) I^*}{p\beta_3 (1 + \alpha S^*) I^*}; \quad A^* = \frac{c_1(-\beta_4 d_3 d_2 + \beta_3 k d_4)}{(\beta_4 d_3 \beta_2 d_5 + \beta_4 d_3 p_1 c_1 + \beta_5 k d_5 \beta_3)}$$

It is observed that for the positivity of endemic equilibrium point (E^*), the following two conditions are satisfied:

$$\frac{\beta_1 k d_4}{\beta_4 d d_2} > \frac{\beta_1 d_3}{\beta_3 d} \tag{17}$$

$$R_0 > W\left(1 + \frac{\beta_1 d_3}{\beta_3 d}\right) \tag{18}$$

$$W = \frac{(\beta_5 d_2 d_5 + \beta_2 d_4 d_5 + c_1 d_4 p_1) k \beta_3 d}{(\beta_5 \beta_3 k d_5 + \beta_2 \beta_4 d_3 d_5 + \beta_4 c_1 d_3 p_1) d_1 d_2}$$

Due to complex expressions for elements of 6×6 Jacobian matrix, using Mathematica, the local stability analysis of E^* is not conclusive. Therefore, numerical simulations have been performed for the following relevant data [8]:

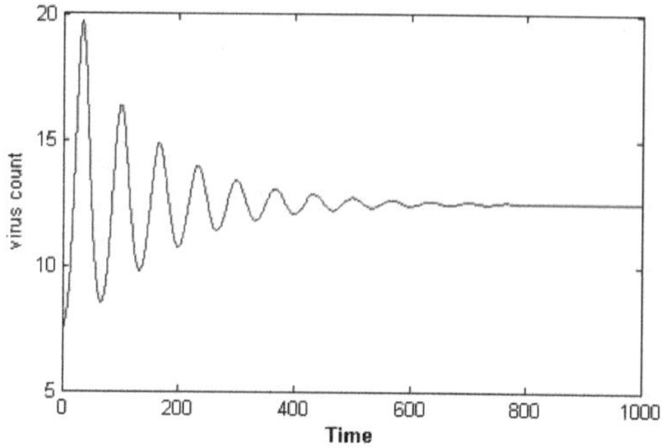

Fig. 3 Time series for the virus particles

$$\omega = 10, \beta_1 = 0.001, \beta_2 = 0.001, \beta_3 = 0.004, \beta_4 = 0.005, \beta_5 = 0.001, k = 3.5, d = 0.05, d_1 = 0.5, d_2 = 0.5, d_3 = 0.05, d_4 = 0.02, d_5 = 0.051, p_1 = 0.001, p = 0.001, c_1 = 0.08$$

For the above choice of data, endemic state exists. It is found that starting with the neighborhood of $E^*(160,\ 2.291272629,\ 12.5,\ 8.543636855,\ 372.8773585,\ 13.40178330)$, the solution converges to endemic state (Fig. 3).

4 Existence of Bi-Stability

With the help of existence and stability conditions of various equilibrium points, the bifurcation diagrams with respect to R_0 have been drawn for $\tilde{V}(= \frac{kd_4}{\beta_4 d_2}) > \breve{V}(= V^*)(= \frac{\beta_1 d_3}{\beta_3 d})$ and $\tilde{V} < \breve{V}(= V^*)$ in Figs. 4 and 5 respectively. The following observations are made from bifurcation diagrams:
The regions of existence and stability of states E_0 and E_1 remain same in Figs. 4 and 5.

In Fig. 4, the state E_2 gets unstable when its viral load is higher than that of E_3 or E^* states. Further, for $W < 1$, the state E_3 becomes unstable. However, for $W > 1$, it is found to be stable in the region given in Fig. 4. The endemic state E^* exists, when R_0 is higher than $(1 + \tilde{V})$, the virus particles density in E_2 state. No region of bi-stability could be observed in this case.

In Fig. 5, when virus load in E_3 or E^* states is higher than the E_2 state, the state E_2 gets stable if exists. However, for $W < 1$, the state E_3 remains unstable. For $W > 1$, the state E_3 gets stable in the region $(1 + V^*) < R_0 < W(1 + V^*)$. The bi-stability of states E_2 and E_3 occurs in this region. The endemic state does not exist in this case.

Thus, the bi-stability of the states E_2 and E_3 exist in the region $(1 + V^*) < R_0 < W(1 + V^*)$ for $\tilde{V} > V^*(= \breve{V})$.

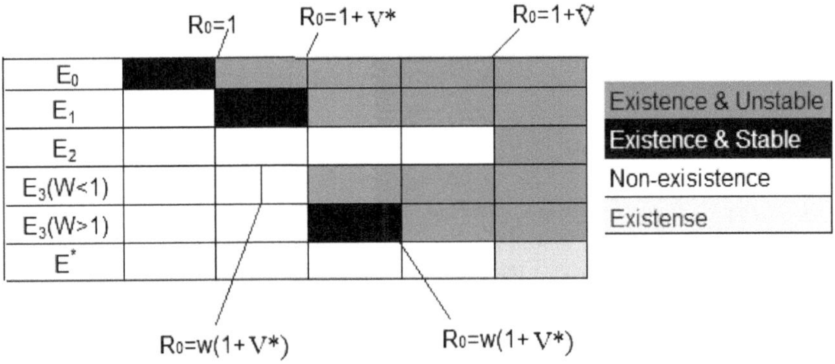

Fig. 4 Bifurcation diagram for $\tilde{V} > V^*(=\breve{V})$

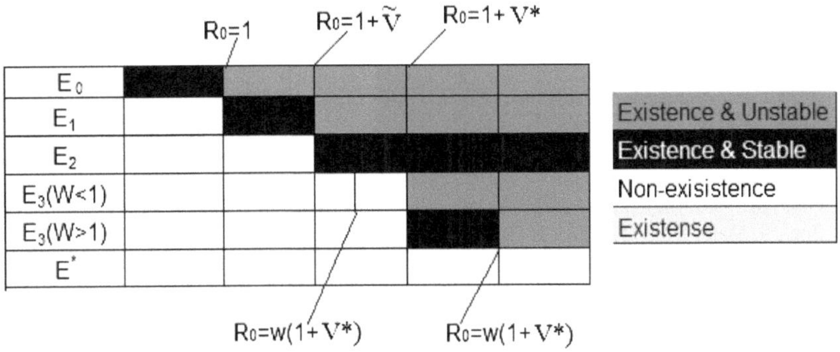

Fig. 5 Bifurcation diagram for $\tilde{V} < V^*(=\breve{V})$

5 Conclusions

In this paper, the cellular dynamics of primary dengue infection has been proposed and analyzed through non-linear mathematical model. The model incorporates the effects of T cells specifically, CD_4 and CD_8 and antibody response to present the clinical features of primary dengue infection. It validates the clinical features of disease that dengue virus gets cleared from the body within 7–14 days. Further, it is found analytically that for $R_0 < 1$, the infection free state remains stable while rest of the other states do not exist. The endemic state is found to be stable for the choice of data. By bifurcation diagrams, it can be concluded that there is a region for the bi-stability of states E_2 and E_3 for $W > 1$ and $V^*(= \breve{V}) > \tilde{V}$. It is also observed that there is a critical level of infected cells and virus particles inside the body above which T cell response activates. Once the state E_2 gets stable, the virus load is found to be low as compared to E_3 and E^* states. The states E_2 and E_3 will be unstable if the E_1 state gets stable. If the state E_3 stabilizes and $W > 1$, infection will not persist as will violate the existence condition of the endemic state E^*. Therefore, humoral response is more strong as compared to immune response by CD_8 T cells.

References

1. Ansari, H., Hesaaraki, M.: A with-in host dengue infection model with immune response and beddington-deangelis incidence rate. App. Math. **3**, 177–184 (2012)
2. Ben-Shachar, R., Koelle, K.: Minimal within-host dengue models highlight the specific roles of the immune response in primary and secondary dengue infections. J. R. Soc. Interface **12**, 20140886 (2015)
3. Bisen, P.S., Raghuvanshi, R.: Emerging Epidemics: Management and Control. Wiley Blackwell, Hoboken (2013)
4. Clapham, H.E., Tricou, V., et. al.: Within-host viral dynamics of dengue serotype 1 infection. J. R. Soc. Interface **11**, 20140094 (2014)
5. Esteva, L., Vargas, C.: Analysis of a dengue disease transmission model. Math. Biosci. **150**, 131–151 (1998)
6. Esteva, L., Vargas, C.: Coexistence of different serotypes of dengue virus. J. Math. Biol. **150**, 31–47 (2003)
7. Gubler, D.J.: Resurgent vector-borne, diseases as a global health problem. Emerg. Infect. Dis. **4**, 442–450 (1998)
8. Gujarati, T.P., Ambika, G.: Virus antibody dynamics in primary and secondary dengue infections. J. Math. Biol. **69**, 1773–1800 (2014)
9. Jindadamrongwech, S., Thepparit, C., Smith, D.R.: Identification of GRP78(BiP)as a liver cell expressed receptor element for dengue virus serotype 2. Arch. Virol. **149**, 915–927 (2004)
10. Kliks, S.C., Nisalak, A., Brandt, W.E., Wahl, L., Burke, D.S.: Antibody-dependent enhancement of dengue virus growth in human monocytes as a risk factor for dengue haemorrhagic fever. Am. J. Trop. Med. Hyg. **40**, 444–451 (1989)
11. Lasalle, J.P.: Stability theory for ordinary differential equations. J. Differ. Equ. **4**, 57–65 (1968)
12. Magdalini, M., Huynh-Hoa, B., et al.: Vaccinia virus-specific CD_4 T cell responses target a set of antigens largely distinct from those targeted by CD_8 T cell responses. J. Immunol. **178**, 6814–6820 (2007)
13. Nowak, M.A., May, R.M.: Virus Dynamics: Mathematical Principles of Immunology and Virology. Oxford University Press, Oxford (2000)
14. Nuraini, N., Soewono, E., Sidarto, K.A.: Mathematical model of dengue disease transmission with severe DHF compartment. Bull. Malays. Math. Sci. Soc. **30**(2), 143–157 (2007)
15. Nuraini, N., Tasman, H., et al.: A with-in host Dengue infection model with immune response. Math. Comput. Model. **49**, 1148–1155 (2009)
16. Rivino, L., Emmanuelle, A.P.K., Vojislav, J., et al.: Differential targeting of viral components by CD_4 versus CD_8 T lymphocytes in dengue virus infection. J. Virol. **87**(5), 2693–2706 (2013)
17. Supriatna, A.K., Soewono, E., van Gils, S.A.: A two-age-classes dengue transmission model. Math. Biosci. **216**, 114–121 (2008)
18. Tewa, J.J., Dimi, J.L., Bowong, S.: Lyapunov functions for a dengue disease transmission model. Chaos Soliton Fract. **39**, 936–941 (2009)
19. Willey, J.M., Sherwood, L.M., Woolverton, C.J.: Microbiology, 7th edn. McGraw Hill, New York (2008)
20. Yauch, L.E., Prestwood, T.R., et al.: CD_4 + T cells are not required for the induction of dengue virusspecific CD_8 + T cell or antibody responses but contribute to protection after vaccination. J Immunol. **185**(9), 5405–5416 (2010)

Stabilization of Prey Predator Model via Feedback Control

Anuraj Singh

Abstract In this paper, the effect of feedback linearization in Leslie–Gower type prey-predator model with Holling-type IV functional response is investigated. It is shown that the closed loop system may be stabilized using either approximate or exact linear approach. The former approach uses a linear control variable to provide a feedback linearization law whereas in latter approach, state space coordinates are suitably changed. Using this feedback control, a complex non-linear system is reduced to a linear controlled system that yields a globally asymptotically stable equilibrium point. Finally Analytical findings are validated through numerical simulations.

Keywords Feedback control · Exact linearization · Closed loop · Asymptotic stable

2010 Mathematics Subject Classification 34H15 · 93D15 · 93C15

1 Introduction

The prey predator dynamics has been extensively discussed by several investigators [4, 15, 16]. Leslie and Gower proposed a prey-predator model under the assumption that there is a correlation between reduction in population of predator and its preferred food [10]. Even, another prey-predator model has been introduced by Leslie in which carrying capacity of predator population depends commensurately upon prey population [10, 11]. Later, May incorporated Holling type functional response in this model [15]. On account of the functional responses of types I, II and III, the model produced a wide range of dynamics and investigators explored global stability of equilibrium point, occurrence of chaos and periodicity in the system [2, 5, 7, 9]. Sokol and Howell [19] introduced a Holling type IV functional response in the model which fitted their experimental data significantly better. The functional

A. Singh (✉)
Graphic Era University, Dehrdaun, India
e-mail: anurajiitr@gmail.com

J.M. Cushing et al. (eds.), *Applied Analysis in Biological and Physical Sciences*,
Springer Proceedings in Mathematics & Statistics 186,
DOI 10.1007/978-81-322-3640-5_10

response IV is characterized by inflation in predation rate with prey population to utmost at a threshold prey density beyond which predation rate drops [6]. Whereas, in case of other functional responses the predation rate inflates with prey density. In another study [3], a model with functional response IV numerically demonstrated different dynamics at significant levels of prey interference in comparison to the other functional responses. Li and Xiao investigated Leslie–Gower model with functional response IV which manifested limit cycles and bifurcation [12]. Models incorporating time delays with Holling type IV functional response are extensively studied by Jiang and Lian [8, 13]. They explored complex dynamics in the system and demonstrated stability, periodic orbits and direction of bifurcating periodic orbits.

The employment of feedback control in complex systems generated significant interest after influential work by Ott [17]. The method is so simple and convenient that it appears to be remarkable for biological problem. Not much of research has been done in this area pertaining to ecological systems [14, 18].

By studying a Leslie–Gower prey-predator model with Holling's functional response of type IV, it is shown in this paper that an appropriately chosen control approach can render an unstable system into one that is globally stable.

2 The Mathematical Model

Let the density of prey and predator population be $X(t)$ and $Y(t)$ respectively. It is assumed that prey population is growing logistically with Holling's functional response of type IV

$$\begin{aligned} \frac{dX}{dT} &= rX\left(1-\frac{X}{K}\right)-\frac{mXY}{X^2+a} \\ \frac{dY}{dT} &= s\left(1-\frac{Y}{nX}\right)Y. \end{aligned} \tag{1}$$

where r is the intrinsic growth of prey species with carrying capacity K. m denotes per capita consumption rate of the predator and the constant a denotes the number of prey required to make maximum rate just half. s is the growth rate of the logistically growing population Y and n is magnitude of food quality of prey for reproduction in predator population. All the parameters are assumed to be taken positive. The following set of non-dimensional variables and parameters helps reduce the number of parameters from 6 to 3:

$$t = rT,\ x = X/K,\ y = mY/rK^2 \ \text{ and}$$

$$\alpha = a/K^2,\ \beta = mn/Kr,\ \gamma = s/r$$

This leads to non-dimensional form of the system

$$\frac{dx}{dt} = x(1-x) - \frac{xy}{x^2+\alpha}$$
$$\frac{dy}{dt} = \gamma\left(1 - \frac{y}{\beta x}\right)y. \tag{2}$$

The initial conditions for the system (2) are:

$$x \geq 0, y \geq 0. \tag{3}$$

From the biological point of view, the equilibrium point lying in the positive quadrant R^2 is of interest for the dynamics of the system. The interior equilibrium point $E^* = (x^*, y^*)$ can be obtained from the equations:

$$1 - x = \frac{y}{x^2+\alpha}, \quad y = \beta x.$$

It is seen that the system (2) exhibits Hopf bifurcation and admits a limit cycle under certain conditions (refer [12]).

The main objective of this paper is to show that a dynamic balance can be reached and the system (2) can be stabilized both locally and globally by suitably chosen feedback linearization design.

3 Feedback Linearization

Feedback linearization fully or partly transforms the primary nonlinear system into an equivalent linear system. This approach is entirely different from the traditional Jacobian approach.

3.1 Approximate Linearization

In the present section, an effort is made to stabilize the system (2) by employing approximate linearization.

Theorem 1 *The feedback control law u stabilizes the closed-loop system (2), where u is*

$$u = k_1(x - x^*) + k_2(y - y^*)$$

$$k_1 > \frac{(k_2 - \gamma)\left(1 - 2x^* - \frac{(\alpha - x^{*2})\beta x^*}{(x^{*2}+\alpha)^2}\right)}{\frac{x^*}{x^{*2}+\alpha}} - \beta\gamma$$

$$k_2 < \gamma + 2x^* + \frac{(\alpha - x^{*2})\beta x^*}{(x^{*2}+\alpha)^2} - 1.$$

Proof A linear control u is exerted on the system (2) as

$$\begin{aligned} \frac{dx}{dt} &= x(1-x) - \frac{xy}{x^2+\alpha} \\ \frac{dy}{dt} &= \gamma\left(1 - \frac{y}{\beta x}\right) y + u. \end{aligned} \tag{4}$$

Transformations $v = x - x^*$ and $w = y - y^*$ reduce the system (4) to

$$\begin{aligned} \dot{v} &= (v + x^*)\left(1 - (v + x^*)\right) - \frac{(v + x^*)(w + y^*)}{(v + x^*)^2 + \alpha} \\ \dot{w} &= (w + y^*)\left(\gamma - \frac{\gamma(w + y^*)}{\beta(v + x^*)}\right) + u. \end{aligned} \tag{5}$$

where x^* and y^* are equilibrium values. The linearized form of (5) can be written as

$$\dot{\mathbf{U}} = A\mathbf{U} + Bu \tag{6}$$

where

$$\mathbf{U} = \begin{pmatrix} v \\ w \end{pmatrix}, \quad A = \begin{pmatrix} 1 - 2x^* - \frac{(\alpha - x^{*2})\beta x^*}{(x^{*2}+\alpha)^2} & -\frac{x^*}{x^{*2}+\alpha} \\ \beta\gamma & -\gamma \end{pmatrix}, \quad B = \begin{pmatrix} 0 \\ 1 \end{pmatrix}.$$

In linear feedback, each control variable takes as a linear combination of state variables. In our case

$$u = K\mathbf{U}. \tag{7}$$

where row vector $K = \begin{pmatrix} k_1 & k_2 \end{pmatrix}$ represents a constant feedback. Using (7), (6) can be written as

$$\dot{\mathbf{U}} = (A + BK)\mathbf{U} = C\mathbf{U} \tag{8}$$

and

$$C = A + BK = \begin{pmatrix} 1 - 2x^* - \frac{(\alpha - x^{*2})\beta x^*}{(x^{*2}+\alpha)^2} & -\frac{x^*}{x^{*2}+\alpha} \\ \beta\gamma + k_1 & -\gamma + k_2 \end{pmatrix}.$$

The trace and determinant of matrix C are

$$Trace\ C = 1 - 2x^* - \frac{(\alpha - x^{*2})\beta x^*}{(x^{*2} + \alpha)^2} - \gamma + k_2$$
$$det\ C = \left(1 - 2x^* - \frac{(\alpha - x^{*2})\beta x^*}{(x^{*2} + \alpha)^2}\right)(-\gamma + k_2) + (\beta\gamma + k_1)\left(\frac{x^*}{x^{*2} + \alpha}\right).$$

It follows from the Routh–Hurwitz criterion that the controlled system (8) is stable iff

$$Trace\ C < 0,\ det\ C > 0.$$

Thus suitably chosen k_1 and k_2 such that

$$k_2 < \gamma + 2x^* + \frac{(\alpha - x^{*2})\beta x^*}{(x^{*2} + \alpha)^2} - 1 < 0 \tag{9}$$

and

$$k_1 > \frac{(k_2 - \gamma)\left(1 - 2x^* - \frac{(\alpha - x^{*2})\beta x^*}{(x^{*2}+\alpha)^2}\right)}{\frac{x^*}{x^{*2}+\alpha}} - \beta\gamma. \tag{10}$$

would make the system (8) stable.

3.2 Exact Linearization

In the previous section, a control law was obtained, using approximate linearization approach, locally. In this section, another approach known as exact linearization, is employed which makes the system locally as well as globally stable.

The nonlinear system is assumed to be of the form

$$\begin{aligned} \dot{\mathbf{X}} &= f(\mathbf{X}) + g(X)u', \quad \mathbf{X}(0) = \mathbf{X}_0 \\ \tilde{X} &= h(\mathbf{X}) \end{aligned} \tag{11}$$

where $\mathbf{X} \in R^n$ and $u' \in R^m$ are state vector and input vector respectively. $\tilde{X} \in R^m$is output vector having continuous derivatives where f, g are continuous vector on R^n with $f(0) = 0$.

The feedback control is employed as

$$u' = \alpha(\mathbf{X}) + \beta(\mathbf{X})v \tag{12}$$

where v is an external reference input. Further, a change of variable $z = \Phi(\mathbf{X})$ is introduced that transforms the nonlinear system into a linear controllable system.

The original system (2) with control term can be written as

$$\dot{\mathbf{X}} = \begin{pmatrix} x\,(1-x) - \frac{xy}{x^2+\alpha} \\ y\left(\gamma - \frac{\gamma y}{\beta x}\right) \end{pmatrix} + \begin{pmatrix} 0 \\ 1 \end{pmatrix} u^{'} \tag{13}$$

Here $u^{'}$ is an exerted control introducing an output $\tilde{X} = x - x^*$ stands for chasing of prey population. Theorem 2 is the main result of this section.

Theorem 2 *The closed loop system (13) is globally asymptotically stable provided the feedback control law is*

$$u^{'} = -x(x^2+\alpha) - y + \frac{x^2+\alpha}{x} - \frac{(\alpha - x^2)\beta}{(x^2+\alpha)}. \tag{14}$$

Proof Using the transformations be $\bar{x} = x - x^*$, $\bar{y} = y - y^*$, the system (13) can be rewritten as

$$\begin{aligned} \dot{\bar{\mathbf{X}}} &= f(\bar{\mathbf{X}}) + g(\bar{\mathbf{X}})u^{'} \\ \tilde{X} &= h(\bar{\mathbf{X}}) = \bar{x} \end{aligned} \tag{15}$$

where

$$f(\bar{\mathbf{X}}) = \begin{pmatrix} (\bar{x}+x^*)\,(1-(\bar{x}+x^*)) - \frac{(\bar{x}+x^*)(\bar{y}+y^*)}{(\bar{x}+x^*)^2 + \, alpha} \\ (\bar{y}+y^*)\left(\gamma - \frac{\gamma(\bar{y}+y^*)}{\beta(\bar{x}+x^*)}\right) \end{pmatrix}, \quad g(\bar{\mathbf{X}}) = \begin{pmatrix} 0 \\ 1 \end{pmatrix}$$

$$\text{and} \quad \bar{\mathbf{X}} = \begin{pmatrix} \bar{x} \\ \bar{y} \end{pmatrix}.$$

As $h(\bar{\mathbf{X}}) = \bar{x}$, then

$$L_f h(\bar{\mathbf{X}}) = \dot{\bar{x}} = x\,(1-x) - \frac{xy}{x^2+\alpha}, \tag{16}$$

obviously,

$$L_g L_f^{r-1} h(\bar{\mathbf{X}}) = L_g L_f^{2-1} h(\bar{\mathbf{X}}) = -\frac{x}{x^2+\alpha} \neq 0. \tag{17}$$

Here r is relative degree. For the particular system $r = 2$.

Letting

$$z = \Phi(\bar{\mathbf{X}}) = \begin{pmatrix} h(\bar{\mathbf{X}}) \\ L_f h(\bar{\mathbf{X}}) \end{pmatrix} = \begin{pmatrix} \bar{x} \\ \dot{\bar{x}} \end{pmatrix} \tag{18}$$

which denotes change of variables and $h(\bar{\mathbf{X}})$ and $L_f h(\bar{\mathbf{X}})$ being linearly independent, it is global diffeomorphism. Describing

$$\begin{aligned} \dot{z}_1 &= z_2 \\ \dot{z}_2 &= v \end{aligned} \tag{19}$$

in the new z-coordinate system, where v is a input relating to actual input u' by

$$v = L_f^2 h(\bar{\mathbf{X}}) + L_g L_f h(\bar{\mathbf{X}}) u', \quad L_g L_f h(\bar{\mathbf{X}}) \neq 0$$

The Brunovsky canonical form is written as (see [1])

$$u' = \frac{1}{L_g L_f h(\bar{\mathbf{X}})}(-z_1 - z_2 - L_f^2 h(\bar{\mathbf{X}})). \tag{20}$$

Accordingly, the system (2) converts into the linear system and the Brunovsky linear system is absolutely controllable [1]. Hence the system (13) is globally stable (ref. [18]).

From (20), the control law u' can be written as

$$u' = -x(x^2 + \alpha) - y + \frac{x^2 + \alpha}{x} - \frac{(\alpha - x^2)\beta}{(x^2 + \alpha)}. \tag{21}$$

Hence the proof.

4 Numerical Simulation

In the current section, numerical simulations are given to validate the analytic results for the stabilization of the system (2). Let us consider the following set of parameters [8]:

$$\alpha = 0.2, \beta = 1.2, \gamma = 0.02. \tag{22}$$

For this choice of parameters, system (2) shows limit cycle and corresponding oscillating time series, respectively for initial values $(0.1578, 0.18935)$. Analysis suggests that system (2) exhibits periodic orbits and thrashing time series (see Figs. 1 and 2).

By employing feedback control with approximate linearization (see Sect. 3.1), a control law $u = K\mathbf{U}$ is achieved with $K = \begin{pmatrix} 1 & 0.27 \end{pmatrix}$ through which system attains asymptotic stability. In Fig. 3, the time series converges to equilibrium point of the system (2).

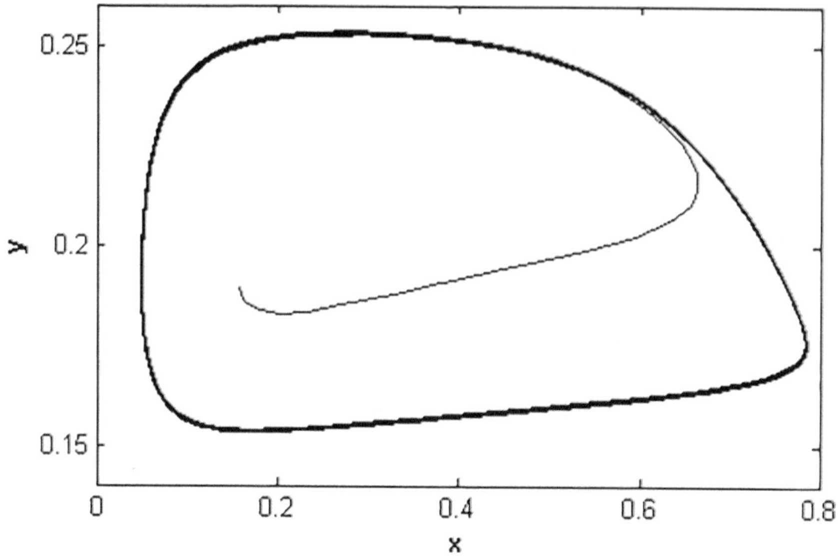

Fig. 1 Closed loop of original system (2)

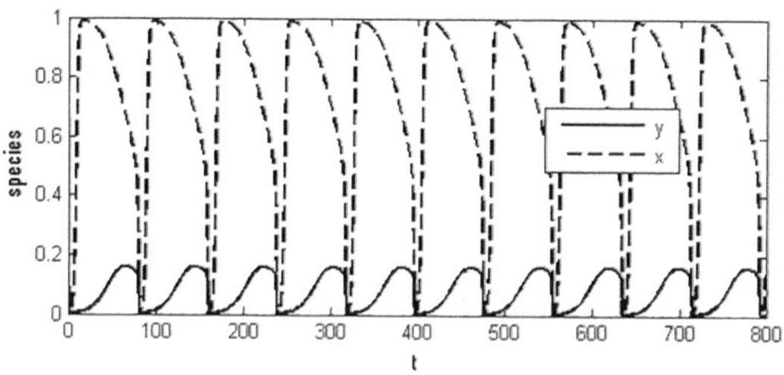

Fig. 2 Oscillating time series of original system (2)

In exact linearization (see Sect. 3.2), the control law (14) as given in Theorem 2, i.e.

$$u' = -x(x^2 + 0.2) - y + \frac{(x^2 + 0.2)}{x} + \frac{(0.2 - x^2)1.2}{(x^2 + 0.2)}$$

makes system (2) globally stable. Time series plotted in Fig. 4 shows that solution trajectory approaches to equilibrium point $E^*(0.0025, 0.2005)$.

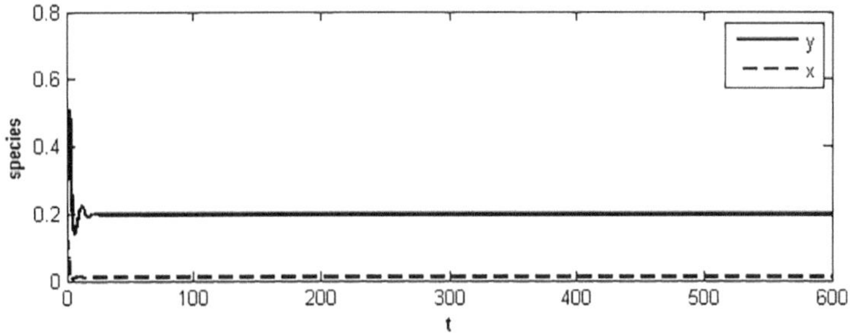

Fig. 3 Approximate linearization

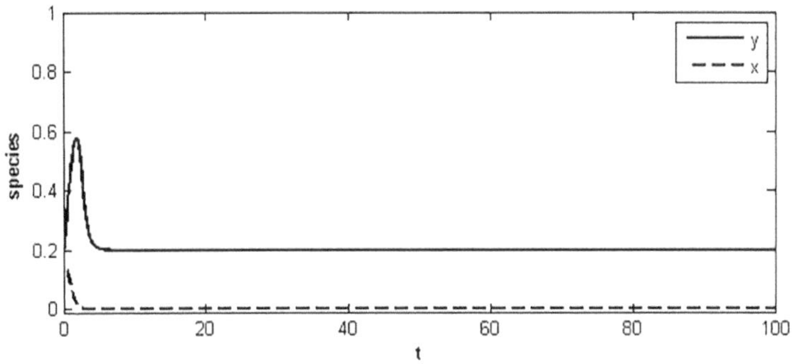

Fig. 4 Exact linearization

5 Conclusions

The Leslie–Gower prey-predator model with Holling type IV functional response shows periodic solutions and bifurcations [8, 12]. In this paper, it is shown that desired control laws can be determined using feedback (approximate and exact) linearizations that can bring the system into order such that its positive equilibrium solution becomes locally and globally stable.

Feedback linearization approaches are important methods to deal with nonlinear dynamical systems. Results of this study further show that control theory has a vital role to play even in biological sciences.

References

1. Andrievskii, B.R., Fradkov, A.L.: Control of chaos: methods and applications. I. Methods. Autom. Remote Control **64**, 673–713 (2003)
2. Braza, P.A.: The bifurcation structure of the Holling–Tanner model for predator-prey interactions using two-timing. SIAM J. App. Math. **63**, 889–904 (2003)

3. Collings, J.B.: The effects of the functional response on the bifurcation behavior of a mite predator-prey interactionmodel. J. Math. Biol. **36**, 149–168 (1997)
4. Freedman, H.I.: Deterministic Mathematical Models in Population Ecology. Dekker, New York (1980)
5. Gakkhar, S., Singh, A.: Complex dynamics in a prey predator system with multiple delays. Commun. Nonlinear Sci. Num. Simul. **17**, 914–929 (2012)
6. Holling, C.S.: Principles of insect predation. Ann. Rev. Entomol. **6**, 163–182 (1961)
7. Hsu, S.B., Huang, T.W.: Global stability for a class of predator-prey systems. SIAM J. App. Math. **55**, 763–783 (1995)
8. Jiang, J., Song, Y.: Stability and bifurcation analysis of a delayed Leslie–Gower predator-prey system with nonmonotonic functional response. Abstract App. Anal. **2013**, Article ID 152459, 19 p
9. Korobeinikov, A.: A Lyapunov function for Leslie–Gower predator-prey models. App. Math. Lett. **14**, 697–699 (2001)
10. Leslie, P.H., Gower, J.C.: The properties of a stochastic model for the predator-prey type of interaction between two species. Biometrica **47**, 219–234 (1960)
11. Leslie, P.H.: Some further notes on the use of matrices in population mathematics. Biometrica **35**, 213–245 (1948)
12. Li, Y., Xiao, D.: Bifurcations of a predator-prey system of Holling and Leslie types. Chaos Sol. Frac. **34**, 606–620 (2007)
13. Lian, F., Xu, Y.: Hopf bifurcation analysis of a predator-prey system with Holling type-IV functional response and time delay. Appl. Math. Comput. **215**, 1484–1495 (2009)
14. Liu, X., Zhang, Q., Zhao, L.: Stabilization of a kind of prey-predator model with holling functional response. J. Syst. Sci. Complex. **19**, 436–440 (2006)
15. May, R.M.: Stability and Complexity in Model Ecosystems. Princeton University Press, Princeton (2001)
16. Murray, J.D.: Mathematical Biology: I. An Introduction. Springer, New York (2002)
17. Ott, E., Grebogi, C., Yorke, J.A.: Controlling chaos. Phys. Rev. Lett. **64**, 1196–1199 (1990)
18. Singh, A., Gakkhar, S.: Stabilization of modifed Leslie–Gower prey-predator model. Differ. Equ. Dyn. Syst. **22**, 239–249 (2014)
19. Sokol, W., Howell, J.A.: Kinetics of phenol exidation by washed cells. Biotechnol. Bioeng. **23**, 2039–2049 (1980)

Graph Theoretic Concepts in the Study of Biological Networks

M. Indhumathy, S. Arumugam, Veeky Baths and Tarkeshwar Singh

Abstract The theory of complex networks has a wide range of applications in a variety of disciplines such as communications and power system engineering, the internet and worldwide web (www), food webs, human social networks, molecular biology, population biology and biological networks. The focus of this paper is on biological applications of the theory of graphs and networks. Graph theory and several graph theoretic properties serve as an ideal mathematical tool in the analysis of complex networks. We present the basic concepts and notations from graph theory which is widely used in the study of biological networks. Various biological networks such as Protein interaction networks, Metabolome based reaction network, Gene regulatory network, Gene coexpression network, Protein structure network, Structural brain network, Phylogenetic networks, Ecological networks and Food web networks are described. We also deal with various centrality measures which provide deep insight in the study of biological networks. Applications of biological network analysis in several areas are also discussed.

M. Indhumathy · S. Arumugam (✉)
National Centre for Advanced Research in Discrete Mathematics,
Kalasalingam University, Anand Nagar, Krishnankoil 626126, Tamil Nadu, India
e-mail: s.arumugam.klu@gmail.com

M. Indhumathy
e-mail: indhumathy.bio@gmail.com

S. Arumugam
Department of Computer Science, Liverpool Hope University, Liverpool, UK

S. Arumugam
Department of Computer Science, Ball State University, Muncie, IN, USA

V. Baths
Department of Biological Sciences, Birla Institute of Technology
and Science Pilani, K K Birla Goa Campus, NH-17B, Zuarinagar, Goa, India
e-mail: veeky@goa.bits-pilani.ac.in

T. Singh
Department of Mathematics, Birla Institute of Technology and Science Pilani,
K K Birla Goa Campus, NH-17B, Zuarinagar, Goa, India
e-mail: tksingh@goa.bits-pilani.ac.in

J.M. Cushing et al. (eds.), *Applied Analysis in Biological and Physical Sciences*,
Springer Proceedings in Mathematics & Statistics 186,
DOI 10.1007/978-81-322-3640-5_11

Keywords Biological networks · Centrality measures · Graph · Motifs

1 Introduction

The theory of complex networks has a wide range of applications in a variety of disciplines such as communications and power system engineering, the internet and worldwide web (www), food webs, human social networks, molecular biology, population biology and biological networks. The focus of this paper is on biological applications of the theory of graphs and networks. Network analysis leads to a better understanding of the critical role of these networks in many key questions. For instance the complex interplay between the structure of social networks and the spread of disease is a topic of critical importance and hence in recent years the researches in ecology and epidemiology have focused attention in network analysis.

Graph theory and several graph theoretic properties serve as an ideal mathematical tool in the analysis of complex networks. In Sect. 2, we present the basic concepts and notations from graph theory which is widely used in the study of biological networks. Various biological networks such as Protein interaction networks, Metabolome based reaction network, Gene regulatory network, Gene coexpression network, Protein structure network, Structural brain network, Phylogenetic networks, Ecological networks and Food web networks are described in Sect. 3. Section 4 deals with the various centrality measures which provide deep insight in the study of biological networks. Topology of biological networks and network motifs have been stated in Sects. 5 and 6 respectively. Sections 7 and 8 deal with network databases and network visualizing tools. Applications of biological network analysis in several areas and concluding remarks are given in Sects. 9 and 10 respectively.

2 Graph Theoretic Concepts

In this section we present a few basic concepts in graph theory which are essential for the study of biological networks. For graph theoretic terminology we refer to Chartrand and Lesniak [8].

A graph G is a finite nonempty set of objects called *vertices* or *nodes* together with a set of unordered pairs of distinct vertices of G called *edges* or *links*. The vertex set and the edge set of G are denoted by $V(G)$ and $E(G)$ respectively. The edge $e = \{u, v\}$ is said to join the vertices u and v. We write $e = uv$ and say that u and v are adjacent vertices; u and e are incident, as are v and e. If e_1 and e_2 are distinct edges of G incident with a common vertex, then e_1 and e_2 are adjacent edges.

The number of vertices in G is called the *order* of G and the number of edges in G is called the *size* of G. A graph of order n and size m is called a (n, m)-*graph*. A graph is *trivial* if its vertex set is a singleton. A graph $G = (V, E)$ is called a weighted graph if for every edge e of G a weight $w(e)$ is assigned. The weight is usually a positive number. The graph G is called a signed graph if every edge e of G is assigned a positive or a negative sign.

A vertex u is called a *neighbor* of a vertex v in G, if uv is an edge of G. The set of all neighbors of v is the *open neighborhood* of v and is denoted by $N(v)$; the set $N[v] = N(v) \cup \{v\}$ is the *closed neighborhood* of v in G.

A graph H is called a *subgraph* of G if $V(H) \subseteq V(G)$ and $E(H) \subseteq E(G)$. A subgraph H of a graph G is a *proper subgraph* of G if either $V(H) \neq V(G)$ or $E(H) \neq E(G)$. A *spanning subgraph* of G is a subgraph H of G with $V(H) = V(G)$.

For a set S of vertices of G, the *induced subgraph* is the maximal subgraph of G with vertex set S and is denoted by $\langle S \rangle$. Thus two vertices of S are adjacent in $\langle S \rangle$ if and only if they are adjacent in G. The induced subgraph $\langle S \rangle$ is also denoted by $G[S]$.

Let v be a vertex of a graph G and $|V(G)| \geq 2$. Then the induced subgraph $\langle V(G) \backslash \{v\} \rangle$ is denoted by $G - v$ and it is the subgraph of G obtained by the removal of v and the edges incident with v. If $e \in E(G)$, the spanning subgraph with edge set $E(G) \backslash \{e\}$ is denoted by $G - e$ and it is the subgraph of G obtained by the removal of the edge e. The graph obtained from G by adding an edge e is denoted by $G + e$.

The *degree* of a vertex v in a graph G is defined to be the number of edges incident with v and is denoted by $\deg(v)$.

The minimum of $\{\deg(v) : v \in V(G)\}$ is denoted by $\delta(G)$ or simply δ and the maximum of $\{\deg(v) : v \in V(G)\}$ is denoted by $\Delta(G)$ or simply Δ.

A graph G is *complete* if every pair of distinct vertices of G are adjacent in G. A complete graph on n vertices is denoted by K_n.

A *clique* in G is a complete subgraph of G. The maximum order of a clique in G is called the *clique number* of G and is denoted by $\omega(G)$ or simply ω. A clique H in G with $|V(H)| = \omega$ is called a *maximum clique* in G.

A *bipartite* graph is a graph G whose vertex set $V(G)$ can be partitioned into two nonempty subsets X and Y such that each edge of G has one end in X and the other end in Y. The pair (X, Y) is called a *bipartition* of G. If further, every vertex in X is adjacent to all the vertices of Y, then G is called a *complete bipartite graph*. The complete bipartite graph with bipartition (X, Y) such that $|X| = r$ and $|Y| = s$ is denoted by $K_{r,s}$. In particular, the graph $K_{1,n-1}$ is called a *star* and the graph $K_{1,3}$ is called a *claw*.

A *walk* in a graph G is an alternating sequence $u_0, e_1, u_1, \ldots, u_{n-1}, e_n, u_n$ of vertices and edges of G, beginning and ending with vertices such that $e_i = u_{i-1}u_i$, for $1 \leq i \leq n$. This walk joins u_0 and u_n and may also be denoted $(u_0, u_1, u_2, \ldots, u_{n-1}, u_n)$; it is sometimes called a u_0 - u_n walk. It is *closed* if $u_0 = u_n$ and is *open* otherwise.

A *path* P *of length* n in a graph G is a sequence $(u_0, u_1, u_2, \ldots, u_{n-1}, u_n)$ of distinct vertices such that for $0 \leq i \leq n-1$, the vertices u_i and u_{i+1} are adjacent. We say that P is an u_0–u_n *path*. The vertices u_0 and u_n are called the *origin* and *terminus* of P respectively. The vertices $u_1, u_2, \ldots, u_{n-1}$ are called the *internal vertices* of P. A path on n vertices is denoted by P_n.

A *cycle of length* $n \geq 3$ in a graph G is a sequence $(u_0, u_1, u_2, \ldots, u_{n-1}, u_0)$ of vertices of G such that for $0 \leq i \leq n-2$, the vertices u_i and u_{i+1} are adjacent, u_{n-1} and u_0 are adjacent and $u_0, u_1, u_2, \ldots, u_{n-1}$ are distinct. A cycle on n vertices is denoted by C_n. A cycle C_n of length n is called *even* or *odd* according as n is even or odd.

A graph G is said to be *connected* if every pair of vertices of G are joined by a path. A maximal connected subgraph of G is called a *component* of G.

A graph G having more than one component is called a *disconnected* graph. An edge e of a connected graph G is called a *cut-edge* if $G - e$ is disconnected. A vertex v of a connected graph G is called a *cut-vertex* if $G - v$ is disconnected.

The *distance* $d(u, v)$ between two vertices u and v of a connected graph G is defined to be the length of any shortest path joining u and v. A shortest u–v path is often called a *geodesic*.

The *diameter* of a connected graph G is the length of any longest geodesic and is denoted by diam(G). We call an u - v path in G for which $d(u, v) = \text{diam}(G)$ as a *diametrical path.*

3 Biological Networks

In this section we present some of the popular biological networks which have been investigated by several authors.

Protein-Protein Interaction network (PPI-Network) is a graph $G = (V, E)$ where V is a set of proteins and two proteins are joined by an edge if they interact physically. The interaction between viral proteins and human proteins can be represented as a bipartite graph G. The vertex set of G is $V_1 \cup V_2$, where V_1 is the set of viral proteins and V_2 is the set of all human proteins. A viral protein $v \in V_1$ is joined to a human protein $w \in V_2$ if v interacts with w. This bipartite graph is called viral-human protein interaction network and this network has been investigated by Mukhopadhyay and Maulik [26].

Human protein and disease association network is a bipartite graph G whose vertex is $V_1 \cup V_2$, where V_1 is the set of human proteins and V_2 is the set of diseases and $v_1 \in V_1$ is joined by an edge to $v_2 \in V_2$, if the human protein v_1 is associated with the disease v_2. This network has been investigated by Mukhopadhyay and Maulik [26].

Metabolome based reaction network is a directed graph $D = (V, A)$ where V is a set of metabolites and a vertex v is joined to a vertex w by an arc (v, w) if there is a reaction or interaction which transforms the metabolite v to the metabolite w. This network has been investigated by Veeky Baths et al. [5].

Gene regulation is a general term for cellular control of the synthesis of protein at the transcription step. Often one gene is regulated by another gene via the corresponding protein. Thus gene regulation leads to the concept of gene regulatory network, which has been investigated by Yue and Chunmei [36]. Gene regulatory network is a directed graph $D = (V, A)$ where V is the set of genes and two genes $g_1, g_2 \in V$ are joined by an arc if there is a regulatory relationship between g_1 and

g_2, or more precisely g_1 regulates g_2. The regulatory relationship between two genes may be either positive direct regulatory influence or inverse causality or no correlation. Hence gene regulatory network can also be represented as a directed weighted graph, where the weight of an arc is an estimate of the probability of relationship between the genes in the network. This nctwork has been investigated by Raza and Jaiswal [29]. Positive regulatory relationship represents activation and negative regulatory relationship represents inhibition. This leads to the representation of a gene regulatory network as a signed directed graph where an arc (g_1, g_2) is assigned a positive sign if the corresponding regulatory relationship is activation and is assigned a negative sign if the corresponding relationship is inhibition. A study of gene regulatory network leads to a better understanding of the regularity mechanism of the genes and prediction of the behavior of some unknown genes. This network has been studied in Christensen et al. [9].

A gene coexpression network is a graph $G = (V, E)$ where V is a set of genes and two genes are connected by an edge if there is a significant coexpression relationship between them. There are several methods for constructing the gene coexpression network and this network has been studied in Perkins et al. [28]. A coexpression measure is selected and a similarity score is calculated for each pair of genes using this measure. Two genes which have a similarity score higher than a selected threshold value are joined by an edge (Azuaje [2]).

To understand the protein structures, a graph representation of protein structure has recently been introduced. Protein structure is modeled as residue interaction graph (RIG) in which nodes represent the amino acid residuals and an edge represents a pairwise contact between residuals. A contact between two residuals is defined if the distance between any pair of their heavy atoms is within a specified distance cut-off and the cut-off is normally taken in the range (4,5). Understanding RIGs may provide deeper insights into protein structures binding and folding mechanisms as well as inter protein stability and function. Properties of this network are given in [24].

Structural brain network can be represented as a graph whose vertex set is the set of neural elements (neurons or brain regions) and edges representing physical connections (Synapses or axonal projections). This is used to understand the complex structure of the brain and brain associated diseases such as Alzheimers disease, brain tumor and epilepsy. This network has been studied in [7].

Consider an ecological community V consisting of predators and preys where a predator eats a prey. The food web network is a directed graph D whose vertex set is V and if $u, v \in V$, then (u, v) is an arc in D if u is a predator, v is a prey and u eats v. The competition graph is a graph whose vertex set is the set of predators and there is an edge between two predators if they have a common prey. If further we associate a weight with edge in the competition graph, where the weight is the number of common preys then we obtain a weighted competition graph. These concepts have been investigated by Dunne et al. [10].

A Phylogenetic network is a graph which is used to visualize evolutionary relationships between nucleotide sequences, genes, chromosomes or genomes. They are used when reticulated events such as hybridization, horizontal gene transfer, recombination or gene duplication and loss are believed to be involved. For further details we refer to [17].

An Ecological network is a representation of the biotic interactions in an ecosystem. This is an undirected graph whose vertex set is a set of species and two species are joined by an edge if they interact. These interactions can be trophic or symbiotic. Applications of ecological networks include exploration of how the community context affects pairwise interactions. Other related studies include Metapopulations, epidemiology and evolution of cooperation. For some basic results in Ecological network we refer to Sole and Montoya [32].

4 Centrality Measures

Graph theoretic concepts are being extensively used in the analysis of biological networks. In this section we present several centrality measures which are used to rank the nodes of a network in the order of their performance.

4.1 Stress

The stress is a node centrality index, which has been studied by Shannon et al. [31]. The stress of a node v is the number of shortest paths passing through v. A node with a high stress is traversed by a high number of shortest paths. However, a node has high stress value does not imply that it is critical to maintain communication. Indeed two nodes may be connected by other shortest paths not passing through v. Hence "high" and "low" stress are more meaningful when stress of a node v is compared with the average stress value of the graph G. In biological terms, the stress of a node indicates its relevance in holding together communicating nodes. Hence, if the stress is higher, then the relevance of the node connecting other node is higher. Due to the nature of this centrality it may also be possible that the stress indicates how a molecule is heavily involved in cellular processes but not necessarily in maintaining the communication between the other nodes.

4.2 Betweenness

Betweenness is another node centrality index which is similar to stress, but provides more information. Let v_1, v_2 and v be three distinct nodes. Let $\sigma_{v_1v_2}(v)$ denote the number of shortest v_1–v_2 paths which pass through v. Let $\sigma_{v_1v_2}$ denote the total number of shortest v_1–v_2 paths. The betweenness centrality index of v is defined by

$$C_B(v) = \sum_{\substack{v_1, v_2 \in V \\ v_1 \neq v_2}} \left(\frac{\sigma_{v_1 v_2}(v)}{\sigma_{v_1 v_2}} \right).$$

Since for computing the betweenness centrality index, the summation is taken overall pairs of nodes, we divide it by $\frac{(n-1)(n-2)}{2}$ for graphs, where n is the total number of nodes; so that the betweenness index of v lies in the range $[0, 1]$. A high betweenness index indicates that the node for certain paths is crucial to maintain node connections (Scardoni and Laudana [30]). The betweenness index of a node in a protein-signaling network indicates the relevance of a protein as functionally capable of holding together communicating proteins. The higher the value the higher the relevance of the protein as organizing regularity molecule.

4.3 Edge Betweenness

Edge betweenness centrality is the edge version of the node betweenness centrality. Let v_1, v_2 be two distinct nodes and let e be an edge. Let $\sigma_{v_1 v_2}(e)$ denote the number of shortest v_1–v_2 paths which pass through the edge e. Let $\sigma_{v_1 v_2}$ denote the total number of shortest v_1–v_2 paths. Then the edge betweenness centrality index of e is defined by

$$C_B(e) = \sum_{\substack{v_1, v_2 \in V \\ v_1 \neq v_2}} \left(\frac{\sigma_{v_1 v_2}(e)}{\sigma_{v_1 v_2}} \right).$$

The edge betweenness index is normalized as in the case of node betweenness index. In the context of a protein-signaling network, edge betweenness centrality indicates that a specific biochemical reaction has a central role in the network functional organizations.

4.4 Diameter

The diameter of a graph G is defined by $diam(G) = max\ \{d(u, v) : u, v \in V\}$, where $d(u, v)$ is the distance between the vertices u and v. It is a simple general parameter which indicates the compactness of the network. If G has high diameter, then G has two vertices whose distance is high. However, a graph with high diameter may have subgraphs which are compact. If a graph has low diameter, then it surely indicates that all the nodes are close to each other and the graph is compact.

For example, the diameter of a protein-signaling network can be interpreted as the overall easiness of the proteins to communicate or influence their reciprocal function.

4.5 Average Distance

The average distance of a graph G is defined by

$$Ad(G) = \frac{\left(\sum_{\substack{u,v\in V\\ u\neq v}} d(u,v)\right)}{\frac{n(n-1)}{2}}$$

where n in the number of nodes in G.

In general $Ad(G)$ is not an integer. In most cases, $Ad(G)$ is more informative than the diameter. High average distance indicates that the nodes are distant, implying that the network is not compact. A low average distance indicates that the nodes are close to each other and the network is compact. For example if a big protein signaling network has low average distance, then the proteins within the network have the tendency to generate functional complexes or modules.

4.6 Closeness

Closeness is another node centrality index. The closeness centrality of a node v is defined by

$$CC(v) = \frac{n-1}{\sum\limits_{u\in V-\{v\}} d(u,v)},$$

where n is the number of nodes in the network. Here high and low values are more meaningful when compared to the average closeness of G. High value of closeness of v indicates that all the nodes are in proximity to v and low value of closeness of v indicates that all other nodes are distant from v.

For example, the closeness of a node in a protein-signaling network can be interpreted as the probability of a protein to be functionally relevant for several other proteins. Thus a protein with high closeness will be easily central to the regularity of other proteins. This concept has been used in [1] for the study of protein structures.

4.7 Eigenvector Centrality

The adjacency matrix A of a graph G with vertex set $V(G) = \{v_1, v_2, \ldots\ldots, v_n\}$ is the $n \times n$ matrix defined by

$$a_{ij} = \begin{cases} 1 & \text{if } v_i \text{ and } v_j \text{ are adjacent} \\ 0 & \text{otherwise.} \end{cases}$$

A number λ is called an eigen value of A, if there exists a vector e such that $Ae = \lambda e$ and e is called the eigen vector corresponding to the eigen value λ. Since the matrix A is symmetric, all its eigen values are real. Let e_1 be the eigen vector corresponding to the largest eigen value λ_1 of A. Then the ith component of the vector e_1 is the eigen vector centrality of the node v_i. In biological terms, a node with high eigen vector centrality value is adjacent to the other nodes that themselves have high eigenvector centrality value. This concept is given in [6].

4.8 *Eccentricity*

The eccentricity of a node v is defined by $e(v) = \max\{d(u, v) : u \in V - \{v\}\}$. Thus eccentricity of v is the distance between v and a node which is farthest from v. Let $f(v) = \frac{1}{e(v)}$, the reciprocal of the eccentricity. If $f(v)$ has a high value, then all other nodes are in proximity with v. On the other hand, if $f(v)$ is low, then there is at least one node which is far from v. For example in a protein-signaling network a protein with high value of f will be more easily influenced by activity of other proteins.

4.9 *Subgraph Centrality*

Subgraph centrality was introduced by Estrada in 2005. This centrality measure helps in finding hidden subgraph within a network. Here, smaller subgraph has given more weightage than the larger one as smaller subgraph can reveal the network motifs [13].

5 Topology of Biological Network

5.1 *Watts and Strogatz Small World Network*

Small world network was proposed by Watts and Strogatz [34]. This network is a random network with high clustering coefficient value indicating that most pairs of nodes contain at least one shortest path of small length between them. Therefore, in this type of network, mean length of shortest path is always small and in a given time, it is possible to reach from one node to another with few steps. Internet connectivity, social network, gene network are examples of small world network. This type of topology is only applied to the network where single nodes have few neighbors. If links between nodes are grown in huge number, then this theory fails as there may not be a shortest path of small length between two distant nodes.

5.2 Erdos Renyi Random Network

Paul Erdos and Alfred Renyi proposed this random graph network for non regular complex networks. Here edges are randomly added between pairs of randomly selected nodes with an initial condition of N nodes without any edges between them. This network follows poissonian distribution assuming added edges $<< N^2$ [11]. Real world network like internet, social network, biological network etc. do not follow this network.

5.3 Barabasi–Albert Scalefree Network

Scalefree network was proposed by Barabasi–Albert [3]. This model opposes the idea that all complex networks are random in nature. According to this network, there are some special kind of mechanisms which shape this randomness of complex network. In scalefree network, structure and evolution are closely related and it is constantly changing by addition of new node or link to the existing network. When a new node comes in the existing network, it will tend to link with the node having maximum number of connections in a given network. This type of attachment is known as preferential attachment, and the node with maximum number of connections is known as hub. Here, degrees are distributed following the power law distribution resulting a few nodes with maximum links(hubs) while many nodes with a very few links.

6 Network Motifs

In biological network, it has been observed that a particular group of nodes with a fixed structural pattern, are involved in specific functions and these are known as motifs. Motifs are often called simple building blocks of complex network [25]. In graphs, motifs are basically repeating units of small subgraphs within a single network or among many networks. As motifs are involved in the particular functions, it is possible to predict the function of unknown proteins by comparing with the known motifs and then with its function. There are many motifs finder algorithm like Mavisto, FANMOD used to identify motifs within a network.

7 Network Databases

To date, vast amount of biological data has been created with the help of high throughput techniques like yeast two hybrid screening systems, DNA microarray and next generation sequencing. These data can be accessed through databases. There are many such online databases housing these data and are freely accessible. Following are some important databases of biological network.

- **KEGG**. Kyoto Encyclopedia of Genes and Genomes [21].
- **STRING**. Search Tool for the Retrieval of Interacting Genes/Proteins [18].
- **HPRD**. Human Protein Reference Database [23].
- **MINT**. Molecular Interaction Database [37].
- **DIP**. Database of Interacting Proteins [35].
- **Reactome**. It is database for reaction pathways and biological processes. The pathways represented here are species specific [20].
- **BioGRID**. Biological General Repository for Interaction Datasets [33].
- **SPIKE**. Signaling Pathway Integrated Knowledge Engine [27].
- **IntAct**. InAct Molecular interaction database [22].

8 Network Visualizing and Analyzing Tools

There are many open source tools routinely used for network visualization and also for calculating different centrality values along with many network parameters like diameter, degree, shortest pathlength, clustering coefficient etc. Following three tools are widely used in biological network visualizing and analysis.

- **Cytoscape** [31]
- **Pajek** [4]
- **Visant** [16]

9 Applications

Applications of graph theory in the fields of biology and medicine include identification of drug targets, determination of the role of proteins or genes of unknown function [12, 14], design of effective containment strategies for infectious diseases and early diagnosis of neurological disorders by detecting abnormal patterns of neural synchronization. The knowledge of the topologies of biological networks and their impact on biological processes is needed to develop more sophisticated treatment strategies for complex diseases such as cancer [19]. Protein-Protein interaction networks have recently been combined with the networks describing the relationships between the diseases and disease gene causing them as well as between drugs and their protein targets, thus giving insights into pharmacology. Another application is the study of genetic disorders. A diseasome is represented as a bipartite graph, whose vertex set is the set of genetic disorders and disease genes. A genetic disorder X is joined to the disease gene Y if there is a mutation in the gene Y which gives genetic disorder X [15]. Two other graphs in this connection are constructed as follows. The human disorder network has its vertex set the set of genetic disorders and two disorders are joined by an edge if they are both caused by at least one common gene. The disease gene network has its vertex set the set of all disease genes and two genes are joined by a link if they are associated with at least one common disorder. These networks are used to examine and understand human disease gene and phenotype associations. Measures of centrality are used to identify structurally important genes or proteins in interaction networks. Network motifs help in finding structural pattern along with unknown protein functions.

10 Conclusion

In this survey article a detailed account of various biological networks, basic graph theoretic concepts, various centrality measures which play a crucial role in the analysis of biological networks, biological databases which are essential for the construction of biological networks and software tools required for the network visualizing and analyzing have been presented. Analysis of the biological networks using graph theoretic tools lead to the identification of influential proteins or genes, which can be confirmed experimentally.

References

1. Amitai, G., Shemesh, A., Sitbon, E., Shklar, M., Netanely, D., Venger, I., Pietrokovski, S.: Network analysis of protein structures identifies functional residues. J. Mol. Biol. **344**(4), 1135–1146 (2004)
2. Azuaje, F.J.: Selecting biologically informative genes in coexpression networks with a centrality score. Biol. Direct **9**(12), 1–23 (2014)
3. Barabási, A.L., Albert, R.: Emergence of scaling in random networks. Science **286**(5439), 509–512 (1999)
4. Batagelj, V., Mrvar, A.: Pajek - program for large network analysis. Connections **21**, 47–57 (1998)
5. Baths, V., Roy, U., Singh, T.: Disruption of cell wall fatty acid biosynthesis in Mycobacterium tuberculosis using a graph theoretic approach. Theor. Biol. Med. Model. **8**(5), 1–13 (2011)
6. Bonacich, P.: Factoring and weighting approaches to status scores and clique identification. J. Math. Sociol. **2**, 113–120 (1972)
7. Bullmore, E.D., Sporns, O.: Complex brain networks graph theoretical analysis of structural and functional systems. Nature Rev. Neurosci. **10**, 186–198 (2009)
8. Chartrand, G., Lesniak, L.: Graphs & Digraphs, 4th edn. Chapman and Hall, CRC, Boca Raton (2005)
9. Christensen, C., Gupta, A., Maranas, C.D., Albert, R.: Large scale inference and graph theoretical analysis of gene-regulatory networks in B. Subtilis. Physica A **373**, 796–810 (2007)
10. Dunne, J.A., Williams, R.J., Martinez, N.D.: Food-web structure and network theory: the role of connectance and size. PNAS **99**(20), 12917–12922 (2002)
11. Erdös, P., Rényi, A.: On the strength of connectedness of a random graph. Acta Mathematica Academiae Scientiarum Hungarica **12**, 261–267 (1964)
12. Estrada, E.: Virtual identification of essential protein within the protein interaction network of yeast. Proteomics **6**(1), 35–40 (2006)

13. Estrada, E., Rodríguez-Velázquez, J.A..: Subgraph centrality in complex networks. Phys. Rev. **71** (2005)
14. Fuller, T.F., Ghazalpour, A., Aten, J.E., Drake, T.A., Lusis, A.J., Horrath, S.: Weighted gene coexpression network analysis strategies applied to mouse weight. Mamm Genome **18**(6–7), 463–472 (2007)
15. Goh, K.I., Cusick, M.E., Valle, D., Childs, B., Vidal, M., Barabási, A.: The human disease network. Proc. National Acad. Sci. **104**(21), 8685–8690 (2007)
16. Hu, Z., Mellor, J., Wu, J., Yamada, T., Holloway, D., DeLisi, C.: VisANT: data-integrating visual framework for biological networks and modules. Nucleic Acids Res. **33**, 352–357 (2005)
17. Huson, D.H., Rupp, R., Scornavacca, C.: Phylogenetic Networks. Cambridge University Press, Cambridge (2010)
18. Jensen, L.J., Kuhn, M., Stark, M., Chaffron, S., Creevey, C., Muller, J., Doerks, T., Julien, P., Roth, A., Simonovic, M., Bork, P., von Mering, C.: STRING 8-a global view on proteins and their functional interactions in 630 organisms. Nucleic Acids Res. **37**, 412–416 (2009)
19. Jiang, W., Li, X., Rao, S., Wang, L., Du, L., Li, C., Wu, C., Wang, H., Wang, Y., Yang, B.: Constructing disease-specific gene networks using pair-wise relevance metric: application to colon cancer identifies interleukin 8, desmin and enolase 1 as the central elements. BMC Syst. Biol. **2**(1), 72 (2008)
20. Joshi-Tope, G., Gillespie, M., Vastrik, I., D'Eustachio, P., Schmidt, E., de Bono, B., Jassal, B., Gopinath, G.R., Wu, G.R., Matthews, L., Lewis, S., Birney, E., Stein, L.: Reactome: a knowledgebase of biological pathways. Nucleic Acids Res. **33**, 428–432 (2005)
21. Kanehisa, M., Goto, S., Furumichi, M., Tanabe, M., Hirakawa, M.: KEGG for representation and analysis of molecular networks involving diseases and drugs. Nucleic Acids Res. **38**, 355–360 (2010)
22. Kerrien, S., Alam-Faruque, Y., Aranda, B., Bancarz, I., Bridge, A., Derow, C., Dimmer, E., Feuermann, M., Friedrichsen, A., Huntley, R., Kohler, C., Khadake, J., Leroy, C., Liban, A., Lieftink, C., Montecchi-Palazzi, L., Orchard, S., Risse, J., Robbe, K., Roechert, B., Thorneycroft, D., Zhang, Y., Apweiler, R., Hermjakob, H.: IntAct-open source resource for molecular interaction data. Nucleic Acids Res. **35**, 561–565 (2007)
23. Keshava Prasad, T.S., Goel, R., Kandasamy, K., Keerthikumar, S., Kumar, S., Mathivanan, S., Telikicherla, D., Raju, R., Shafreen, B., Venugopal, A., Balakrishnan, L., Marimuthu, A., Banerjee, S., Somanathan, D.S., Sebastian, A., Rani, S., Ray, S., Harrys Kishore, C.J., Kanth, S., Ahmed, M., Kashyap, M.K., Mohmood, R., Ramachandra, Y.L., Krishna, V., Abdul Rahiman, B., Mohan, S., Ranganathan, P., Ramabadran, S., Chaerkady, R., Pandey, A.: Human protein reference database. Nucleic Acids Res. **37**, 767–772 (2009)
24. Memisevic, V., Milenkovic, T., Przulj, N.: An integrative approach to modeling biological networks. J. Integr. Bioinform. **7**(3), 1–22 (2010)
25. Milo, R., Shen-Orr, S., Itzkovitz, S., Kashtan, N., Chklovskii, D., Alon, U.: Network motifs: simple building blocks of complex networks. Science **298**(5594), 824–827 (2002)
26. Mukhopadhyay, A., Maulik, U.: Network-based study reveals potential infection pathways of Hepatitis-C leading to various diseases. PLOS One **9**(4), 1–12 (2014)
27. Paz, A., Brownstein, Z., Ber, Y., Bialik, S., David, E., Sagir, D., Ulitsky, I., Elkon, R., Kimchi, A., Avraham, K.B., Shiloh, Y., Shamir, R.: SPIKE: a database of highly curated human signaling pathways. Nucleic Acids Res. **39**, 793–799 (2011)
28. Perkins, A.D., Langston, M.A.: Threshold selection in gene coexpression networks using spectral graph theory techniques. BMC Bioinform. **10**(54), 1–11 (2008)
29. Raza, K., Jaiswal, R.: Reconstruction and analysis of cancer-specific gene regulatory networks from gene expression profiles. Int. J. Bioinform. Biosci. **3**(2), 25–34 (2013)
30. Scardoni, G., Laudana, C.: Centralities based analysis of complex networks. In: Zhang, Y. (ed.) New Frontiers in Graph Theory, InTech (2012)
31. Shannon, P., Markiel, A., Ozier, O., Baliga, N.S., Wang, J.T., Ramage, D., Amin, N., Schwikowski, B., Idekar, T.: Cytoscape: a software environment for integrated models of biomolecular interaction networks. Genome Res. **13**(11), 2498–2504 (2003)

32. Sole, R.V., Montoya, J.M.: Complexity and fragility in ecological networks. Proc. R. Soc. Lond. B **268**, 2039–2045 (2001)
33. Stark, C., Breitkreutz, B.J., Reguly, T., Boucher, L., Breitkreutz, A., Tyers, M.: BioGRID: a general repository for interaction datasets. Nucleic Acids Res. **34**, 535–539 (2006)
34. Watts, D.J., Strogatz, S.H.: Collective dynamics of 'small-world' networks. Nature **393**, 440–442 (1998)
35. Xenarios, I., Rice, D.W., Salwinski, L., Baron, M.K., Marcotte, E.M., Eisenberg, D.: DIP: the database of interacting proteins. Nucleic Acids Res. **28**(1), 289–291 (2000)
36. Yue, H., Chunmei, L.: Study of Gene regulatory network based on graph. In: 4th International Conference on Biomedical Engineering and Informatics, pp. 2236–2240. IEEE (2011)
37. Zanzoni, A., Montecchi-Palazzi, L., Quondam, M., Ausiello, G., Helmer-Citterich, M., Cesareni, G.: MINT: a Molecular INTeraction database. FEBS Lett. **513**(1), 135–140 (2002)

Some Algebraic Aspects and Evolution of Genetic Code

Tazid Ali and Nisha Gohain

Abstract The genetic code is the set of rules by which DNA stores genetic information of organisms. In this paper we discuss an algebraic structure of the genetic code in terms of the four DNA bases (A, C, G, T). Some relations between transition/transversion mutation of codons and algebraic properties of respective codons of the group structure are obtained. A distance matrix of the amino acids is constructed. We establish some relations between the distance matrix and physico-chemical properties of amino acids. Further we argue that the distance matrix reflects evolutionary pattern of amino acids.

Keywords DNA · Genetic code · Mutation · Algebra · Distance matrix

1 Introduction

The genetic code gives the information about the formation of protein molecule. It is the set of rules by which information encoded in genetic material (DNA or RNA sequences) is translated into proteins. A linear chain of amino acids forms the protein molecule. There are 20 different amino acids found till now that occur in proteins. Amino acids are synthesized by RNAs and RNAs are obtained from DNAs. Each amino acid is a triplet code (codon) of four possible bases (A, C, G, T) of DNA. The base Thymine (T) is replaced by Uracil (U) in RNA.

The transmission of information from DNA to protein goes through transcription and translation. Due to mutation, the sequence of bases is not copied precisely in replicating the strand of DNA. This affects protein formation. Codons are muted in different ways such as deletion, insertion, inversion, point and frame-shift mutation. In this paper only the case of point mutation is considered. The point mutation is a

T. Ali (✉) · N. Gohain (✉)
Department of Mathematics, Dibrugarh University, Dibrugarh 786004, India
e-mail: tazidali@yahoo.com

N. Gohain
e-mail: gohainnisha@gmail.com

J.M. Cushing et al. (eds.), *Applied Analysis in Biological and Physical Sciences*, Springer Proceedings in Mathematics & Statistics 186,
DOI 10.1007/978-81-322-3640-5_12

simple change in one base of the gene sequence. Point mutation may be at a single point or at more than one point. Transition is the point mutation either from purine (A, G) to purine or from pyrimidine (C, T) to pyrimidine. Point mutation from purine to pyrimidine or vice-versa is called transversion. Point mutations usually occur during DNA replication.

The 20 amino acids are coded by the 64 codons, which makes the genetic code. That is same amino acid may be coded by different codons. Mathematically, it can be considered as a many to one function from codons to amino acids. Balakrishnan [3] pointed out that as the total number of codons are almost triple than the total number of amino acids, so there may be some mathematical structures on genetic code.

A codon is formed of three bases and the importance of a base differs according to its position on the codon. The error frequency (accepted mutations) of codons plays an important role on the importance of different positions of bases in codons. The frequency of errors decreases from the third base to the first base and then next to the second base. That is the second base of a codon is biologically most relevant base. Also, the hydrophobicity property of an amino acid is connected with the most significant base i.e., the second base of codons. According to Watson and Crick [17], the hydrophilic amino acids are coded by the codons having Adenine (A) as second base and in case of hydrophobic amino acids, the codons has Uracil (U) as second base.

Many researchers such as, Hornos and Hornos [8], Beland and Allen [6], Schuster et al. [15], Bashford et al. [5], Robin et al. [10], Bashford et al. [4], Lehmann [9], Stadler et al. [16], Antoneli et al. [2] and Ali and Phukan [1] tried to study the genetic code algebraically. Sanchez et al. [11–14] discussed some algebraic structures of DNA sequences. They proposed a partial order on codon set and studied the genetic code through Boolean deductions. They also pointed out that based on some biological properties of codons it was possible to deduce and study different structures of the genetic code algebraically. Gohain et al. [7] also worked on the genetic code and found some interesting relations between the lattice structure and the biological aspects.

In this paper, some algebraic concepts in genetic code are investigated and the evolutionary pattern of amino acids based on hamming distance are explored.

2 Algebraic Structure on Genetic Code

Sanchez et al. [14] discussed two orderings of the RNA bases. The ordering is based on the chemical types (that is purine and pyrimidine) and the number of hydrogen bonds. The two orderings are: {A, C, G, U} and {U, G, C, A} and a sum operation (Table 1) is defined on these two base sets. The two sets are isomorphic to the cyclic group Z_4 (group Z_4 of integer module 4).

Working on the same field, by considering the same order of bases, Ali and Phukan [1] define a product operation (Table 1) on the base set $X = \{A, C, G, U\}$. With these two binary operations (sum and product) the set X fulfils the axioms of a

Table 1 Sum and product operations on X = {A, C, G, U}

Sum	+	A	C	G	U	Product	•	A	C	G	U
	A	A	C	G	U		A	A	A	A	A
	C	C	G	U	A		C	A	C	G	U
	G	G	U	A	C		G	A	G	A	G
	U	U	A	C	G		U	A	U	G	C

ring structure. In the ring $(X, +, \bullet)$, A is additive identity and C is the multiplicative identity. Also, X has commutative ring structure with identity element.

Ali and Phukan [1] arranged all the codons in the genetic code table (Table 2) by using the Cartesian product of the ring X i.e., $X \times X \times X$and denoted it as C_G, where

$$X \times X \times X = \{(P, Q, R) : P, Q, R \in \{A, C, G, U\}\} \tag{2.1}$$

$$i.e., C_G = \{PQR : P, Q, R \in \{A, C, G, U\}\} \tag{2.2}$$

Each codon of the form PQR is associated with the element (P, Q, R) of $X \times X \times X$ and thus a one to one correspondence can be established between set $X \times X \times X$and C_G. Next, sum and product operations are defined between the codons as follows:

$$(PQR) + (P'Q'R') = (P + P')(Q + Q')(R + R') \tag{2.3}$$

$$PQR.P'Q'R' = P.P'Q.Q'R.R' \tag{2.4}$$

With these two operations, C_G possesses ring structure and is isomorphic to $Z_4 \times Z_4 \times Z_4$.

For example, the element $GAC \in C_G$ has correspondence with the element $(2, 0, 1) \in Z_4 \times Z_4 \times Z_4$. The genetic code with corresponding amino acids is shown in Table 2.

We propose the following

Definition 1 Codon in which all bases are purines is an even codon and the codon in which at least one base is a pyrimidine is an odd codon.

It is observed that the set of all even codons, that is $\{AAA, AAG, GAA, GAG, AGA, AGG, GGA, GGG\}$ is a subgroup of the group $(C_G, +)$.

The order of the elements of the group $(C_G, +)$ divides the group into three classes. The following Table 3 gives the order of the codons.

The transition mutation (purine to purine or pyrimidine to pyrimidine) and transversion mutation of codons are connected with changes in parity (change from odd codon to even or vice-versa) and the order of codons (order as element of the ring). Following are a few connections that we have observed:

Table 2 The $Z_4 \times Z_4 \times Z_4$ table of the Genetic code

	A			C			G			U			
	(1)	(2)	(3)	(1)	(2)	(3)	(1)	(2)	(3)	(1)	(2)	(3)	
A	000	AAA	K	010	ACA	T	020	AGA	R	030	AUA	I	A
	001	AAC	N	011	ACC	T	021	AGC	S	031	AUC	I	C
	002	AAG	K	012	ACG	T	022	AGG	R	032	AUG	M	G
	003	AAU	N	013	ACU	T	023	AGU	S	033	AUU	I	U
C	100	CAA	Q	110	CCA	P	120	CGA	R	130	CUA	L	A
	101	CAC	H	111	CCC	P	121	CGC	R	131	CUC	L	C
	102	CAG	Q	112	CCG	P	122	CGG	R	132	CUG	L	G
	103	CAU	H	113	CCU	P	123	CGU	R	133	CUU	L	U
G	200	GAA	E	210	GCA	A	220	GGA	G	230	GUA	V	A
	201	GAC	D	211	GCC	A	221	GGC	G	231	GUC	V	C
	202	GAG	E	212	GCG	A	222	GGG	G	232	GUG	V	G
	203	GAU	D	213	GCU	A	223	GGU	G	233	GUU	V	U
U	300	UAA	–	310	UCA	S	320	UGA	–	330	UUA	L	A
	301	UAC	Y	311	UCC	S	321	UGC	C	331	UUC	F	C
	302	UAG	–	312	UCG	S	322	UGG	W	332	UUG	L	G
	303	UAU	Y	313	UCU	S	323	UGU	C	333	UUU	F	U

(1) → Corresponding elements of $Z_4 \times Z_4 \times Z_4$
(2) → The codons
(3) → The amino acids, – → The stop codons

Table 3 Partition of the group into three classes w.r.t. their orders

Order	Related codon
1	AAA
2	AAG, GAA, GAG, AGA, AGG, GGA, GGG
3	AAC, CAC, UAC, UUC, CCC, GCC, UCC, AGC, CGC, UCU, GGC, UGC, AUC, CUC, ACC, AAU, CAU, GAU, UAU, CAA, UAA, GUC, CAG, UAG, CGG, ACA, ACG, ACU, CCA, CCG, GAC, CCU, GCA, UUG, GUU, GUA, GCG, GCU, UCA, UCG, AGU, CGA, CGU, GGU, UGA, UGG, UGU, AUA, AUG, AUU, CUA, CUG, CUU, GUG, UUA, UUU

1. One-point transition of any base keeps the codon parity as well as codon order unchanged. No extreme changes are introduced in the properties of amino acids due to these mutations.
2. Transversion of bases changes codon parity as well as codon order.
3. Transversion of codons having a pyrimidine as second base (biologically most significant position) keeps the codon parity as well as codon order unchanged.
4. All odd codons have maximal order and all even codons have order less than that.
5. During single base transversion, even codons are always muted to odd codons and for each codon, the resulting muted codons are algebraically inverse of one

Table 4 Substitution of the bases of all even codons w.r.t Watson–Crick base pairs

Even codon $\longleftrightarrow$ not zero-divisor codon
AAA $\longleftrightarrow$ UUU
AAG $\longleftrightarrow$ UUC
GAA $\longleftrightarrow$ CUU
GAG $\longleftrightarrow$ CUC
AGA $\longleftrightarrow$ UCU
AGG $\longleftrightarrow$ UCC
GGA $\longleftrightarrow$ CCU
GGG $\longleftrightarrow$ CCC

another. For example, first base transversion of the even codon AAG are CAG and UAG, which are algebraically inverse of one another.

6. In the first base transversion, the even codons are changed to a codon that code to a polar amino acid, for the second base it is to hydrophobic and for the third base it is to a small amino acid. Also, due to third base transversion, the hydrophilic (hydrophobic) codon changes to a hydrophilic (hydrophobic) codon.

We have eight even codons and the substitution of the bases of the codons with respect to the Watson–Crick base pairing ($A \longleftrightarrow U, G \longleftrightarrow C$) gives another eight codons (see Table 4) which are not zero-divisors of the group $(C_G, +)$.

The even codons with their muted codons (transversion) and the not zero-divisor codons with their muted codons (transversion) partition the whole set of codons into two equal, disjoint subsets. Table 5 gives the even and not zero-divisor codons with their muted codons.

Thus, we can define a function $f : C_G \to C_G$ such that for $\alpha \in C_G$,

$$f(\alpha) = f(\alpha_1, \alpha_2, \alpha_3) = (\alpha_1 + x_1, \alpha_2 + x_2, \alpha_3 + x_3)$$

$$x_i = \begin{cases} U & \text{if } \alpha_i \text{ is a purine} \\ C & \text{if } \alpha_i \text{ is a pyrimidine, for every } i = 1, 2, 3. \end{cases}$$

An alternative way of defining the function f is $f : C_G \to C_G$ such that for all $\alpha \in C_G$

$$f(\alpha) = f(\alpha_1\alpha_2\alpha_3) = (UUU - \alpha_1\alpha_2\alpha_3).$$

It is observed that all the elements having order less than 4 map to an element of order 4 and will give us the set of all not zero-divisors of C_G. The function f represents the triple base mutation of all even codons in terms of Watson–Crick base pairs.

The set obtained by transversion of domain of f (even codons) together with the domain set and the range set together with the set obtained by the transversion of the

Table 5 Transversion of even codons and not zero-divisor codons

Even codons	Muted codons(transversion)	Not zero-divisor codons	Muted codons(transversion)
AAA	CAA, UAA, ACA, AUA, AAC, AAU	UUU	AUU, GUU, UAU, UGU, UUA, UUG
AAG	CAG, UAG, ACG, AUG, AAC, AAU	UUC	AUC, GUC, UAC, UGC, UUA, UUG
GAA	CAA, UAA, GCA, GUA, GAC, GAU	CUU	AUU, GUU, CAU, CGU, CUA, CUG
GAG	CAG, UAG, GCG, GUG, GAC, GAU	CUC	AUC, GUC, CAC, CGC, CUA, CUG
AGA	CGA, UGA, ACA, AUA, AGC, AGU	UCU	ACU, GCU, UAU, UGU, UCA, UCG
AGG	CGG, UGG, ACG, AUG, AGC, AGU	UCC	ACC, GCC, UAC, UGC, UCA, UCG
GGA	CGA, UGA, GCA, GUA, GGC, GGU	CCU	ACU, GCU, CAU, CGU, CCA, CCG
GGG	CGG, UGG, GCG, GUG, GGC, GGU	CCC	ACC, GCC, CAC, CGC, CCA, CCG

range set (set of all not zero-divisors) partitions the whole set C_G into two disjoints sets. In other words, if M is the set of even codons and their one-point transversions, N is the set of all not zero-divisors and their one point transversions, then,

$$C_G = M \cup N \text{ and } M \cap N = \phi$$

3 Distances Between Amino Acids and Their Biological Significance

We define a distance matrix using the Hamming distance between each pair of codons. Further, this matrix is used to determine the distances between amino acids. For that purpose, the average distances between the coded codons for the respective amino acids are considered.

For example, if we consider the codons GGA and CGG, then the number of base positions at which the corresponding codons are different gives the hamming distance between the respective codons. Next, we consider the amino acids H and T then the distance between them can be calculated by using the hamming distances between their corresponding coded codons. The amino acid H is coded by the codons CAC, CAU and the amino acid T is coded by the codons ACA, ACC, ACG, ACU.

Table 6 The distance matrix of 20 amino acids obtained from the hamming distances between codons. The distance between amino acids pairs computed as the mean distance between their respective codons

	G	W	C	R	S	V	L	F	M	I	E	D	Y	K	N	Q	H	A	T	P
G	0.00	1.75	1.75	1.75	2.46	1.81	2.75	2.75	2.75	2.75	1.75	1.75	2.75	2.88	2.75	2.63	2.75	1.75	2.75	2.75
W	1.75	0.00	1.00	1.67	1.83	2.75	2.33	2.00	3.00	3.00	3.00	3.00	2.00	2.50	3.00	2.50	3.00	2.75	2.75	2.75
C	1.75	1.00	0.00	1.83	1.67	2.75	2.50	1.50	3.00	2.67	3.00	2.50	2.00	3.00	2.50	3.00	2.50	2.75	2.75	2.75
R	1.75	1.67	1.83	0.00	2.36	2.75	2.28	2.83	2.33	2.44	2.67	2.83	2.83	2.33	2.50	2.00	2.17	2.75	2.42	2.08
S	2.46	1.83	1.67	2.36	0.00	2.79	2.56	2.00	2.50	2.39	2.83	2.67	2.00	2.50	2.33	2.83	2.67	2.08	1.75	2.08
V	1.81	2.75	2.75	2.75	2.79	0.00	1.75	1.75	1.75	1.75	1.75	1.75	2.75	2.75	2.75	2.75	2.75	1.75	2.75	2.75
L	2.75	2.33	2.50	2.28	2.56	1.75	0.00	1.50	1.67	1.78	2.67	2.83	2.50	2.67	2.83	2.00	2.17	2.75	2.75	2.08
F	2.75	2.00	1.50	2.83	2.00	1.75	1.50	0.00	2.00	1.67	3.00	2.50	1.50	3.00	2.50	3.00	2.50	2.75	2.75	2.75
M	2.75	3.00	3.00	2.33	2.50	1.75	1.67	2.00	0.00	1.00	2.50	3.00	3.00	1.50	2.00	2.50	3.00	2.75	1.75	2.75
I	2.75	3.00	2.67	2.44	2.39	1.75	1.78	1.67	1.00	0.00	2.83	2.67	2.67	1.83	1.67	2.83	2.67	2.75	1.75	2.75
E	1.75	3.00	3.00	2.67	2.83	1.75	2.67	3.00	2.50	2.83	0.00	1.00	2.00	1.50	2.00	1.50	2.00	1.75	2.75	2.75
D	1.75	3.00	2.50	2.83	2.67	1.75	2.83	2.50	3.00	2.67	1.00	0.00	1.50	2.00	1.50	2.00	1.50	1.75	2.75	2.75
Y	2.75	2.00	2.00	2.83	2.00	2.75	2.50	1.50	3.00	2.67	2.00	1.50	0.00	2.00	1.50	2.00	1.50	2.75	2.75	2.75
K	2.88	2.50	3.00	2.33	2.50	2.75	2.67	3.00	1.50	1.83	1.50	2.00	2.00	0.00	1.00	1.50	2.00	2.75	1.75	2.75
N	2.75	3.00	2.50	2.50	2.33	2.75	2.83	2.50	2.00	1.67	2.00	1.50	1.50	1.00	0.00	2.00	1.50	2.75	1.75	2.75
Q	2.63	2.50	3.00	2.00	2.83	2.75	2.00	3.00	2.50	2.83	1.50	2.00	2.00	1.50	2.00	0.00	1.00	2.75	2.75	1.75
H	2.75	3.00	2.50	2.17	2.67	2.75	2.17	2.50	3.00	2.67	2.00	1.50	1.50	2.00	1.50	1.00	0.00	2.75	2.75	1.75
A	1.75	2.75	2.75	2.75	2.08	1.75	2.75	2.75	2.75	2.75	1.75	1.75	2.75	2.75	2.75	2.75	2.75	0.00	1.75	1.75
T	2.75	2.75	2.75	2.42	1.75	2.75	2.75	2.75	1.75	1.75	2.75	2.75	2.75	1.75	1.75	2.75	2.75	1.75	0.00	1.75
P	2.75	2.75	2.75	2.08	2.08	2.75	2.08	2.75	2.75	2.75	2.75	2.75	2.75	2.75	2.75	1.75	1.75	1.75	1.75	0.00

Therefore the distances between the codons of H and T are

	ACA	ACC	ACG	ACU
CAC	3	2	3	3
CAU	3	3	3	2

Hence the distance between the amino acids H and T is 2.75.

In this way the distance matrix of the 20 amino acids is calculated as shown in Table 6.

It is observed that as the distance values increase, the differences of physico-chemical properties of amino acids increase in most of the cases. Also, the distance values are higher between most of the hydrophilic and hydrophobic amino acids. For example, the strong hydrophilic amino acid Lysine and the strong hydrophobic amino acid Phenylalanine have maximum distance value 3. When there is small difference in the distance values between two amino acids, there is also little difference in their properties. Further it is observed that the distance obtained above defines a metric on the set of amino acids.

It may be noted that similar results were also obtained by Sanchez et al. [11] wherein he used a different approach to obtain the distance table.

From the distance matrix of Table 6 we have obtained graph of the amino acids as explained below. Vertices are represented by the amino acids, and two vertices (amino acids) α and β are connected by an edge if their distance is less than some given threshold value $\epsilon > 0$. First we consider the average distance (2.21) as threshold value. The corresponding graph is depicted below in Fig. 1. The graph structures for threshold values $\epsilon = 2.00$, 1.75 and 1.5 are given in Figs. 2, 3 and 4 respectively.

From these graph structures we observe that with the increase in the threshold value, the accessibility of getting an amino acid from other decreases simultaneously. The graphs in Figs. 1 and 2 are connected while the others are disconnected. Also in Fig. 3, the amino acids A, G, P, T, V are isolated. These five amino acids are

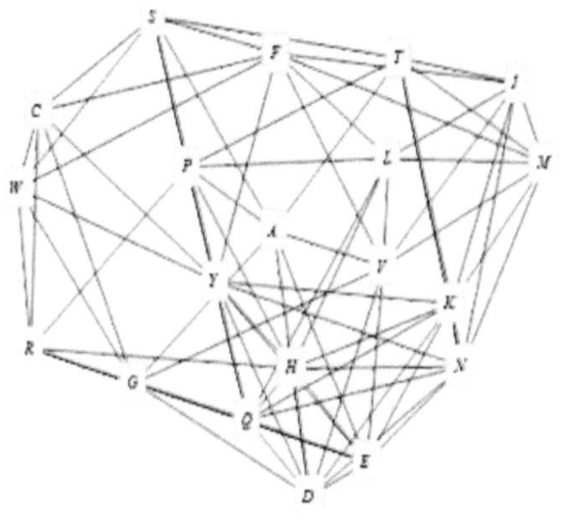

Fig. 1 Graph of amino acids for the threshold value 2.21

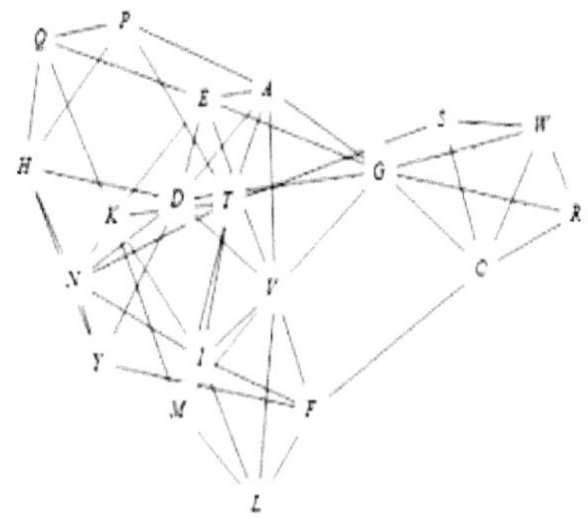

Fig. 2 Graph of amino acids for the threshold value 2.00

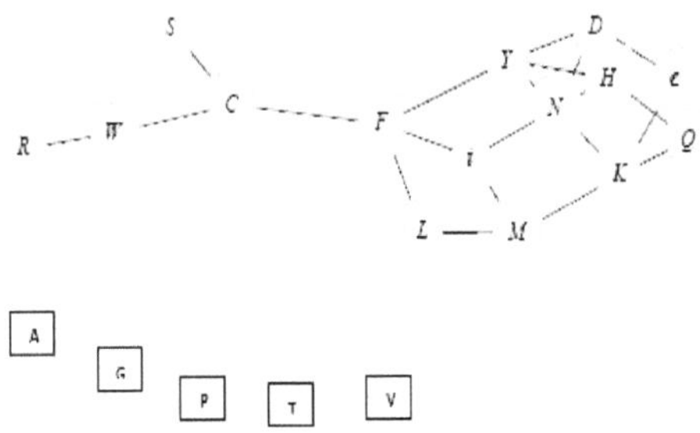

Fig. 3 Graph of amino acid for threshold value 1.75

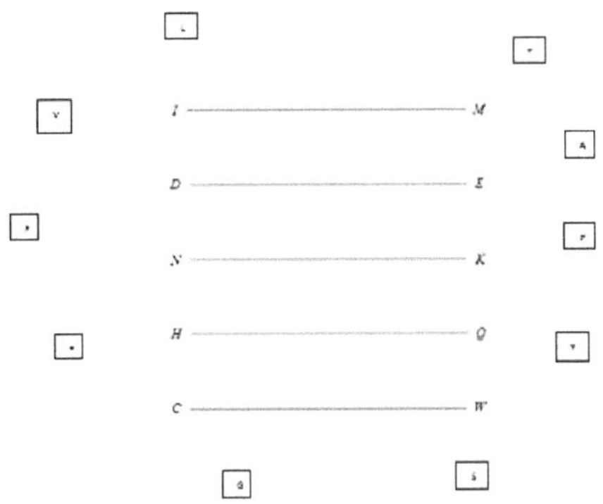

Fig. 4 Graph of amino acid for the threshold value 1.5

Table 7 The hamming distance of the mutations observed in the HIV protease gene. It confers drug resistance with related to the wild type-HIV-HXB2

Amino acid Mutations	Codon-Mutation	Distance value	Antiviral drug	Amino acid Mutations	Codon-Mutation	Distance value	Antiviral drug
A711	GCU→AUU	2	ABT-378	L10Y	CUC→UAC	2	BMS 232632
A71T	GCU→ACU	1	Indinavir, Crixivan	L24I	UUA→AUA	1	Indinavir, Crixivan
A71L	GCU→CUC	3	ABT-378	L23I	CUA→AUA	1	BILA 2185 BS
A71V	GCU→GUU	1	Nelfinavir, Viracept	L24V	UUA→GUA	1	Telinavir
D60E	GAU→GAA	1	DMP 450	L63P	CUC→CCC	1	ABT-378, AG1343
D30N	GAU→AAU	1	Nelfinavir, Viracept	L33F	UUA→UUC	1	ABT-538
G16E	GGG→GAG	1	ABT-378	L90M	UUG→AUG	1	Nelfinavir, Viracept
G52S	GGU→AGU	1	AG1343	M36I	AUG→AUA	1	Nelfinavir, Viracept
G48V	GGG→GUG	1	Telinavir, MK-639	L97V	UUA→GUA	1	DMP-323
G73S	GGU→AGU	1	AG1343 MK-639	M46F	AUG→UUC	2	A-77009
I47V	AUA→GUA	1	ABT-378	M46L	AUG→UUC	2	Indinavir, Crixivan
H69Y	CAU→UAU	1	Aluviran, Lopinavir	M46I	AUG→AUA	1	Indinavir, Crixivan
I50L	AUU→CUU	1	BMS 232632	M46V	AUG→GUG	1	A-77006
I54M	AUU→AUG	1	BILA 2185 BS	N88S	AAU→AGU	1	BMS 232632
I54L	AUC→CUC	1	ABT-378	N88D	AAU→GAU	1	Nelfinavir, Viracept
I54T	AUC→ACC	1	ABT-378	P81T	CCU→ACU	1	Telinavir
I82T	AUC→ACC	1	A-77003	R8Q	CGA→CAA	2	A-77004
I54V	AUC→GUC	1	ABT-378, MK-639	R8K	CGA→AAA	2	A-77003
I84A	AUA→GCA	2	BILA 1906 BS	R57K	AGA→AAA	1	AG1343
K20M	AAG→AUG	1	Indinavir, Crixivan	V32I	GUA→AUA	1	A-77005, Telinavir
I84V	AUA→GUA	1	Nelfinavir, Viracept	T91S	ACU→UCU	1	ABT-378

(continued)

Table 7 (continued)

Amino acid Mutations	Codon-Mutation	Distance value	Antiviral drug	Amino acid Mutations	Codon-Mutation	Distance value	Antiviral drug
K20R	AAG→AGG	1	Indinavir, Crixivan	V75I	GUA→AUA	1	Telinavir
K55R	AAA→AGA	1	AG1343	V82A	GUC→GCC	1	Ritonovir, Norvir
K45I	AAA→AUA	1	DMP-323	V77I	GUA→AUA	1	AG1343
L10I	CUC→AUC	1	Indinavir, Crixivan	V82F	GUC→UUC	1	Ritonovir, Norvir
L10V	CUC→GUC	1	Indinavir, Crixivan	V82S	GUC→UCC	2	Ritonovir, Norvir
L10R	CUC→CGC	1	Indinavir, Crixivan	V82I	GUC→AUC	1	A-77011
L10F	CUC→UUC	1	Lopinavir	V82T	GUC→ACC	2	Ritonovir, Norvir

Table 8 The hamming distance of the mutations observed in the human beta-globin gene

Amino acid Mutations	Codon Mutation	Distance value	Biological effect	Reference
T123I	ACC→ATC	1	Asymptomatic	[11300351] Hemoglobin. 2001, 25, 67–78
P36H	CCT→CAT	1	High oxygen affinity	[11939509] Hemoglobin. 2002, 26, 21–31
V20E	GTG→GAG	1	High oxygen affinity	[7914875] Eur J Haematol. 1994, 53, 21–25
V126L	GTG→CTG	1	Neutral	[11939515] Hemoglobin. 2002, 26, 7–12
V20M	GTG→ATG	1	High oxygen affinity	[7914875] Eur J Haematol. 1994, 53, 21–25
V111F	GTC→TTC	1	Low oxygen affinity	[10975442] Hemoglobin. 2000, 24, 227–237
V34F	GTC→TTC	1	High oxygen affinity	[10846826] Int J Hematol. 2000, 71, 221–226
H97Q	CAC→CAA	1	High oxygen affinity	[8571935] Am J Hematol. 1996, 51, 32–36
E121Q	GAA→CAA	1		[8095930] Hemoglobin. 1993, 17, 9–17
A128V	GCT→GTT	1	Mild instability	[11300349] Hemoglobin. 2001, 25, 45–56
L114P	CTG→CCG	1	Non-functional	[11300352] Hemoglobin. 2001, 25, 79–89
D99E	GAT→GAA	1	High oxygen affinity	[1814856] Hemoglobin. 1991, 15, 487–496
H97Q	CAC→CAG	1	High oxygen affinity	[8890707] Ann Hematol. 1996, 73, 183–188
D21N	GAT→AAT	1		[8507722] Ann Hematol. 1993, 66, 269–272
V34D	GTC→GAC	1	Unstable	[1260309] Hemoglobin. 2003, 27, 31–35
N139Y	AAT→TAT	1	High oxygen affinity	[8718692] Hemoglobin. 1995, 19, 335–341
E121K	GAA→AAA	1		[7908281] Hemoglobin. 1993, 17, 523–535
K82E	AAG→GAG	1	Altered oxygen affinity	[9028820] Hemoglobin. 1997, 21, 17–26
A140V	GCC→GTC	1	Mild polycythemia	[7908281] Hemoglobin. 1993, 17, 523–535
G83D	GGC→GAC	1	Hb Pyrgos (Normal)	[9255613] Hemoglobin. 1997; 21, 345–361
G15R	GGT→CGT	1	Neutral	[11939517] Hemoglobin. 2002, 26, 77–81
D99N	GAT→AAT	1	High oxygen affinity	[11843288] Int J Hematol. 2002, 75, 35–39
V111L	GTC→CTC	1	Fannin-Lubbock variant	[7852084] Hemoglobin. 1994, 18, 297–306

(continued)

Table 8 (continued)

Amino acid Mutations	Codon Mutation	Distance value	Biological effect	Reference
E26K	GAG→AAG	1		[9140717] Hemoglobin. 1997, 21, 205–218
G119D	GGC→GAC	1	Fannin-Lubbock variant	[7852084] Hemoglobin. 1994, 18, 297–306
N108I	AAC→ATC	1	Low affinity	[12010673] Haematologica. 2002, 87, 553–554
H146P	CAC→CCC	1	High oxygen affinity	[11475152] Ann Hematol. 2001, 80, 365–367
C112W	TGT→TGG	1	Silent and unstable	[8936462] Hemoglobin. 1996, 20, 361–369
H92Y	CAC→TAC	1	Cyanosis	[9494043] Hemoglobin. 1998, 22, 1–10
A111V	GCC→GTC	1	Silent	[7615398] Hemoglobin. 1995, 19, 1–6
D52G	GAT→GGT	1	Silent	[9730366] Hemoglobin. 1998, 22, 355–371
A123S	GCC→TCC	1	Silent	[7615398] Hemoglobin. 1995, 19, 1–6
V126G	GTG→GGG	1	Mild beta-thalassaemia	[1954392] Blood. 1991, 78, 3070–3075
F42L	TTT→TTG	1	Hemolytic anemia	[11920235] Hematol J. 2001, 2(1):61–66
W15Stop	TGG→TAG	1	Beta-thalassaemia	[10722110] Hemoglobin. 2000 Feb, 24(1):1–13

different from the remaining amino acids in the sense that all of them are coded by four codons, having same first and second bases. In Fig. 4, we have observed that the amino acids V, L, F, R, G, S, A, T, P, Y are isolated and the amino acids I, W, K, E, Q are connected with M, C, N, D, H respectively. Here the non-isolated amino acids differ from the other 10 isolated amino acids in the sense that each is obtained from any of the other by third base mutation of a codon and the corresponding codons of the connected amino acids have same first and second bases. Also, for the isolated amino acids, the third base mutation of the corresponding codons of an amino acid produces synonymous codons. That is the muted codon codes the same amino acid.

Next we discuss a real life example and observe that the distance value is usually small in between frequently occurring codon mutations. At first we check the distance between the single point drug resistance mutations in HIV-1 protease gene (see Table 7). Next we go through the respective gene of the HXB2 strain and human beta globin gene (refer. Table 8). The distance value obtained is 1 between most of the codons in both the cases. And it is also noted that if a small change occurs in the physico-chemical properties of the amino acids in human beta-globin gene, then there is a change in the biological function of hemoglobin.

Therefore we can conclude that the physico-chemical properties of the amino acids are connected with the hamming distances determined in the genetic code.

4 Conclusions

In this paper, we discussed an algebraic structure of the genetic code which exhibited some interesting connections of physico-chemical properties of amino acids with the algebraic structure. We observed that there is a close connection between the order of the codons and transition/transversion mutations. We have shown that the set of all codons which are not zero divisors can be obtained from the even codons and the transversion of these two sets partitioned the whole set of codons into disjoint subsets. Next, a distance matrix of codons is obtained from which a distance matrix of amino acids is constructed. The distance matrix reflects the fact that the difference of physico-chemical properties of amino acids is related to the distance between amino acids. A graph of the amino acids is generated from the distance matrix. This graph structure roughly depicts the evolutionary pathway of the amino acids.

References

1. Ali, T., Phukan, C.K.: Topology in genetic code algebra. Math. Sci. Int. Res. J. **2**(2), 179–182 (2013)
2. Antoneli, F., Braggion, L., Forger, M., Hornos, J.E.M.: Extending the search for symmetries in the genetic code. Int. J. Mod. Phys. B **17**, 3135–3204 (2003)
3. Balakrishnan, J.: Symmetry scheme for amino acid codons. Phys. Rev. E **65**, 021912–5 (2002)

4. Bashford, J.D., Jarvis, P.D.: The genetic code as a periodic table. Biosystems **57**, 147–161 (2000)
5. Bashford, J.D., Tsohantjis, I., Jarvis, P.D.: A supersymmetric model for the evolution of the genetic code. Proc. Natl. Acad. Sci. USA **95**, 987–992 (1998)
6. Beland, P., Allen, T.F.: The origin and evolution of the genetic code. J. Theor. Biol. **170**, 359–365 (1994)
7. Gohain, N., Ali, T., Akhtar, A.: Lattice structure and distance matrix of genetic code. J. Biol. Sys. **23**(3), 485–504 (2015)
8. Hornos, J.E.M., Hornos, Y.M.M.: Algebraic model for the evolution of the genetic code. Phys. Rev. Lett. **71**(26), 4401–4404 (1993)
9. Lehmann, J.: Physico-chemical constraints connected with the coding properties of the genetic system. J. Theor. Biol. **202**, 129144 (2000)
10. Robin, D., Knight, R.D., Freeland, S.J., Landweber, L.F.: Selection, history and chemistry: the three faces of the genetic code. Trends Biochem. Sci. **24**, 241247 (1999)
11. Sanchez, R., Morgado, E., Grau, R.: The genetic code boolean lattice. MATCH Commun. Math. Comput. Chem. **52**, 29–46 (2004)
12. Sanchez, R., Morgado, E., Grau, R.: A genetic code boolean structure. I. The meaning of boolean deductions. Bull. Math. Biol. **67**, 1–14 (2005)
13. Sanchez, R., Perfetti, L.A., Morgado, E., Grau, R.: A new DNA sequences vector space on a genetic code galois field. MATCH Commun. Math. Comput. Chem. **54**(1), 3–28 (2005)
14. Sanchez, R., Morgado, E., Grau, R.: Gene algebra from a genetic Code algebra structure. J. Math. Biol **51**(431), 457 (2005)
15. Schuster, P., Fontana, W., Hofacker, I.L.: From sequences to shapes and back: a case study in RNA secondary structures. Proc. Biol. Sci. **255**, 279–284 (1994)
16. Stadler, B., Stadler, P., Wagner, G., Fontana, W.: The topology of the possible: formal spaces underlying patterns of evolutionary change. J. Theor. Biol. **213**, 241–274 (2001)
17. Watson, J.D., Crick, F.H.C.: A structure for deoxyribose nucleic acid. Nature **171**(3), 737–738 (1953)

Part II
Applications to Physical Sciences

On Ramsey $(2K_2, 2H)$-Minimal Graphs

Kristiana Wijaya and Edy Tri Baskoro

Abstract The Ramsey set $\mathscr{R}(G, H)$ consists of all graphs F satisfying that any red-blue coloring of edges of F contains a red copy of G or a blue copy of H but if any edge in F is deleted then this condition is not fulfilled. For any integer $m \geq 2$, let mG be a disjoint union of m copies of graph G. In this paper, we construct graphs in $\mathscr{R}(2K_2, 2H)$ by using graph theoretical operations over graphs in $\mathscr{R}(2K_2, H)$ if H is either a cycle, a path, or a star.

Keywords Ramsey minimal graph · Matching · Cycle · Path · Star

2010 Mathematics Subject Classification. 05D10 · 05C55

1 Introduction

Let G and H be two graphs. A graph F is said to *arrow* the pair (G, H), written $F \rightarrow (G, H)$, if every red-blue coloring of G results in a subgraph isomorphic to G every edge of which is colored red (a red G) or a subgraph isomorphic to H every edge of which is colored blue (a blue H). Therefore, if $F \nrightarrow (G, H)$, then there exists a red-blue coloring of F for which there is neither a red G nor a blue H. Such a red-blue coloring of F is called a *(G, H)-coloring*. A graph F is called a *Ramsey* (G, H)*-minimal* if $F \rightarrow (G, H)$ and for every edge $e \in F$, we have that $F - e \nrightarrow (G, H)$. Next, the set of all Ramsey (G, H)-minimal graphs is denoted by $\mathscr{R}(G, H)$.

K. Wijaya · E.T. Baskoro (✉)
Combinatorial Mathematics Research Group, Faculty of Mathematics and Natural Sciences, Institut Teknologi Bandung, Jalan Ganesa 10, Bandung 40132, Indonesia
e-mail: ebaskoro@math.itb.ac.id

K. Wijaya
e-mail: kristiana.w@students.itb.ac.id

J.M. Cushing et al. (eds.), *Applied Analysis in Biological and Physical Sciences*, Springer Proceedings in Mathematics & Statistics 186,
DOI 10.1007/978-81-322-3640-5_13

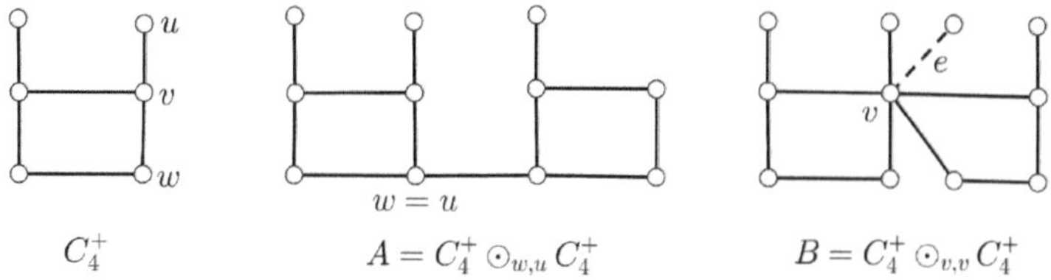

Fig. 1 The graphs $A \in \mathscr{R}(2K_2, 2P_4)$ and $B \notin \mathscr{R}(2K_2, 2P_4)$

Investigation on Ramsey (G, H)-minimal graphs was initiated by Burr et al. [3]. The set $\mathscr{R}(G, H)$ can be finite or infinite. For any integer $m \geq 2$, let mG be a disjoint union of m copies of graph G. Burr et al. [3] proved that if G is a matching, namely $G = mK_2$, then $\mathscr{R}(G, H)$ is finite for any graph H. In this paper, we focus on $G = 2K_2$. The study of Ramsey $(2K_2, H)$-minimal graphs has received much attention for various graphs H. The set $\mathscr{R}(2K_2, P_3) = \{2P_3, C_4, C_5\}$ was showed by Mengersen and Oeckermann [4]. In the same paper, they characterized all graphs in $\mathscr{R}(2K_2, K_{1,3})$. Moreover, Wijaya et al. [6] determined the graphs in $\mathscr{R}(2K_2, C_4)$. The sets $\mathscr{R}(2K_2, P_4) = \{2P_4, C_4^+, C_5, C_6, C_7\}$ and $\mathscr{R}(2K_2, P_5)$ were characterized by Baskoro and Yulianti [1], where C_4^+ is the graph in Fig. 1. Later, Tatanto and Baskoro [5] gave the graphs in the set $\mathscr{R}(2K_2, 2P_4)$ and $\mathscr{R}(2K_2, 2P_5)$. Recently, Baskoro and Wijaya et al. [2] showed that a graph obtained from two disjoint graphs in $\mathscr{R}(2K_2, K_4)$ by identifying vertices and edges is a member of $\mathscr{R}(2K_2, 2K_n)$.

Motivated by these last results, in this paper, we will propose some constructions for a member of $\mathscr{R}(2K_2, 2H)$ by using the graphs in $\mathscr{R}(2K_2, H)$, provided H is a connected graph. We only consider if H is a cycle, a path, or a star. Let G and H be graphs. Let $u \in V(G)$, $v \in V(H)$, $a \in E(G)$ and $b \in E(H)$. The graph $G \odot_{u,v} H$ is defined as a graph obtained from two disjoint graphs G and H by identifying vertices u and v. So, the graph $G \odot_{u,v} H$ has $(|V(G)| + |V(H)| - 1)$ vertices and $(|E(G)| + |E(H)|)$ edges. Next, the graph $G \ominus_{a,b} H$ is defined a graph obtained from two disjoint graphs G and H by identifying edges a and b. The graph $G \ominus_{a,b} H$ has $(|V(G)| + |V(H)| - 2)$ vertices and $(|E(G)| + |E(H)| - 1)$ edges.

2 Previous Results

In this section, we give some previous results from Baskoro and Wijaya [2] used in this paper.

Theorem 1 ([2]) *For any integers $s \geq 2$, $m \geq 1$ and any connected graph H, the graph $(s + m - 1)H$ is in $\mathscr{R}(sK_2, mH)$.*

Theorem 2 ([2]) *If $F_1, F_2 \in \mathscr{R}(2K_2, K_n)$ and both are connected, then $G \odot_{u,v} H \in \mathscr{R}(2K_2, 2K_n)$, for any $u \in V(F_1)$ and $v \in V(F_2)$.*

Theorem 3 ([2]) *If $F_1, F_2 \in \mathscr{R}(2K_2, K_n)$ and both are connected, then $(F_1 \ominus_{a,b} F_2) \in \mathscr{R}(2K_2, 2K_n)$, for $a \in E(F_1)$ and $b \in E(F_2)$.*

3 Union of Two Ramsey Minimal Graphs

By Theorem 1, for any connected graph H, one of disconnected graphs belonging to $\mathscr{R}(2K_2, 2H)$ is given by the following corollary.

Corollary 1 *Let H be a connected graph. Then, $3H \in \mathscr{R}(2K_2, 2H)$.*

Theorem 4 *Let H_1 and H_2 be connected graphs. Let $F_1 \in \mathscr{R}(2K_2, H_1)$, $F_2 \in \mathscr{R}(2K_2, H_2)$ be two connected graphs. If each of $\{F_1, F_2\}$ contains no $H_1 \cup H_2$, then $F_1 \cup F_2 \in \mathscr{R}(2K_2, H_1 \cup H_2)$.*

Proof We consider $F = F_1 \cup F_2$. Then, any red-blue coloring of edges of F with no red $2K_2$ will contain a blue H_1 in F_1 and H_2 in F_2. Hence, in total, we have a blue $H_1 \cup H_2$ in F. So, $F_1 \cup F_2 \rightarrow (2K_2, H_1 \cup H_2)$.

Now, we will prove that for every $e \in E(F_1 \cup F_2)$, $(F_1 \cup F_2) - e \nrightarrow (2K_2, H_1 \cup H_2)$. Without loss of generality, we can only consider when $e \in F_1$. Then, there exits a $(2K_2, H_1)$-coloring ϕ_1 of edges of $F_1 - e$. We now define ϕ as a red-blue coloring of edges of $(F_1 \cup F_2) - e$ such that $\phi(x) = \phi_1(x)$ for every $x \in E(F_1 - e)$ and $\phi(x) =$ blue for every $x \in E(F_2)$. Since F_2 contains no $H_1 \cup H_2$, under the coloring ϕ, $(F_1 \cup F_2) - e$ contains neither a red $2K_2$ nor a blue $H_1 \cup H_2$. We obtain a $(2K_2, H_1 \cup H_2)$-coloring of $(F_1 \cup F_2) - e$. □

Theorem 5 *Let H be a connected graph. If $F_1, F_2 \in \mathscr{R}(2K_2, H)$, then*

(a) $F_1 \cup F_2 \in \mathscr{R}(4K_2, H)$,

(b) $F_1 \cup F_2 \in \mathscr{R}(2K_2, 2H)$.

Proof Let $F_1, F_2 \in \mathscr{R}(2K_2, H)$.

(a) We consider $F = F_1 \cup F_2$. Then, any red-blue coloring of edges of F with no blue H will contain a red $2K_2$ in F_1 and a red $2K_2$ in F_2. Therefore, in total, we have a red $4K_2$ in F. So, $F_1 \cup F_2 \rightarrow \mathscr{R}(4K_2, H)$. Next, let $e \in E(F)$. Then, e is in either $E(F_1)$ or $E(F_2)$. Consider $e \in E(F_1)$ (the case of $e \in E(F_2)$ is similar). Then, there is a $(2K_2, H)$-coloring ϕ_1 of edges of $F_1 - e$. Let φ be a red-blue coloring of edges of F_2 with a red $2K_2$ and no blue H. Now, we define ϕ as a red-blue coloring of F such that $\phi(x) = \phi_1(x)$ for $x \in E(F_1)$ and $\phi(x) = \varphi(x)$ for $x \in E(F_2)$. Under the coloring ϕ, the graph F contains neither a red $4K_2$ nor a blue H. Hence, ϕ is a $(4K_2, H)$-coloring of edges of $F_1 \cup F_2$. This means that $F_1 \cup F_2 \in \mathscr{R}(4K_2, H)$.

(b) The minimality of F_1 and F_2 implies that each of $\{F_1, F_2\}$ must contain no $2H$. Furthermore, by Theorem 4, we obtain $F_1 \cup F_2 \in \mathscr{R}(2K_2, 2H)$. □

Note that the second statement of Theorem 5 has been appeared in [2].

4 Identifying Two Vertices of Two Ramsey Minimal Graphs

Motivated by Theorem 2, we will apply a similar method on cycles, paths, and stars. Now, we consider the graphs in $\mathcal{R}(2K_2, C_n)$. If $F_1, F_2 \in \mathcal{R}(2K_2, C_n)$ and $u \in V(F_1)$, $v \in V(F_2)$, then every cycle in $F_1 \odot_{u,v} F_2$ must inclusively use the vertices in either F_1 or F_2, but not both. It means that if there exists a cycle C containing the identified vertex $u = v$ in $F_1 \odot_{u,v} F_2$, then either $V(C) - u \subseteq F_1$ or $V(C) - u \subseteq F_2$. Therefore, we obtain that $3C_n \nsubseteq F_1 \odot_{u,v} F_2$. Consequently, we have the following theorem.

Theorem 6 *Let $n \geq 3$ be an integer. If connected graphs $F_1, F_2 \in \mathcal{R}(2K_2, C_n)$, $u \in V(F_1)$ and $v \in V(F_2)$, then $F_1 \odot_{u,v} F_2 \in \mathcal{R}(2K_2, 2C_n)$.*

Proof First, observe that $F_1 \odot_{u,v} F_2 \rightarrow (2K_2, 2C_n)$. Now, we will prove that for every $e \in E(F_1 \odot_{u,v} F_2)$, $(F_1 \odot_{u,v} F_2) - e \nrightarrow (2K_2, 2C_n)$. Without loss of generality, we need only consider the case $e \in F_1$. We have a $(2K_2, C_n)$-coloring ϕ_1 of edges of $F_1 - e$. Now, we define ϕ as a red-blue coloring of edges of $(F_1 \odot_{u,v} F_2) - e$ such that $\phi(x) = \phi_1(x)$ for $x \in E(F_1 - e)$ and $\phi(x) =$ blue otherwise. Thus, we obtain $(F_1 \odot_{u,v} F_2) - e$ containing neither a red $2K_2$ nor a blue $2C_n$. □

On the other hand, we cannot replace the complete graph K_n in Theorem 2 directly by a path P_n or a star $K_{1,n}$. If $F_1, F_2 \in \mathcal{R}(2K_2, P_n)$ and $u \in V(F_1)$, $v \in V(F_2)$, then $3P_n \subseteq F_1 \odot_{u,v} F_2$ is possibly true. Therefore, we need some additional conditions so that $F_1 \odot_{u,v} F_2 \in \mathcal{R}(2K_2, 2P_n)$. One such condition is that $F_1 \odot_{u,v} F_2$ contains no $3P_n$. But, this is not enough to guarantee $F_1 \odot_{u,v} F_2 \in \mathcal{R}(2K_2, 2P_n)$. For instance, consider the graphs in Fig. 1. The graphs A and B are obtained from $C_4^+ \odot C_4^+$ but two different identifying vertices. Even though both graphs A and B do not contain a graph $3P_n$, we can show that $A \in \mathcal{R}(2K_2, 2P_4)$ but $B \notin \mathcal{R}(2K_2, 2P_4)$, since there is no $(2K_2, 2P_4)$-coloring in $B - e$. We have the following theorem.

Theorem 7 *Let $n \geq 3$ be an integer. Let F_1, F_2 be connected graphs in $\mathcal{R}(2K_2, P_n)$, $u \in V(F_1)$ and $v \in V(F_2)$. If $F_1 \odot_{u,v} F_2$ contains no $3P_n$, then $F_1 \odot_{u,v} F_2 \rightarrow (2K_2, 2P_n)$.*

Proof Suppose $F_1 \odot_{u,v} F_2 \nrightarrow (2K_2, 2P_n)$. Then, there exists a $(2K_2, 2P_n)$-coloring ϕ of edges of $F_1 \odot_{u,v} F_2$. So, under the coloring ϕ, $F_1 \odot_{u,v} F_2$ contains at most one red K_2 and at most one blue P_n. It means that the red subgraph of $F_1 \odot_{u,v} F_2$ forms either a triangle or a star. If the red subgraph of $F_1 \odot_{u,v} F_2$ forms a triangle K_3, then this K_3 can be either in F_1 or F_2. Now, if this red K_3 is in F_1, then, this blue P_n can be in F_1 or F_2. If P_n is in F_1, then it contradicts $F_2 \rightarrow (2K_2, P_n)$. If P_n is in F_2, then it contradicts $F_1 \rightarrow (2K_2, P_n)$. If P_n is in both F_1 and F_2, then it contradicts either $F_1 \rightarrow (2K_2, P_n)$ or $F_2 \rightarrow (2K_2, P_n)$. Therefore, $F_1 \odot_{u,v} F_2 \rightarrow (2K_2, 2P_n)$. A similar argument can be applied if the red subgraph of $F_1 \odot_{u,v} F_2$ forms a star. □

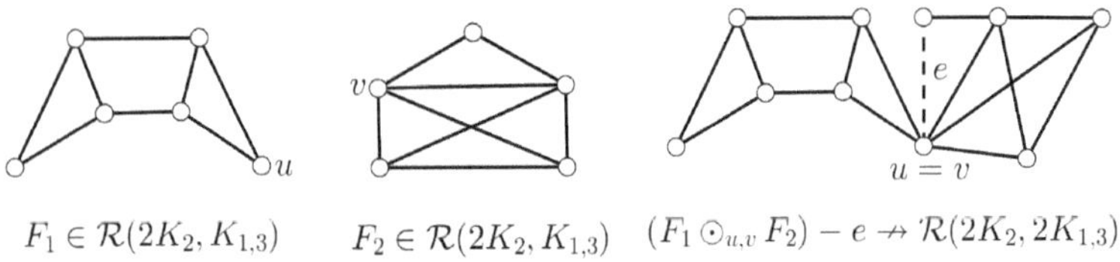

Fig. 2 The graph $F_1 \odot_{u,v} F_2 \not\supseteq 3K_{1,3}$ but $F_1 \odot_{u,v} F_2 \notin \mathcal{R}(2K_2, 2K_{1,3})$

Similarly, there are two graphs $F_1, F_2 \in \mathcal{R}(2K_2, K_{1,n})$, $u \in V(F_1)$, $v \in V(F_2)$, and $F_1 \odot_{u,v} F_2 \not\supseteq 3K_{1,n}$ but $F_1 \odot_{u,v} F_2$ is not in $\mathcal{R}(2K_2, 2K_{1,n})$. An example can be seen in Fig. 2. Thus, we have the following theorem. The proof is similar.

Theorem 8 *Let $n \geq 2$ be an integer. Let F_1, F_2 be connected graphs in $\mathcal{R}(2K_2, K_{1,n})$, $u \in V(F_1)$ and $v \in V(F_2)$. If $F_1 \odot_{u,v} F_2$ contains no $3K_{1,n}$, then $F_1 \odot_{u,v} F_2 \rightarrow (2K_2, 2K_{1,n})$.*

5 Identifying Two Edges of Two Ramsey Minimal Graphs

In this section, we replace the complete graph in Theorem 3 by a cycle, a path, or a star. First, we let F_1 and F_2 be connected graphs in $\mathcal{R}(2K_2, C_n)$. Then, we have the following theorem for the graph $F_1 \ominus_{a,b} F_2$ for any edges $a \in E(F_1)$ and $b \in E(F_2)$.

Theorem 9 *Let $n \geq 3$ be an integer. Let F_1, F_2 be connected graphs in $\mathcal{R}(2K_2, C_n)$, $a \in E(F_1)$ and $b \in E(F_2)$. If $F_1 \ominus_{a,b} F_2$ contains no $3C_n$, then $F_1 \ominus_{a,b} F_2 \in \mathcal{R}(2K_2, 2C_n)$.*

Proof Let $F_1, F_2 \in \mathcal{R}(2K_2, C_n)$ be connected graphs, $a \in E(F_1)$, and $b \in E(F_2)$. It is easily seen that $F_1 \ominus_{a,b} F_2 \rightarrow \mathcal{R}(2K_2, 2C_n)$. Next, we show that for every edge $e \in E(F_1 \ominus_{a,b} F_2)$, $(F_1 \ominus_{a,b} F_2) - e \nrightarrow (2K_2, 2C_n)$. Without loss of generality, we consider $e \in E(F_1)$. Then, there exists a $(2K_2, C_n)$-coloring ϕ_1 of edges of F_1. Let ϕ be a red-blue coloring of edges of $(F_1 \ominus_{a,b} F_2) - e$ such that $\phi(x) = \phi_1(x)$ for $x \in E(F_1 - e)$ and $\phi(x) =$ blue for all $x \in E(F_2)$. It can be immediately shown that ϕ is a $(2K_2, 2C_n)$-coloring of $(F_1 \ominus_{a,b} F_2) - e$. Hence, for every $a \in E(F_1)$ and $b \in E(F_2)$, $F_1 \ominus_{a,b} F_2 \in \mathcal{R}(2K_2, 2C_n)$. □

Before we observe a path P_n and a star $K_{1,n}$, we first give one class of graph belonging in $\mathcal{R}(2K_2, 2P_n)$.

Theorem 10 *Let $n \geq 3$ be an integer. The cycle $C_s \in \mathcal{R}(2K_2, 2P_n)$ if and only if $2n + 1 \leq s \leq 3n - 1$.*

Proof Let s and n be integers such that $7 \leq 2n + 1 \leq s \leq 3n - 1$. Consider a red-blue coloring of C_s containing no red $2K_2$. Then, all edges of C_s are blue or the subgraph induced by red edges form a path P_3. Hence, the remaining edges form a path P_{s-1}. It means that there exists a blue $2P_n$. Now, let e be an edge of C_s. Then, $C_s - e = P_s$. Let $V(P_s) = \{v_1, v_2, \ldots, v_s\}$. Let ϕ be a red-blue coloring of edges of $C_s - e$ such that $\phi(x) =$ red for x incident to v_n and $\phi(x) =$ blue otherwise. We obtain a $(2K_2, 2P_n)$-coloring of edges of $C_s - e$. So, $C_s \in \mathscr{R}(2K_2, 2P_n)$.

Next, let $C_s \in \mathscr{R}(2K_2, 2P_n)$ for $n \geq 3$. For a contradiction, if $s \leq 2n$, then we have a $(2K_2, 2P_n)$-coloring ϕ of edges of C_s where $\phi(x) =$ red for x incident to v_1 and the remaining edges are colored by blue. Under the coloring ϕ, there is neither a red $2K_2$ nor a blue $2P_n$. If $s \geq 3n$, then C_s is not minimal, since C_s contains $3P_n$. □

Now, we will replace the cycle C_n in Theorem 9 by a path P_n. Let F_1, F_2 be connected graphs in $\mathscr{R}(2K_2, P_n)$. For every $a \in E(F_1)$ and $b \in E(F_2)$, a graph $F_1 \ominus_{a,b} F_2$ can contain either $3P_n$ or a cycle C_s with $2n + 1 \leq s \leq 3n - 1$. Even though, we exclude the graph $3P_n$ and cycle C_s with $2n + 1 \leq s \leq 3n - 1$, we cannot guarantee that $F_1 \ominus_{a,b} F_2 \in \mathscr{R}(2K_2, 2P_n)$. For example, by [1], we know that $C_7, F_3 \in \mathscr{R}(2K_2, P_5)$, where F_3 is the graph in Fig. 3. For the edge $a \in E(C_7)$ and $b \in E(F_3)$ as depicted in Fig. 2, we have $C_7 \ominus_{a,b} F_3$ containing neither $3P_n$ nor C_{11}. Since for $e = a = b$, $(C_7 \ominus_{a,b} F_3) - e \rightarrow \mathscr{R}(2K_2, 2P_5)$, then $C_7 \ominus_{a,b} F_3 \notin \mathscr{R}(2K_2, 2P_5)$. A similar argument holds for a star $K_{1,n}$ (see Fig. 4). So, we only have the following theorems. The proofs of these theorems can be done in a similar way as the proof of Theorem 7.

Theorem 11 *Let $n \geq 3$ be an integer. Let F_1, F_2 be two connected graphs in $\mathscr{R}(2K_2, P_n)$, $a \in E(F_1)$ and $b \in E(F_2)$. If $F_1 \ominus_{a,b} F_2$ contains neither $3P_n$ nor C_s with $2n + 1 \leq s \leq 3n - 1$, then $F_1 \ominus_{a,b} F_2 \rightarrow \mathscr{R}(2K_2, 2P_n)$.*

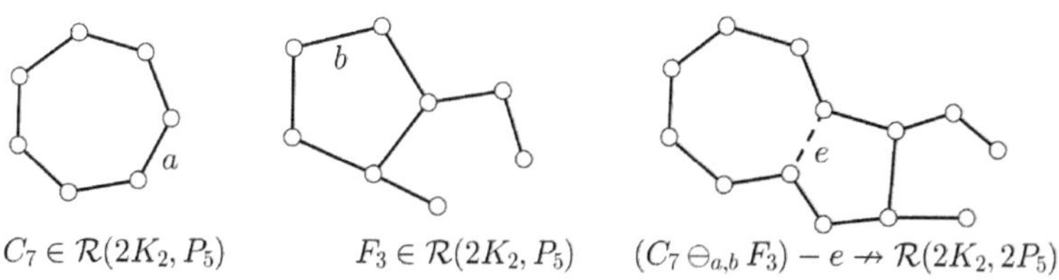

Fig. 3 The graph $C_7 \ominus_{a,b} F_3 \not\supseteq 3P_5$ and $C_7 \ominus_{a,b} F_3 \not\supseteq C_{11}$ but $C_7 \ominus_{a,b} F_3 \notin \mathscr{R}(2K_2, 2P_5)$

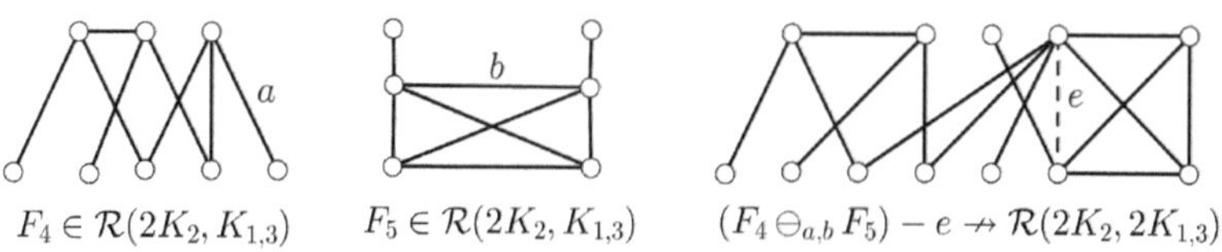

Fig. 4 The graph $F_4 \ominus_{a,b} F_5 \not\supseteq 3K_{1,3}$ but $F_4 \ominus_{a,b} F_5 \notin \mathscr{R}(2K_2, 2K_{1,3})$

Theorem 12 *Let $n \geq 3$ be an integer. Let F_1, F_2 be two connected graphs in $\mathscr{R}(2K_2, K_{1,n})$, $a \in E(F_1)$ and $b \in E(F_2)$. If $F_1 \ominus_{a,b} F_2$ contains no $3K_{1,n}$, then $F_1 \ominus_{a,b} F_2 \to \mathscr{R}(2K_2, 2K_{1,n})$.*

Acknowledgments This research was supported by Research Grand "Program Hibah Riset Unggulan ITB-DIKTI", Ministry of Research, Technology and Higher Education, Indonesia.

References

1. Baskoro, E.T., Yulianti, L.: On Ramsey minimal graphs for $2K_2$ versus P_n. Adv. Appl. Discret. Math. **8**(2), 83–90 (2011)
2. Baskoro, E.T., Wijaya, K.: On Ramsey $(2K_2, K_4)$-minimal graphs, Mathematics in the 21st Century. Springer Proceedings in Mathematics and Statistics, vol. 98, pp. 11–17 (2015)
3. Burr, S.A., Erdös, P., Faudree, R.J., Schelp, R.H.: A class of Ramsey-finite graphs. In: Proceeding of the Ninth Southeastern Conference on Combinatorics, Graph Theory and Computing, pp. 171–180 (1978)
4. Mengersen, I., Oeckermann, J.: Matching-star Ramsey sets. Discret. Appl. Math. **95**, 417–424 (1999)
5. Tatanto, D., Baskoro, E.T.: On Ramsey $(2K_2, 2P_n)$-minimal graphs. AIP Conf. Proc. **1450**, 90–95 (2012)
6. Wijaya, K., Yulianti, L., Baskoro, E.T., Assiyatun, H., Suprijanto, D.: All Ramsey $(2K_2, C_4)$-minimal graphs. J. Algorithms Comput. **46**, 9–25 (2015)

Solution of Viscous Burgers Equation Using a New Flux Based Scheme

Mohammad Belal and Nadeem Hasan

Abstract In this work, a new scheme (PVU-M+) is used for obtaining the numerical solution of viscous Burgers equation (VBE) in 1-D which is widely regarded as a 1D cartoon of the Navier–Stokes Equation. The scheme is a variant of PVU family of schemes. The analytical/exact solutions of problems governed by the equation are utilized for comparing the numerical solutions obtained by PVU-M+ scheme.

Keywords Viscous Burgers equation · Discontinuities · Waves

1 Introduction

While computing the compressible flows which are governed by Euler/Navier–Stokes equations, the discontinuities pose a stiff challenge. There should be sufficient numerical diffusion in the numerical scheme to stably resolve the discontinuities without any spurious oscillations. It is generally believed that schemes relying on the wave dynamics would capture the flow physics of compressible flows much better. There has been much effort in development of schemes that utilize the wave dynamics. However, schemes relying on direct discretization of the conservation equations didnt get much attention.

1.1 Background

Laney [1] classified the time marching methods for Euler/Navier–Stokes equations as:

M. Belal · N. Hasan (✉)
Department of Mechanical Engineering, Aligarh Muslim University, Aligarh, India
e-mail: nadhasan@gmail.com

M. Belal
e-mail: belalbh38@gmail.com

J.M. Cushing et al. (eds.), *Applied Analysis in Biological and Physical Sciences*, Springer Proceedings in Mathematics & Statistics 186,
DOI 10.1007/978-81-322-3640-5_14

1. Flux based methods
2. Wave based methods of Reconstruction-Evolution type
3. Wave based methods of flux splitting type

Qamar et al. 2006 [2] gave a flux based scheme for computation of compressible flows. The scheme is a two-step, predictor-corrector algorithm utilizing the basic ideas of upwinding. The inter cell convective fluxes were estimated on the basis of upwind interpolation of convective transport vector based on particle convective velocities and thus the scheme was named as Particle Velocity Upwind (PVU) scheme. The scheme employed first-order upwind procedure without explicitly using artificial viscosity/damping to stabilize the solution in the vicinity of shocks.

Later, Qamar et al. 2010 [3] presented an extension of the PVU scheme. Here first order upwind interpolations was only utilized in the regions of discontinuities while second order upwind interpolation was employed for both inter cell velocity and convective transport vector in smooth regions of solution. The use of first order upwind interpolations in the neighborhood of shocks resulted in high dissipation.

In its third phase Hasan et al. [4] extended the scheme to an efficient and robust higher order scheme named as PVU-M+. In this version higher order (central/upwind) estimates for inter-cell particle velocity and inter-cell convective property vector are obtained and blended through a suitably defined weight function W_f in the smooth varying solution region. In the vicinity of shocks, the higher order estimates (central/upwind) are combined with the lower order estimates (upwind biased) through suitably defined solution sensitive weight functions or limiter functions in order to compute the inter-cell numerical convective flux. This reduces the large dissipation associated with first order upwind interpolation estimate employed in earlier versions reported in [2, 3].

In the present work, the solutions of 1D viscous Burgers equation are obtained using PVU-M+ scheme. VBE is taken as a model, not only to test the numerical methods but also to obtain the numerical solution of the equation for small values of the viscosity, largely due to its similarity with Navier–Stokes equation (NSE). In the present study, the 1D viscous Burgers equation (VBE) has been selected as a test bed as analytical solutions can be obtained for initially continuous as well as discontinuous spatial data. Consequently, the properties like numerical dissipation and numerical dispersion for a nonlinear convection combined with linear diffusion model problems can be easily identified and studied. Such properties play an important role in capturing the chaotic/turbulent solutions of Navier–Stokes equations. The compressible flow test cases for the PVU-M+ scheme reported earlier in [4] comprised mostly of inviscid model problems so that the ability of the scheme to capture the shock dynamics under action of non-linear convection and physical diffusion is not known. Also the ability of the scheme in resolving the wave dynamics for a non-linear convection combined with linear diffusion scenario, as represented by VBE is also not established and warrants an investigation.

1.2 Viscous Burgers Equation

The classical VBE in 1D is given as,

$$\frac{\partial u}{\partial t} + u\frac{\partial u}{\partial x} = \nu\frac{\partial^2 u}{\partial x^2} \tag{1}$$

where u is a characteristic velocity in x direction, t is the time and ν is the damping or viscosity parameter.

VBE is a fundamental partial differential equation that occurs in various areas of physical sciences. It is named after Johannes Martinus Burgers (1895–1981). It can be considered as a simplified form of the Navier–Stokes equation [5] due to the presence of nonlinear convection term and the linear viscosity term. The main feature of VBE is the simultaneous existence of a nonlinear convection term and a linear diffusion term. If the diffusion term is dominant over convection, the solution of VBE approaches the solution of the diffusion equation. On the other hand, if the nonlinear term dominates, there will be formation of shock-like discontinuities. The conservation form of VBE is given as,

$$\frac{\partial u}{\partial t} + \frac{\partial}{\partial x}\left(\frac{u^2}{2} - \nu\frac{\partial u}{\partial x}\right) = 0 \tag{2}$$

The VBE is an important model which appears in various fields of physical science [6–12]. Analytic solutions of VBE can be obtained by the Cole–Hopf transformation, which transforms the VBE to a linear heat equation [9, 10].

1.3 Exact Solutions of Viscous Burgers Equation

To transform VBE into a linear heat equation via the Cole–Hopf transformation [9, 10], we consider Eq. (2). The transformation can be done in two steps, first introduce ζ such that,

$$u = \zeta_x \tag{3}$$

$$\frac{u^2}{2} - \nu\frac{\partial u}{\partial x} = -\zeta_t \tag{4}$$

Now, taking $\xi(x,t) = e^{-\frac{\zeta}{2\nu}}$

$$\zeta = -2\nu log\xi \tag{5}$$

Using the above relations, Eq. (2) transforms to a linear parabolic equation (pure diffusion).

$$\xi_t = \nu\xi_{xx} \tag{6}$$

Now, when the VBE has been transformed into a linear partial differential equation, the initial and boundary conditions on u can be transformed via

$$u(x,t) = -2\nu\frac{\xi_x}{\xi} \tag{7}$$

The **initial condition** can be given as,

$$u(x,0) = f(x) = -2\nu\frac{\xi_x}{\xi} \tag{8}$$

On integrating this equation, we get

$$\xi(x,0) = \xi(0,0)exp\left[-\frac{\int_0^x u(x',0)dx'}{2\nu}\right] \tag{9}$$

Now, using Eq. (7) the **boundary conditions** can be given as,

$$u(0,t) = -2\nu\frac{\xi_0}{\xi} \Rightarrow \xi_x + \frac{u(0,t)}{2\nu}\xi = 0 \tag{10}$$

$$u(l,t) = -2\nu\frac{\xi_l}{\xi} \Rightarrow \xi_x + \frac{u(l,t)}{2\nu}\xi = 0 \tag{11}$$

1.3.1 Wave Solutions

For wave solution over the interval $-\infty < x < +\infty$, the boundary conditions are not required. A common example of a solution with space-time translational symmetry is the travelling wave solution. Suppose the Eq. (6) admits a travelling wave solution [13] of the form given as,

$$\xi(x,t) = Ae^{\mu(x+ct)}, \tag{12}$$

where c and μ are arbitrary complex constants. On substituting in Eq. (6), c and μ can be shown to be related as,

$$c = \nu\mu. \tag{13}$$

Since Eq. (6) is linear, by superposition,

$$\xi(x,t) = D + \sum_{1}^{p} B_i e^{\mu_i(x+\nu\mu_i t)}, \tag{14}$$

is also a solution. Expressing $\mu_i = \mu_{R_i} + j\mu_{I_i}$, the real part of the solution is given as,

$$\xi(x, t) = D + \sum_{1}^{p} B_i e^{\mu_{R_i}x+\nu\left(\mu_{R_i}^2-\mu_{I_i}^2\right)t} \cos\left(\mu_{I_i}x + 2\mu_{R_i}\mu_{I_i}\nu t\right). \tag{15}$$

This equation represents a general wave solution to the heat equation.

1. Standing waves: Standing waves are characterized by a zero phase speed. This implies that for standing waves, $\mu_{R_i}\mu_{I_i} = 0$.
   For $\mu_{R_i} = 0$, $\mu_i = \mu_{I_i}$. One obtains solution using Eq. (15) of the form given as,

$$\xi(x, t) = D + \sum_{1}^{p} B_i e^{-\nu\mu_{I_i}^2 t} \cos\left(\mu_{I_i}x\right). \tag{16}$$

   This equation represents a superposition of individual standing waves whose amplitude damps out in time for $\nu > 0$. The corresponding solution for the VBE can be obtained using Eq. (7) as,

$$\mu(x, t) = 2\nu \frac{\sum_1^p \left[C_i\mu_{I_i} \exp\left\{-\nu\mu_{I_i}^2 t\right\} sin\left(\mu_{I_i}x\right)\right]}{1 + \sum_1^p \left[C_i\mu_{I_i} \exp\left\{-\nu\mu_{I_i}^2 t\right\} cos\left(\mu_{I_i}x\right)\right]} \tag{17}$$

   where $C_i = B_i/D$.
2. Travelling waves: Travelling waves are mathematically represented as $\xi = func\ (x \pm at)$, where a is a real constant representing the phase velocity of the wave. Therefore, if we choose $\mu_{I_i} = 0$, using Eq. (15) one obtains a solution of the form given as,

$$\xi(x, t) = D + \sum_{1}^{p} B_i e^{\mu_{R_i}(x+\nu\mu_{R_i}t)}. \tag{18}$$

   This equation represents a spatially non-periodic travelling wave solution to the heat equation.
3. Spatially modulated travelling wave: If $\mu_i = \mu_{R_i} = \pm(\mu_{I_i})$ one obtains a solution of the form given as,

$$\xi(x, t) = D + \sum_{1}^{p} B_i e^{\mu_i x} cos\mu_i(x \pm 2\mu_i\nu t). \tag{19}$$

   The above solution is readily recognized as a spatially modulated travelling wave. For $\nu > 0$, $\mu_i > 0$, it represents a left travelling wave, while for $\mu_i < 0$ one obtains a right travelling wave in the (t-x) plane. In both cases the waves are damped along their direction of propagation.

1.3.2 Exact Solutions for Specified Initial and Boundary Conditions

Consider the boundary and initial conditions as,

$$u(0,t) = 0, u(l,t) = 0, t > 0, \tag{20}$$

$$u(x,0) = f(x), 0 \le x \le l. \tag{21}$$

The methodology for obtaining the exact solution has been taken from Lokenath Debnath [14]. It follows from Eq. (9) that

$$\xi(x,0) = exp\left[-\frac{1}{2\nu}\int_0^x f(x')dx'\right] \tag{22}$$

The standard solution of the linear diffusion equation is given by

$$\xi(x,t) = a_o + \sum_{n=1}^{\infty} a_n exp\left(-\frac{n^2\pi^2\nu t}{l^2}\right) cos\left(\frac{n\pi x}{l}\right) \tag{23}$$

where

$$a_o = \frac{1}{l}\int_0^l \xi(x,0)dx \tag{24}$$

and

$$a_n = \frac{2}{l}\int_0^l \xi(x,0)cos\left(\frac{n\pi x}{l}\right) dx \tag{25}$$

The corresponding solution for the VBE can be obtained using Eq. (7).

2 PVU-M+ Scheme

The general equation for the compressible form of the Navier–Stokes equations strong conservative form [15] in three dimensions is

$$\frac{\partial U}{\partial t} + \frac{\partial F}{\partial x} + \frac{\partial G}{\partial y} + \frac{\partial H}{\partial z} = J \tag{26}$$

As the present work deals with one-dimensional viscous Burgers equation, the equation in dimensionless form is expressed as,

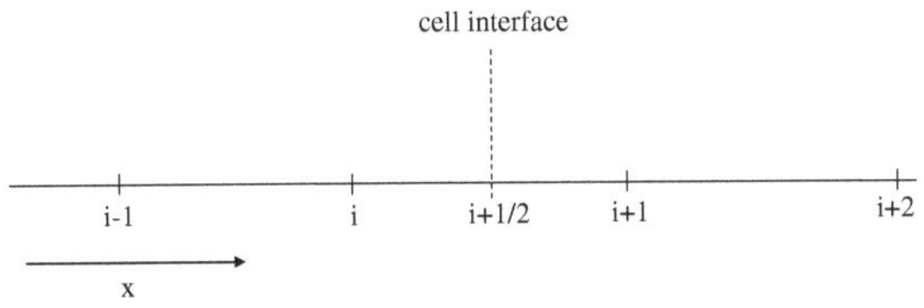

Fig. 1 Diagram showing a computational molecule at a grid point (i)

$$\frac{\partial U}{\partial t}+\frac{\partial (F^{c}+F^{nc})}{\partial x}=0 \tag{27}$$

where U is the solution vector, F is the flux vector, F^c is the convective flux vector and F^{nc} is the non-convective flux vector. Figure 1 shows a typical uniform 1-D mesh with a computational segment surrounding the ith node or grid point. The PVU-M+ scheme is a two-step predictor-corrector scheme [4]. The solution vector at the ith grid point can be obtained at the new time level (nt + 1) through the predictor and corrector steps given as,

Predictor Step:

$$U^{*}=U^{nt}-\Delta t\left(\frac{F^{c}_{i+1/2}\left(U^{nt}\right)-F^{c}_{i-1/2}\left(U^{nt}\right)}{x_{i+1/2}-x_{i-1/2}}+\frac{F^{nc}_{i+1}\left(U^{nt}\right)-F^{nc}_{i}\left(U^{nt}\right)}{x_{i+1}-x_{i}}\right) \tag{28}$$

Corrector Step:

$$U^{nt+1}=\frac{U^{*}+U^{nt}}{2}-\frac{\Delta t}{2}\left(\frac{F^{c}_{i+1/2}\left(U^{*}\right)-F^{c}_{i-1/2}\left(U^{*}\right)}{x_{i+1/2}-x_{i-1/2}}+\frac{F^{nc}_{i}\left(U^{*}\right)-F^{nc}_{i-1}\left(U^{*}\right)}{x_{i}-x_{i-1}}\right) \tag{29}$$

The inter-cell numerical convective flux given in the above steps is expressed as,

$$F^{c}_{i+1/2}=u_{i+1/2}\Phi_{i+1/2} \tag{30}$$

where $u_{i+1/2}$ and $\Phi_{i+1/2}$ is an estimate of inter-cell particle velocity and the inter-cell convective property vector Q respectively.

2.1 Estimation of $u_{i+1/2}$ and $\Phi_{i+1/2}$ in Smooth Varying Solution Region

The higher order (cubic central interpolation) and lower order (quadratic upwind biased and first order upwind biased interpolation) estimates of any discrete function f_i at midway location i + 1/2 in the interval $[x_i, x_{i+1}]$ is given by Hasan et al. [4]. The cubic central and quadratic upwind estimates for both inter-cell numerical particle velocity and the inter-cell convective property vector are blended through a suitably defined weight function W_f. The weighted estimates, thus obtained are employed in the smooth regions. The weight function has been designed in such a way that it approaches unity with increasing magnitude of the convection velocities, thereby giving more weight to the upwind estimate [4].

To maintain the numerical stability in the vicinity of shocks and other types of discontinuities, the artificial viscosity or numerical dissipation is enhanced by employing lower order estimates. In order to achieve this, the weighted higher order estimates obtained above are combined with the lower order estimates, via solution sensitive weight functions ψ and η, to finally estimate the inter-cell values. The solution sensitive normalized weight functions ψ and η perform the dual role of identifying the discontinuities in the solution and automatically adjusting the proportions of the higher/lower order interpolation estimates to the inter-cell values. Finally, the estimates of inter-cell particle velocity and the inter-cell convective property vector are examined against a range boundedness criteria. If any of the estimates does not satisfy the range boundedness criteria, then the value of the component is taken to be the mean of the value on either side of the interface.

2.2 Estimation of $u_{i+1/2}$ and $\Phi_{i+1/2}$ in the Vicinity of Shocks or High Solution Gradients

The procedures described earlier, to obtain the estimates of $u_{i+1/2}$ and $\Phi_{i+1/2}$, lead to oscillations/overshoots/undershoots in the vicinity of high solution gradients or shocks on the failure of range boundedness criteria. A value of ψ or η in excess of a threshold value is used to identify a shock or a non-smooth solution feature. Threshold limits in the range of [0.7–0.9] are found to be suitable for the detection of shock.

In order to specifically identify the shock, the wave speeds on either side of the interface (i+1/2) are utilized. It is known that for the formation of shocks, the characteristics must converge. Once a shock is detected the inter-cell particle velocity and convective transport property vector can be determined as,

$$u_{i+1/2} = u_i \text{ and } \Phi_{i+1/2} = Q_i \tag{31}$$

3 Numerical Solution of Viscous Burgers Equation Using PVU-M+ Scheme

The numerical solution for VBE using PVU-M+ scheme has been obtained for six cases. Three cases involve wave solutions while the other three cases involve solutions with continuous and discontinuous initial conditions and fixed boundary conditions. The methodology for computing the analytical solution of the VBE in all these cases has already been presented in Sect. 1.3. Table 1 summarizes the cases employed for

Table 1 Summary of the Test cases employed for the wave solutions

Test case	Wave	Domain	Time
1	Standing wave	[−5, 5]	3.0, 6.0
2	Travelling wave	[−5, 5]	3.0, 6.0
3	Spatially modulated travelling wave	[−1, 1]	1.0, 2.0

Table 2 Summary of the Test cases employed for the wave solutions

Test case	Initial conditions u(x,0)	Boundary condition	Domain	Time
4	$u_O sin\left(\frac{j\pi x}{l}\right) sin\left(\frac{k\pi x}{l}\right)$	u(−2.5, t) = 0 u(2.5, t) = 0	[−2.5, 2.5]	3.0, 6.0
5	$0, 0 \le x \le \frac{1}{3}$ $1, 0 < x < \frac{2}{3}$ $0, \frac{2}{3} \le x \le 1$	u(0, t) = 0 u(1, t) = 0	[0, 1]	0.6
6	$-1, 0 \le x \le \frac{1}{3}$ $1, 0 < x < \frac{2}{3}$ $-1, \frac{2}{3} \le x \le 1$	u(0, t) = −1 u(1, t) = −1	[0, 1]	0.3

wave solutions and Table 2 summarizes the cases involving solutions with continuous and discontinuous initial conditions and fixed boundary conditions.

For cases 1–3, the parameter ν has been taken as 0.1 while for cases 5–6 it is taken as 0.005. For case 4 the parameter ν is chosen as 0.01. For all cases uniform grids having 160 points have been utilized except for case 4 where a uniform 400 point grid has been utilized in order to capture the small length scales in the solution. A time step of 5×10^{-4} is employed for test cases 1–4 while a smaller time step of 1×10^{-4} is employed for test cases involving initial discontinuities (cases 5 and 6).

Case 1: A three wave (p = 3) exact standing wave solution (Eq. (17)) is constructed. The values of exact solution parameters C_i have been taken as: $C_1 = 0.6$, $C_2 = 0.65$, $C_3 = 0.7$, $\mu_1 = 1.1$, $\mu_2 = 1.2$, and $\mu_3 = 1.3$. Figure 2 compares the solution obtained by the PVU-M+ scheme with the exact solution. The waves are damped with time. The shape and amplitude are well captured by the PVU-M+ scheme. This is a direct evidence of the fact that the physical rates of convection and diffusion are very faithfully captured.

Case 2: In this case also, a three wave (p = 3) exact travelling wave solution is employed. The values of parameters C_i and that of μ_{R_i} have been taken identical to those considered for case 1. Figure 3 compares the solution obtained by the PVU-M+ scheme with the exact solution. The wave motion (towards left) is resolved quite accurately without any dispersion and noticeable numerical (excessive) diffusion.

Case 3: This is a test case based on spatially modulated left travelling wave solutionof the transformed heat equation (i.e. $\mu_i > 0$). A single wave (p = 1) exact

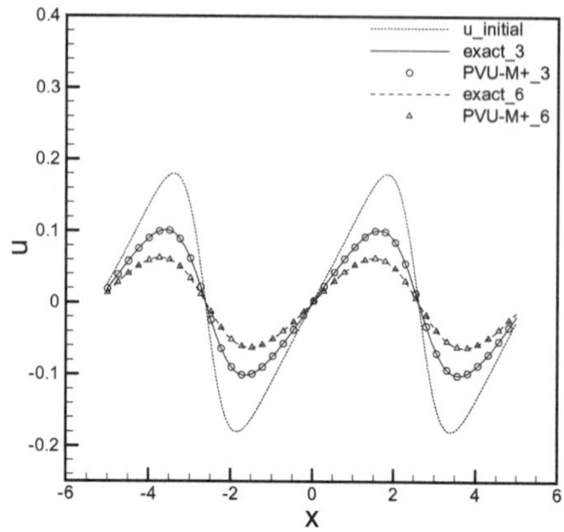

Fig. 2 Comparison of solution obtained by the PVU-M+ scheme with the exact solution for case 1 taking 160 grid points

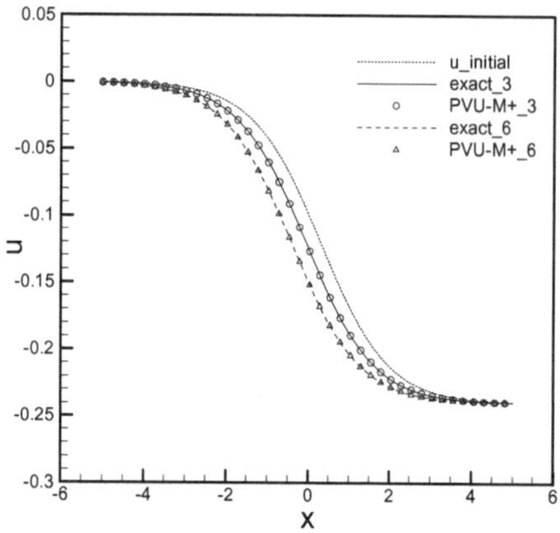

Fig. 3 Comparison of solution obtained by the PVU-M+ scheme with the exact solution for case 2 taking 160 grid points

solution is obtained by setting the solution parameters as $C_1 = 0.001$ and $\mu_1 = 0.4$, respectively. Figure 4 compares the numerical solutions with the exact solutions. The wave movement towards the left is faithfully captured by the scheme.

Cases 4–6 involve exact solutions of VBE subjected to specified initial and boundary conditions. The integrals involved in Eqs. (22), (24) and (25) are estimated numerically using one-third Simpsons rule. For case 4, uniformly spaced 151 points over the entire domain are utilized while for case 5 and case 6 a data sample of 201 uniformly spaced points are employed. A convergence study of the integral values ensures that the number of sample points is appropriate for each case. The exact solutions are constructed (Eqs. (23) and (7)) by considering leading 40 terms in case 4 while 50 terms are utilized in cases 5 and 6. The number of terms is chosen so as to yield a convergence upto 5th decimal place in the analytical series solutions.

Case 4: This case involves a smooth initial data having the superposition of sine wave. The value of u_o has been taken as 0.2 which controls the amplitude of the waveform. The values of j and k have been taken as 6 and 26 respectively and l is

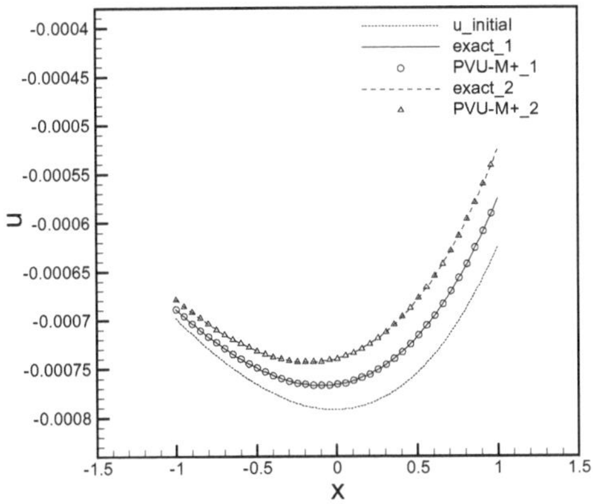

Fig. 4 Comparison of solution obtained by the PVU-M+ scheme with the exact solution for case 3 taking 160 grid points

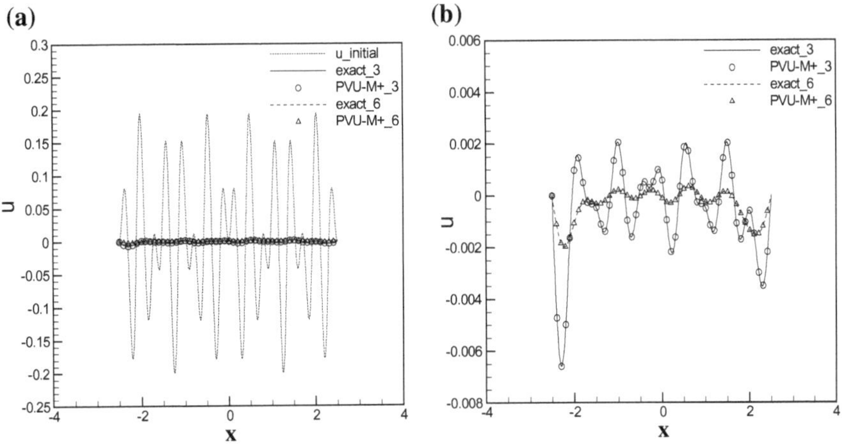

Fig. 5 Comparison of solution obtained by the PVU-M+ scheme with the exact solution on 400 grid points for case 4 **a** showing initial waveform **b** without initial waveform

the length of the domain which is 5. Figure 5 compares the solutions. For clarity, the exact and numerical solutions (using PVU-M+) at time t = 3 and t = 6 have been shown separately in Fig. 5b without the initial data. The shape of complex or superposed waves and their amplitude are well captured by the PVU-M+ scheme without any noticeable effects of numerical dissipation and dispersion.

Case 5: This case involves a jump discontinuity in the initial data. Further, with passage of time, the solution involves a moving shock towards right. Figure 6 compares the numerical solution with the exact solution solutions. The spreading and movement of the initial discontinuity under the combined effects of non-linear convection and linear viscous diffusion is captured very faithfully by the scheme.

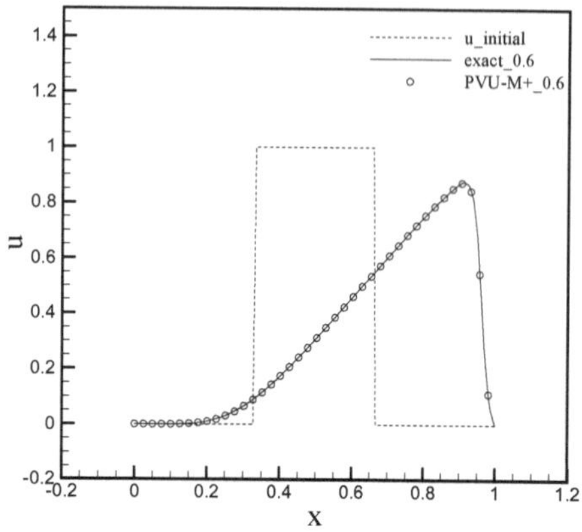

Fig. 6 Comparison of solution obtained by the PVU-M+ scheme with the exact solution for case 5 taking 160 grid points

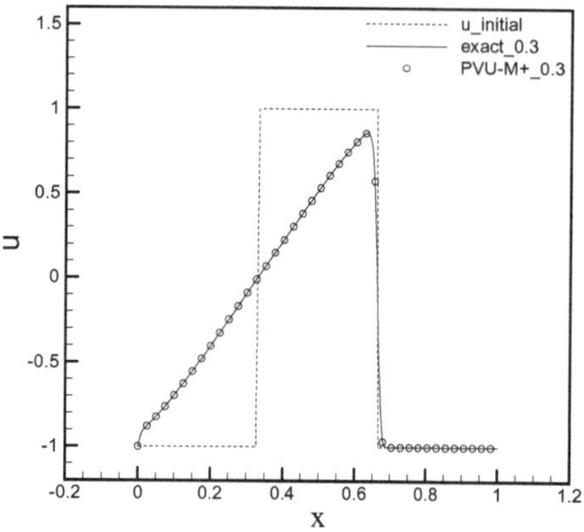

Fig. 7 Comparison of solution obtained by the PVU-M+ scheme with the exact solution for case 6 taking 160 grid points

Case 6: This test case involves a stronger jump discontinuity in the initial data in comparison to the previous case. In contrast with case 5, the discontinuity is stationary in time as the medium flows through it. Figure 7 compares the numerical solution with the exact solution solutions. The scheme captures the evolution of initial discontinuity very accurately and thus, can easily capture the steep gradients of stationary strong shocks without any numerical contamination in the form of excessive diffusion and dissipation.

3.1 Global Convergence Rates

The global convergence rates are a measure of the rate at which the numerical solution converges to the exact solution as the numbers of grid points are increased. For a discrete approximate solution $u_N(x_i, t)$ and exact solution $u_{EX}(x, t)$ on a uniform mesh of length scale h and total grid points N, the global error can be expressed as,

$$E(h) = \left\{ \frac{1}{N} \sum_{i=1}^{N} [u_N(x_i, t) - u_{EX}(x_i, t)]^2 \right\}^{\frac{1}{2}}. \quad (32)$$

For all the test cases, except case 4 and case 2, uniform grids with 40, 80 and 160 points have been taken. For test case 4, the grid points have been taken as 100, 200 and 400 while for case 2, the numerical solutions converge to the exact solutions at much coarser grids and solutions are considered at 10, 20 and 40 grid points. Figure 8 depicts a typical convergence characteristic for test case 3. Similar linear characteristics were obtained for other test cases.

For sufficiently large N (depending on the problem), the convergence characteristics exhibit a linear trend which implies that the behaviour of E with N can be mathematically described by the power law,

$$E = d\,(N)^m, \quad (33)$$

where, d and m are constants.

The slope m of the characteristic is a measure of the global convergence rate. The value of the constant m for the various test cases found from the convergence characteristics is summarized in Table 3. The PVU-M+ scheme exhibits a nearly quadratic

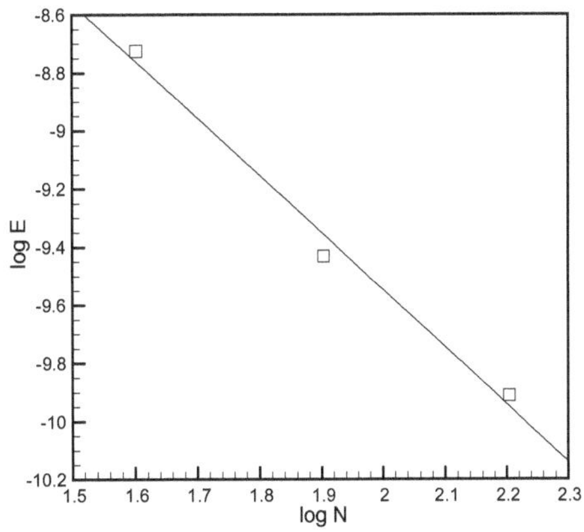

Fig. 8 Convergence characteristics for test case 3 employing 40, 80 and 160 grid points

Table 3 Convergence statistics for the different test cases

Test case	m
1	−2.27
2	−2.04
3	−1.97
4	−2.48
5	−2.42
6	−2.44

and even super-quadratic global convergence rate. This is in close agreement with the formal local spatial order of accuracy (between 2 and 3) of the various discretizations involved.

4 Conclusions

The cases involving wave solutions clearly indicate that the shape and amplitude are well captured by the PVU-M+ scheme without any significant numerical dissipation and dispersion. This indicates good potential of the scheme for capturing instabilities in non-linear viscous flow processes. The scheme also captures the solutions even on lowering the viscosity coefficient. Even on taking coarser grids, the scheme has been found to behave faithfully.

The case involving smooth initial data with superposition of sine waves and characterized by large number of length scales was easily captured by the scheme. Thus, the scheme shows good promise in capturing spatio-temporal evolution of widely varying length scales under the influence of non-linear interactions and linear physical diffusion as in the case of turbulent motion of fluids.

The evolution of initial discontinuities in the last two cases was easily captured by the scheme without any spurious oscillations or overshoots/undershoots. For all the test cases considered, the scheme delivers a quadratic/super-quadratic global convergence rate. Thus, the capturing of smooth as well as non-smooth or discontinuous features of the solutions without any spurious oscillations or overshoots/undershoots essentially establishes TVD character of the PVU-M+ scheme for VBE. The PVU-M+ scheme can further be used for the solution of 2D viscous Burgers equation (VBE). In a multi-dimensional scenario the interaction between different variables adds a new dimension and a new challenge for numerical methods.

References

1. Laney, C.: Computational Gasdynamics. Cambridge University Press, Cambridge (1998)
2. Qamar, A., Hasan, N., Sanghi, S.: New scheme for the computation of compressible flows. AIAA J. **44**, 1025–1039 (2006)
3. Qamar, A., Hasan, N., Sanghi, S.: A new spatial discretization strategy of the convective flux term for the hyperbolic conservation laws. Eng. Appl. Comput. Fluid Mech. **4**(4), 593–611 (2010)
4. Hasan, N., Khan, S.M., Shameem, F.: A new flux-based scheme for compressible flows. Comput. Fluids **119**, 58–86 (2015)
5. Polyanin, A.D., Zaitsev, V.F.: Handbook of Nonlinear Partial Differential Equations. Chapman & Hall/CRC Press, Boca Raton-London-New York (2003)
6. Burgers, J.: A mathematical model illustrating the theory of turbulence. Adv. Appl. Mech. **1**, 171–199 (1948)

7. J. M. J. M. Burgers, *The nonlinear diffusion equation : asymptotic solutions and statistical problems*. Dordrecht-Holland ; Boston : D. Reidel Pub. Co, 1974. First published in 1973 under title: Statistical problems connected with asymptotic solutions of the one-dimensional nonlinear diffusion equation
8. Bec, J., Khanin, K.: Burgers turbulence. Phys. Rep. **447**(12), 1–66 (2007)
9. Cole, J.D.: On a quasi-linear parabolic equation occurring in aerodynamics. Q. Appl. Math. **9**(3), 225–236 (1951)
10. Hopf, E.: The partial differential equation ut + uux=xx. Commun. Pure Appl. Math. **3**(3), 201–230 (1950)
11. Batchelor, G.K., Davies, R.M.: Surveys in Mechanics. Cambridge University Press, Cambridge (1956)
12. Woyczynski, W.A.: Burgers-KPZ Turbulence : Gttingen Lectures. Springer, Berlin (1998)
13. Salas, A.H.: Symbolic computation of solutions for a forced burgers equation. Appl. Math. Comput. **216**, 18–26 (2010)
14. Debnath, L.: Nonlinear Partial Differential Equations for Scientists and Engineers. Birkhauser, Basel (2012)
15. Anderson, J.: Computational Fluid Dynamics. The Basics with Applications. McGraw-Hill Education, New York (1995)

Effect of Slip Velocity on the Performance of a Magnetic Fluid Based Transversely Rough Porous Narrow Journal Bearing

Snehal Shukla and Gunamani Deheri

Abstract Efforts have been made to study and analyze the effect of slip velocity on the performance of a magnetic fluid based transversely rough porous narrow journal bearing. The Neuringer-Rosensweig model governs the fluid flow while the velocity slip is modeled by the method of Beavers and Joseph. The stochastic model of Christensen and Tonder has been adopted to evaluate the effect of transverse surface roughness. With the adding of suitable boundary conditions, the associated stochastically averaged Reynolds' equation is solved to obtain the fluid pressure, in turn, which results in the calculation of load carrying capacity. It is found that the combined effect of slip velocity and surface roughness is to decrease the load carrying capacity significantly, in general. Of course, in augmenting the performance of the bearing system, the eccentricity ratio plays a central role even if the slip parameter is at minimum. It is established that the bearing can support a load even in the absence of flow, unlike the case of a conventional lubricant.

Keywords Journal bearing · Load carrying capacity · Magnetic fluid · Porosity · Roughness · Slip velocity

Nomenclature

c radial clearance
e eccentricity ratio ($e = \frac{e_1}{c}$)
h fluid film thickness at any point
y Co-ordinate in axial dirction
e_1 eccentricity
H magnitude of magnetic field

S. Shukla (✉)
Department of Mathematics, Shri R.K. Parikh Arts and Science College,
Petlad, Gujarat, India
e-mail: snehaldshukla@gmail.com

G. Deheri
Department of Mathematics, Sardar Patel University,
Vallabh Vidyanagar, Anand, Gujarat, India
e-mail: gm.deheri@rediffmail.com

J.M. Cushing et al. (eds.), *Applied Analysis in Biological and Physical Sciences*,
Springer Proceedings in Mathematics & Statistics 186,
DOI 10.1007/978-81-322-3640-5_15

L length of the bearing
P lubricant pressure
R radius of the bearing
S slip parameter
U shaft surface speed
W load carrying capacity
P^* dimensionless pressure
S^* non-dimensional slip velocity
W^* dimensionless load carrying capacity
σ standard deviation
ε skewness
α variance
θ circumferential co-ordinate
φ Porosity
Ψ attitude angle of bearing
η dynamic viscosity of fluid
σ^* non-dimensional standard deviation
ε^* non-dimensional skewness
α^* non-dimensional variance
φ^* non-dimensional porosity
μ^* magnetization parameter
$\bar{\mu}$ magnetic susceptibility
μ_0 permeability of the free space
$\frac{dy}{dx}$ gradient of the film thickness in the direction of motion

1 Introduction

Journal bearings support load acting in the direction normal to the rotating shaft. It is the most common bearing among sliding bearings. Journal bearings are used even today as indispensable bearings in many rotating machines such as stream turbines, generators, blowers, compressors and ship propulsion shafts. Moreover, journal bearings have a strong impact on the vibration characteristics of machinery. The type of machinery we are concered with range from small high speed spindles to motors, fans, amd pumps to large turbines to some paper mill rolls and other large slow speed rotors. Journal-type models are often best for situations where there is a lot of motion. People often commit to regular oil changes for car and truck engines in part to keep these moving pieces in good working order. Additionally, they are widely use in gasoline and fueled piston engine in motor vehicle. Porous journal bearings impregnated with oils are widely used in industrial applications. Also, they are more advantageous because they do not need continuous lubrication, therefore, their structure is simple and also reduce cost is reduced.

The hydrodynamic theory of lubrication of porous journal bearing has been discussed in Cameron [1], who obtained a way out for oil film pressure and load carrying capacity of finite, full bearing using the short bearing assumptions. Baka [2] computed the pressure function using short bearing as well as long bearing approximations and calculated the load capacity and coefficient of friction and compared to one another. Desai and Patel [3] conducted an experimental analysis of pressure distribution in hydrodynamic journal bearing for various loading conditions with several operating parameters. Agostino et al. [4] presented an approximate model for unsteady finite porous journal bearings and proposed a model which provided a quick analytical calculation of the fluid film force for different values of the permeability factor, aspect ratio and eccentricity ratio. Durany et al. [5] considered the numerical solution of a transient thermo-hydrodynamical model for a journal bearing device.

Recently, considerable attention has been paid to the use of magnetic fluid as a lubricant modifying the performance of bearing system. The properties of magnetic fluids are well controlled by external magnetic field that gives broad possibilities for technical and biomedical applications. The use of magnetic fluid as a lubricant modifying the performance of a bearing system has been discussed in various investigations, Yan [6] deliberated on the performance of dynamically loaded journal bearings lubricated with couple stress fluids considering the elasticity of the liner. Lin [7] presented a study dealing with the effects of couple stress fluids based upon the Stock micro-continuum theory together with the Hopf bifurcation. Zakharov [8] reviewed the progress in the theory of hydrodynamic lubrication. Guha [9] dealt with the steady-state performance of hydrodynamic flexible journal bearings of finite width considering micro-polar ferrofluid lubrication. Shah and Patel [10] analyzed the performance of porous layered axially journal bearing lubricated with ferrofluid considering the effects of permeability of porous facing, slip velocity at the interface of the porous matrix and squeeze velocity. Lin et al. [11] investigated the non-Newtonian effects on the nonlinear stability boundary of short journal bearings. Here, it was found that the non-Newtonian effects provided a large stability boundary within the clearance circle as compared to the bearing lubricated with a Newtonian fluid.

When gap between two mating surfaces becomes smaller, the effects of roughness become more important. In most of the applications, the smoothness of bearing surfaces would not be valid for the accurate prediction of the performance and life of the bearings. Thus, surface roughness has been studied with much interest in the recent years because all bearing surfaces are rough to some extent. The effect of roughness in hydrodynamic bearings was first studied by Tzeng and Saibel [12] and subsequently by Christensen and Tonder [13–15]. Tzeng and Saibel [12] found that the load carrying capacity of such bearings was high and frictional force in an infinitely long bearing was low. The performance characteristics of transverse rough bearing were compared with that of smooth bearing. Later, Christensen and Tonder [13–15], extended the model of Tzeng and Saibel [12] for the longitudinal roughness case. Tala-Ighil [16] dealt with a numerical study, based on finite difference method to find the tendency for the importance of tribological properties of a journal bearing such as minimum film thickness, maximum pressure, and axial oil flow and friction torque. Buuren [17] solved the problem of porous journal bearing using Galerkin's method with a novel approach. The proposed method was verified by analytical expressions and new results were shown for porous journal bearings including the influence of rough surfaces.

Hsu et al. [18] investigated the performance of a ferrofluid under the combined influence of surface roughness and a magnetic field. Deheri and Patel [19] analyzed the performance of a squeeze film in an infinitely long rough journal bearing using the ferrofluid flow model of Jenkins. Shukla and Deheri [20] dealt with the performance of a magnetic fluid based squeeze film in porous rough infinitely long parallel plates under the influence of slip velocity.

Thus, it was deemed appropriate to make an investigation on the performance of a magnetic fluid based transversely rough porous journal bearing taking the slip velocity in to account.

2 Analysis

The geometry and configuration of the bearing system is provided in Fig. 1.

Under usual assumptions of hydrodynamic lubrication theory the governing Reynolds' equation for pressure, Cameron [1], by employing the Beavers and Joseph [21] model for smooth bearing system takes the form

$$dP/dx = (3U\eta/(h^3 S))(dh/dx)((L^2/4) - y^2) \quad (1)$$

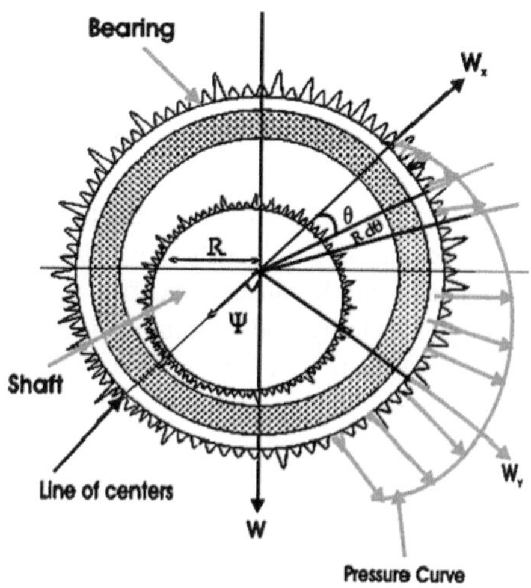

Fig. 1 The configuration of the bearing system

where

$$S = (1 + sh)/(2 + sh)$$

Following Christensen and Tonder [13–15] the expression for film thickness is considered in the form of

$$h(x) = \bar{h}(x) + h_s \tag{2}$$

where $\bar{h}$ is the mean film thickness, while h_s is a randomly varying portion measured from the mean level characterizing the random roughness. Here h_s is assumed to be governed by the probability density function $f(h_s)$ (Tzeng and Seibel [12])

$$f(h_s) = \begin{cases} \frac{32}{35b}(1 - (h_s^2/b^2))^3 \ ; -b \le h_s \le b \\ 0 \qquad\qquad\qquad\qquad , \text{elsewhere}, \end{cases} \tag{3}$$

The details regarding the mean α, standard deviation σ, and skewness ε associated with the characterization of roughness, can be seen from [13–15].

The magnetic field is oblique to the stator and its magnitude is given by

$$H^2 = kc^2 sin(2\pi - \theta) \tag{4}$$

where $k = 10^{14}\ A^2 m^{(-4)}$ chosen so as to hence a magnetic field of strength over 10^5 (Bhat [22]).

Stochastically averaging equation (1) by the method of Christensen and Tonder, and using the standard narrow bearing theory and assumptions of hydrodynamic lubrication as well as the method adopted by Bhat [22], the Reynolds type equation governing the pressure distribution is found to be

$$(d/dx)(P - (\mu_0 \bar{\mu} H^2)/2) = (3U\eta/a(h)S)(dh/dx)(L^2/4 - y^2) \tag{5}$$

where

$$a(h) = h^3 + 3\alpha h^2 + 3(\sigma^2 + \alpha^2)h + \varepsilon + 3\sigma^2\alpha + \alpha^3 + 12\varphi H$$

which is just $h^3 + 12\varphi H$ in the case of smooth bearings.

Considering

$$x = R\theta$$

and following the discussions of (Cameron [1]), the film thickness is expressed as

$$h = c(1 + ecos\theta) \tag{6}$$

The concerned boundary conditions are

$$P = 0, x = -L/2 \text{ and } x = L/2 \tag{7}$$

Introducing the dimensionless quantities

$$(y^*) = y/L,\ P^* = (Rc^2)/(U\eta L^2)P,\ \mu^* = (k\mu_0\bar{\mu}Rc^2)/U\eta,\ \varepsilon^* = \varepsilon/c^3,\ \sigma^* = \sigma/c,$$

$$\alpha^* = \alpha/c,\ \varphi^* = 12\varphi H/c^3,\ A_1 = \varepsilon^* + 3((\sigma^*)^2\alpha^*) + (\alpha^*)^3,\ A_2 = 12\varphi^*,\ A_4 = 3\alpha^*,$$

$$A_3 = 3((\sigma^*)^2 + (\alpha^*)^2),\ S^* = Sh \tag{8}$$

and fixing the following symbols

$$Q_1 = 1 + A_1 + A_2 + A_3 + A_4;\ Q_2 = eA_3 + 2eA_4 + 3e,\ Q_3 = e^2A_4 + 3e^2,\ Q_4 = e^3;$$

$$a_1 = Q_1 + Q_2 + Q_3 + Q_4;\ a_2 = (3/2)Q_4 + Q_3 + (1/2)Q_2$$

One obtains the expression for pressure distribution in dimensionless form as

$$P^* = ((0.25 - (y^*)^2)(\mu^*/6) + ((0.25 - (y^*)^2)esin\theta)/(Q_1 + Q_2cos\theta + Q_3cos^2\theta + Q_4cos^3\theta))(1/S^*) \tag{9}$$

The load carried by the narrow journal bearing can be found by integrating the pressure around it, taking the directions in to account and using half Sommerfeld conditions, which state that when the film diverges (at $\theta = \pi$), the pressure is uniformly zero. What is of course needed is the resulting force which is balanced by the load applied to the shaft. Following the discussions in Cameron, [1] the dimensionless load carrying capacity per unit width is given by

$$W^* = \sqrt{(W_X)^2 + (W_Y)^2} \tag{10}$$

where in
W_X: The total component in the direction of the line of center
W_Y: The total force at right angles to the line of center, determined by the relations

$$W_X = \int_0^{\pi}\int_{-L/2}^{L/2} (P - (\mu_o\bar{\mu}H^2)/2)Rcos\theta d\theta dx,\ W_Y = \int_0^{\pi}\int_{-L/2}^{L/2} (P - (\mu_o\bar{\mu}H^2)/2)Rsin\theta d\theta dy$$

This leads to the dimensionless load carrying capacity calculated as

$$W^* = (c^2/(U\eta L^3S^*))W = W_1 + W_2 \tag{11}$$

where

$$W_1 = \pi^2/2 + ((a_1/(2a_2)) - 1)\log((\pi^2a_2 - a_1)/a_1),$$

and

$$W_2 = \pi - 1/2\sqrt{(a_1/a_2)}log((\pi + \sqrt{(a_1/a_2)})/(\pi - \sqrt{(a_1/a_2)})) + ((\mu^* a_2)/3e).$$

3 Results and Discussions

It is clearly seen that Eq. (9) determines the dimensionless pressure distribution; while the distribution of load carrying capacity in non-dimensional form is given by Eq. (11). It is seen from Eqs. (9) and (11) that the increase in the dimensionless pressure and load carrying capacity as compared to the case of conventional lubricant respectively, turns out to be,

$$(0.25 - ((y^*)^2)\mu^*/6) \text{ and } (a_2/3e)\mu^* \tag{12}$$

It is manifested that the expression for the load carrying capacity is linear with respect to the magnetization parameter and hence the dimensionless load carrying capacity increases with increasing magnetization parameter as shown in Fig. 2. The magnetization results in an improved performance because it increases the viscosity of the lubricant leading to an increase in pressure, thus, providing increased load carrying capacity (Bhat [22]).

The profile for the non-dimensional load carrying capacity with respect to standard deviation is presented in Figs. 3 and 4. It is clearly seen that the standard deviation adversely affects the bearing system in the sense that the load carrying capacity decreases due to the standard deviation. Figure 3 says that the rate of decrease in the load carrying capacity due to σ^*gets decreased due to negatively skewed roughness. This is because the surface roughness of the bearing system retards the motion of lubricant, resulting in decreased load carrying capacity. However, this decrease is relatively less in the case of porosity and negligible up to $\varphi^*= 0.001$, which can be seen from Fig. 4.

One can observe the variation of non-dimensional load-carrying capacity with respect to variance from Figs. 5, 6 and 7. α^*(positive) decreases the load-carrying capacity. It is seen that α^*(negative) induces an increase in the load-carrying capacity. One can easily find that the effect of slip velocity and eccentricity ratio is more sharp, which can be seen from Figs. 6 to 7. It is interesting to note that the effect of the porosity on the distribution of the load-carrying capacity with respect to variance is negligible up to 0.001.

The profile for the distribution of the load-carrying capacity with respect to skewness is depicted in Figs. 8, 9, 10 and 11.

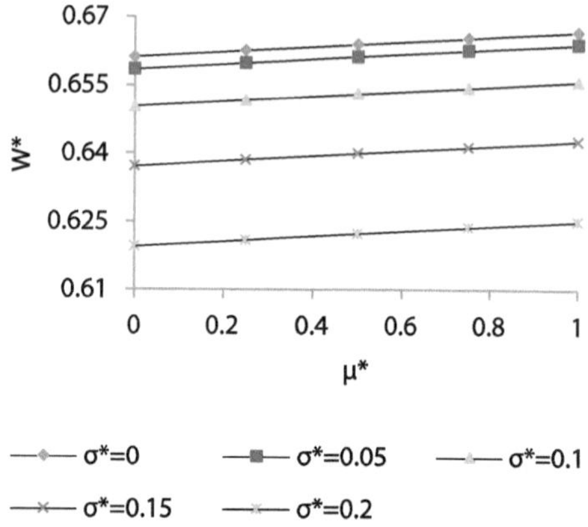

Fig. 2 The Variation of load carrying capacity with respect to μ^* and σ^*

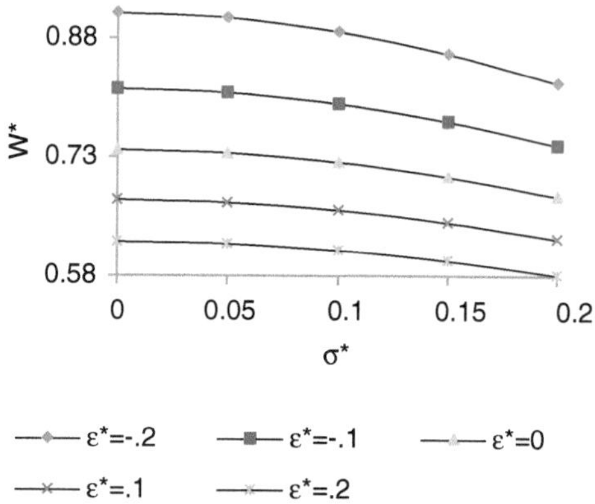

Fig. 3 The variation of load carrying capacity with respect to σ^* and ε^*

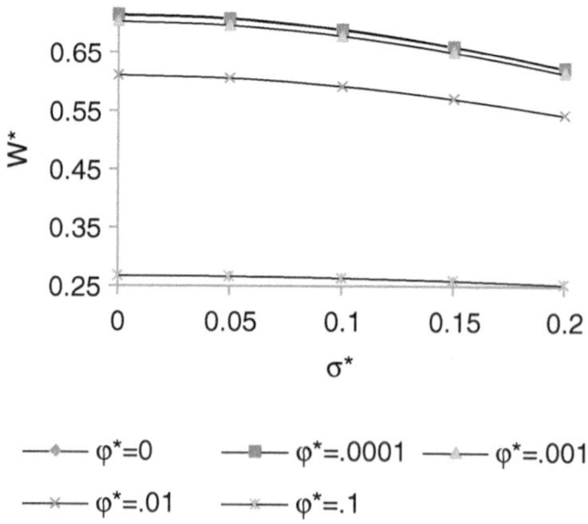

Fig. 4 The variation of load carrying capacity with respect to σ^* and φ^*

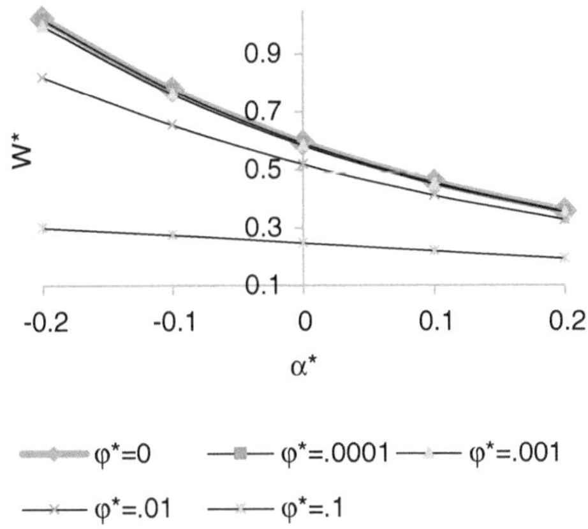

Fig. 5 The variation of load carrying capacity with respect to α^* and φ^*

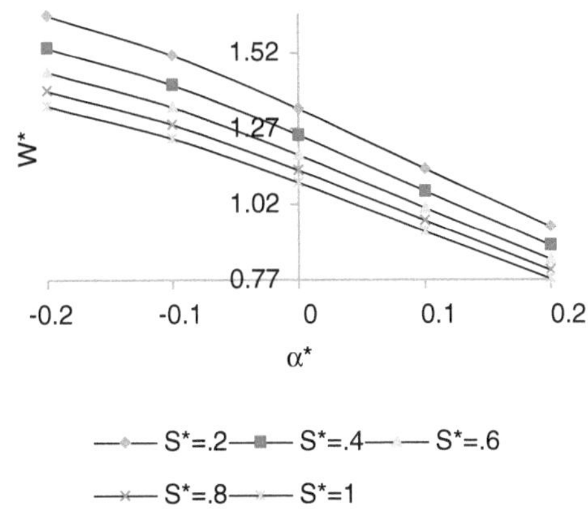

Fig. 6 The variation of load carrying capacity with respect to α^* and S^*

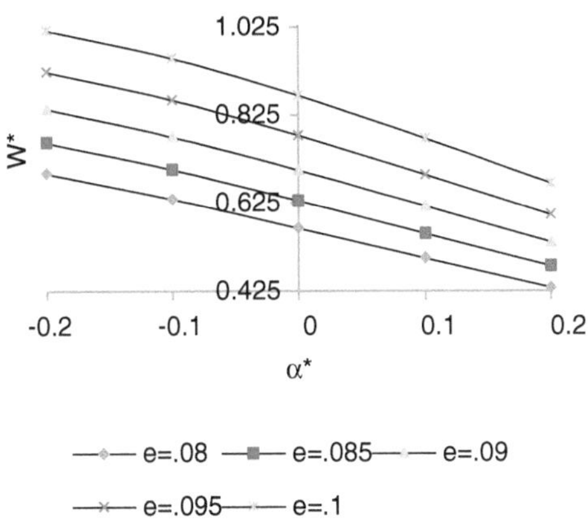

Fig. 7 The variation of load carrying capacity with respect to α^* and e

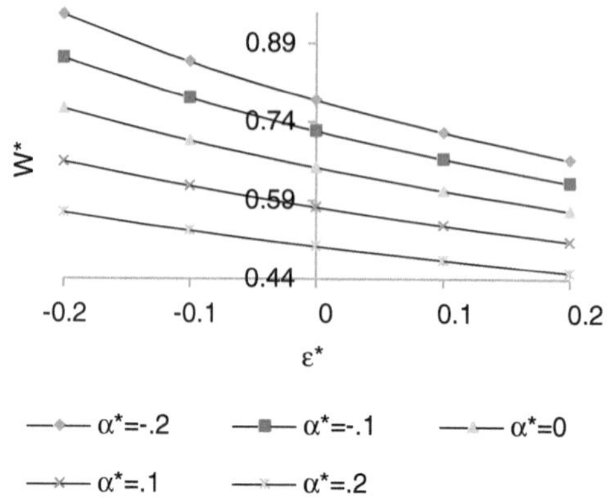

Fig. 8 The variation of load carrying capacity with respect to ε^* and α^*

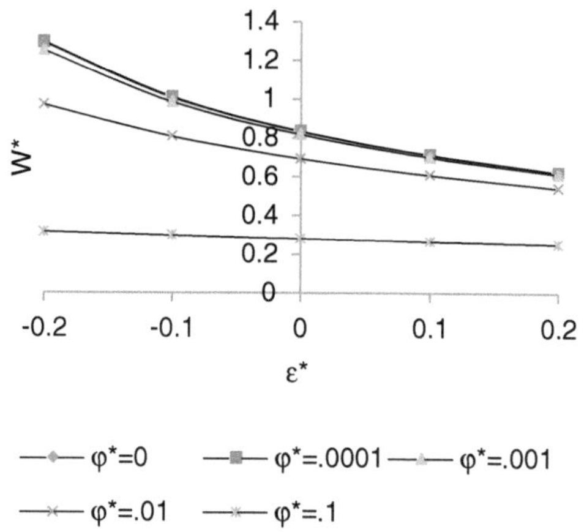

Fig. 9 The variation of load carrying capacity with respect to ε^* and φ^*

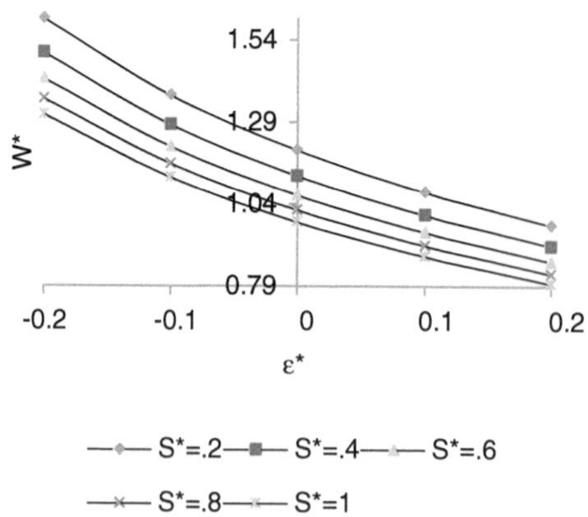

Fig. 10 The variation of load carrying capacity with respect to ε^* and S^*

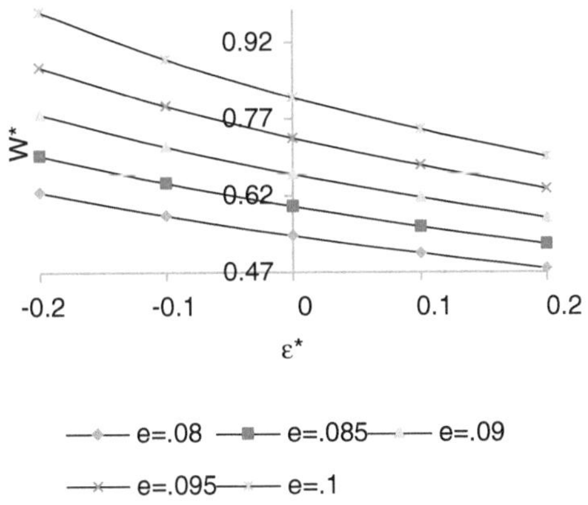

Fig. 11 The variation of load carrying capacity with respect to ε^* and e

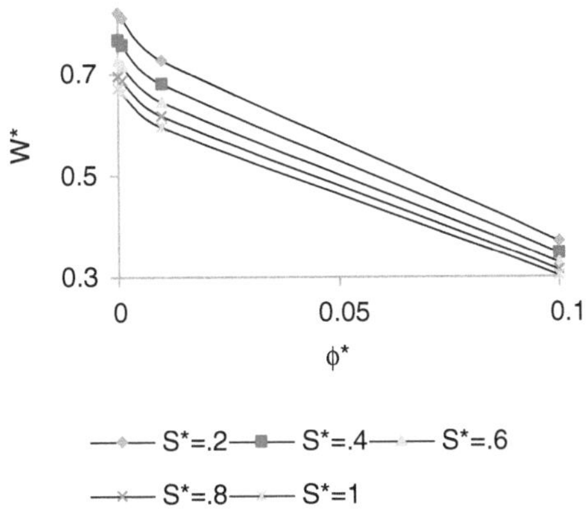

Fig. 12 The variation of load carrying capacity with respect to φ^* and S^*

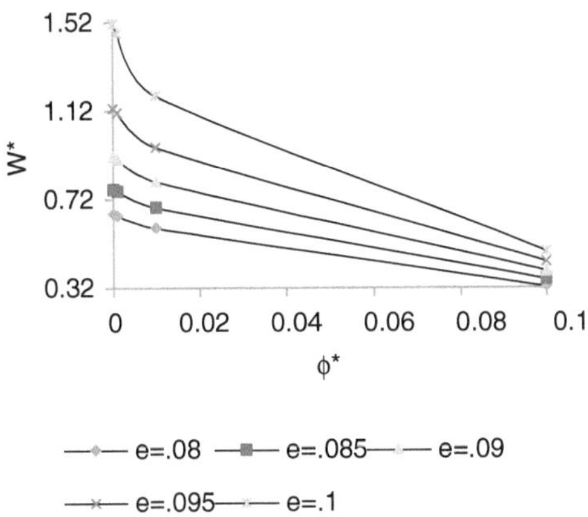

Fig. 13 The variation of load carrying capacity with respect to φ^* and e

It is found that the skewness follows the trends of the variance. The effect of slip velocity as well as eccentricity ratio is sharp which can be seen from Figs. 10 and 11. It is noticed that the effect of the porosity on the distribution of the load-carrying capacity with respect to skewness is negligible up to 0.001. However, one can easily notice that the positive effect of negative variance is relatively sharp as compared to the effect of negatively skewed roughness. Furthermore, one can visualize that the combined effect of negatively skewed roughness and negative variance is significantly positive in most of the situations.

The effect of the porosity is considerably adverse as can be seen from Figs. 12 and 13. One can easily conclude that indeed, the load carrying capacity decreases sharply due to porosity.

A close glance at the graphs reveals that the negative effect of porosity can be overcome to a large extent by the positive effect of magnetization in the case of negatively skewed roughness for small to moderate values of the slip parameter. The adverse effect of standard deviation can be compensated to some extent by the magnetization for small values of slip parameter at least in the case when α^* (-ve) is in place. The combined negative effect of positively skewed roughness and variance (+ve) can be reduced only to a small extent, by the positive effect of magnetization, keeping slip at minimum, for small values of standard deviation suitably choosing the eccentricity ratio.

These observations confirm that in spite of the fact that, the porosity, standard deviation and slip parameter combine decreases the load carrying capacity, this article offers some measures to improve the situation.

4 Validation

A comparison of the result found here with that of Cameon [1], indicates that the load is decreased by 1.35%. It is not surprising as roughness, slip velocity and porosity combine decreases the load carrying capacity. Only due to the effect of surface roughness, does the load capacity gets decreased by 3.29%.

In order to validate our results, the following sets of of comparison have been made with well-known published work of Cameon [1].

Quantity		$\bar{\varepsilon} = -0.05$				
		μ^*	α^*	σ^*	φ^*	S^*
Load carrying capacity in this Manuscript	With consideration	1.11955	0.66635	0.37117	1.06792	0.37876
	With consideration	1.10829	0.67170	0.44588	1.09453	0.38049
Article in Cameron	With consideration	1.16890	0.91033	0.68983	1.16894	0.75377
Quantity		$\varphi^* = 0.25$				
		μ^*	$\bar{\alpha}$	$\bar{\sigma}$	ε^*	S^*
Load carrying capacity in this Manuscript	With consideration	1.09112	0.32586	0.36186	0.42283	0.35659
	With consideration	1.07073	0.34374	0.37306	0.43791	0.40434
Article in Cameron	With consideration	1.16894	0.57110	0.60973	0.68983	0.64928
Quantity		$S^* = 0.3$				
		μ^*	$\bar{\alpha}$	$\bar{\sigma}$	φ^*	ε^*
Load carrying capacity in this Manuscript	With consideration	0.75952	0.73401	0.57497	0.57497	0.45948
	With consideration	0.74701	0.81488	0.60570	0.60570	0.46836
Article in Cameron	With consideration	0.91032	0.9558	0.81825	0.81825	0.68984

5 Conclusion

This investigation makes it clear that from bearing's life period point of view the roughness aspects must be considered carefully while designing the bearing system, even if there is the presence of a suitable magnetic strength and slip is at the minimum. To mitigate the adverse effect of porosity and standard deviation, even the eccentricity ratio may offer some help in the case of negatively skewed roughness when variance ($-ve$) occurs. In spite of the fact that there is a hoast of factors bringing down the load carrying capacity of the bearing system supports a good amount of load even when there is no flow. If developed this, can be a useful piece of work for space craft vehicles and nuclear power plant.

Acknowledgements The authors gratefully acknowledge with thanks the fruitful comments and suggestions of the Reviewer/Editor leading to an improvement in the presentation of the paper.

References

1. Cameron, A.: Basic theory of lubrication. Willy, New York (1972)
2. Baka, E.: Calculation of the hydrodynamic load carrying capacity of porous journal bearings. Periodica Polytechnica Ser. Mech. Eng. **46**, 03–14 (2002)
3. Desai, C.K., Patel, D.C.: Experimental analysis of pressure distribution of hydrodynamic journal bearing: a parametric study. In: Proceedings of the International Conference on Mechanical Engineering, Bangladesh, ICME 05-AM-30 (2005)
4. Agostino, V., Ruggiero, A., Senatore, A.: Approximate model for unsteady finite porous journal bearings fluid film force calculation. J. Eng. Tri. Part-J **220**, 227–234 (2006)
5. Durany, J., Pereira, J., Varas, F.: Dynamic stability of journal bearing devices through simulation of thermo hydrodynamic models. Tri. Inter. **9**, 1703–1718 (2010)
6. Yan-yan, M.A.: Performance of dynamically loaded journal bearings lubricated with couple stress fluids considering the elasticity of the liner. J. Zhejiang University SCIENCE A. **9**, 916–921 (2008)
7. Lin, J.: Weakly non-linear bifurcation analysis of a short journal bearing lubricated with Non-Newtonian couple stress fluids. J. Chin. inst. Eng **31**, 721–727 (2008)
8. Zakharov, S.: Hydrodynamic lubrication research: current situation and future prospects. J. Friction Wear **31**, 56–67 (2010)
9. Guha, S.: On the steady-state performance of hydrodynamic flexible journal bearings of finite width lubricated by ferro fluids with micro-polar effect. Int. J. Mech. Eng. Robotics Res. **1**, 32–49 (2012)
10. Shah, R.C., Patel, D.B.: Mathematical analysis of newly designed ferro fluid lubricated double porous layered axially undefined journal bearing with anisotropic permeability, slip velocity and squeeze velocity. Int. J. fluid. mech. **40**, 446–454 (2013)
11. Lin, J., Li, P., Hung, T., Liang, L.: Nonlinear stability boundary of journal bearing systems operating with non-Newtonian couple stress fluids. Tri. Int. **71**, 114–119 (2014)
12. Tzeng, S., Seibel, E.: Surface roughness effect on slider bearing lubrication. Trans. ASME J. Lubr. Technol. **10**, 334–338 (1967)
13. Christensen, H.,Tonder, K.C.: Tribology of rough surface: Stochastic models of hydrodynamic lubrication. SINTEF, Report No. 10/69-18 (1969)
14. Christensen, H.,Tonder, K.C.: Tribology of rough surface: parametric study and comparison of lubrication models. SINTEF, Report No. 22/69-18 (1969)
15. Christensen, H., Tonder, K.C.: Hydrodynamic lubrication of rough bearing surfaces of finite width. ASME-ASLE Lubrication Conference, paper No. 70-Lub (1970)
16. Tala-Ighil, N., Fillon, M., Maspeyrot, P., Bounif, A.: Hydro dynamic effects of texture geometries on journal bearing surfaces. In: Proceedings of the Tenth International Conference on Tribology, Romania, pp. 47–52 (2007)
17. Buuren, S., Hetzier, H., Hinterkausen, H., Seemann, W.: Novel approach to solve the dynamical porous journal bearing problem. Tri. Inter. **46**, 30–40 (2012)
18. Hsu, T., Chen, J., Chiang, H., Chou, T.: Lubrication performance of short journal bearings considering the effects of surface roughness and magnetic field. Tri. Inter. **61**, 169–175 (2013)

19. Deheri, G.M., Patel, N.D.: Ferrofluid lubrication of an infinitely long rough porous journal bearing. J. Ser. Soc. Commun. Mech **7**, 36–58 (2013)
20. Shukla, S.D., Deheri, G.M.: Effect of slip velocity on the performance of a magnetic fluid based squeeze film in porous rough infinitely long parallel plates. Friction. Wear. Res. **1**, 06–15 (2014)
21. Beavers, G.S., Joseph, D.D.: Boundary conditions at a naturally permeable wall. J. Fluid. Mech. **1**, 197–207 (1967)
22. Bhat, M.V.: Lubrication with a magnetic fluid. Team Spirit (India) Pvt. Ltd. (2003)

On the Wave Equations of Kirchhoff–Narasimha and Carrier

Pratik Suchde and A.S. Vasudeva Murthy

Abstract A nonlinear nonlocal wave equation modelling the coupling between transverse and longitudinal vibrations was derived by Carrier in 1945. In 1968, using careful asymptotics, Narasimha derived a similar equation but with a different nonlinearity (nowadays referred as Kirchhoff type nonlinearity). In this study we solve both the equations numerically and compare them. Since there are no experimental data available it is not possible to suggest which is the better model. However Kurmyshev has pointed out that Carrier's model cannot be valid for rubber or soft nylon strings.

Keywords Nonlinear wave equations · Kirchhoff

2010 Mathematics Subject Classification 74J30 · 74H45

1 Introduction

It is well known that the classical wave equation

$$\frac{\partial^2 u}{\partial t^2} = c^2 \frac{\partial^2 u}{\partial x^2} \tag{1}$$

does not accurately describe the transverse motion of a vibrating elastic string [5, 7, 10, 13, 17]. This was known to Kirchhoff and Rayleigh over a century ago [15]. The modern study to correct this anomaly was started by Carrier [5, 6] who derived the equation

P. Suchde · A.S. Vasudeva Murthy (✉)
TIFR Centre for Applicable Mathematics, Bangalore, India
e-mail: vasu@math.tifrbng.res.in

P. Suchde
e-mail: pratiksuchde@gmail.com

J.M. Cushing et al. (eds.), *Applied Analysis in Biological and Physical Sciences*, Springer Proceedings in Mathematics & Statistics 186,
DOI 10.1007/978-81-322-3640-5_16

$$\mathbf{w}_{tt}(x,t) = \left(1 + \frac{1}{2\pi}\int_0^{\pi} \mathbf{w}^2(x,t)\,dx\right)\mathbf{w}_{xx}(x,t), \quad 0 < x < \pi, \quad 0 < t \le T \quad (2)$$

followed by several others (see [15]) including Narasimha [17] who derived the equation

$$\mathbf{w}_{tt}(x,t) + 2R\mathbf{w}_t(x,t) = \left(1 + \frac{\Gamma}{2}\int_0^{1} \mathbf{w}_x^2(x,t)\,dx\right)\mathbf{w}_{xx}(x,t) + \mathbf{f}(x,t),$$
$$0 < x < 1, \quad 0 < t \le T \quad (3)$$

Although the nonlinearity in (2) and (3) are different, it was only recently that this difference has been highlighted. The nonlinearity in (3) is of the same form as that in Kirchhoff's wave equation [10] for slender beams. Even though mathematical studies of Kirchhoff's equation dates back to the 1940s [3], there has been no reference to Kirchhoff's work in the studies of Rayleigh [22], Carrier, Narasimha [17], Pohozaev [21] and Lions [12]. In fact, till 1985, (3) was referred as the Carrier–Narasimha equation [2]. Kirchhoff's work was only rediscovered in the 1990s [23].

Numerical study of (3) for planar ($w(x,t) = (y(x,t), 0)$), undamped ($R = 0$) and free ($f(x,t) \equiv 0$) vibrations have been done by Bilbao [4], Peradze [20], Liu and Rincon [13] and Christie and Sanz-Serna [7]. However, the same as not been done for the nonplanar, damped and forced case or for Eq. (2). Furthermore, there has been no comparison between the solutions of (2) and (3). The aim of the present study is to make this comparison. Since exact solutions are difficult to obtain, we seek numerical solutions.

2 The Equations of Narasimha and Carrier

While Kirchhoff's equation represented planar motion, Narasimha and Carrier's equations represent non-planar motion for a vibrating elastic string. In Eqs. (2) and (3), $\mathbf{w}(x,t)$ is the transverse displacement of the string where $\mathbf{w}(x,t) = (y(x,t), z(x,t))$, $y(x,t)$ and $z(x,t)$ being the transverse motion in the y and z direction respectively. Subscripts t and x refer to temporal and spatial derivatives respectively. $\mathbf{w}^2(x,t) = y^2(x,t) + z^2(x,t)$ and $\mathbf{w}_x^2(x,t) = y_x^2(x,t) + z_x^2(x,t)$ is the squared x derivative of $\mathbf{w}$. $\mathbf{f}$ is the external force acting on the string, R the damping coefficient and Γ the nonlinearity parameter (in (3)). Here, R and Γ are fixed positive constants, same in both directions, which holds true for a uniform string of circular cross-section.

Adding a damping and forcing term to (2) and changing the non-dimensional values to facilitate easy comparison with (3), we get the damped Carrier's equation

$$\mathbf{w}_{tt}(x,t) + 2R\mathbf{w}_t(x,t) = \left(1 + \frac{\Gamma}{2}\int_0^1 \mathbf{w}^2(x,t)\,dx\right)\mathbf{w}_{xx}(x,t) + \mathbf{f}(x,t),$$
$$0 < x < 1, \quad 0 < t \le T \qquad (4)$$

Now, consider (3) and (4) with the initial boundary conditions

$$\begin{aligned} \mathbf{w}(x,0) &= \phi(x) \\ \mathbf{w}_t(x,0) &= \psi(x) \\ \mathbf{w}(0,t) &= \mathbf{w}(1,t) = (0,0) \end{aligned} \qquad (5)$$

boundary conditions corresponding to fixed ends.

3 Numerical Algorithm

We now present a numerical scheme to solve (3), (5) and (4), (5) using a spectral spatial approximation similar to that used by Peradze [20] for a special case of (3) with time integration similar to that proposed by Christie and Sanz-Serna [7]. We consider approximate solutions of the form

$$\begin{aligned} \mathbf{W}(x,t) &= \sum_{j=1}^{n} \mathbf{W_j}(t)\sin(j\pi x) \\ \mathbf{W_j}(t) &= \big(W_{j,y}(t), W_{j,z}(t)\big) \;\; j = 1, 2, \ldots, n \end{aligned} \qquad (6)$$

for both (3) and (4). Henceforth, for a fixed j, the summand in (6) will be referred to as the jth mode and the jth harmonic interchangeably. Further, $j\pi$ will be referred to as the (spatial) frequency. Now, we assume that the initial conditions and forcing function are of the form:

$$\begin{aligned} \phi(x) &= \sum_{j=1}^{n} \alpha_{\mathbf{j}} \sin(j\pi x) \\ \psi(x) &= \sum_{j=1}^{n} \beta_{\mathbf{j}} \sin(j\pi x) \\ \mathbf{f}(x,t) &= \sum_{j=1}^{n} \mathbf{F_j}(t)\sin(j\pi x) \end{aligned} \qquad (7)$$

where

$$\begin{aligned}\alpha_{\mathbf{j}} &= (\alpha_{j,y}, \alpha_{j,z})\\ \beta_{\mathbf{j}} &= (\beta_{j,y}, \beta_{j,z})\\ \mathbf{F_j}(t) &= (F_{j,y}(t), F_{j,z}(t))\end{aligned}$$

Further details of the algorithm can be found in the Appendix.

4 Difference Between the Two Equations

For the same initial conditions, forcing function, R and Γ. The difference in the two solutions can be considerable even for very small values of time. Figure 1 shows the displacements at $t = 0.15$ for

$$\begin{aligned}\phi(x) &= (\sin(\pi x), \sin(2\pi x))\\ \psi(x) &= (5\sin(\pi x), \sin(\pi x) + \sin(2\pi x))\\ \mathbf{f}(x,t) &= (\sin(\pi x), \sin(2\pi x))\\ R &= 0.1\\ \Gamma &= 0.5\\ \Delta t &= 0.05\end{aligned}$$

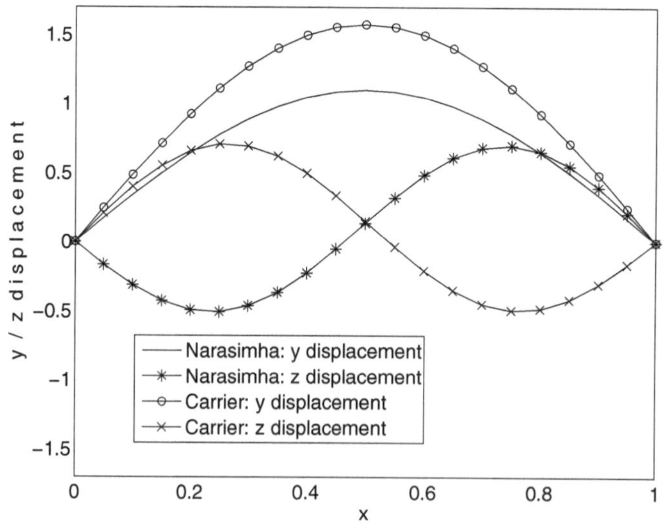

Fig. 1 Displacement at t = 0.15

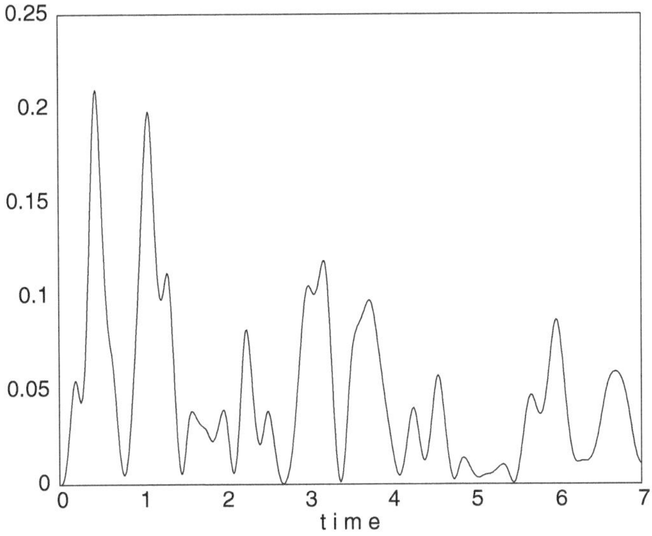

Fig. 2 Square of L^2 norm of the difference of the two solutions

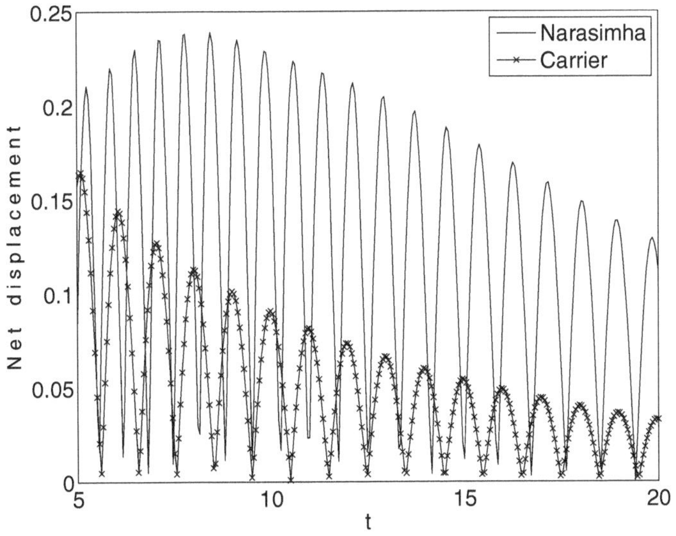

Fig. 3 Net displacement of centre of the string

Figure 2 shows the square of L^2 norm of the difference of the solutions of the two equations for the same conditions as above. Figure 3 shows the net displacement of the centre of the string $((y(0.5,t)^2 + z(0.5,t)^2)^{\frac{1}{2}})$ following both Narasimha's and Carrier's wave equation for the same conditions as above with zero forcing function. This illustrates that the solutions of the two equations oscillate with different (temporal) frequencies. It disagrees with Anand's [1] conclusion that under free vibrations the amplitude of total transverse displacement does not exhibit oscillatory behaviour.

4.1 Nonlinearity

Comparing the nonlinearity in the Eqs. (9) and (23) for the same values of W_k,

$$1+\frac{\Gamma}{2}\pi^2\sum_{k=1}^{n}\frac{k^2\mathbf{W_k}^2(t)}{2} \geq 1+\frac{\Gamma}{2}\sum_{k=1}^{n}\frac{\mathbf{W_k}^2(t)}{2} \tag{8}$$

The inequality will be strict for non-trivial solutions. Thus, we can conclude that the resulting nonlinearity for same initial conditions will be more in the Narasimha's equation compared to Carrier's wave equation (for very small time). Physically, this can interpreted as the solution of (3) starts with a higher phase velocity and thus higher temporal frequency than the solution of (4) under the same initial conditions. This is illustrated in Fig. 4 for the conditions

$$\begin{aligned}\phi(x) &= (\sin(\pi x)+\sin(2\pi x), \sin(\pi x))\\ \psi(x) &= (5\sin(\pi x), \sin(\pi x)+\sin(2\pi x))\\ \mathbf{f}(x,t) &= (\sin(\pi x)+, \sin(2\pi x))\end{aligned}$$

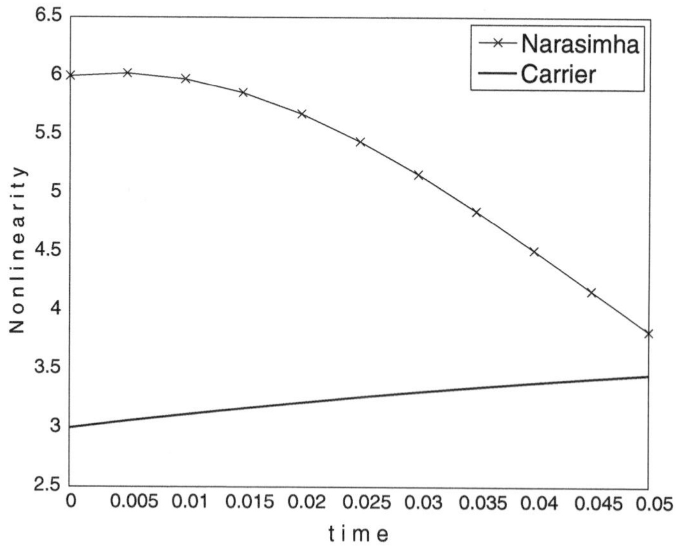

Fig. 4 Evolution of nonlinearity, for small time

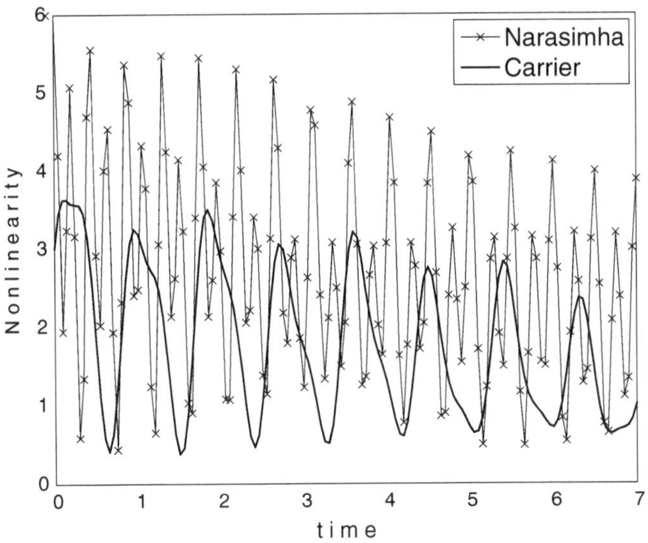

Fig. 5 Evolution of nonlinearity

$$R = 0.05$$
$$\Gamma = 0.5$$
$$\Delta t = 0.005$$

Figure 5 shows the variation of the nonlinearity (the coefficient of w_{xx}) with time for (3) and (4) for the above conditions with $\Delta t = 0.05$.

4.2 Precession

In physics and astronomy, precession refers to the motion or simply the change in orientation of the axis of a rotating body. It is this phenomenon which causes the wobble of a spinning top and the shift in the orientation of the Earth's axis of rotation. For a string undergoing nonplanar vibration, each point of the string follows a slightly distorted elliptic orbit. For a linear string, the orientation of the elliptical orbit followed by each point is constant as shown in Fig. 6. It only decreases in size due to damping.

For time-independent forcing $\mathbf{f}(x, t) = \mathbf{g}(x)$, the nonlinearity causes a change in orientation of this elliptical orbit, which is similar to the phenomenon of precession if the whirling motion of each point of the string was to be imagined as a rotating body. Figures 6 and 7a, b show the path traced by the centre point of a string following

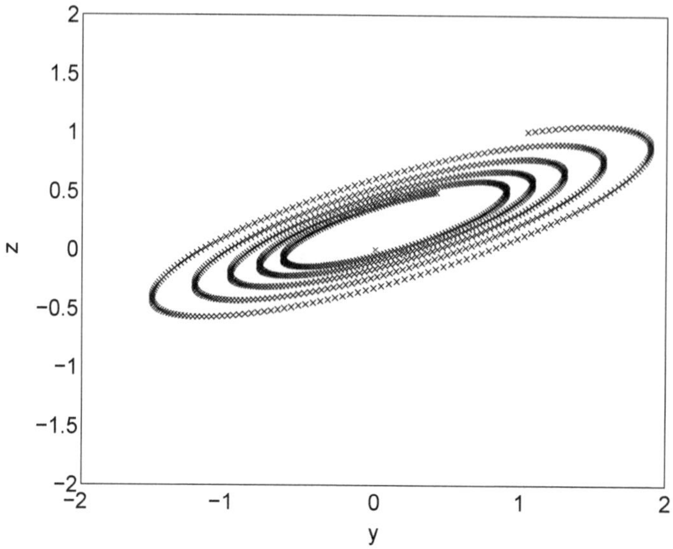

Fig. 6 Path traced by the centre point of a string following the linear wave equation

a linear wave equation, Carrier's wave equation and Narasimha's wave equation respectively up to time $t = 10\,\mathrm{s}$ under the conditions

$$\begin{aligned} \mathbf{f}(x,t) &= (\sin(\pi x), 2\sin(\pi x)) \\ \mathbf{w}(x,0) &= (\sin(\pi x), \sin(\pi x)) \\ \mathbf{w}_t(x,0) &= (5\sin(\pi x), \sin(\pi x)) \\ R &= 0.1 \\ \Gamma &= 0\,(\text{Fig. 6}) \\ \Gamma &= 1\,(\text{Fig. 7a, b}) \\ \Delta t &= 0.01 \end{aligned}$$

This phenomenon of precession occurs in both the nonlinear wave equation, but it does so at a much higher rate in Narasimha's wave equation than in Carrier's. The rate of precession is proportional to the amplitude of motion for both equations, which agrees with the results of Gough [9]. Thus the rate of precession decreases with time as damping reduces the amplitude. Precession occurs even for time-dependent forcing but the orbit followed is extremely distorted.

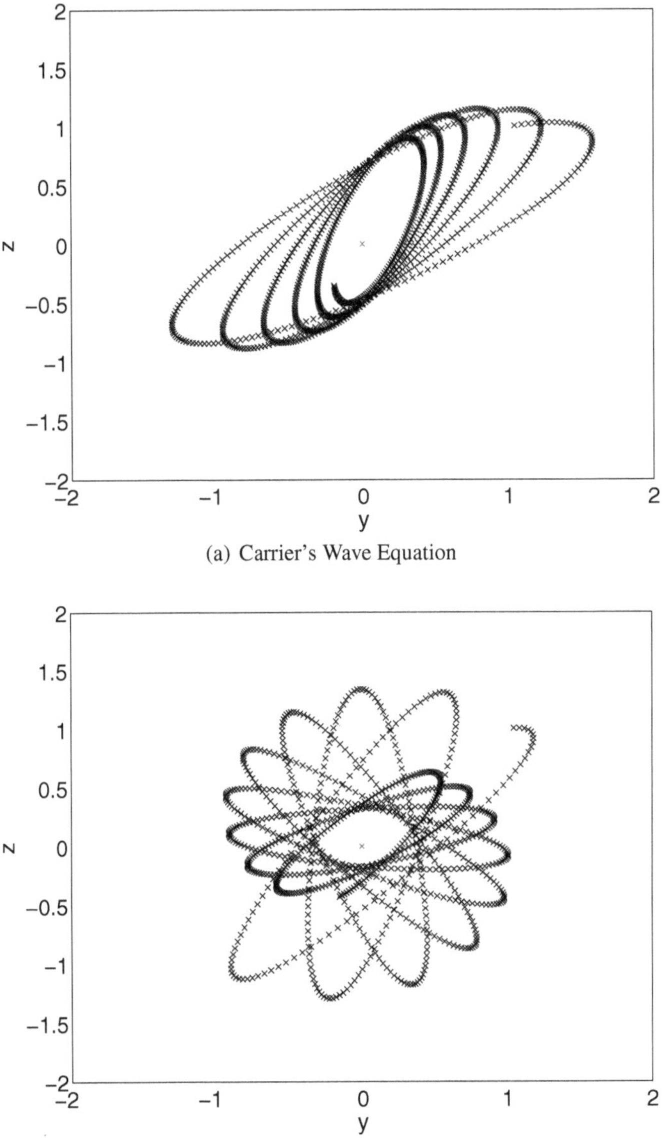

(a) Carrier's Wave Equation

(b) Narasimha's Wave Equation

Fig. 7 Precession in a string following the nonlinear Wave Equations

4.3 Onset of Nonplanar Motion

Determining the conditions under which a force in one directions (say y) can cause motion in the perpendicular transverse direction z has been an area of interest for a long time. This has been studied analytically for (3) by Murthy and Ramakrishna

[16], Miles [14] and Narasimha [17]. O'Reilly and Holmes [19] studied the same and concluded that in a 'hysteresis region' near the fundamental frequency planar motion is unstable for high enough amplitudes and ballooning or whirling motion develops. Our numerical simulations agree with this conclusion and show that the same also holds for (4). Physically, the effective natural frequency changes with damping and nonlinearity.

Numerically we study the onset of nonplanar motion by answering the following problem. If a small disturbing z motion is introduced in planar xy motion, under what conditions does this disturbance grow with time. Narasimha studied onset of nonplanar motion by trying to answer the same analytically. For only one transverse mode of motion and $\mathbf{F_1} = (c_1 \cos(\omega t), 0)$ for a constant c_1 and varying ω, we observe that in a neighbourhood around $\omega = 1.5$ for Carrier's wave equation and $\omega = 2$ for Narasimha's wave equation, z motion is ensured to be excited. This shows the difference in the effective natural frequency for the two equations. For both equations, increasing R and increasing Γ(keeping the other fixed), shifts this neighbourhood to the right. Increasing R also decreases the length of this neighbourhood. Further, increasing R and decreasing Γ suppresses the onset of nonplanar motion. This disagrees with Narasimha's conclusion that increasing nonlinearity suppresses nonplanar motion. Further, for fixed R, there exists a minimum value of Γ for which this neighbourhood is observed which increases with increasing R beyond a certain value. This minimum value is shown in Fig. 8 for $c_1 = 0.5\pi^3$, initial conditions $\phi(x) = (\sin(\pi x), 0.05 \sin(\pi x))$ and $\phi(x) = (0, 0)$.

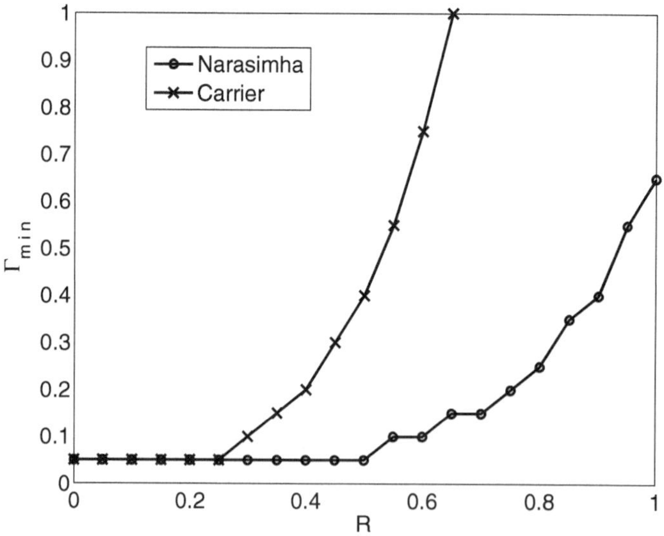

Fig. 8 Minimum Γ for onset of nonplanar motion

4.4 Damping of Free Motion

Each mode of the solution to both (3) and (4) undergoes harmonic oscillations for small R. For free motion $\mathbf{f}(x,t) \equiv (0,0)$, the peaks of these oscillations damp to zero exponentially at a rate almost equal to R for both the equations.

5 Conclusion

The difference in the integrands of the wave equations of Narasimha and Carrier show considerable difference. Extensive experimental results are needed to determine which of them provides a better model to the vibrating string.

Appendix: Algorithm

Narasimha's Wave Equation

Using (6) with Narasimha's wave Eq. (3), we get[1]:

$$
\begin{aligned}
&\mathbf{W_j}''(t) + 2R\mathbf{W_j}'(t) + \left(1 + \frac{\Gamma}{2}\pi^2 \sum_{k=1}^{n} \frac{k^2 \mathbf{W_k}^2(t)}{2}\right) \mathbf{W_j}(t) j^2 \pi^2 = \mathbf{F_j}(t) \\
&\mathbf{W_j}(0) = \alpha_j \\
&\mathbf{W_j}'(0) = \beta_j \quad j = 1, 2, \ldots, n
\end{aligned}
\tag{9}
$$

Now, we introduce $\mathbf{a_j}(t) = j\mathbf{W_j}(t)$ and $\mathbf{b_j}(t) = \mathbf{W_j}'(t)$ to get the equivalent first order system

$$
\begin{aligned}
&\mathbf{b_j}'(t) = \mathbf{F_j} - 2\mathbf{R}\mathbf{b_j}(t) - \left(1 + \frac{\Gamma}{4}\pi^2 \sum_{k=1}^{n} \mathbf{a_j}^2(t)\right) \pi^2 j \mathbf{a_j}(t) \\
&\mathbf{a_j}'(t) = j\mathbf{b_j}(t) \\
&\mathbf{a_j}(0) = j\alpha_j \\
&\mathbf{b_j}(0) = \beta_j \qquad j = 1, 2, \ldots, n
\end{aligned}
\tag{10}
$$

[1]Different modes are independent. It is also observed experimentally [16] that when the forcing function is in the vicinity of a spatial frequency $\omega_n = n\pi$, the mode of vibration can be closely described by the function $\sin(n\pi x)$. See also [8].

Now, we form the vectors

$$
\begin{aligned}
\mathbf{a}(t) &= (\mathbf{a_j}(t))_{j=1}^{n} \\
\mathbf{b}(t) &= (\mathbf{b_j}(t))_{j=1}^{n} \\
\mathbf{F}(t) &= (\mathbf{F_j}(t))_{j=1}^{n} \\
\alpha &= (\alpha_{\mathbf{j}})_{j=1}^{n} \\
\beta &= (\beta_{\mathbf{j}})_{j=1}^{n}
\end{aligned}
$$

Further we define the norm of a constant or time-dependent vector $\mathbf{r} = (\mathbf{r_j})_{j=1}^{n} = ((r_{j,y}, r_{j,z}))_{j=1}^{n}$ as

$$
\|\mathbf{r}\|^2 = \sum_{j=1}^{n} \mathbf{r_j}^2 \quad \text{with} \quad \mathbf{r_j}^2(t) = \mathbf{r_j} \cdot \mathbf{r_j} = r_{j,y}^2 + r_{j,z}^2
$$

We also define the (linear bounded) operator K acting on the vector $\mathbf{r}$

$$
K(\mathbf{r}) = K(((r_{j,y}, r_{j,z}))_{j=1}^{n}) = ((j r_{j,y}, j r_{j,z}))_{j=1}^{n} \tag{11}
$$

Using these our system (10) becomes

$$
\begin{aligned}
\mathbf{b}'(t) &= \mathbf{F} - 2R\mathbf{b}(t) - \left(1 + \frac{\Gamma}{4}\pi^2 \|\mathbf{a}(t)\|^2\right) \pi^2 K(\mathbf{a}(t)) \\
\mathbf{a}'(t) &= K(\mathbf{b}(t)) \\
\mathbf{a}(0) &= K(\alpha) \\
\mathbf{b}(0) &= \beta
\end{aligned} \tag{12}
$$

Due to the presence of the nonlocal term $\|\mathbf{a}(t)\|^2$ we will solve (12) by the use of a predictor corrector algorithm similar to that proposed by Christie and Sanz-Serna [7]. First we discretize the time t by introducing a uniform grid $\{t_m | 0 = t_0 < t_1 < \cdots < t_M = T\}$ with a time step Δt. The approximate values obtained at the time $t = t_m$ will be denoted by $\mathbf{a}^m$ and $\mathbf{b}^m$ respectively and the forcing function by $\mathbf{F}^m$.

Predictor Stage:

$$
\frac{\mathbf{b}^{[0]} - \mathbf{b}^{m-1}}{\Delta t} + 2R\frac{\mathbf{b}^{[0]} + \mathbf{b}^{m-1}}{2} + \left(1 + \frac{\Gamma}{4}\pi^2 \|\mathbf{a}^{m-1}\|^2\right) \pi^2 K(\mathbf{a}^{m-1}) = \frac{\mathbf{F}^m + \mathbf{F}^{m-1}}{2} \tag{13}
$$

$$
\frac{\mathbf{a}^{[0]} - \mathbf{a}^{m-1}}{\Delta t} = K(\mathbf{b}^{m-1})
$$

Corrector Stage:

$$\frac{\mathbf{b}^{[s]} - \mathbf{b}^{m-1}}{\Delta t} + 2R\frac{\mathbf{b}^{[s]} + \mathbf{b}^{m-1}}{2} + \left(1 + \frac{\Gamma}{4}\pi^2 \frac{\|\mathbf{a}^{m-1}\|^2 + \|\mathbf{a}^{[s-1]}\|^2}{2}\right)\pi^2 \times K\left(\frac{\mathbf{a}^{m-1} + \mathbf{a}^{[s-1]}}{2}\right) = \frac{\mathbf{F}^m + \mathbf{F}^{m-1}}{2} \tag{14}$$

$$\frac{\mathbf{a}^{[s]} - \mathbf{a}^{m-1}}{\Delta t} = K\left(\frac{\mathbf{b}^{m-1} + \mathbf{b}^{[s-1]}}{2}\right) \quad s = 1, 2, \ldots \tag{15}$$

The corrector stage is repeated till a value of s is found such that the approximations of $\mathbf{a}$ and $\mathbf{b}$ obtained at the s and $s-1$ stage are identical, up to a certain level of tolerance. Then $\mathbf{a}^m$ and $\mathbf{b}^m$ are taken to be $\mathbf{a}^{[s]}$ and $\mathbf{b}^{[s]}$ respectively. Thus, the predictor-corrector algorithm gives an approximation to the modified Crank–Nicolson scheme

$$\frac{\mathbf{b}^{m} - \mathbf{b}^{m-1}}{\Delta t} + 2R\frac{\mathbf{b}^{m} + \mathbf{b}^{m-1}}{2} + \left(1 + \frac{\Gamma}{4}\pi^2 \frac{\|\mathbf{a}^{m}\|^2 + \|\mathbf{a}^{m-1}\|^2}{2}\right)\pi^2 \times K\left(\frac{\mathbf{a}^{m} + \mathbf{a}^{m-1}}{2}\right) = \frac{\mathbf{F}^m + \mathbf{F}^{m-1}}{2}$$

$$\frac{\mathbf{a}^{m} - \mathbf{a}^{m-1}}{\Delta t} = K\left(\frac{\mathbf{b}^{m} + \mathbf{b}^{m-1}}{2}\right)$$

At time $t = t_m$ our approximate solution can be obtained by

$$\mathbf{W}(x, t_m) = \sum_{j=1}^{n} \frac{\mathbf{a_j}^m}{j} \sin(j\pi x) \tag{16}$$

Further space and time derivatives can be obtained by

$$\mathbf{W}_x(x, t_m) = \sum_{j=1}^{n} \mathbf{a_j}^m \cos(j\pi x)\pi$$

$$\mathbf{W}_t(x, t_m) = \sum_{j=1}^{n} \mathbf{b_j}^m \sin(j\pi x)$$

To determine a criteria for convergence of these corrector steps, we shall assume that $\int_0^1 \mathbf{w}_x^2(x,t)\,dx$ is bounded. We shall also assume that the corresponding values, $\|\mathbf{a}^m\|$ and $\|\mathbf{a}^{[s]}\|$ are also bounded. Further, say

$$\|\mathbf{a}^m\|, \|\mathbf{a}^{[s]}\| \le C_{max} \quad \forall m, s \tag{17}$$

It is easy to show that

$$\|K(\mathbf{r})\| \leq n\|\mathbf{r}\| \tag{18}$$

Upon replacing s by $s-1$ in (14), subtracting the resulting equation from (14), taking the norm, using (17), (18) and simplifying, for $s \geq 3$ we get:

$$\left(\frac{1+R\Delta t}{\Delta t}\right)\|\mathbf{b}^{[s]}-\mathbf{b}^{[s-1]}\| \leq \left(\frac{3}{8}\Gamma\pi^4 nC_{max}^2+\frac{\pi^2 n}{2}\right)\|\mathbf{a}^{[s-1]}-\mathbf{a}^{[s-2]}\| \tag{19}$$

Using a similar procedure on (15), we get

$$\|\mathbf{a}^{[s-1]}-\mathbf{a}^{[s-2]}\| \leq \frac{\Delta tn}{2}\|\mathbf{b}^{[s-3]}-\mathbf{b}^{[s-4]}\| \tag{20}$$

Using (19) and (20) we get

$$\|\mathbf{b}^{[s]}-\mathbf{b}^{[s-1]}\| \leq \frac{(\Delta t)^2}{1+R\Delta t}n^2\frac{\pi^2}{4}\left[1+\frac{3\Gamma}{4}\pi^2 C_{max}^2\right]\|\mathbf{b}^{[s-2]}-\mathbf{b}^{[s-3]}\| \tag{21}$$

Thus,

$$\frac{(\Delta t)^2}{1+R\Delta t}n^2\frac{\pi^2}{4}\left[1+\frac{3\Gamma}{4}\pi^2 C_{max}^2\right] < 1 \tag{22}$$

is a necessary condition for the convergence of the corrector stages at every time step. Under typical conditions when the corrector steps converge, they do so in within 5 corrector iterations for a tolerance of 10^{-2} and within 35 corrector steps with an tolerance of 10^{-8}.

Carrier's Wave Equation

The same method is used to obtain approximate solutions to Carrier's wave Eq. (4). The first equation of the system (9) will get modified to

$$\mathbf{W_j}''(t)+2R\mathbf{W_j}'(t)+\left(1+\frac{\Gamma}{2}\sum_{k=1}^{n}\frac{\mathbf{W_k}^2(t)}{2}\right)\mathbf{W_j}(t)j^2\pi^2=\mathbf{F_j}(t) \tag{23}$$

This is converted to an equivalent first order system by the introduction of $\mathbf{a_j}(t)=\mathbf{W_j}(t)$ (note the difference in the definition of $\mathbf{a_j}(t)$) and $\mathbf{b_j}(t)=\mathbf{W_j}'(t)$. Upon further introducing the same vector notation, norm and operator as done before, our system becomes

$$\begin{aligned}\mathbf{b}'(t)&=\mathbf{F}-2R\mathbf{b}(t)-\left(1+\frac{\Gamma}{4}\|\mathbf{a}(t)\|^2\right)\pi^2K^2(\mathbf{a}(t))\\ \mathbf{a}'(t)&=\mathbf{b}(t)\\ \mathbf{a}(0)&=\alpha\\ \mathbf{b}(0)&=\beta\end{aligned} \tag{24}$$

Now using a similar predictor-corrector algorithm to get an approximation to a modified Crank–Nicolson scheme.

Predictor Stage:

$$\frac{\mathbf{b}^{[0]} - \mathbf{b}^{m-1}}{\Delta t} + 2R\frac{\mathbf{b}^{[0]} + \mathbf{b}^{m-1}}{2} + \left(1 + \frac{\Gamma}{4}\|\mathbf{a}^{m-1}\|^2\right)\pi^2 K^2(\mathbf{a}^{m-1}) = \frac{\mathbf{F}^m + \mathbf{F}^{m-1}}{2} \tag{25}$$

$$\frac{\mathbf{a}^{[0]} - \mathbf{a}^{m-1}}{\Delta t} = \mathbf{b}^{m-1}$$

Corrector Stage:

$$\frac{\mathbf{b}^{[s]} - \mathbf{b}^{m-1}}{\Delta t} + 2R\frac{\mathbf{b}^{[s]} + \mathbf{b}^{m-1}}{2} + \left(1 + \frac{\Gamma}{4}\frac{\|\mathbf{a}^{m-1}\|^2 + \|\mathbf{a}^{[s-1]}\|^2}{2}\right)\pi^2 \times K^2\left(\frac{\mathbf{a}^{m-1} + \mathbf{a}^{[s-1]}}{2}\right) = \frac{\mathbf{F}^m + \mathbf{F}^{m-1}}{2} \tag{26}$$

$$\frac{\mathbf{a}^{[s]} - \mathbf{a}^{m-1}}{\Delta t} = \frac{\mathbf{b}^{m-1} + \mathbf{b}^{[s-1]}}{2} \qquad s = 1, 2, \ldots$$

At time $t = t_m$ our approximate solution can be obtained by

$$\mathbf{W}(x, t_m) = \sum_{j=1}^{n} \mathbf{a_j}^m \sin(j\pi x) \tag{27}$$

The criteria for convergence of the corrector steps can be obtained in a manner similar to that done in section "Narasimha's Wave Equation". For

$$\|\mathbf{a}^m\|, \|\mathbf{a}^{[s]}\| \le C_{max} \quad \forall m, s$$

$$\frac{(\Delta t)^2}{1 + R\Delta t} n^2 \frac{\pi^2}{4}\left[1 + \frac{3\Gamma}{4} C_{max}^2\right] < 1 \tag{28}$$

is a necessary condition for the convergence of the corrector steps. Further the corrector steps take a similar number of steps to converge as those for Narasimha's wave equation.

For both equations, the accuracy of the algorithm and the code was verified by comparing the numerical solutions with various manufactured solutions. One such manufactured solution used was

$$\mathbf{w}(x, t) = (\sin(\pi x) + 0.5\sin(2\pi x))\cos(\pi t), (2\sin(\pi x) + \sin(2\pi x))\cos(\pi t))$$

for which, we obtain

$$\begin{aligned}\mathbf{f}_1(x,t) &= \pi \sin(\pi x)\left(\frac{5}{2}\Gamma\pi^3 cos^3(\pi t) - 2R\sin(\pi t)\right)\\ &\quad + \pi \sin(2\pi x)\left(\frac{3}{2}\pi\cos(\pi t) + 5\Gamma\pi^3\cos^3(\pi t) - R\sin(\pi t)\right)\\ \mathbf{f}_2(x,t) &= \pi \sin(\pi x)\left(\frac{25}{16}\Gamma\pi\cos^3(\pi t) - 2R\sin(\pi t)\right)\\ &\quad + \pi \sin(2\pi x)\left(\frac{3}{2}\pi\cos(\pi t) + \frac{25}{8}\Gamma\pi\cos^3(\pi t) - R\sin(\pi t)\right)\end{aligned}$$

where $\mathbf{f}_1$ and $\mathbf{f}_2$ are the corresponding y components of the forcing functions for (3) and (4) respectively, with the z components equal to twice the y components.

By setting $\Gamma = 0$, we obtain the well known linear wave equation, the closed form solution of which is known. In this case, the numerical solution matched the d'Alembert's solutions very well.

References

1. Anand, G.V.: Large-amplitude damped free vibration of a stretched string. J. Accoust. Soc. Am. **45**, 1089–1096 (1969)
2. Arosio, A.: Global (in time) solution of the approximate nonlinear string equation of G.F. Carrier and R. Narasimha. Comment. Math. Univ. Carolin. **26**, 166–172 (1985)
3. Bernstein, S.: Sur une classe dequations fonctionnelles aux derivees partielles. Izv. Akad. Nauk SSSR Ser. Mat. **4**, 17–26 (1940)
4. Bilbao, S.: Energy-conserving finite difference schemes for tension-modulated strings. In: Proceedings of IEEE International Conference on Acoustics, Speech, and Signal Processing, 2004 (ICASSP '04), vol. 4, pp. 285–288. Montreal, Canada (2004)
5. Carrier, G.F.: On the nonlinear vibration problem of the elastic string. Q. J. Appl. Math. **3**, 157–165 (1945)
6. Carrier, G.F.: A note on the vibrating string. Q. J. Appl. Math. **7**, 97–101 (1949)
7. Christie, I., Sanz-Serna, J.M.: A Galerkin method for a nonlinear integro-differential wave system. Comput. Methods Appl. Mech. Eng. **44**, 229–237 (1984)
8. Dickey, R.W.: Stability of periodic solutions of the nonlinear string. Q. Appl. Math. **38**, 253–259 (1980)
9. Gough, C.: The nonlinear free vibration of a damped elatic string. J. Accout. Soc. Am. **75**, 1770–1776 (1984)
10. Kirchhoff, G.: Vorlesungen ber Mathematische Physik: Mechanik, vol. 28. Druck und Verlag von B.G. Teubner, Leipzig (1876)
11. Kurmyshev, E.V.: Transverse and longitudinal mode coupling in a free vibrating soft string. Phys. Lett. (A) **310**, 148–160 (2003)
12. Lions, J.L.: On some questions in boundary value problems of mathematical physics. In: G.M. de La Penha, L.A. Medeiros (eds.), Contemporary Developments in Continuum Mechanics and Partial Differential Equation, pp. 284–346. North-Holland, msterdam (1978)
13. Liu, I.S., Rincon, M.A.: Effect of moving boundaries on the vibrating elastic string. Appl. Numer. Math. **47**, 159–172 (2003)

14. Miles, J.W.: Stability of forced oscialltions of a vibrating string. J. Accoust. Soc. Am. **38**, 855–861 (1965)
15. Vasudeva Murthy, A.S.: On the String Equation of Narasimha. Connected at Infinity II: A Selection of Mathematics by Indians, Hindustan Book Agency (2013)
16. Murthy, G.S.S., Ramakrishna, B.S.: Nonlinear character of resonance in stretched strings. J. Acoust. Soc. Am. **38**, 461–471 (1965)
17. Narasimha, R.: Nonlinear vibration of an elastic string. J. Sound Vib. **8**, 134–155 (1968)
18. Nayfeh, S.A., Nayfeh, A.H., Mook, D.T.: Nonlinear response of a taut string to longitudinal and transverse end excitation. J. Vib. Control **1**, 291–305 (1995)
19. O'Reilly, O., Holmes, P.J.: Non-Linear, non-planar and non-periodic vibrations of a string. J. Sound Vib. **153**, 413–435 (1992)
20. Peradze, J.: A numerical algorithm for the nonlinear Kirchhoff string equation. Numer. Math. **102**, 311–342 (2005)
21. Pohozaev, S.I.: On a class of quasilinear hyperbolic equation. Mat. Sb. **96**, 152–165 (1975)
22. Rayleigh, J.W.S.: On maintained vibrations. Philos. Mag. **XV**, 229–235 (1883)
23. Spagnolo, S.: The Cauchy problem for the Kirchhoff equations. In: Proceedings of the Second International Conference on Partial Differential Equations (Italian) (Milan, 1992). Rend. Sem. Fis. Mat. Milano, vol. 62, pp. 17–51 (1992)
24. Suchde, P., Vasudeva Murthy, A.S.: From Carrier-Narasimha to Kirchhoff–Narasimha (communicated)

Mathematical Model of Flow in a Channel with Permeability - Combined Effect of Straight and Curved Boundaries

P. Muthu and M. Varunkumar

Abstract We investigate the effect of straight and curved boundary on the viscous, incompressible fluid flow in a channel with absorbing walls. The effect of fluid absorption through permeable wall is considered by taking flux as a function of axial distance. The nonlinear equations of motion are linearized by perturbation method by assuming δ (ratio of inlet half-width to length of channel) as a small parameter and are solved by numerical methods. The effects of double constriction the velocity profiles and mean pressure drop are observed and are presented graphically. The model may be considered for possible application to flow in renal tubule.

Keywords Renal tubules · Straight and curved boundary channel · Perturbation method · Permeable walls

1 Introduction

The study of flow through tubes of non-uniform cross-section has attracted by many researchers mainely due to its relevance to flow in renal tubules. Mathematical models of flow in renal tubule has been studied by various authors. Macey [1] seems to be the first to study the flow in the proximal renal tubule and modeled the problem as the flow of an incompressible viscous fluid through a circular tube with linear rate of reabsorption at the wall. A mathematical model for the bulk flow in the proximal tubule decays exponentially with axial distance was introduced by Kelman [2]. Macey [3] extended his first model using Kelman's condition and solved the equations of motion to find the average pressure drop.

P. Muthu (✉) · M. Varunkumar
Department of Mathematics, National Institute of Technology, Warangal 506004, Telangana, India
e-mail: muthuatbits@gmail.com

M. Varunkumar
e-mail: varun.nitw@gmail.com

J.M. Cushing et al. (eds.), *Applied Analysis in Biological and Physical Sciences*, Springer Proceedings in Mathematics & Statistics 186,
DOI 10.1007/978-81-322-3640-5_17

Marshal and Trowbridge [4] considered a flow of a newtonian fluid through a rigid permeable tube and transmural seepage is assumed to obey Darcy's law. Pallat et al. [5] have assumed that fluid loss from a porous tube is a function of pressure gradients across the tubule wall.

All these above studies treated renal tubule is assumed as cylindrical tube of uniform cross-section, while in general, such tubes may not have uniform cross-section through out their length. The hydrodynamical studies of an incompressible viscous fluid in a circular tube of varying cross-section with reabsorption at the wall have been investigated by Radhakrishnamacharya et al. [6] The effects of wall permeability on the velocity and wall shear stress in tubes(renal tubules) have been studied by Chathurani and Ranganatha [7], who treated the tubule as non-uniform.

Recently, Muthu and Tesfahun [8] developed a mathematical model for a viscous, incompressible fluid flow in a channel with slowly varying cross-section with permeable walls.

The objective of this is to investigate the flow through the renal tubule by considering an incompressible viscous fluid in a straight and curved boundary channel with reabsorbing walls.

2 Geometry

The boundary of the channel walls are assumed to be symmetric about x-axis and vary with x. It is taken as [9] (Fig. 1),

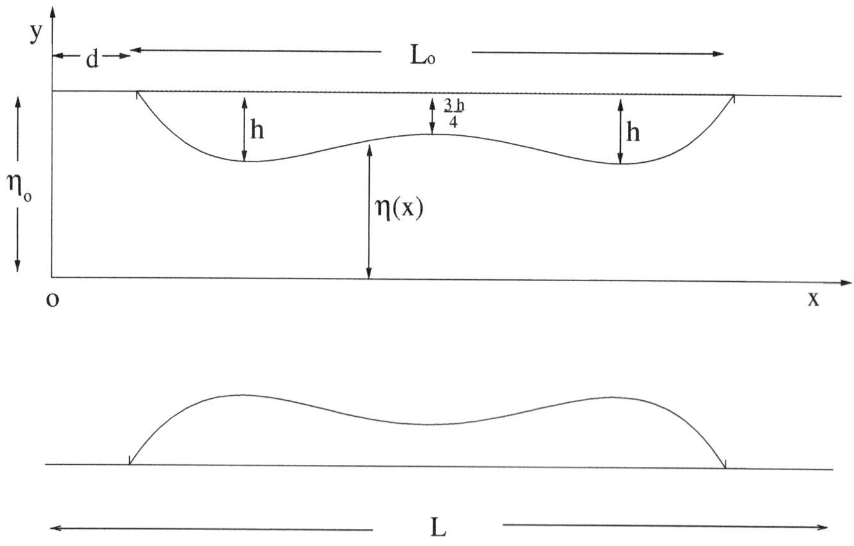

Fig. 1 Geometry of the straight and curved boundary channel

$$\eta(x) = \begin{cases} \eta_0 - \frac{3}{2}h\left[11(\frac{x-d}{L_0}) - 47(\frac{x-d}{L_0})^2 + 72(\frac{x-d}{L_0})^3 - 36(\frac{x-d}{L_0})^4\right], & d \le x \le d + L_0 \\ \eta_0, & \text{otherwise.} \end{cases} \tag{1}$$

where η_0 is the half width of the channel inlet (at $x = 0$), h is the maximum height of the constriction, d is the location of constriction, L_0 is the length of the spread of constriction and L is the length of the channel.

3 Mathematical Formulation

Consider an incompressible Newtonian fluid flow through a straight and curved boundary channel as given by Eq. (1). The motion of the fluid is assumed to be laminar, steady and symmetric. The channel is assumed to be long enough to neglect both the entrance and end effects. The governing equations of such fluid motion are given by

$$\frac{\partial u}{\partial x} + \frac{\partial v}{\partial y} = 0 \tag{2}$$

$$u\frac{\partial u}{\partial x} + v\frac{\partial u}{\partial y} = -\frac{1}{\rho}\frac{\partial p}{\partial x} + \frac{\mu}{\rho}\left(\frac{\partial^2 u}{\partial x^2} + \frac{\partial^2 u}{\partial y^2}\right) \tag{3}$$

$$u\frac{\partial v}{\partial x} + v\frac{\partial v}{\partial y} = -\frac{1}{\rho}\frac{\partial p}{\partial y} + \frac{\mu}{\rho}\left(\frac{\partial^2 v}{\partial x^2} + \frac{\partial^2 v}{\partial y^2}\right) \tag{4}$$

where u and v are the velocity components along the x and y axes, respectively, p is the pressure, $(\frac{\mu}{\rho})$ is the kinematic viscosity of the fluid, μ is the kinematic viscosity and ρ is the density.

The boundary conditions are taken as follows:

(i) The tangential velocity at the wall is zero. That is,

$$u + \frac{d\eta}{dx}v = 0 \quad \text{at} \quad y = \eta(x) \tag{5}$$

(ii) The regularity condition requires,

$$v = 0 \quad \text{and} \quad \frac{\partial u}{\partial y} = 0 \quad \text{at} \quad y = 0 \tag{6}$$

(iii) The reabsorption has been accounted by considering the bulk flow as a decreasing function of x. That is, the flux across a cross section is given as

$$Q(x) = \int_0^{\eta(x)} u(x,y)dy = Q_0 F(\alpha x) \tag{7}$$

where $F(\alpha x) = 1$ when $\alpha = 0$ and decreases with x. Further, $\alpha \geq 0$ is the reabsorption coefficient, a constant and Q_0 is the flux across the cross section at $x = 0$.

Eliminating pressure p from Eqs. (3) and (4) and introducing stream function $\psi(x, y)$

$$u = \frac{\partial \psi}{\partial y} \quad \text{and} \quad v = -\frac{\partial \psi}{\partial x} \tag{8}$$

and the following non-dimensional quantities,

$x' = \frac{x}{L}, y' = \frac{y}{\eta_0}, \eta' = \frac{\eta}{\eta_0}, \alpha' = \alpha L_0, p' = \frac{p\eta_0^2}{\mu Q_0}$.

The non-dimensional form of the boundary, the governing equations and the boundary conditions are (after dropping primes)

$$\eta(x) = \begin{cases} 1 - \frac{3}{2}\frac{\delta_1}{\eta_0}\left[11(\frac{x-a}{\epsilon}) - 47(\frac{x-a}{\epsilon})^2 + 72(\frac{x-a}{\epsilon})^3 - 36(\frac{x-a}{\epsilon})^4\right], & a \leq x \leq a+\epsilon \\ 1, & \text{otherwise.} \end{cases} \tag{9}$$

where $a = \frac{d}{L}$ and $\epsilon = \frac{L_0}{L}$.

And the equations of motion (3)–(4) transform to the following form, which is written after dropping the primes,

$$\nabla^2 \psi = \delta Re \left[\frac{\partial \psi}{\partial y} \nabla \frac{\partial \psi}{\partial x} - \frac{\partial \psi}{\partial x} \nabla \frac{\partial \psi}{\partial y}\right] \tag{10}$$

where $\nabla = \left[\delta^2 \frac{\partial^2}{\partial x^2} + \frac{\partial^2}{\partial y^2}\right]$, $\delta = \frac{\eta_0}{L}$ and $Re = \frac{Q_0 \rho}{\mu}$.

And the boundary conditions (3)–(7) become

$$\frac{\partial \psi}{\partial y} = -\frac{3}{2}A\delta\left[11 - 94\left(\frac{x-a}{\epsilon}\right) + 216\left(\frac{x-a}{\epsilon}\right)^2 - 144\left(\frac{x-a}{\epsilon}\right)^3\right]\frac{\partial \psi}{\partial x} \quad \text{at} \quad y = \eta(x) \tag{11}$$

$$\psi = 0 \quad \text{and} \quad \frac{\partial^2 \psi}{\partial y^2} = 0 \quad \text{at} \quad y = 0 \tag{12}$$

$$\psi = F(\alpha x) \quad \text{at} \quad y = \eta(x) \tag{13}$$

where $A = \frac{h}{L_0}$, $\delta = \frac{\eta_0}{L}$, $\epsilon = \frac{L_0}{L}$ and $a = \frac{d}{L}$.

In this problem, we consider exponentially decaying bulk flow, that is, in Eq. (7), F is taken as,

$$F(\alpha x) = e^{-\alpha x} \tag{14}$$

4 Method of Solution

The flow is complex because of the nonlinearity of the governing equations and boundary conditions (10)–(13). Thus, to solve (10) for velocity components in the present analysis, assuming the geometrical parameter $\delta \ll 1$, we shall seek a solution for stream function $\psi(x, y)$ in the form of a power series in terms of δ, as

$$\psi(x, y) = \psi_0(x, y) + \delta\psi_1(x, y) + \cdots \quad (15)$$

Substituting (15) in Eqs. (10)–(13) and collecting coefficients of various like powers of δ, we get the following sets of equations for $\psi_0(x, y)$, $\psi_1(x, y)$,…

δ^0 case:

$$\frac{\partial^4\psi_0}{\partial y^4} = 0 \quad (16)$$

The boundary conditions are

$$\frac{\partial\psi_0}{\partial y} = 0 \quad \text{at} \quad y = \eta(x) \quad (17)$$

$$\psi_0 = 0 \quad \text{and} \quad \frac{\partial^2\psi_0}{\partial y^2} = 0 \quad \text{at} \quad y = 0 \quad (18)$$

$$\psi_0 = e^{-\alpha x} \quad \text{at} \quad y = \eta(x) \quad (19)$$

δ^1 case:

$$\frac{\partial^4\psi_1}{\partial y^4} = Re\left[\frac{\partial\psi_0}{\partial y}\frac{\partial^3\psi_0}{\partial y^2\partial x} - \frac{\partial\psi_0}{\partial x}\frac{\partial^3\psi_0}{\partial y^3}\right] \quad (20)$$

The boundary conditions are

$$\frac{\partial\psi_1}{\partial y} = -\frac{3}{2}A\left[11 - 94\left(\frac{x-a}{\epsilon}\right) + 216\left(\frac{x-a}{\epsilon}\right)^2 - 144\left(\frac{x-a}{\epsilon}\right)^3\right]\frac{\partial\psi_0}{\partial x} \quad \text{at} \quad y = \eta(x) \quad (21)$$

$$\psi_1 = 0 \quad \text{and} \quad \frac{\partial^2\psi_1}{\partial y^2} = 0 \quad \text{at} \quad y = 0 \quad (22)$$

$$\psi_1 = 0 \quad \text{at} \quad y = \eta(x) \quad (23)$$

Similar expressions can be written for higher orders of δ. However, since we are looking for an approximate analytical solution for the problem, we consider upto order of δ^1 equations. The solution of Eq. (16) together with Eqs. (17)–(19) is

$$\psi_0 = A_1(x)y + A_2(x)y^3 \tag{24}$$

where $A_1(x) = \frac{3}{2\eta}e^{-\alpha x}$ and $A_2(x) = -\frac{1}{2\eta^3}e^{-\alpha x}$.

The solution of Eq. (20) together with Eqs. (21)–(23) is

$$\psi_1(x, y) = A_3 y + A_4 y^3 + Re\left[\left(A_1\frac{dA_2}{dx} - A_2\frac{dA_1}{dx}\right)\frac{y^5}{20} + A_2\frac{dA_2}{dx}\frac{y^7}{70}\right] \tag{25}$$

where

$$A_3(x) = Re\left[\left(A_1\frac{dA_2}{dx} - A_2\frac{dA_1}{dx}\right)\frac{\eta^4}{20} + A_2\frac{dA_2}{dx}\frac{\eta^6}{35}\right]$$
$$+\frac{3}{4}A\left[11 - 94\left(\frac{x-a}{\epsilon}\right) + 216\left(\frac{x-a}{\epsilon}\right)^2 - 144\left(\frac{x-a}{\epsilon}\right)^3\right]\left[\frac{dA_1}{dx}\eta + \frac{dA_2}{dx}\eta^3\right]$$

$$A_4(x) = -Re\left[\left(A_1\frac{dA_2}{dx} - A_2\frac{dA_1}{dx}\right)\frac{\eta^2}{10} + 3A_2\frac{dA_2}{dx}\frac{\eta^4}{70}\right]$$
$$-\frac{3}{4}A\left[11 - 94\left(\frac{x-a}{\epsilon}\right) + 216\left(\frac{x-a}{\epsilon}\right)^2 - 144\left(\frac{x-a}{\epsilon}\right)^3\right]\left[\frac{dA_1}{dx}\frac{1}{\eta} + \frac{dA_2}{dx}\eta\right]$$

Hence, substituting ψ_0 and ψ_1 in Eq. (15), we get

$$\psi(x, y) = A_1 y + A_2 y^3 + \delta\left(A_3 y + A_4 y^3 + A_5\frac{y^5}{120} + A_6\frac{y^7}{840}\right) \tag{26}$$

where $A_5(x) = 6Re\left[A_1\frac{dA_2}{dx} - A_2\frac{dA_1}{dx}\right]$ and $A_6(x) = 12ReA_2\frac{dA_2}{dx}$.

The velocities along x- and y-directions respectively, are obtained by substituting Eq. (25) in (8), as

$$u = \frac{\partial\psi}{\partial y} = A_1 + 3A_2y^2 + \delta\left(A_3 + 3A_4y^2 + A_5\frac{y^4}{4} + A_6\frac{y^6}{10}\right) \tag{27}$$

$$v = -\frac{\partial\psi}{\partial x} = -\frac{dA_1}{dx}y - \frac{dA_2}{dx}y^3 - \delta\left[\frac{dA_3}{dx}y + \frac{dA_4}{dx}y^3 + \frac{dA_5}{dx}\frac{y^5}{20} + \frac{dA_6}{dx}\frac{y^7}{70}\right] \tag{28}$$

Now, the non dimensional pressure $p(x, y)$ can be obtained by using Eqs. (25), (8) and (3). It is given as

$$p(x, y) = \delta\frac{\partial u}{\partial x} + \frac{1}{\delta}\int\frac{\partial^2 u}{\partial y^2}dx - Re\int\left[u\frac{\partial u}{\partial x} + v\frac{\partial u}{\partial y}\right]dx \tag{29}$$

The mean pressure is given as

$$\bar{p}(x) = \frac{1}{\eta(x)} \int_0^{\eta(x)} p(x, y)dy \tag{30}$$

Further, the mean pressure drop between $x = 0$ and $x = x_0$

$$\triangle \bar{p}(x_0) = \bar{p}(0) - \bar{p}(x_0), \quad 0 \le x_0 \le 1 \tag{31}$$

The wall shear stress $\tau_w(x)$ is defined as,

$$\tau_w(x) = \frac{(\sigma_{yy} - \sigma_{xx})\frac{d\eta}{dx} + \sigma_{xy}(1 - (\frac{d\eta}{dx})^2)}{1 + (\frac{d\eta}{dx})^2} \quad at \quad y = \eta(x) \tag{32}$$

where $\sigma_{xx} = 2\mu\frac{\partial u}{\partial x}$, $\sigma_{yy} = 2\mu\frac{\partial v}{\partial y}$ and $\sigma_{xy} = \mu(\frac{\partial u}{\partial y} + \frac{\partial v}{\partial x})$

Using the non-dimensional quantity $\tau'_w = \frac{\eta_0^2}{\mu Q_0}\tau_w$, the wall shear stress becomes,

$$\tau_w(x) = \frac{2\delta^2(\frac{\partial v}{\partial y} - \frac{\partial u}{\partial x})\frac{d\eta}{dx} + (\frac{\partial u}{\partial y} + \delta^2\frac{\partial v}{\partial x})(1 - \delta^2(\frac{d\eta}{dx})^2)}{1 + \delta^2(\frac{d\eta}{dx})^2} \tag{33}$$

It may be noted that in Eq. (29), the integrals are difficult to evaluate analytically to get the expression for $p(x, y)$. Therefore, they are calculated by numerical integration.

5 Results and Discussion

The aim of this analysis is to observe the behaviour of an incompressible fluid flow through a straight and curved boundary channel with absorbing walls. The parameter A characterizes the double constriction of the walls and α represents reabsorption coefficient of walls.

We discuss the effects of these parameters on the transverse velocity $v(x, y)$ and mean pressure drop ($\Delta\bar{p}$), given in Eqs. (28) and (31). In all our calculations, the following parameters are fixed as $\delta = 0.1$, $Re = 1.0$ and $a = 0.1$.

Velocity v:

In this case, the velocity profile of the flow is obtained by taking different values of constriction parameter A at different cross-sections $x = 0.25$, $x = 0.50$ and $x = 0.75$ of the channel. Here, the constriction is narrow at locations $x = 0.25$ and $x = 0.75$ and is wide at $x = 0.50$. The reabsorption coefficient is taken as $\alpha = 1.0$.

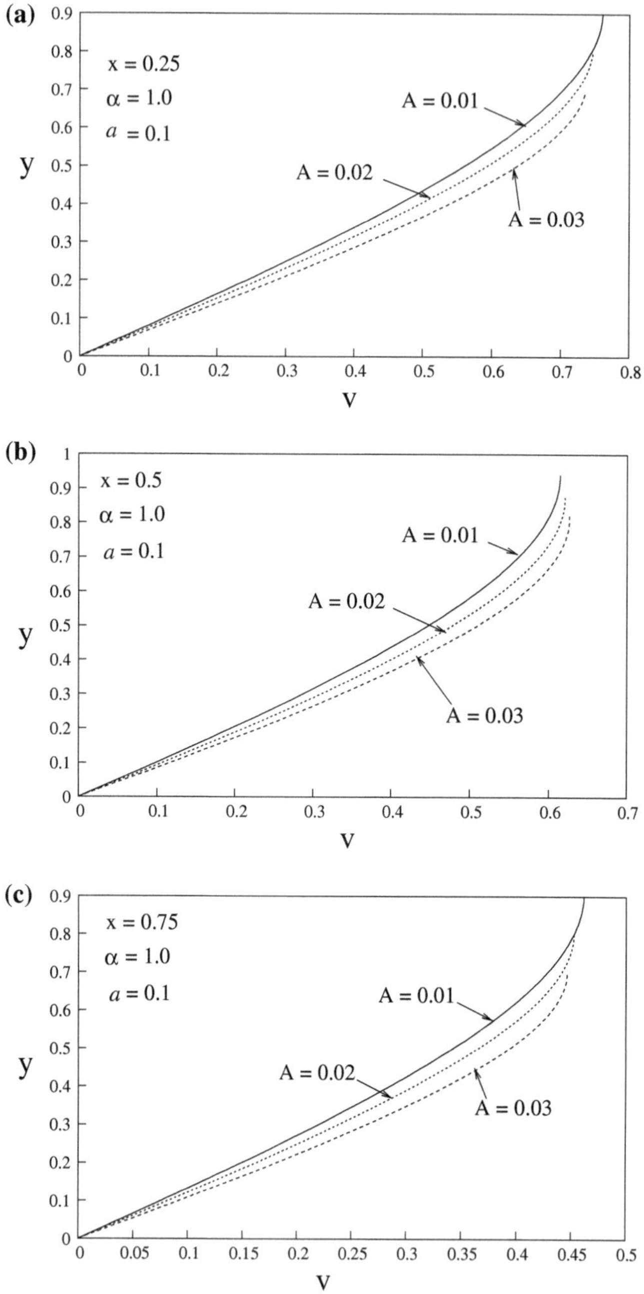

Fig. 2 Distribution of transverse velocity (v) with y

As the value of A increases from 0.01 to 0.03, the narrowness of double constriction increases. From the Fig. 2a–c, the value of transverse velocity at $x = 0.25$ is more than at $x = 0.50$ and $x = 0.75$. That is, the transverse velocity decreases as x increases at different cross-sections, even though the degree of narrowness is same at locations $x = 0.25$ and $x = 0.75$. In the both cases, the downstream of the flow, though there is no significant change in the behaviour of transverse velocity, the quantity of the velocity decreases.

Mean Pressure Drop $\Delta\bar{p}$:

The value of the mean pressure drop over the length of the channel is calculated for different values of A, a and α. It can be observed, from Fig. 3a, that the mean pressure drop increases as A increases. Further, from Fig. 3b–d, as the reabsorption coefficient α increases the mean pressure drop decreases.

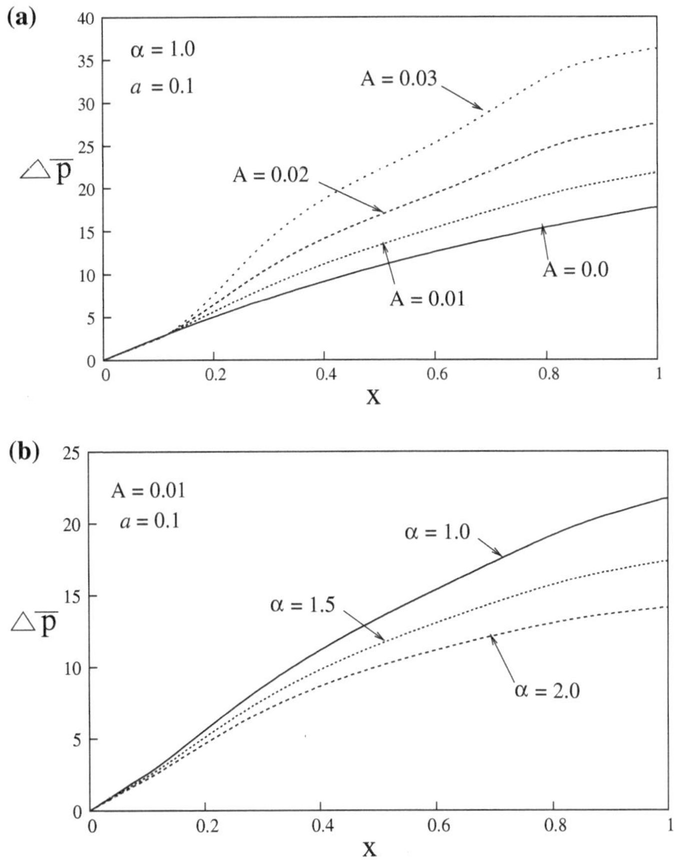

Fig. 3 Distribution of mean pressure drop $\Delta\bar{p}$ with x

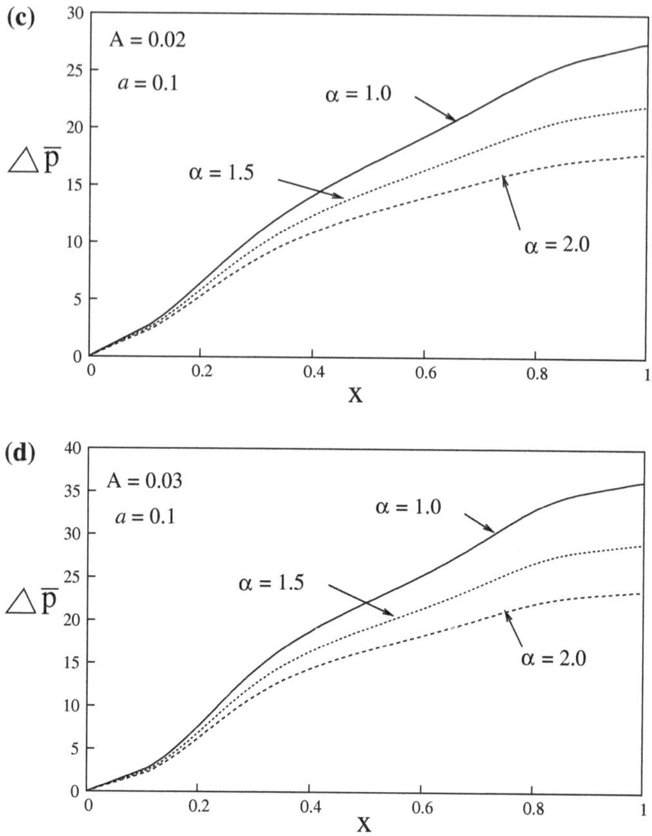

Fig. 3 (continued)

Wall Shear Stress $|\tau_w|$:

Figure 4a–f gives the magnitude of wall shear stress ($|\tau_w|$) at various values of the axial distance. As the reabsorption coefficient α increases, the magnitude of wall shear stress decreases (Fig. 4a–c). Also noted that, as constriction parameter A increases, the magnitude of the wall shear stress increases.

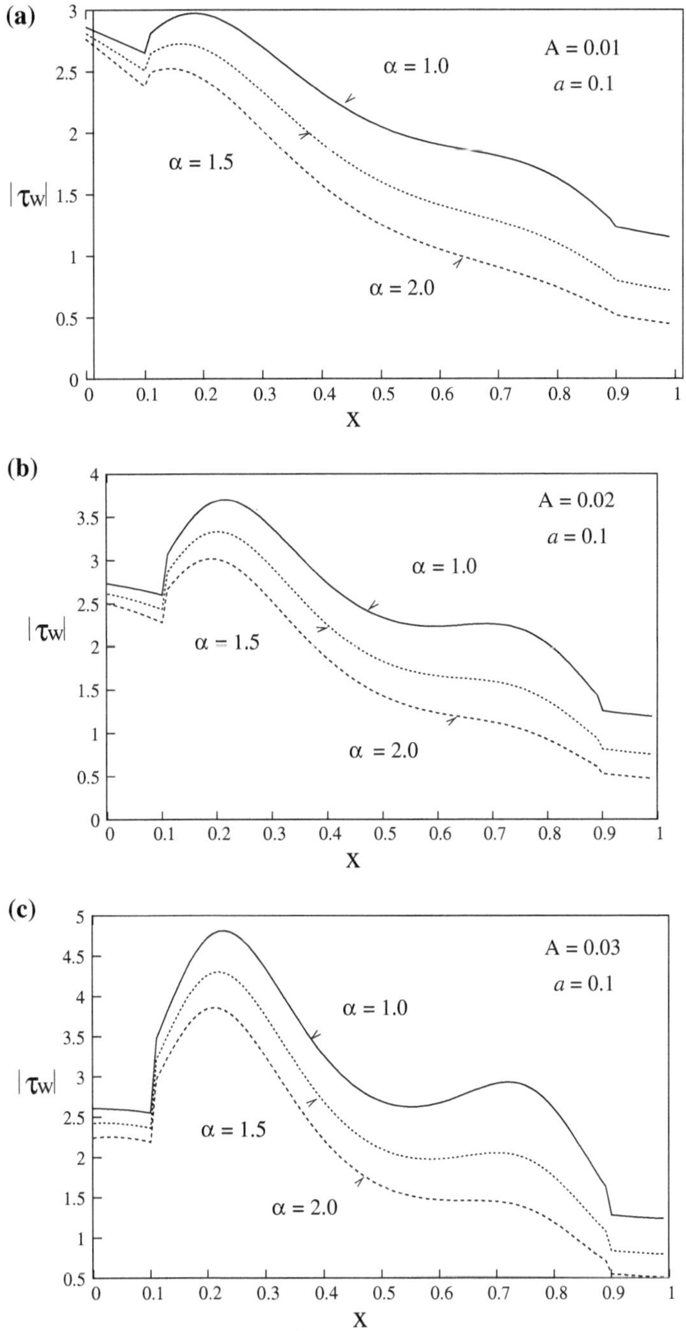

Fig. 4 Distribution of wall shear stress $|\tau_w|$ with x

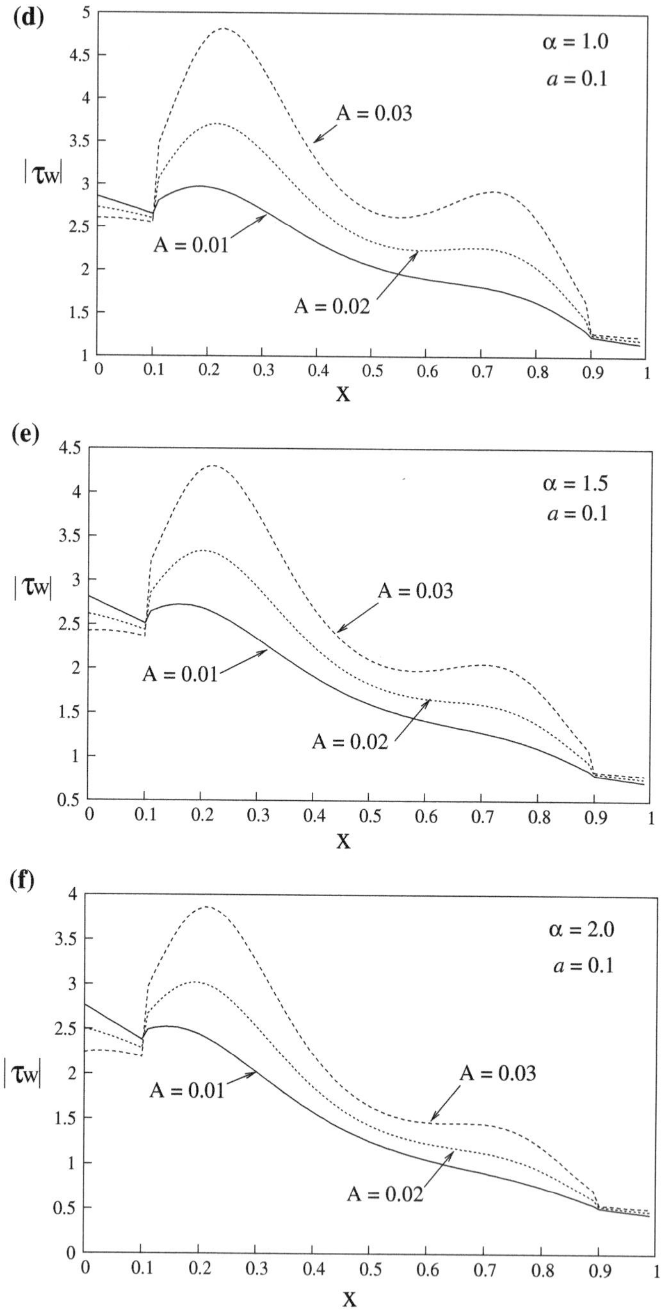

Fig. 4 (continued)

References

1. Robert, I.M.: Pressure flow patterns in a cylinder with reabsorbing walls. Bull. Math. Biophys. **25**, 1–9 (1963)
2. Kelman, R.B.: A theoretical note on exponential flow in the proximal part of the mammalian nephron. Bull. Math. Biophys. **24**, 303–317 (1962)
3. Robert, I.M.: Hydrodynamics of renal tubule. Bull. Math. Biophys. **27**, 117–124 (1965)
4. Marshall, E.A., Trowbridge, E.A.: Flow of a newtonian fluid through a permeable tube: the application to the proximal renal tubule. Bull. Math. Biol. **36**, 457–476 (1974)
5. Palatt, Paul J., Sackin, H., Tanner, R.I.: A hydrodynamical model of a permeable tubule. J. Theor. Biol. **44**, 287–303 (1974)
6. Radhakrisnamacharya, G., Chandra, P., Kaimal, M.R.: A hydrodynamical study of the flow in renal tubules. bull. math. biol. **43**, 151–163 (1981)
7. Chaturani, P., Ranganatha, T.R.: Flow of newtonian fluid in non-uniform tubes with variable wall permeability with application to flow in renal tubules. Acta Mech. **88**, 11–26 (1991)
8. Muthu, P., Berhane, T.: Mathematical model of flow in renal tubules. Int. J. App. Math. Mech. **6**, 94–107 (2010)
9. Chakravarthy, S., Mandal, P.K.: Mathematical modelling of blood flow through an overlapping stenosis. Math. Comput. Model. **19**, 59–73 (1994)

Part III
Applied Analysis

Approximate Controllability of Nonlocal Fractional Integro-Differential Equations with Finite Delay

Kamaljeet, D. Bahuguna and R.K. Shukla

Abstract Our purpose in this paper is to derive the approximate controllability of nonlocal integro-differential system of fractional order with finite delay. The main result is proved by using Schauder'r fixed point theorem along with semigroup theory and fractional calculus. Finally, we endow an example to show the application of the main result.

Keywords Approximate controllability · Fractional differential equations · Finite delay · Semigroup theory

2010 Mathematics Subject Classification 34A08 · 34G20 · 34K30 · 93B05

1 Introduction

Let Y and V be Hilbert spaces. Consider the following nonlocal fractional integro-differential system

$$\begin{cases} {}^cD^p y(t) = & -Ay(t) + Bw(t) + F(t, y_t, \int_0^t \zeta(t, r, y_r)dr), \quad t \in J = [0, b], \\ \quad y_0(v) = & \varphi(v) + h(y)(v), \quad v \in [-\sigma, 0], \end{cases} \tag{1}$$

where $\sigma, b > 0$, $0 < p < 1$, ${}^cD^p$ denotes the Caputo fractional derivative; $-A : D(A) \to Y$ generates an analytic and compact semigroup $\{T(t),\ t \geq 0\}$ of bounded linear operators in Y; the control $w(\cdot) \in L^2(J, V)$; the operator $B : V \to Y$ is linear and bounded; the nonlinear operators $F : J \times D \times Y_\alpha \to Y_\beta$, $\zeta : \Sigma \times D \to Y_\alpha$

Kamaljeet · D. Bahuguna (✉)
Department of Mathematics & Statistics, Indian Institute of Technology Kanpur, Kanpur 208016, India
e-mail: dhiren@iitk.ac.in

R.K. Shukla
Invertis University, Bareilly 243123, India

J.M. Cushing et al. (eds.), *Applied Analysis in Biological and Physical Sciences*, Springer Proceedings in Mathematics & Statistics 186,
DOI 10.1007/978-81-322-3640-5_18

are given continuous functions, here $0 < \alpha \le 1, \beta \in [\alpha, 1]$, $D = C([-\sigma, 0], Y_\alpha)$ and $\Sigma = \{(t, r) | 0 \le r \le t \le b\}$; the operator $h : C([-\sigma, b], Y_\alpha) \to D$ is completely continuous; $\varphi \in D$ and the history function y_t is defined by $y_t(\upsilon) = y(t + \upsilon), \upsilon \in [-\sigma, 0]$ and belongs to D. The terms Y_α, Y_β are explained in Sect. 2.

Nowadays, numerous researchers have shown more interest in fractional calculus (see [1–6] and references therein). It has a lot of applications in the field of control, electromagnetic, porous media, electrochemistry, etc. Also fractional derivatives and integrals help us to describe various phenomena appearing in economics, engineering and science, more accurately. In [7–11], the authors have discussed the existence results of fractional delay systems. Several authors have worked on nonlocal fractional differential systems; see, for instance, [3, 11–13]. In applications of physical problems, many times the nonlocal initial condition gives better effects and is more suitable than the usual initial value of the form $y(0) = y_0$.

Kalman was first one who introduced the concept of controllability. Controllability problem allows us to drive the state of dynamical system to the desired state with help of a control parameter existing in the system. In recent years, the controllability of the various nonlinear system has been investigated by many authors (see [10, 11, 14–24]). Exact controllability makes us able to drive the system to any final state, but in case of approximate controllability, the system can be driven to arbitrarily small neighbourhood of a final state. Thus the approximately controllability is more suitable to a dynamical system and quite appropriate in applications. In [11, 15], authors have derived the controllability of abstract functional differential equations with the help of measures of noncompactness and Mönch fixed point theorem. Surendra and Sukavanam [10] have established the approximate controllability of semilinear system of fractional order with bounded delay. In the papers [17, 19, 20, 22–24], the authors have assumed that associated linear system is approximately controllable for proving the approximate controllability of nonlinear systems.

This paper is motivated by recent works [10, 11, 20, 22, 23]. We establish some sufficient conditions to investigate the approximate controllability of the system (1). Schauder's fixed point theorem, Semigroup theory and fractional calculus are applied to derive the main results. To the best of our information, approximate controllability of nonlocal integro-differential system (1) has not yet been studied.

The outline of the paper is as: In the coming Sect. 2, we provide some notations, definitions and preliminary results. The existence result and the approximate controllability of (1) are discussed in Sect. 3. In last Sect. 4, we endow an example to show the application of the main result.

2 Preliminaries

Since $-A$ generates an analytic and compact semigroup $\{T(t),\ t \ge 0\}$ of uniformly bounded linear operator in Y. Therefore $\sup_{t \ge 0} \|T(t)\| \le M$ for some $M > 1$. Without loss of generality, let $0 \in \rho(A)$ (a resolvent set of A). Now it is feasible to define $A^\alpha, 0 < \alpha \le 1$, which is closed linear operator with domain $D(A^\alpha) \subseteq Y$. Moreover,

$D(A^\alpha)$ is dense in Y. Now, we denote by Y_α the Hilbert space $D(A^\alpha)$ endowed with norm $\|\cdot\|_\alpha = \|A^\alpha \cdot\|$. More details are given in the book by Pazy [25]. The following lemma gives the basic properties of A^α.

Lemma 1 (see [25]) *The operator A^α has satisfied the following:*

(i) $D(A^\beta) \subset D(A^\alpha)$ *for any* $\beta \geq \alpha > 0$.
(ii) *If* $\alpha, \beta \in \mathbb{R}$ *and* $\gamma = \max\{\alpha+\beta, \alpha, \beta\}$ *then*

$$A^{\alpha+\beta}y = A^\alpha.A^\beta y, \quad y \in D(A^\gamma).$$

(iii) *For any* $t > 0$, $\alpha \geq 0$ *and* $y \in Y$, *then* $T(t)y \in D(A^\alpha)$.
(iv) $T(t)A^\alpha y = A^\alpha T(t)y, \quad \forall\, y \in D(A^\alpha)$.
(v) $A^\alpha T(t)$ *is bounded in* Y *for each* $t > 0$ *and*

$$\|A^\alpha T(t)\| \leq M_\alpha t^{-\alpha}, \ \textit{for some } M_\alpha > 0.$$

(vi) *The linear operator* $A^{-\alpha}$, $0 \leq \alpha \leq 1$, *is bounded and*

$$\|A^{-\alpha}\| \leq C_\alpha, \textit{ for some } C_\alpha > 0.$$

Definition 1 (*see* [1]) The fractional integral for a function $f \in L^1([0,b], Y)$ is given by

$$I^p f(t) = \frac{1}{\Gamma(p)} \int_0^t (t-r)^{p-1} f(r)\,dr, \quad t > 0, \ p > 0,$$

where $\Gamma(\cdot)$ is the gamma function.

Definition 2 (*see* [1]) The Caputo fractional derivative of a function f is defined as

$${}^cD^p f(t) = \frac{1}{\Gamma(k-p)} \int_0^t \frac{f^{(k)}(r)}{(t-r)^{p+1-k}}\,dr, \quad t > 0,$$

where $0 \leq k-1 < p < k$.

Definition 3 ([2, 10]) A function $y(\cdot) \in C([-\sigma, b], Y_\alpha)$ is said to be mild solution of the Eq. (1) if it satisfies

$$y(t) = \begin{cases} \mathcal{U}_p(t)(\varphi(0) + h(y)(0)) + \int_0^t (t-r)^{p-1}\mathcal{V}_p(t-r) \\ \times\Big[F(r, y_r, \int_0^r \zeta(r, \eta, y_\eta)d\eta) + Bw(t)\Big]dr, \quad t \in J, \\ \varphi(t) + h(y)(t), \quad [-\sigma, 0], \end{cases} \tag{2}$$

where

$$\mathcal{U}_p(t) = \int_0^\infty \psi_p(\lambda) T(t^p\lambda)d\lambda, \quad \mathcal{V}_p(t) = p\int_0^\infty \lambda\psi_p(\lambda)T(t^p\lambda)d\lambda, \tag{3}$$

$$\psi_p(\lambda) = \frac{1}{p}\lambda^{-1-1/p}\rho_p(\lambda^{-1/p}) \quad and$$

$$\rho_p(\lambda) = \frac{1}{\pi}\sum_{n=1}^{\infty}(-1)^{n-1}\lambda^{-np-1}\frac{\Gamma(np+1)}{n!}\sin(n\pi p), \quad \lambda \in (0,\infty).$$

Note that $\psi_p(\lambda)$ is a probability density function on $(0,\infty)$, that is $\psi_p(\lambda) \geq 0$, $\int_0^{\infty}\psi_p(\lambda)d\lambda = 1$ and $\int_0^{\infty}\lambda\psi_p(\lambda) = \frac{1}{\Gamma(1+p)}$.

Lemma 2 (see [2]) *The operators $\mathcal{U}_p(t)$ and $\mathcal{V}_p(t)$ satisfy the following:*

(i) For any fixed $t \geq 0$,

$$\|\mathcal{U}_p(t)y\|_\alpha \leq M\|y\|_\alpha, \quad \|\mathcal{V}_p(t)y\|_\alpha \leq \frac{M}{\Gamma(\alpha)}\|y\|_\alpha, \quad \forall y \in Y_\alpha.$$

(ii) $\mathcal{U}_p(t)$ and $\mathcal{V}_p(t)$ are strongly continuous for each $t \geq 0$.
(iii) If the semigroup $T(t)$ is compact in Y, then $\mathcal{U}_p(t)$ and $\mathcal{V}_p(t)$ are norm-continuous and compact in Y for each $t > 0$. Also the restriction of $\mathcal{U}_p(t)$ and $\mathcal{V}_p(t)$ to Y_α are compact in Y_α for each $t > 0$.
(iv) $\mathcal{U}_p(t)A^\alpha y = A^\alpha\mathcal{U}_p(t)y$, $\mathcal{V}_p(t)A^\alpha y = A^\alpha\mathcal{V}_p(t)y$ for each $t \geq 0$ and $y \in Y_\alpha$.
(v) If $\beta \in (0,1]$, then for any $y \in Y$, $t > 0$

$$\|A^\beta\mathcal{V}_p(t)y\| \leq \frac{N_\beta}{t^{p\beta}}\|y\|, \quad where \ \frac{pM_\beta\Gamma(2-\beta)}{\Gamma(1+p(1-\beta))}.$$

Definition 4 The system (1) is called approximately controllable on J if for each final state $z \in Y_\alpha$ and $\varepsilon > 0$, there is a control $w(\cdot) \in L^2(J,V)$ such that the mild solution $y(\cdot,w)$ of the system (1) satisfies that $\|y(b,w) - z\| < \varepsilon$.

Consider the linear control system of fractional order

$$\begin{cases} {}^cD^p y(t) = & -Ay(t) + Bw(t), \quad t \in J, \\ \quad y(0) = \varphi(0). \end{cases} \tag{4}$$

The controllability operator $\Gamma_0^b : Y_\alpha \to Y_\alpha$ and the resolvent operator $R(\varepsilon,\Gamma_0^b) : Y_\alpha \to Y_\alpha$ associated with (4) are defined as

$$\Gamma_0^b = \int_0^b (b-r)^{p-1}\mathcal{V}_p(b-r)BB^*\mathcal{V}_p^*(b-r)\,dr,$$
$$R(\varepsilon,\Gamma_0^b) = (\varepsilon I + \Gamma_0^b)^{-1}, \quad \varepsilon > 0,$$

where $\mathcal{V}_p^*$ and B^* are the adjoint of $\mathcal{V}_p$ and B respectively.

Theorem 3 (see [22]) *Let Y^* be a dual space of a separable reflexive Banach space Y. Also let $\Gamma : Y^* \to Y$ be a symmetric operator, i.e. $\langle y_1^*, \Gamma y_2^*\rangle = \langle y_2^*, \Gamma y_1^*\rangle$ for all $y_1^*, y_2^* \in Y^*$. Then the statements given below are equivalent:*

(i) $\langle y, \Gamma y\rangle > 0$ *for each nonzero* $y \in Y^*$, *i.e.,* Γ *is a positive operator.*
(ii) *For each* $z \in Y$, $y_\varepsilon(z) = \varepsilon(\varepsilon I + \Gamma P)^{-1}(z) \to 0$ *strongly as* $\varepsilon \to 0^+$. *Here* $P: Y \to Y^*$ *is a duality map.*

Lemma 4 (see [20]) *The linear control system (4) is approximately controllable on* J *if and only if* $\varepsilon R(\varepsilon, \Gamma_0^b) \to 0$ *strongly as* $\varepsilon \to 0^+$.

3 Main Results

To prove the result of approximate controllability of the problem (1), we make the following assumptions:

(H1) The function $\zeta : \Sigma \times D \to Y_\alpha$ satisfies the following:

(i) For every $(t, r) \in \Sigma$, the function $\zeta(t, r, \cdot) : D \to Y_\alpha$ is continuous and for $\varphi \in D$, $\zeta(\cdot, \cdot, \varphi)$ is strongly measurable on Σ.
(ii) There exists $m(\cdot, \cdot) \in L^1(\Sigma, \mathbb{R}^+)$ such that

$$\|\zeta(t, r, y_r)\|_\alpha \le m(t, r)\|y_r\|_D,$$

where $\|\varphi\|_D = \sup\{\|\varphi(t)\|_\alpha \,:\, t \in [-\sigma, 0]\},\ \varphi \in D$. For convenience, we write $L_0 = \max \int_0^t m(t, r)dr$.

(H2) The function $F\colon J \times D \times Y_\alpha \to Y_\beta$ satisfies the following:

(i) The function $F(t, \cdot, \cdot) : D \times Y_\alpha \to Y_\beta$ is continuous for each $t \in J$ and for all $(\varphi, y) \in D \times Y_\alpha$, $F(\cdot, \varphi, y)$ is strongly measurable.
(ii) For $k > 0$ and $t \in J$, there exist $\frac{a_k(\cdot)}{(t-\cdot)^{1-p}} \in L^1([0, t], \mathbb{R}^+)$ such that

$$\sup\{\|F(t, \varphi, y)\|_\beta \,:\, \|\varphi\| \le k,\ \ \|y\|_\alpha \le L_0 bk\} \le a_k(t) \quad \text{for a.e. } t \in J,$$

and

$$\liminf_{k\to\infty} \frac{1}{k}\int_0^t \frac{a_k(r)}{(t-r)^{1-p}} = \mu < +\infty.$$

(H3) $h\colon C([-\sigma, b], Y_\alpha) \to C([-\sigma, 0], Y_\alpha)$ is a completely continuous operator such that

$$\lim_{\|y\|_\infty\to\infty} \frac{\|h(y)\|_D}{\|y\|_\infty} = 0,$$

where $\|y\|_\infty = \sup\{\|y(t)\|_\alpha \,:\, t \in [-\sigma, b]\},\ y \in C([-\sigma, b], Y_\alpha)$.

For any $\varepsilon > 0$ and $z \in Y_\alpha$, we define a control $w_\varepsilon(t, y)$ as

$$w_\varepsilon(t,y) = B^*\mathcal{V}_p^*(b-t)R(\varepsilon,\Gamma_0^b)\Big[z-\mathcal{U}_p(b)(\varphi(0)+h(y)(0)) - \int_0^b (b-r)^{p-1}\mathcal{V}_p(b-r)F\Big(r,y_r,\int_0^r \zeta(r,\eta,y_\eta)d\eta\Big)dr\Big]. \quad (5)$$

For sake of the convenience, we write

$$R = \frac{M}{\varepsilon\Gamma(p)}\|B\|^2\Big[C_\alpha\|z\|_\alpha + MC_\alpha(\|\varphi(0)\|_\alpha + \|h(y)(0)\|_\alpha) + \frac{MC_\beta}{\Gamma(p)}\int_0^b (b-r)^{p-1}a_k(r)dr\Big].$$

Theorem 5 *Assume that the hypothesis (H1)–(H3) hold. Then, for each $\varepsilon > 0$, the nonlocal integro-differential system* (1) *corresponding to the control function $w_\varepsilon(t,y)$ has a mild solution if*

$$\frac{M}{\Gamma(p)}\mu\Big[C_{\beta-\alpha} + \frac{Mb^{p(1-\alpha)}N_\alpha C_\beta}{\varepsilon(1-\alpha)\Gamma(p+1)}\|B\|^2\Big] < 1. \quad (6)$$

Proof Let $B_k = \{y \in C([-\sigma,b],Y_\alpha) : \|y\|_\infty \le k\}$, where $k > 0$. Define an operator $Q_\varepsilon : C([-\sigma,b],Y_\alpha) \to C([-\sigma,b],Y_\alpha)$ by

$$(Q_\varepsilon y)(t) = \begin{cases} \mathcal{U}_p(t)\,[\varphi(0)+(h(y))(0)] + \int_0^t (t-r)^{p-1}\mathcal{V}_p(t-r) \\ \quad\times\big[F\big(r,y_r,\int_0^r \zeta(r,\eta,y_\eta)d\eta\big) + Bw_\varepsilon(r,y)\big]dr \quad t\in J, \\ \varphi(t)+(h(y))(t), \quad t\in[-\sigma,0]. \end{cases} \quad (7)$$

It is obvious that any fixed point of Q_ε is a mild solution of the system (1). So we have to prove that the operator Q_ε has a fixed point for each $\varepsilon > 0$. Firstly, for each $\varepsilon > 0$, we show that $Q_\varepsilon(B_k) \subset B_k$ for some $k = k(\varepsilon) > 0$. If it is not true for each $k > 0$, then there would exist $y^{(k)} \in B_k$ and $t \in J$ such that $\|(Q_\varepsilon y^{(k)})(t)\|_\alpha > k$. By Lemma 2, (H1) and (H2), we get

$$\|F\Big(r,y_r^{(k)},\int_0^r \zeta(r,\eta,y_\eta^{(k)})d\eta\Big)\|_\beta \le a_k(r), \quad (8)$$

$$\begin{aligned}\|Bw_\varepsilon(r,y^{(k)})\| &\le \frac{1}{\varepsilon}\|BB^*\mathcal{V}_p^*(b-r)\|\Big[\|A^{-\alpha}A^\alpha z\| \\ &\quad + \|A^{-\alpha}A^\alpha\mathcal{U}_p(b)(\varphi(0)+h(y)(0))\| + \|\int_0^b (b-r)^{p-1} \\ &\quad \times A^{-\beta}A^\beta\mathcal{V}_p(b-r)F\Big(r,y_r^{(k)},\int_0^r \zeta(r,\eta,y_\eta^{(k)})d\eta\Big)dr\|\Big] \\ &\le \frac{M}{\varepsilon\Gamma(p)}\|B\|^2\Big[C_\alpha\|z\|_\alpha + MC_\alpha\big(\|\varphi(0)\|_\alpha\end{aligned}$$

$$+ \|h(y)(0)\|_\alpha\big) + \frac{MC_\beta}{\Gamma(p)} \int_0^b (b-r)^{p-1} a_k(r)dr\Big]$$
$$= R \tag{9}$$

and

$$\begin{aligned} k < \|Q_\varepsilon y^{(k)}(t)\|_\alpha &\le \|\mathcal{U}_p(t)(\varphi(0) + h(y)(0)\|_\alpha + \Big\| \int_0^t (t-r)^{p-1} \mathcal{V}_p(t-r) \\ &\times \Big[F\Big(r, y_r^{(k)}, \int_0^r \zeta(r, \eta, y_\eta^{(k)})d\eta\Big) + Bw_\varepsilon(r, y^{(k)})\Big]dr\Big\|_\alpha \\ &\le M(\|\varphi\|_\alpha + \|h(y)\|_\alpha) + \int_0^t (t-r)^{p-1} \Big\| A^{\alpha-\beta}\mathcal{V}_p(t-r) \\ &\times A^\beta F\Big(r, y_r^{(k)}, \int_0^r \zeta(r, \eta, y_\eta^{(k)})d\eta\Big)\Big\| dr \\ &+ \int_0^t (t-r)^{p-1} \|A^\alpha \mathcal{V}_p(t-r) Bw_\varepsilon(r, y^{(k)})\| dr \\ &\le M(\|\varphi\|_\alpha + \|h(y)\|_\alpha) + \frac{MC_{\beta-\alpha}}{\Gamma(p)} \int_0^t (t-r)^{p-1} a_k(r)dr \\ &+ \frac{Mb^{p(1-\alpha)}N_\alpha}{\varepsilon(1-\alpha)\Gamma(p+1)} \|B\|^2 \Big[C_\alpha \|z\|_\alpha + MC_\alpha \\ &\times \big(\|\varphi(0)\|_\alpha + \|h(y)(0)\|_\alpha\big) + \frac{MC_\beta}{\Gamma(p)} \int_0^b (b-r)^{p-1} a_k(r)dr\Big]. \end{aligned} \tag{10}$$

Dividing both side of (10) by k and taking $k \to \infty$, we get

$$1 < \frac{M}{\Gamma(p)} \mu \Big[C_{\beta-\alpha} + \frac{Mb^{p(1-\alpha)}N_\alpha C_\beta}{\varepsilon(1-\alpha)\Gamma(p+1)} \|B\|^2\Big].$$

This contradicts (6). Hence $Q_\varepsilon(B_k) \subseteq B_k$ for some $k > 0$.

To prove the continuity of Q_ε on B_k, we let $\{y^{(n)}\} \subset B_k$ with $y^{(n)} \to y \in B_k$ as $n \to \infty$. If any $t \in [-\sigma, 0]$, then $\|Q_\varepsilon y^{(n)}(t) - Q_\varepsilon y(t)\|_\alpha = \|h(y^{(n)})(t) - h(y)(t)\|_\alpha \to 0$ as $n \to 0$. If $t \in J = [0, b]$, then, by assumptions (H1)–(H3), we have

(I) $\zeta(t, \tau, y_\tau^{(n)}) \to \zeta(t, \tau, y_\tau)$.
(II) $F(t, y_t^{(n)}, \int_0^t \zeta(t, \tau, y_\tau^{(n)})d\tau) \to F(t, y_t, \int_0^t \zeta(t, \tau, y_\tau)d\tau)$.
(III) $h(y^{(n)}) \to h(y)$.
(IV) $w_\varepsilon(r, y^{(n)}) \to w_\varepsilon(r, y)$.
(V) $\|F(t, y_t^{(n)}, \int_0^t \zeta(t, \tau, y_\tau^{(n)})d\tau) - F(t, y_t, \int_0^t \zeta(t, \tau, y_\tau)d\tau)\|_\beta \le 2a_k(t)$.
(VI) $\|Bw_\varepsilon(r, y^{(n)}) - Bw_\varepsilon(r, y)\| \le 2R$.

In view of Lebesgue's dominated convergence theorem, we get

$$\begin{aligned}
&\|Q_\varepsilon y^{(n)}(t) - Q_\varepsilon y(t)\|_\alpha \\
&\quad \le M\|h(y^{(n)})(0) - h(y)(0)\|_\alpha + \frac{MC_{\beta-\alpha}}{\Gamma(p)} \int_0^t (t-r)^{p-1} \\
&\quad \times \left\| F\Big(r, y_r^{(n)}, \int_0^r \zeta(r,\tau,y_\tau^{(n)})d\tau\Big) - F\Big(r, y_r, \int_0^r \zeta(r,\tau,y_\tau)d\tau\Big)\right\|_\beta dr \\
&\quad + N_\alpha \int_0^t (t-r)^{p(1-\alpha)-1} \|Bw_\varepsilon(r,y^{(n)}) - Bw_\varepsilon(r,y)\| dr \\
&\quad \to 0 \text{ as } n \to \infty.
\end{aligned}$$

Thus $Q_\varepsilon \colon B_k \to B_k$ is continuous.

Further we prove the equicontinuity of $Q_\varepsilon(B_k)$ on $[-\sigma, b]$. Take arbitrary $y \in B_k$ and $t_1, t_2 \in [-\sigma, b]$ with $t_1 \le t_2$. If $t_1,\ t_2 \in [-\sigma, 0]$, then $\|Qy(t_2) - Qy(t_1)\| \le \|\varphi(t_2) - \varphi(t_1)\| + \|h(y)(t_2) - h(y)(t_1)\| \to 0$ as $t_1 \to t_2$ independent of $y \in B$ because $\varphi \in D$ and h is completely continuous. Further, if $t_1, t_2 \in J$, we get

$$\begin{aligned}
&\|Q_\varepsilon y(t_2) - Q_\varepsilon y(t_1)\|_\alpha \le \|\mathcal{U}_p(t_2)(\varphi(0) + h(y)(0)) - \mathcal{U}_p(t_1)(\varphi(0) + h(y)(0))\|_\alpha \\
&\quad + \Big\| \int_0^{t_1} (t_2-r)^{p-1}[\mathcal{V}_p(t_2-r) - \mathcal{V}_p(t_1-r)] F\Big(r, y_r, \int_0^r \zeta(r,\tau,y_\tau)d\tau\Big) dr \Big\|_\alpha \\
&\quad + \Big\| \int_0^{t_1} \Big[(t_1-r)^{p-1} - (t_2-r)^{p-1}\Big] \mathcal{V}_p(t_1-r) F\Big(r, y_r, \int_0^r \zeta(r,\tau,y_\tau)d\tau\Big) dr \Big\|_\alpha \\
&\quad + \Big\| \int_{t_1}^{t_2} (t_2-r)^{p-1} \mathcal{V}_p(t_2-r) F\Big(r, y_r, \int_0^r \zeta(r,\tau,y_\tau)d\tau\Big) dr \Big\|_\alpha \\
&\quad + \Big\| \int_0^{t_1} (t_2-r)^{p-1}[\mathcal{V}_p(t_2-r) - \mathcal{V}_p(t_1-r)] Bw_\varepsilon(r,y)\, dr \Big\|_\alpha \\
&\quad + \Big\| \int_0^{t_1} \Big[(t_1-r)^{p-1} - (t_2-r)^{p-1}\Big] \mathcal{V}_p(t_1-r) Bw_\varepsilon(r,y)\, dr \Big\|_\alpha \\
&\quad + \Big\| \int_{t_1}^{t_2} (t_2-r)^{p-1} \mathcal{V}_p(t_2-r) Bw_\varepsilon(r,y)\, dr \Big\|_\alpha \\
&\le \|\mathcal{U}_p(t_2)(\varphi(0) + h(y)(0)) - \mathcal{U}_p(t_1)(\varphi(0) + h(y)(0))\|_\alpha \\
&\quad + C_{\beta-\alpha} \int_0^{t_1} (t_2-r)^{p-1} \|\mathcal{V}_p(t_2-r) - \mathcal{V}_p(t_1-r)\| a_k(r)\, dr \\
&\quad + \frac{MC_{\beta-\alpha}}{\Gamma(p)} \int_0^{t_1} |(t_1-r)^{p-1} - (t_2-r)^{p-1}|\, a_k(r)\, dr \\
&\quad + \frac{MC_{\beta-\alpha}}{\Gamma(p)} \int_{t_1}^{t_2} (t_2-r)^{p-1} a_k(r)\, dr \\
&\quad + R \int_0^{t_1} (t_2-r)^{p-1} \|A^\alpha(\mathcal{V}_p(t_2-r) - \mathcal{V}_p(t_1-r))\|\, dr \\
&\quad + N_\alpha R \int_0^{t_1} |(t_1-r)^{p(1-\alpha)-1} - (t_2-r)^{p(1-\alpha)-1}|\, dr
\end{aligned}$$

$$
\begin{aligned}
&+ N_\alpha R \int_{t_1}^{t_2} (t_2 - r)^{p(1-\alpha)-1} \, dr \\
&= I_1 + I_2 + I_3 + I_4 + I_5 + I_6 + I_7. \qquad (11)
\end{aligned}
$$

From expression of I_1, I_3, I_4, I_6 and I_7, we can easily see that $I_1 \to 0$, $I_3 \to 0$, $I_4 \to 0$, $I_6 \to 0$ and $I_7 \to 0$ as $t_2 \to t_1$ independent of $y \in B_k$ and for any $\epsilon \in (0, t_1)$, we have

$$
\begin{aligned}
I_2 Z &\le C_{\beta-\alpha} \int_0^{t_1-\epsilon} (t_2 - r)^{p-1} \|\mathcal{V}_p(t_2 - r) - \mathcal{V}_p(t_1 - r)\| a_k(r) \, dr \\
&\quad + C_{\beta-\alpha} \int_{t_1-\epsilon}^{t_1} (t_2 - r)^{p-1} \|\mathcal{V}_p(t_2 - r) - \mathcal{V}_p(t_1 - r)\| a_k(r) \, dr \\
&\le C_{\beta-\alpha} \int_0^{t_1-\epsilon} (t_2 - r)^{p-1} a_k(r) \, dr. \sup_{r \in [0, t_1-\epsilon]} \|\mathcal{V}_p(t_2 - r) - \mathcal{V}_p(t_1 - r)\| \\
&\quad + \frac{2MC_{\beta-\alpha}}{\Gamma(p)} \int_{t_1-\epsilon}^{t_1} (t_2 - r)^{p-1} a_k(r) \, dr \qquad (12)
\end{aligned}
$$

and

$$
\begin{aligned}
I_5 &\le R \int_0^{t_1-\epsilon} (t_2 - r)^{p-1} \big\| A^\alpha \big[\mathcal{V}_p(t_2 - r) - \mathcal{V}_p(t_1 - r)\big] \big\| dr \\
&\quad + R \int_{t_1-\epsilon}^{t_1} (t_2 - r)^{p-1} \big\| A^\alpha \big[\mathcal{V}_p(t_2 - r) - \mathcal{V}_p(t_1 - r)\big] \big\| dr \\
&\le pR \int_0^{t_1-\epsilon} (t_2 - r)^{p-1} \Big\| \int_0^\infty \lambda \psi_p(\lambda) A^\alpha \big[T\big((t_2 - r)^p \lambda\big) - T\big((t_1 - r)^p \lambda\big)\big] d\lambda \Big\| dr \\
&\quad + N_\alpha R \int_{t_1-\epsilon}^{t_1} \big[(t_2 - r)^{p(1-\alpha)-1} + (t_1 - r)^{p(1-\alpha)-1}\big] dr \\
&\le pR \int_0^{t_1-\epsilon} (t_2 - r)^{p-1} \int_0^\infty \lambda \psi_p(\lambda) \big\| A^\alpha T\big((t_1 - r)^p \lambda\big) \big\| \\
&\quad \times \big\| T\big((t_2 - r)^p \lambda - (t_1 - r)^p \lambda\big) - I \big\| d\lambda dr \\
&\quad + N_\alpha R \int_{t_1-\epsilon}^{t_1} \big[(t_2 - r)^{p(1-\alpha)-1} + (t_1 - r)^{p(1-\alpha)-1}\big] dr \\
&\le pM_\alpha R \int_0^{t_1-\epsilon} (t_1 - r)^{p(1-\alpha)-1} dr \\
&\quad \times \sup_{r \in [0, t_1-\epsilon]} \int_0^\infty \lambda^\alpha \psi_p(\lambda) \big\| T\big((t_2 - r)^p \lambda - (t_1 - r)^p \lambda\big) - I \big\| d\lambda \\
&\quad + N_\alpha R \int_{t_1-\epsilon}^{t_1} \big[(t_2 - r)^{p(1-\alpha)-1} + (t_1 - r)^{p(1-\alpha)-1}\big] dr. \qquad (13)
\end{aligned}
$$

Since the semigroup $\{T(t),\ t \geq 0\}$ is analytic and compact, then, by Lemma 2, we obtain that $I_2, I_5 \to 0$ as $t_2 \to t_1$ and $\epsilon \to 0$ independent of $y \in B_k$. Thus the equicontinuity of $Q_\varepsilon(B_k)$ on $[-\sigma, b]$ has been proved.

Finally, we have to prove that $G(t) = \{(Q_\varepsilon y)(t) :\ y \in B_k\}, t \in [-\sigma, b]$, is relatively compact set in Y_α. Take a fixed number $t \in (0, b]$ and then take $\kappa \in (0, t)$. For $y \in B_k$ and $\delta > 0$, we define

$$\begin{aligned}
(Q_\varepsilon^{\kappa,\delta} y)(t) &= \int_\delta^\infty \psi_p(\lambda)T(t^p\lambda)d\lambda\Big(\varphi(0) + h(y)(0)\Big) \\
&\quad + p\int_0^{t-\kappa} (t-r)^{p-1} \int_\delta^\infty \lambda\psi_p(\lambda)T((t-r)^p\lambda) \\
&\quad \times \Big[F\Big(r, y_r, \int_0^r \zeta(r,\tau,y_\tau)d\tau\Big) + Bw_\varepsilon(r,y)\Big]d\lambda dr \\
&= T(\kappa^p\delta)\int_\delta^\infty \psi_p(\lambda)T(t^p\lambda - \kappa^p\delta)d\lambda\Big(\varphi(0) + h(y)(0)\Big) \\
&\quad + pT(\kappa^p\delta)\int_0^{t-\kappa} (t-r)^{p-1} \int_\delta^\infty \lambda\psi_p(\lambda)T((t-r)^p\lambda - \kappa^p\delta) \\
&\quad \times \Big[F\Big(r, y_r, \int_0^r \zeta(r,\tau,y_\tau)d\tau\Big) + Bw_\varepsilon(r,y)\Big]d\lambda dr.
\end{aligned}$$

By the compactness of the semigroup $T(t)$ in Y_α, the set $\{(Q_\varepsilon^{\kappa,\delta} y)(t) :\ y \in B_k\}$ is relatively compact in Y_α. But we also get that

$$\begin{aligned}
&\|(Q_\varepsilon y)(t) - (Q_\varepsilon^{\kappa,\delta} y)(t)\|_\alpha \\
&\leq \Big\|\int_0^\delta \psi_p(\lambda)T(t^p\lambda)d\lambda\Big(\varphi(0) + h(y)(0)\Big)\Big\|_\alpha \\
&\quad + p\Big\|\int_0^t (t-r)^{p-1}\int_0^\delta \lambda\psi_p(\lambda)T((t-r)^p\lambda) \\
&\quad \times \Big[F\Big(r, y_r, \int_0^r \zeta(r,\tau,y_\tau)d\tau\Big)d\lambda + Bw_\varepsilon(r,y)\Big]dr\Big\|_\alpha \\
&\quad + p\Big\|\int_{t-\kappa}^t (t-r)^{p-1}\int_\delta^\infty \lambda\psi_p(\lambda)T((t-r)^p\lambda) \\
&\quad \times \Big[F\Big(r, y_r, \int_0^r \zeta(r,\tau,y_\tau)d\tau\Big)d\lambda + Bw_\varepsilon(r,y)\Big]dr\Big\|_\alpha \\
&\leq M\|\big(\varphi(0) + h(y)(0)\big)\|\int_0^\delta \psi_\alpha(\lambda)d\lambda \\
&\quad + pMC_{\beta-\alpha}\int_0^t (t-r)^{p-1}a_k(r)\,dr\int_0^\delta \lambda\psi_p(\lambda)d\lambda
\end{aligned}$$

$$
\begin{aligned}
&+ \frac{MC_{\beta-\alpha}}{\Gamma(p)} \int_{t-\kappa}^{t} (t-r)^{p-1} a_k(r)\, dr \\
&+ pM_\alpha R \int_0^t (t-r)^{p(1-\alpha)-1}\, dr \int_0^\delta \lambda^{1-\alpha} \psi_p(\lambda) d\lambda \\
&+ N_\alpha R \int_{t-\kappa}^{t} (t-r)^{p(1-\alpha)-1}\, dr \\
&\to 0 \quad \text{as } \kappa, \delta \to 0^+.
\end{aligned}
$$

Thus, for each $t \in (0, b]$, $G(t)$ is relatively compact in Y_α. If $t \in [-\sigma, 0]$, then $G(t) = \{\varphi(t) + h(y)(t) : \ y \in B\}$ is relatively compact in Y_α as given that $h : C([-\sigma, b], Y_\alpha) \to D$ is completely continuous. Hence, for each $t \in [-\sigma, b]$, the set $G(t)$ is relatively compact in Y_α.

By the Ascoli–Arzela theorem, the set $\{Q_\varepsilon y : y \in B_k\}$ is relatively compact in $C([-\sigma, b] : Y_\alpha)$. Hence the operator $Q_\varepsilon \colon B_k \to B_k$ is completely continuous for each $\varepsilon > 0$. So by Schauder's fixed point theorem, Q_ε has a fixed point on B_k for each $\varepsilon > 0$. This consummate the proof. □

Theorem 6 *Assume that* $F : J \times D \times Y_\alpha \to Y_\beta$ *and* $h : C([-\sigma, b], Y_\alpha) \to C([-\sigma, 0], Y_\alpha)$ *are uniformly bounded, and also the hypotheses (H1)–(H3) are satisfied. If the linear system (4) is approximately controllable on* J*, then the nonlocal integro-differential system* (1) *is also approximately controllable on* J.

Proof Since $F : J \times D \times Y_\alpha \to Y_\beta$ and $h : C([-\sigma, b], Y_\alpha) \to C([-\sigma, 0], Y_\alpha)$ are uniformly bounded, then all hypotheses of Theorem 5 are verified. Therefore, for each $\varepsilon > 0$ the system (1) has a mild solution $y^{(\varepsilon)}$ in some $B_k \subset C([-\sigma, b], Y_\alpha)$ under the control

$$
\begin{aligned}
w_\varepsilon(t, y^{(\varepsilon)}) &= B^* \mathcal{V}_p^*(b-t) R(\varepsilon, \Gamma_0^b) p(y^{(\varepsilon)}), \\
p(y^{(\varepsilon)}) &= z - \mathcal{U}_p(b)\big(\varphi(0) + h(y^{(\varepsilon)})(0)\big) \\
&\quad - \int_0^b (b-r)^{p-1} \mathcal{V}_p(b-r) F\Big(r, y_r^{(\varepsilon)}, \int_0^r \zeta(r, \tau, y_\tau^{(\varepsilon)}) d\tau\Big) dr.
\end{aligned}
$$

Also note that

$$
\begin{aligned}
y^{(\varepsilon)}(b) &= \mathcal{U}_p(b)\big[\varphi(0) + (h(y^{(\varepsilon)}))(0)\big] + \int_0^b (b-r)^{p-1} \mathcal{V}_p(b-r) \\
&\quad \times \Big[F\Big(r, y_r^{(\varepsilon)}, \int_0^r \zeta(r, \tau, y_\tau^{(\varepsilon)}) d\tau\Big) + B w_\varepsilon(r, y^{(\varepsilon)})\Big] dr \\
&= z - p(y^{(\varepsilon)}) + \Gamma_0^b R(\varepsilon, \Gamma_0^b) p(y^{(\varepsilon)}) \\
&= z - \varepsilon R(\varepsilon, \Gamma_0^b) p(y^{(\varepsilon)}). \qquad (14)
\end{aligned}
$$

The sequence $\big\{F\big(r, y_r^{(\varepsilon)}, \int_0^r \zeta(r, \tau, y_\tau^{(\varepsilon)}) d\tau\big)\big\}$ is bounded as $F : J \times D \times Y_\alpha \to Y_\beta$ is bounded. So there is a subsequence denoted by itself, that converges weakly to say $F(r)$ in $L_2([0,b], Y_\beta)$. The set $\left\{\mathcal{U}_p(b)h(y^{(\varepsilon)})(0)\right\}$ is relatively compact as h is bounded. Hence there is a subsequence denoted by $\left\{\mathcal{U}_p(b)h(y^{(\varepsilon)})(0)\right\}$, that converges to say $\overline{h}$ in Y_α. Now we define

$$v = z - \mathcal{U}_p(b)\varphi(0) - \overline{h} - \int_0^b (b-r)^{p-1}\mathcal{V}_p(b-r)F(r)dr.$$

So we have

$$\begin{aligned}
&\|p(y^{(\varepsilon)}) - v\|_\alpha \le \|\mathcal{U}_p(b)h(y^{(\varepsilon)})(0) - \overline{h}\|_\alpha \\
&\quad + \left\| \int_0^b (b-r)^{p-1}\mathcal{V}_p(b-r)\Big[F\Big(r, y_r^{(\varepsilon)}, \int_0^r \zeta(r,\tau,y_\tau^{(\varepsilon)})d\tau\Big) - F(r)\Big]dr \right\|_\alpha \\
&\quad \le \|\mathcal{U}_p(b)h(y^{(\varepsilon)})(0) - \overline{h}\|_\alpha \\
&\quad + \sup_{0\le t\le b}\left\| \int_0^t (t-r)^{p-1}A^{\alpha-\beta}\mathcal{V}_p(t-r)A^\beta\Big[F\Big(r, y_r^{(\varepsilon)}, \int_0^r \zeta(r,\tau,y_\tau^{(\varepsilon)})d\tau\Big) - F(r)\Big]dr \right\| \\
&\quad \to 0 \text{ as } \varepsilon \to 0.
\end{aligned} \tag{15}$$

By Ascoli–Arzela theorem, the operator $l(\cdot) \to \int_0^{\cdot}(\cdot - r)^{p-1}\mathcal{V}_p(\cdot - r)l(r)dr$ is compact. So we get $\|p(y^{(\varepsilon)}) - v\|_\alpha \to 0$ as $\varepsilon \to 0$. Thus from (14), we get

$$\begin{aligned}
\|y^{(\varepsilon)}(b) - z\|_\alpha &\le \|\varepsilon R(\varepsilon,\Gamma_0^b)(v)\|_\alpha + \|\varepsilon R(\varepsilon,\Gamma_0^b)\|\|p(y^{(\varepsilon)}) - v\|_\alpha \\
&\le \|\varepsilon R(\varepsilon,\Gamma_0^b)(v)\|_\alpha + \|p(y^{(\varepsilon)}) - v\|_\alpha.
\end{aligned}$$

It follows from the Lemma 4 and (15) that $\|y^{(\varepsilon)}(b) - z\|_\alpha \to 0$ as $\varepsilon \to 0$. Hence the nonlocal integro-differential system (1) is approximately controllable. □

4 Example

Consider the following finite delay partial differential equation of fractional order:

$$\begin{cases}
{}^cD_t^p y(t,\eta) = \frac{\partial^2}{\partial\eta^2}y(t,\eta) + w(t,\eta) + a\int_0^t (t-r)^{-\frac{1}{2}} \\
\qquad \times \int_0^\pi \varrho(\eta,\psi)\sin\left(\frac{|y(r-\sigma,\psi)|}{r^{\frac{1}{2}}}\right)d\psi dr, \; t\in[0,b], \eta\in[0,\pi], \\
y(t,0) = y(t,\pi) = 0, \quad t\in[0,b], \\
y(v,\eta) = \varphi(v,\eta) + \int_0^b K(r,v)\frac{1}{1+(y(r,\eta))^2}dr, \quad -\sigma \le v \le 0,
\end{cases} \tag{16}$$

where ${}^cD_t^p$ is a caputo fractional partial derivative of order p, $0 < p < 1$; $\varphi \in D = C([-\sigma,0]\times[0,\pi], \mathbb{R}^+)$.

Let $Y = L^2([0,\pi],\mathbb{R})$. Define an operator $A: Y \to Y$ by $A\nu = -\nu''$ with

$$D(A) = \{\nu \in Y : \nu, \nu' \text{ is absolutely continuous } \nu'' \in Y, \nu(0) = \nu(\pi) = 0\}.$$

Obviously $-A$ generates an analytic and compact semigroup $\{T(t), t \geq 0\}$ in Y. Moreover, the operator A is given by

$$A\nu = \sum_{m=1}^{\infty} m^2 \langle \nu, e_m \rangle e_m, \quad \nu \in D(A)$$

and semigroup $\{T(t)\}$ is given by

$$T(t)\nu = \sum_{m=1}^{\infty} \exp(-m^2 t) \langle \nu, e_m \rangle e_m, \quad \nu \in Y,$$

where $e_m(\eta) = \sqrt{\frac{2}{\pi}} \sin(m\eta), \ 0 \leq \eta \leq \pi, \ m = 1, 2, \ldots$ For $\alpha = \frac{1}{2}$, the operator $A^{\frac{1}{2}}$ is given by

$$A^{\frac{1}{2}}\nu = \sum_{m=1}^{\infty} m \langle \nu, e_m \rangle e_m, \quad \nu \in D(A^{\frac{1}{2}}),$$

where $D(A^{\frac{1}{2}}) = \{\nu \in Y : \sum_{m=1}^{\infty} m \langle \nu, e_m \rangle e_m \in Y\}$. In Particular $\|A^{-\frac{1}{2}}\| = 1$.

Now we define $y(t)(\eta) = y(t, \eta)$, $y_r(\upsilon)(\psi) = y(r + \upsilon, \psi)$, $f(t, \varphi, x)(\eta) = ax(\eta)$, $\zeta(t, r, \varphi)(\eta) = (t - r)^{-\frac{1}{2}} \int_0^{\pi} \varrho(\eta, \psi) \sin\left(\frac{|\varphi(-\sigma)(y)|}{r^{\frac{1}{2}}}\right) d\psi$, $h(x)(\upsilon)(\eta) = \int_0^b \frac{K(r,\upsilon)}{1+(y(r,\eta))^2} dr$, and $\varphi(\upsilon)(\eta) = \varphi(\upsilon, \eta)$. Thus above nonlocal fractional partial differential equations (16) can be written as the abstract form (1).

We assume that the following conditions holds.

1. The operator $K(r, \upsilon)$ is continuous on compact rectangle $[0, b] \times [-\sigma, 0]$.
2. The function $\varrho : [0, \pi] \times [0, \pi] \to \mathbb{R}$ is continuously differentiable with $\varrho(0, \psi) = \varrho(\pi, \psi) = 0$ and

$$L = \left[\int_0^{\pi} \int_0^{\pi} \left(\frac{\partial \varrho(\eta, \psi)}{\partial \eta}\right)^2 d\psi d\eta\right]^{\frac{1}{2}} < \infty.$$

For $x \in C([-\sigma, b], Y_{\frac{1}{2}})$, we have

$$\begin{aligned}\langle \zeta(t, r, \varphi), e_m \rangle &= \int_0^{\pi} \left((t - r)^{-\frac{1}{2}} \int_0^{\pi} \varrho(\eta, \psi) \sin\left(\frac{|\varphi(-\sigma)(y)|}{r^{\frac{1}{2}}}\right) d\psi\right) e_m(\eta) d\eta \\ &= \frac{(t - r)^{-\frac{1}{2}}}{m} \left\langle \int_0^{\pi} \frac{\partial \varrho(\eta, \psi)}{\partial \eta} \sin\left(\frac{|\varphi(-\sigma)(\psi)|}{r^{\frac{1}{2}}}\right) d\psi, \sqrt{\frac{2}{\pi}} \cos(m\eta) \right\rangle\end{aligned}$$

and in view of Bessel'r inequality,

$$\begin{aligned}\|\zeta(t,r,\varphi)\|^2_{\frac{1}{2}} &= \sum_{m=1}^{\infty} m^2 |\langle \zeta(t,r,\varphi), e_m\rangle|^2\\ &= (t-r)^{-1}\sum_{m=1}^{\infty}\Big|\Big\langle \int_0^{\pi} \frac{\partial \varrho(\eta,\psi)}{\partial \eta}\sin\left(\frac{|\varphi(-\sigma)(\psi)|}{r^{\frac{1}{2}}}\right)d\psi, \sqrt{\frac{2}{\pi}}\cos(m\eta)\Big\rangle\Big|^2\\ &\le (t-r)^{-1}\int_0^{\pi}\left(\int_0^{\pi}\frac{\partial \varrho(\eta,\psi)}{\partial \eta}\sin\left(\frac{|\varphi(-\sigma)(\psi)|}{r^{\frac{1}{2}}}\right)d\psi\right)^2 d\eta\\ &\le (t-r)^{-1}\int_0^{\pi}\int_0^{\pi}\left(\frac{\partial \varrho(\eta,\psi)}{\partial \eta}\right)^2 d\psi d\eta \times \int_0^{\pi}\sin^2\left(\frac{|\varphi(-\sigma)(\psi)|}{r^{\frac{1}{2}}}\right)d\psi\\ &\le (t-r)^{-1}r^{-1}\int_0^{\pi}\int_0^{\pi}\left(\frac{\partial \varrho(\eta,\psi)}{\partial \eta}\right)^2 d\psi d\eta \times \int_0^{\pi}\big(\varphi(-\sigma)(\psi)\big)^2 d\psi.\end{aligned}$$

This implies

$$\|\zeta(t,r,y_r)\|_{\frac{1}{2}} \le (t-r)^{-\frac{1}{2}} r^{-\frac{1}{2}} L \|y_r(-\sigma)\|_{\frac{1}{2}}.$$

Thus the function ζ satisfies the assumption (H1). We also can easily see that the function f satisfies the assumption (H2) and the function $h : C([0,b], Y_{\frac{1}{2}}) \to C([-\sigma,0], Y_{\frac{1}{2}})$ given by

$$h(y)(\upsilon)(\eta) = \int_0^b K(r,\upsilon)\frac{1}{1+(y(r,\eta))^2}dr.$$

is a completely continuous and h satisfies the assumption (H3).

The function $\mathcal{V}_p(t) : Y_{\frac{1}{2}} \to Y_{\frac{1}{2}}$ is defined as

$$\mathcal{V}_p(t)\nu = p\sum_{m=1}^{\infty}\int_0^{\infty}\lambda\psi_p(\lambda)\exp(-m^2 t^p\lambda)d\lambda\langle \nu, e_m\rangle e_m$$

and

$$B^*\mathcal{V}_p^*(T-t)\nu = p\sum_{m=1}^{\infty}\int_0^{\infty}\lambda\psi_p(\lambda)\exp(-m^2(T-t)^p\lambda)d\lambda\langle \nu, e_m\rangle e_m.$$

If $B^*\mathcal{V}_p^*(T-t)\nu = 0$, $0 \le t < b$, then, by the definition of $B^*\mathcal{V}_p^*(T-t)\nu$, we get $\nu = 0$. In view of Lemma 4 and Theorem 6, the system (16) is approximately controllable.

Acknowledgements The first and the second authors wish to acknowledge the UGC of India and the research project SR/S4/MS:796/2012 of DST, New Delhi for support to this work respectively.

References

1. Podlubny, I.: Fractional Differential Equations. Academic Press, San Diego (1999)
2. Zhou, Y., Jiao, F.: Existence of mild solutions for fractional neutral evolution equations. Comput. Math. Appl. **59**(3), 1063–1077 (2010)
3. Wang, J., Fan, Z., Zhou, Y.: Nonlocal controllability of semilinear dynamic system with fractional derivative in Banach spaces. J. Optim. Theory Appl. **154**(1), 292–302 (2012)
4. El-Borai, M.: Some probability densities and fundamental solutions of fractional evolution equations. Chaos Solitons Fractals **14**(3), 433–440 (2002)
5. Lakshmikantham, V., Vatsala, A.S.: Basic theory of fractional differential equations. Nonlinear Anal. Theory Methods Appl. **69**(8), 2677–2682 (2008)
6. Jiang, H.: Existence results for fractional order functional differential equations with impulse. Comput. Math. Appl. **64**(10), 3477–3483 (2012)
7. Benchohra, M., Henderson, J., Ntouyas, S.K., Ouahab, A.: Existence results for fractional order functional differential equations with infinite delay. J. Math. Anal. Appl. **338**(2), 1340–1350 (2008)
8. Abbas, S.: Existence of solutions to fractional order ordinary and delay differential equations and applications. Electron. J. Diff. Equ. **9**, 1–11 (2011)
9. Dabas, J., Chauhan, A.: Existence and uniqueness of mild solution for an impulsive neutral fractional integro-differential equation with infinite delay. Math. Comput. Model. **57**(3–4), 754–763 (2013)
10. Kumar, S., Sukavanam, N.: Approximate controllability of fractional order semilinear system with bounded delay. J. Differ. Equ. **252**(11), 6163–6174 (2012)
11. Machado, J.A., Ravichandran, C., Rivero, M., Trujillo, J.: Controllability results for impulsive mixed-type functional integro-differential evolution equations with nonlocal conditions. Fixed Point Theory Appl. **2013**(66), 1–16 (2013)
12. Balachandran, K., Park, J.Y.: Nonlocal Cauchy problem for abstract fractional semilinear evolution equation. Nonlinear Anal. Theory Methods Appl. **71**(10), 4471–4475 (2009)
13. Li, F.: Nonlocal Cauchy problem for delay fractional integrodifferential equations of neutral type. Adv. Diff. Equ. **2012**(47), 1–23 (2012)
14. Tai, Z., Lun, S.: On controllability of fractional impulsive neutral infinite delay evolution integrodifferential systems in Banach spaces. Appl. Math. Lett. **25**(2), 104–110 (2012)
15. Ji, S., Li, G., Wang, M.: Controllability of impulsive differential system with nonlocal conditions. Appl. Math. Comput. **217**(16), 6981–6989 (2011)
16. Wang, W., Zhou, Y.: Complete controllability of fractional evolution systems. Commun. Nonlinear Sci. Numer. Simul. **17**(11), 4346–4355 (2012)
17. Yan, Z.: Approximate controllability of partial neutral functional differential systems of fractional order with state-dependent delay. Int. J. Control **85**(8), 1051–1062 (2012)
18. Zhou, H.X.: Approximate controllability for a class of semilinear abstract equations. SIAM J. Control Optim. **21**(4), 551–565 (1983)
19. Mahmudov, N.I., Zorlu, S.: Approximate controllability of fractional integro-differential equations involving nonlocal initial conditions. Bound. Value Probl. **2013**(118) (2013)
20. Mahmudov, N.I., Zorlu, S.: On the approximate controllability of fractional evolution equations with compact analytic semigroup. J. Comput. Appl. Math. 259, part A, pp. 194–204 (2014)
21. George, R.K.: Approximate controllability of nonautonomous semilinear systems. Nonlinear Anal. Theory Methods Appl. **24**(9), 1377–1393 (1995)
22. Mahmudov, N.I.: Approximate controllability of semilinear deterministic and stochastic evolution equations in abstract spaces. SIAM J. Control Optim. **42**(5), 1604–1622 (2003)
23. Fu, X.: Approximate controllability for neutral impulsive differential inclusions with nonlocal conditions. J. Dyn. Control Syst. **17**(3), 359–386 (2011)
24. Sakthivel, R., Ganesh, R., Ren, Y., Anthoni, S.M.: Approximate controllability of nonlinear fractional dynamical systems. Commun. Nonlinear Sci. Numer. Simul. **18**(12), 3498–3508 (2013)
25. Pazy, A.: Semigroup of Linear Operators and Applications to Partial Differential Equations. Springer, New York (1983)

Some Solutions of Generalised Variable Coefficients KdV Equation by Classical Lie Approach

Rajeev Kumar, Anupma Bansal and R.K. Gupta

Abstract We investigate the symmetries of the generalised KdV Equation by using the theory of Lie classical method. The similarities obtained are utilized to reduce the order of nonlinear partial differential equation. Some solutions of reduced differential equation are presented.

Keywords Exact solution · Symmetry analysis · KdV equation

1 Introduction

Mathematical formulation of some of the physical systems is represented by nonlinear partial differential equations. Korteweg–de Vires (KdV) [1, 2]

$$\frac{\partial}{\partial t}u\left(x,t\right)+u\left(x,t\right)\frac{\partial}{\partial x}u\left(x,t\right)+\frac{\partial^{3}}{\partial x^{3}}u\left(x,t\right)=0, \tag{1}$$

is the third order nonlinear partial differential equation where $u = u(x, t)$, is a function of x and t. Equation (1) is derived from the long one-dimensional, surface gravity, small amplitude, waves propagating in a shallow channel of water. But, it also describes various physical systems in science and engineering, like hydromag-

R. Kumar (✉)
Department of Mathematics, Maharishi Markandeshwar University, Mullana, Ambala 133001, Haryana, India
e-mail: rajeevkumarbudhiraja@gmail.com

A. Bansal
Department of Mathematics, DAV College for Women, Ferozepur Cantt 152001, Punjab, India
e-mail: anupma2512@yahoo.co.in

R.K. Gupta
School of Mathematics and Computer Applications, Thapar University, Patiala 147004, Punjab, India
e-mail: rajeshgupta@thapar.edu

J.M. Cushing et al. (eds.), *Applied Analysis in Biological and Physical Sciences*, Springer Proceedings in Mathematics & Statistics 186,
DOI 10.1007/978-81-322-3640-5_19

netic waves, stratified internal waves, rotons, plasma, ion-acoustic waves, physics, lattice dynamics and geophysical fluid dynamics. Under different geophysical conditions different forms of KdV equations are obtained. One of the particular forms of KdV is its variable-coefficients form (VCKdV) [3–7]

$$\frac{\partial}{\partial t}u(x,t) + f(t)\,u(x,t)\,\frac{\partial}{\partial x}u(x,t) + g(t)\,\frac{\partial^3}{\partial x^3}u(x,t) = 0, \tag{2}$$

where f(t) and g(t) are some real coefficients, the actual form of which depend on the physics of the problem. Though there exists no general method for the solution of nonlinear partial differential equations (PDEs), but due to great advancement in software technology, a significant progress has been made in the implementation of various methods for nonlinear PDEs like Lie classical method [8–10], nonclassical method [11], simplest equation method [12] and different forms of $\frac{G'}{G}$ method [13–15] etc. Many authors like Bluman, Kumei [16, 17], Bruzón [18], Gandarias [19] and Kudryashov [20, 21] studied the nonlinear PDEs using these methods. Invariant solutions of PDE are obtained by determining the symmetries. These solutions are obtained by utilizing the group invariants to fewer the number of independent variables. According to the Lie classical approach, a reduction transformation exists whenever a given PDE is invariant under a Lie group of transformations. This paper studies the variable-coefficient generalized KdV (VCGKdV) equation which was studied by Wazwaz [22, 23].

$$\frac{\partial}{\partial t}u(x,t) + f(t)\,(u(x,t))^n\,\frac{\partial}{\partial x}u(x,t) + g(t)\,\frac{\partial^3}{\partial x^3}u(x,t) = 0, \tag{3}$$

$f(t)$ and $g(t)$ are variable time-dependent coefficients. By applying the symmetry reduction procedure, we determine some exact solutions for this equation which provide much information about nonlinear phenomena.

2 Classical Symmetries

For applying Lie classical approach to Eq. (3), we assume the one-parameter Lie group of infinitesimal transformations in (x, t, u) given by

$$x^* \to x + \varepsilon\xi(x,t,u) + O\left(\varepsilon^2\right), \tag{4}$$

$$t^* \to t + \varepsilon\tau(x,t,u) + O\left(\varepsilon^2\right), \tag{5}$$

$$u^* \to u + \varepsilon\eta(x,t,u) + O\left(\varepsilon^2\right), \tag{6}$$

where ε is the group parameter. Under the transformation (6), Eq. (3) remains invariant. On solving the invariance condition, a set of system of linear equations for the infinitesimals $\xi(x, t, u)$, $\tau(x, t, u)$ and $\eta(x, t, u)$ is obtained. The corresponding Lie algebra of infinitesimal symmetries is the set of vector fields of the form

$$X = \xi(x, t, u)\frac{\partial}{\partial x} + \tau(x, t, u)\frac{\partial}{\partial t} + \eta(x, t, u)\frac{\partial}{\partial u}. \tag{7}$$

Invariance of the given PDE (3) under Lie's transformations with infinitesimal generator (7) gives a collection of determining equations.

$$\tau_x = 0, \tau_u = 0, \xi_u = 0, \eta_{uu} = 0, \tag{8}$$

$$3g(t)\eta_{xu} - 3g(t)\xi_{xx} = 0, \tag{9}$$

$$g(t)\eta_{xxx} + \eta_t + f(t)u^n\eta_x = 0, \tag{10}$$

$$\tau g(t)' + \tau_t g(t) - 3g(t)\xi_x = 0, \tag{11}$$

$$\tau_t f(t)u^n - g(t)\xi_{xxx} - f(t)u^n\xi_x - \xi_t + \tau f(t)' u^n + nf(t)u^{n-1}\eta + 3g(t)\eta_{xxu} = 0, \tag{12}$$

where the subscript represents the derivative with respect to the indicated variable. On solving the determining equations, we get the following form of symmetry generator.

$$\xi = xk_1 + k_2, \tau = \tau(t), \eta = uk_3, \tag{13}$$

where k_1, k_2, k_3 are constants.

Using (13) into (12), we obtain the classifying relations as

$$\tau g' + \tau_t g - 3gk_1 = 0, \tag{14}$$

$$\tau_t f - fk_1 + \tau f' + nk_3 f = 0. \tag{15}$$

3 Symmetry Classification

The analysis of the classifying relations by specifying the forms of the arbitrary parameters is presented in this section. In doing so, we utilize the principal Lie algebra and hence, obtain the maximal symmetry Lie algebra by which the invariant or similarity solution is obtained.

The analysis of the classifying relations (14) and (15) can be done for the different cases of $f(t)$ and $g(t)$ by considering $\tau \neq 0$ in order to obtain the principal Lie algebra.

3.1 Case 1

If $f(t) \neq 0$, the Eq. (15) can be written as

$$\frac{f_t}{f} = -\frac{\tau_t - k_1 + nk_3}{\tau} = m, \tag{16}$$

where m is a constant.

For solving (16), two subcases i.e. $m \neq 0$ and $m = 0$ are considered.

Subcase 1.1 Assume that $m \neq 0$, then from Eq. (16), we obtain

$$f(t) = f_0 e^{mt}, \tau = \frac{k_4}{e^{mt}} - \frac{nk_3}{m} + \frac{k_1}{m} \tag{17}$$

where f_0 and k_4 are constants. If $k_1 \neq 0$, then from classifying relation we get $g(t)$ as.

$$g(t) = g_0 e^{mt}\left[(1 - n\alpha) e^{mt} + \beta m\right]^{\left(-\frac{n\alpha+2}{n\alpha-1}\right)}, \tag{18}$$

where $\alpha = \frac{k_3}{k_1}, \beta = \frac{k_4}{k_1}$ and g_0 is a nonzero arbitrary constant. The symmetry generator coefficients of (13) become

$$\tau = \left(\beta e^{-mt} + \frac{1 - n\alpha}{m}\right), \xi = x, \eta = \beta u. \tag{19}$$

If $k_2 = 0$, the principal Lie algebra is given by the vector field

$$X_1 = \left(\beta e^{-mt} + \frac{1 - n\alpha}{m}\right)\frac{\partial}{\partial t} + x\frac{\partial}{\partial x} + u\beta\frac{\partial}{\partial u}. \tag{20}$$

Let $k_1 = 0$, then $\tau = \frac{k_4}{e^{mt}} - \frac{nk_3}{m}$ thus, for $k_4 \neq 0$ we have

$$g(t) = \frac{g_0}{me^{-mt} - n\gamma}, \tag{21}$$

where $k_3 = \gamma k_4$. The extended principal Lie algebra is given by the vector field as

$$X_2 = \left(me^{-mt} - n\gamma\right)\frac{\partial}{\partial t} + u\gamma\frac{\partial}{\partial u}. \tag{22}$$

If $k_4 = 0$ we get $g(t) = g_0$

$$X_3 = -n\frac{\partial}{\partial t} + mu\frac{\partial}{\partial u}. \tag{23}$$

Subcase 1.2
If $m = 0$ in (16), we get

$$f = f_0, \tau = (nk_3 - k_1)t + k_5. \tag{24}$$

Putting $k_5 = \mu k_1$ and $k_3 = \alpha k_1$, we obtain

$$g(t) = g_0\left[(n\alpha - 1)t - \mu\right]^{\frac{-(n\alpha+2)}{(n\alpha-1)}} \tag{25}$$

and

$$X_4 = \left[(n\alpha - 1)t + \mu\right]\frac{\partial}{\partial t} + x\frac{\partial}{\partial x} + \alpha u\frac{\partial}{\partial u}. \tag{26}$$

If $k_1 = 0$, we get

$$\tau = nk_3 t + k_5 \tag{27}$$

and the substitution $k_5 = \lambda k_3$ gives the extended principal Lie algebras as

$$X_5 = (nt + \lambda)\frac{\partial}{\partial t} + u\frac{\partial}{\partial u}. \tag{28}$$

3.2 *Case 2*

Considering $g(t) \neq 0$ and applying the same procedure as in case 1 (see Sect. 3.1), the classifying Eq. (14) can be written as

$$\frac{g_t}{g} = \frac{3k_1 - \tau_t}{\tau} = q, \tag{29}$$

where q is a constant and the classifying results can be obtained as follows:

Subcase 2.1 For $q \neq 0$, we have $g(t) = g_0 e^{qt}$ and thus if $k_1 \neq 0$

$$f(t) = f_1 e^{qt}\left[3e^{qt} + \beta\right]^{\frac{-(2+n\alpha)}{3\alpha}}, \tag{30}$$

where $k_6 = \beta k_1$, $k_3 = \alpha k_1$ and f_1 is non-zero arbitrary constant.

$$X_6 = \left(q\beta e^{-qt} + 3\right)\frac{\partial}{\partial t} + qx\frac{\partial}{\partial x} + q\alpha u\frac{\partial}{\partial u}. \tag{31}$$

Putting $k_1 = 0$, we get

$$f(t) = f_1 e^{-\frac{ne^{qt} - q^2 t\gamma}{q\gamma}} \tag{32}$$

and

$$X_7 = \gamma e^{-qt}\frac{\partial}{\partial t} + u\frac{\partial}{\partial u}. \tag{33}$$

Subcase 2.2
If $q = 0$, we get $g = g_1$ and $f(t) = f_1\,[3k_1 t + k_7]^{\frac{-(2k_1+nk_3)}{3K_1}}$, $\tau = 3k_1 + k_7$, putting $k_7 = \mu k_1$ and $k_3 = \alpha k_1$, we obtain

$$f(t) = f_1\,(3t+\mu)^{\frac{-(2+n\alpha)}{3}} \tag{34}$$

and

$$X_8 = (3t+\mu)\frac{\partial}{\partial t} + x\frac{\partial}{\partial x} + \alpha u\frac{\partial}{\partial u}. \tag{35}$$

If $k_1 = 0$ we have $\tau = k_7$ and

$$f = f_0 e^{-\frac{nk_3}{k_7}t}, \tag{36}$$

this and the extended Lie algebra can then be written as

$$f = f_0 e^{-\frac{n\alpha}{\mu}t} \tag{37}$$

$$X_9 = \mu\frac{\partial}{\partial t} + \alpha u\frac{\partial}{\partial u}. \tag{38}$$

3.3 *Case 3*

From (12), we get

$$\tau = \frac{3k_1(\int g\,(t)\,dt)}{g(t)}. \tag{39}$$

The condition governing the function f(t) is given as

$$\tau_t f - fk_1 + \tau f' + nk_3 f = 0 \tag{40}$$

and the infinitesimal generators or vectors of the corresponding Lie algebra are given by

$$X_{10} = x\frac{\partial}{\partial x} + \frac{3(\int g\,(t)\,dt)}{g(t)}\frac{\partial}{\partial t}, \quad X_{11} = \frac{\partial}{\partial x}, \quad X_{12} = u\frac{\partial}{\partial u}. \tag{41}$$

4 Symmetry Reductions

The main objective of determining the symmetries of a given PDE is to utilize them for symmetry reductions and obtain their solutions. In the subsection, we will utilize the different vector fields and reduce the VCGKdV equation (3) into ODEs of one independent variable. The similarity variables and the invariant solutions of the VCGKdV equation (3) can be determined by solving characteristic equation as given below:

$$\frac{dx}{\xi} = \frac{dt}{\tau} = \frac{du}{\eta}. \tag{42}$$

4.1 Reduction 1.

From $X_{10} + \beta X_{12}$ we obtain, $\xi = \frac{x}{\sqrt[3]{\int g(t)dt}}$ and $u = x^{\beta} F(\xi)$ and $f(t) = c_1 g(t) \left(\int g(t)\, dt\right)^{-(n\beta+2)/3}$ which reduce the Eq. (3) into ode

$$-1/3\,\xi^4 \tfrac{d}{d\xi} F(\xi) + c_1 \xi^{n\beta+2} (F(\xi))^{n+1} + c_1 \xi^{n\beta+3} (F(\xi))^n \tfrac{d}{d\xi} f(\xi) + \beta(\beta-1)(\beta-2) F(\xi)$$
$$+3\beta(\beta-1)\xi \tfrac{d}{d\xi} F(\xi) + 3\beta\xi^2 \tfrac{d^2}{d\xi^2} F(\xi) + \xi^3 \tfrac{d^3}{d\xi^3} F(\xi) = 0. \tag{43}$$

4.2 Reduction 2.

From $X_{11} + \alpha X_{12}$ we obtain $\xi = t$, $u = \mathrm{e}^{\alpha x} F(\xi)$ and $n\alpha f(\xi) = 0$ which implies either $n = 0$ or $\alpha = 0$ or $f(\xi) = 0$.

For $\alpha = 0$, we get

$$\frac{d}{d\xi} F(\xi) = 0 \tag{44}$$

For $f(\xi) = 0$, we obtain

$$\frac{d}{d\xi} F(\xi) + g(\xi)\,\alpha^3.F(\xi) = 0. \tag{45}$$

For $n = 0$, we have

$$\frac{d}{d\xi} F(\xi) + f(\xi) F(\xi) + g(\xi)\,\alpha^3 F(\xi) == 0. \tag{46}$$

4.3 Reduction 3

From X_8 we obtain $\xi = \frac{x}{(3t+\mu)^{\frac{1}{3}}}$, $u = (3t+\mu)^{\frac{\alpha}{3}} f(\xi)$ and the reduced ODE is

$$\alpha f(\xi) - \xi f(\xi)' + f_1 f^n(\xi) f'(\xi) + g_1 f'''(\xi) = 0. \tag{47}$$

4.4 Reduction 4

From $X_9 + cX_{11}$ we obtain $\xi = x - \frac{tc}{\mu}$, $u = e^{\frac{t\alpha}{\mu}} f(\xi)$, and the reduced ODE turns out to be

$$\frac{\alpha}{\mu} f(\xi) - \frac{c}{\mu} f^{'}(\xi) + f_0 f^n(\xi) f^{'}(\xi) + g_1 f^{'''}(\xi) = 0. \tag{48}$$

5 Analysis of Reduced Equations

To obtain the solutions of Eqs. (43) and (47), consider

$$F(\xi) = A\xi^p, \tag{49}$$

where p and A are the constants to be determined. The exponents of ξ are suitably equated such that their respective coefficients become zero. By equating the coefficients of $(p+3)$ and $(n\beta + np + p + 2)$ we get $p = \frac{1-n\beta}{n}$. On substituting (49) into (43), the solution of Eq. (3) is obtained as

$$u = \left(\frac{3c_1\,(n+1-n\beta)}{n}\right)^{-n^{-1}} \left(\frac{x}{\sqrt[3]{\int g\,(t)\,dt}}\right)^{\frac{1-n\beta}{n}} \tag{50}$$

where c_1 is a constant.

On using Eq. (49) into Eq. (47), the solution of Eq. (3) is obtained as,

$$u = (3t+\mu)^{\frac{\alpha}{3}} (\mathrm{e}^{\ln(-g_1(6n^{-1}+2+4n^{-2})f_1{}^{-1})n^{-1}}) \left(\frac{x}{(3t+\mu)^{\frac{1}{3}}}\right)^{\frac{-2}{n}} \tag{51}$$

From Eq. (45), the solution of given PDE (3) is given as

$$u = c_2\, \mathrm{e}^{\int -g(\xi)\alpha d\xi}, \tag{52}$$

where c_2 is constant.

From Eq. (46), the solution of Eq. (3) is obtained as

$$u = c_3\, \mathrm{e}^{\int -f(\xi) - g(\xi)\alpha d\xi} \tag{53}$$

where c_3 is constant.

For the solutions of Eq. (48), assume

$$f\,(\xi) = B \tanh\,(\xi)^p\,. \tag{54}$$

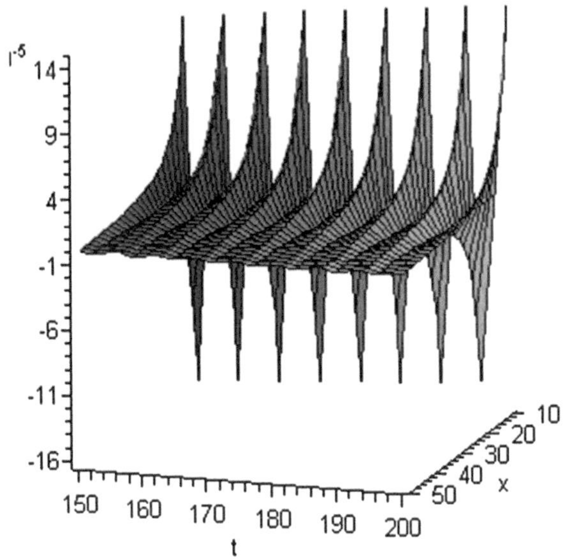

Fig. 1 The figure of solution (50) $\beta = 4, n = 1, c_1 = 1, g(t) = cos(t)$

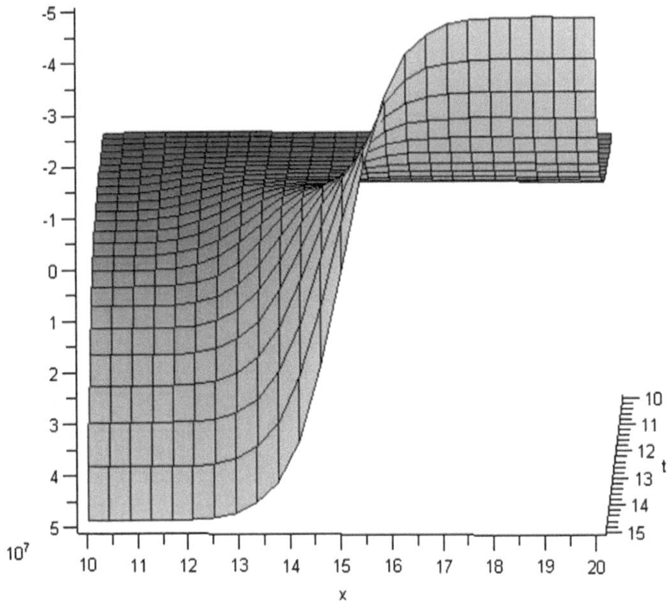

Fig. 2 The figure of solution (52) $n = 1,\ f_0 = 2, c_1 = 3, g_1 = 1, \mu = 1$

The exponents of $\tanh(\xi)$ [24–26] are suitably equated such that their respective coefficients become zero. By equating the exponents pn+p+1 and p-2 respectively, we get $p = \frac{-4}{n}$. Using the Eq. (54) into Eq. (48), the following solitary wave solution of Eq. (3) is obtained as (Figs. 1 and 2):

$$u = e^{\frac{t\alpha}{\mu}} \left(\mathrm{e}^{\ln(-g_1(-4n^{-1}-1)(-4n^{-1}-2)f_0{}^{-1})n^{-1}}\right) \tanh\left(x - \frac{tc}{\mu}\right). \tag{55}$$

6 Conclusion

This paper obtains the Lie symmetries and similarity reductions of variable coefficients generalized KDV equation and further the analysis of the different classifying relations was presented to get the functional forms of the various arbitrary time-dependent parameters for the VCGKDV equation. Also the reduced ODEs have been studied to obtain certain exact solitary wave solutions by using tanh method and other methods. Some graphs of the solution surfaces with some special parameters has been plotted by using the maple software.

References

1. Hong, H., Lee, H.: Korteweg-de Vries equation of ion acoustic surface waves. Phys. Plasmas **6**, 3422–3424 (1999)
2. Satsuma, J., Kaup, D.J.: A B*ä*cklund transformation for a higher order Korteweg-de Vries equation. J. Phys. Soc. Jpn. **43**, 692–726 (1977)
3. Zhang, S.: Application of exp-function method to a KdV equation with variable coefficients. Phys. Lett. A **365**, 448–453 (2007)
4. Singh, K., Gupta, R.K.: On symmetries and invariant solutions of a coupled KdV system with variable coefficients. Int. J. Math. Math. Sci. **23**, 3711–3725 (2005)
5. Gupta, R.K., Singh, K.: Symmetry analysis and some exact solutions of cylindrically symmetric null fields in general relativity. Commun. Nonlinear Sci. Numer. Simul. **16**, 4189–4196 (2011)
6. Abd-el-Malek, M.B., Helal, M.M.: Group method solutions of the generalized forms of Burgers, Burgers-KdV and KdV equations with time-dependent variable coefficients. Acta Mech. **221**, 281–296 (2011)
7. Yan, Z.: A new auto-B*ä*cklund transformation and its applications in finding explicit exact solutions for the general KdV equation. Chin. J. Phys. **40**, 113–120 (2002)
8. Lie, S.: Vorlesungen uber. Differentialgliechurger mit Bekannten Infinite-simalen Transformationen, Teuber, Leipzig (1891)
9. Lie, S.: On integration of a class of linear partial differential equations by means of definite integrals. Arch. Math. **6**, 328–368 (1881)
10. Olver, P.J.: Applications of Lie Groups to Differential Equations. Springer, New York (1993)
11. Arrigo, D.J., Beckham, J.R.: Nonclassical symmetries of evolutionary partial differential equations and compatibility. J. Math. Anal. Appl. **289**, 55–65 (2004)
12. Kudryashov, N.A.: Simplest equation method to look for exact solutions of nonlinear differential equations. Chaos Solitons Fractals **24**, 1217–1231 (2005)
13. Wang, M.L., Zhang, J.L., Li, X.Z.: The $\frac{G'}{G}$-expansion method and travelling wave solutions of nonlinear evolution equations in mathematical physics. Phys. Lett. A **372**, 417–423 (2008)

14. Bansal, A., Gupta, R.K.: Modified $\frac{G'}{G}$-expansion method for finding exact wave solutions of the coupled KleinG-ordon-Schrödinger equation. Math. Methods Appl. Sci. **35**, 1175–1187 (2012)
15. Gupta, R.K., Bansal, A.: Similarity reductions and exact solutions of generalized Bretherton equation with time-dependent coefficients. Nonlinear Dyn. **71**, 1–12 (2013)
16. Bluman, G.W., Anco, S.C.: Symmetries and Differential Equations. Springer, New York (1989)
17. Bluman, G.W., Kumei, S.: Symmetries and Differential Equations. Applied Mathematical Sciences, vol. 81. Springer, Berlin (1989)
18. Bruzón, M.S., Gandarias, M.L., Camacho, J.C.: Classical and nonclassical symmetries for a Kuramoto Sivashinsky equation with dispersive effects. Math. Methods Appl. Sci. **30**, 2091–2100 (2007)
19. Gandarias, M.L., Bruzón, M.S.: Solutions through nonclassical potential symmetries for a generalized inhomogeneous nonlinear diffusion equation. Math. Methods Appl. Sci. **31**, 753–767 (2008)
20. Kudryashov, N.A., Loguinova, N.B.: Extended simplest equation method for nonlinear differential equations. Appl. Math. Comput. **205**, 396–402 (2008)
21. Molati, M., Ramollo, M.P.: Symmetry classification of the Gardner equation with time-dependent coefficients arising in stratified fluids. Commun. Nonlinear Sci. Numer. Simul. **17**, 1542–1548 (2012)
22. Wazwaz, A.M., Triki, H.: Soliton solutions for a generalized KdV and BBM equations with time-dependent coefficients. Commun. Nonlinear Sci. Numer. Simul. **16**, 1122–1126 (2011)
23. Li, W., Zhao, Y.-M.: Exact solutions for a generalized KdV equation with time-dependent coefficients and K(m, n) equation. Appl. Math. Sci. **6**, 2203–2217 (2012)
24. Wazwaz, A.M.: The tanh method for compact and non compact solutions for variants of the KdV-Burger equations. Phys. D: Nonlinear Phenom. **213**, 147–151 (2006)
25. Parkes, E.J., Duffy, B.R.: An automated tanh-function method for finding solitary wave solutions to nonlinear evolution equations. Comput. Phys. Commun. **98**, 288–300 (1996)
26. El-Wakil, S.A., Abdou, M.A.: New exact travelling wave solutions using modified extended tanh-function method. Chaos Solitons Fractals **31**, 840–852 (2007)

Levitin–Polyak Well-Posedness of Strong Parametric Vector Quasi-equilibrium Problems

M. Darabi and J. Zafarani

Abstract We generalize the notion of Levitin–Polyak well-posedness under perturbations for strong version of generalized quasi-equilibrium problems. Some necessary and sufficient conditions for Levitin–Polyak well-posedness and their equivalences between well-posedness of these problems with their scalarizations are given.

Keywords Levitin–Polyak well-posedness · Generalized quasi-equilibrium · Gap function · Scalarization

2010 Mathematics Subject Classification 47J05 · 90C33

1 Introduction and Preliminaries

Well-posedness has played an important role in the theory of optimization. The first concept of well-posedness was initiated by Tykhonov [25] for unconstrained optimization problems. A generalization of Tykhonov well-posedness for an optimization problem (in the scalar case) by the perturbations has been obtained in [7, 30, 31]. Levitin and Polyak [16] extended the well-posedness notion to the constrained case. There are various notions of Levitin–Polyak (in short, LP) well-posedness for scalar and vector optimization problems [10, 12, 15, 26, 29], for scalar and vector equilibrium problems and vector quasi-equilibrium problems [4, 17, 20–22] and

M. Darabi
Department of Basic Sciences, Golpayegan University of Technology, Golpayegan, Isfahan, Iran
e-mail: m-darabi1360@yahoo.com

J. Zafarani (✉)
Department of Mathematics, Sheikhbahaee University, Isfahan, Iran
e-mail: jzaf@zafarani.ir

J. Zafarani
University of Isfahan, Isfahan, Iran

J.M. Cushing et al. (eds.), *Applied Analysis in Biological and Physical Sciences*, Springer Proceedings in Mathematics & Statistics 186,
DOI 10.1007/978-81-322-3640-5_20

for scalar and vector variational inequalities problems [9, 11, 13, 14, 18, 23, 28]. Motivated by the above works, we investigate two generalized vector versions of Levitin–Polyak well-posedness for two class of generalized parametric strong vector quasi-equilibrium problems. We deduce a characterization of Levitin–Polyak well-posedness given in [4] in terms of neighborhoods around solution set.

The outline of the paper is organized as follows: In this section, we present two classes of generalized parametric strong versions of vector quasi-equilibrium problems and some basic definitions and preliminary results. Section 2 deals with notions of LP well-posedness of type I and type II for parametric strong vector quasi-equilibrium problems and conclude their basic properties. In Sect. 3, by introducing two class of gap functions of our problem, we deduce the LP well-posedness of our problems via the LP well-posedness of those gap functions.

Suppose that X, Y and Z are metric topological vector spaces, W is a Hausdorff topological vector space and Λ and Γ are metric spaces. Assume that A, B and D are nonempty closed convex subsets of X, Y and Z, respectively. Assume that $C : X \times \Lambda \times \Gamma \rightrightarrows W$ is a set-valued mapping such that for any $x \in X$, $\lambda \in \Lambda$ and $\gamma \in \Gamma$, $C(x, \lambda, \gamma)$ is a closed, convex and pointed cone in W such that $\operatorname{int} C(x, \lambda, \gamma) \neq \emptyset$. Let $e : X \times \Lambda \times \Gamma \longrightarrow W$ be a continuous vector valued mapping satisfying $e(x, \lambda, \gamma) \in \operatorname{int} C(x, \lambda, \gamma)$. Assume that $K_1 : A \times \Lambda \rightrightarrows A$, $K_2 : A \times \Lambda \rightrightarrows B$, $K_3 : A \times \Lambda \times \Gamma \rightrightarrows D$ and $K_4 : B \times \Lambda \times \Gamma \rightrightarrows D$ are defined. Let the machinery of the problems be expressed by $\Phi : A \times B \times D \times \Gamma \rightrightarrows W$.

For any nonempty sets A and B, we adopt the following notations

$$\begin{aligned} &(x, y)\, r_1\, A \times B \quad &&\text{means} \quad &&\forall x \in A,\ \forall y \in B,\\ &(x, y)\, r_2\, A \times B \quad &&\text{means} \quad &&\forall x \in A,\ \exists y \in B,\\ &(x, y)\, r_3\, A \times B \quad &&\text{means} \quad &&\exists x \in A,\ \forall y \in B,\end{aligned}$$

and

$$\begin{aligned} &\beta_1(A, B) \quad &&\text{means} \quad &&A \subseteq B,\\ &\beta_2(A, B) \quad &&\text{means} \quad &&A \cap B \neq \emptyset.\end{aligned}$$

For $r \in \{r_1, r_2, r_3\}$ and $\beta \in \{\beta_1, \beta_2\}$ and given $\lambda \in \Lambda$ and $\gamma \in \Gamma$, we define the Problem $(P_{3r\beta}(\lambda, \gamma))$ as:

$$\exists\, \bar{x} \in clK_1(\bar{x}, \lambda) : (y, z)\; r\; K_2(\bar{x}, \lambda) \times K_3(\bar{x}, \lambda, \gamma) \Rightarrow \beta\left(\Phi(\bar{x}, y, z, \gamma),\ C(\bar{x}, \lambda, \gamma)\right),$$

and the Problem $(P_{4r\beta}(\lambda, \gamma))$ as:

$$\exists \bar{x} \in clK_1(\bar{x}, \lambda) :,\ (y, z)\; r\; K_2(\bar{x}, \lambda) \times K_4(y, \lambda, \gamma) \Rightarrow \beta\left(\Phi(\bar{x}, y, z, \gamma),\ C(\bar{x}, \lambda, \gamma)\right).$$

We designate the above problems by $(P_{Jr\beta}(\lambda, \gamma))$ where $J \in \{3, 4\}$ and the set of their solutions by $S_{Jr\beta}(\lambda, \gamma)$. In the literature, some important special cases of the above problems were studied; see [1, 3, 8, 19]. For existence result of Problem $(P_{Jr\beta}(\bar{\lambda}, \bar{\gamma}))$ in topological vector spaces and its special cases, one can refer to [5, 6].

Definition 1 A set-valued map $T : X \rightrightarrows Y$ is called

(i) lower semi continuous (l.s.c.), if and only if for each open set $U \subset Y, T^{-}(U) := \{x \in X : T(x) \cap U \neq \emptyset\}$ is open in X.
(ii) closed if and only if $Gr(T) = \{(x, y) \in X \times Y : \ y \in T(x), x \in X\}$ is closed in $X \times Y$.
(iii) upper semi continuous (u.s.c.), if and only if for each closed set $F \subset Y$, $T^{-}(F) := \{x \in X : T(x) \cap F \neq \emptyset\}$ is closed in X.

One can characterize the lower and upper semicontinuity in terms of sequences, as in the following results (see; Theorems 17.16 and 17.19 in [2]).

Lemma 1 ([2]) *A set-valued map $T : X \rightrightarrows Y$ is*

(a) l.s.c. if and only if for any sequence $\{x_n\}$ in X converging to $x \in X$ and each $y \in T(x)$, there is a sequence $\{y_n\}$ converging to y with $y_n \in T(x_n)$, for all n.
(b) u.s.c. if and only if for every sequence $\{x_n\}$ in X converging to $x \in X$ and for any sequence $\{y_n\}$ with $y_n \in T(x_n)$, there is $y \in T(x)$ and a subsequence $\{y_{n_i}\}$ of $\{y_n\}$ converging to y, where T is compact.

2 Levitin–Polyak Well-Posedness

In this section, we define two generalized LP well-posedness of parametric strong vector quasi-equilibrium problems, that improve corresponding definitions in [4, 17, 21].

Definition 2 Let $\{(\lambda_n, \gamma_n)\} \subseteq \Lambda \times \Gamma$ be a convergent sequence to $(\bar{\lambda}, \bar{\gamma})$. A sequence $\{x_n\} \subset A$ is called

(a) LP of type I asymptotically solving sequence corresponding to $\{(\lambda_n, \gamma_n)\}$, for Problem $(P_{3r\beta}(\bar{\lambda}, \bar{\gamma}))$ (resp. $(P_{4r\beta}(\bar{\lambda}, \bar{\gamma})))$, if $x_n \in cl K_1(x_n, \lambda_n)$ and there is a sequence of positive numbers $\{\varepsilon_n\}$ that converges to zero such that

$$(y, z) \ r \ K_2(x_n, \lambda_n) \times K_3(x_n, \lambda_n, \gamma_n)\Big(\text{resp. } (y, z) \ r \ K_2(x_n, \lambda_n) \times K_4(y, \lambda_n, \gamma_n)\Big), \tag{1}$$

$$\beta(\Phi(x_n, y, z, \gamma_n) + \varepsilon_n e(x_n, \lambda_n, \gamma_n), C(x_n, \lambda_n, \gamma_n)). \tag{2}$$

(b) LP of type II asymptotically solving sequence corresponding to $\{(\lambda_n, \gamma_n)\}$, for Problem $(P_{3r\beta}(\bar{\lambda}, \bar{\gamma}))$ (resp. $(P_{4r\beta}(\bar{\lambda}, \bar{\gamma})))$, if there is a sequence of positive numbers $\{\varepsilon_n\}$ that converges to zero such that

$$d(x_n, K_1(x_n, \lambda_n)) \leq \varepsilon_n \tag{3}$$

and (1) and (2) hold.

Example 1 Suppose that $X = Y = Z = \Lambda = \Gamma = \mathbb{R}$, $W = \mathbb{R}^2$. Assume that $K_1(x, \lambda) = K_2(x, \lambda) = K_3(x, \lambda, \gamma) = [0, 1]$ and $\Phi : \mathbb{R} \times \mathbb{R} \times \mathbb{R} \times \mathbb{R} \to \mathbb{R}^2$ is defined by $\Phi(x, y, z, \gamma) = \{y - x\} \times \{y - x, z - x\}$, $C(x, \lambda, \gamma) = {\mathbb{R}^2}_+$ and $e(x, \lambda, \gamma) = (1, 1)$, for all $x, \lambda, \gamma \in \mathbb{R}$. Now for $\beta = \beta_1$, $r = r_1$ and $J = 3$, the sequence $\{\frac{1}{n}\}$ is an example of an LP of type I and type II asymptotically solving sequence for $\varepsilon_n = \frac{1}{n}$, for all n.

Definition 3 (a) The Problem $(P_{Jr\beta}(\bar{\lambda}, \bar{\gamma}))$ is a generalized LP well-posed of type I (resp. type II) if and only if

- (i) there is a solution for Problem $(P_{Jr\beta}(\bar{\lambda}, \bar{\gamma}))$;
- (ii) for any sequence $\{(\lambda_n, \gamma_n)\} \subseteq \Lambda \times \Gamma$ that converges to $(\bar{\lambda}, \bar{\gamma})$, every LP of type I (resp. type II) asymptotically solving sequence for Problem $(P_{Jr\beta}(\bar{\lambda}, \bar{\gamma}))$ corresponding to $\{(\lambda_n, \gamma_n)\}$ contains a subsequence that converges to some points of $S_{Jr\beta}(\bar{\lambda}, \bar{\gamma})$.

(b) The Problem $(P_{Jr\beta}(\bar{\lambda}, \bar{\gamma}))$ is LP well-posed of type I (resp. type II) if and only if

- (i) there is only one solution for Problem $(P_{Jr\beta}(\bar{\lambda}, \bar{\gamma}))$;
- (ii) for any sequence $\{(\lambda_n, \gamma_n)\} \subseteq \Lambda \times \Gamma$, which converges to $(\bar{\lambda}, \bar{\gamma})$, every LP of type I (resp. type II) asymptotically solving sequence for Problem $(P_{Jr\beta}(\bar{\lambda}, \bar{\gamma}))$ corresponding to $\{(\lambda_n, \gamma_n)\}$, converges to $S_{Jr\beta}(\bar{\lambda}, \bar{\gamma})$.

Example 2 In Example 1, we have

$$\begin{aligned} S_{3r_1\beta_1}(\lambda, \gamma) &= \{x \in cl K_1(x, \lambda) : y \in K_2(x, \lambda), z \in K_3(x, \lambda, \gamma),\ \Phi(x, y, z, \gamma) \subseteq {\mathbb{R}^2}_+\} \\ &= \{x \in [0, +\infty[:\ y \in [0, +\infty[,\ z \in [0, +\infty[\ \ \Phi(x, y, z, \gamma) \subseteq {\mathbb{R}^2}_+\} \\ &= \{0\}. \end{aligned}$$

In Example 1, if $\{x_n\} \subseteq cl K_1(x_n, \lambda_n)$ is an LP of type I asymptotically solving sequence corresponding to $(\lambda_n, \gamma_n) \to (\lambda, \gamma)$, then there is a sequence of positive numbers $\{\varepsilon_n\}$ that converges to 0 and

$$y \in K_2(x_n, \lambda_n),\ z \in K_3(x_n, \lambda_n, \gamma_n)\ \ \Phi(x_n, y, z, \gamma_n) + \varepsilon_n e(x_n, \lambda_n, \gamma_n) \subseteq {\mathbb{R}^2}_+,$$

which means that

$$y \in [0, +\infty[,\ z \in [0, +\infty[\ \ (\{y - x_n\} \times \{y - x_n, z - x_n\}) + \varepsilon_n e(x_n, \lambda_n, \gamma_n) \subseteq {\mathbb{R}^2}_+,$$

therefore, $\lim_n x_n = 0$, thus Problem $(P_{3r_1\beta_1}(\lambda, \gamma))$ is LP well-posed of both types.

Example 3 Suppose that $X = Y = Z = \Lambda = \Gamma = \mathbb{R}$, $W = \mathbb{R}^2$. Assume that $K_1(x, \lambda) = K_2(x, \lambda) = K_3(x, \lambda, \gamma) = \mathbb{R}$ and $\Phi : \mathbb{R} \times \mathbb{R} \times \mathbb{R} \times \mathbb{R} \rightrightarrows \mathbb{R}^2$ is defined by

$$\Phi(x, y, z, \gamma) = \begin{cases} \{x\} \times [x, +\infty[& \text{if } x > 1, x < -1, \\ (x, ||x| - 1|) & o.w., \end{cases}$$

for all $x, \lambda, \gamma \in \mathbb{R}$. Let $C(x, \lambda, \gamma) = \mathbb{R}^2{}_+$ and $e(x, \lambda, \gamma) = (1, 1)$. Now for $\beta = \beta_1$, $r = r_1$ and $J = 3$, one has

$$\begin{aligned} S_{3r_1\beta_1} &= \{x \in cl K_1(x, \lambda) : \; y \in K_2(x, \lambda), \; z \in K_3(x, \lambda, \gamma) \; \Phi(x, y, z, \gamma) \subseteq \mathbb{R}^2{}_+\} \\ &= \{x \in \mathbb{R} : \; y \in \mathbb{R}, \; z \in \mathbb{R} \; \Phi(x, y, z, \gamma) \subseteq \mathbb{R}^2{}_+\} \\ &= [0, +\infty[. \end{aligned}$$

If $x_n = n$, for all n, then $\{x_n\}$ is an LP of type I asymptotically solving sequence corresponding to $(\lambda_n, \gamma_n) \to (\lambda, \gamma)$, but $\{x_n\}$ is not convergent. Thus, this problem is not LP well-posed of type I.

Example 4 Suppose that $X = Y = Z$ is the set of all bounded functions $f : [0, 1] \to \mathbb{R}$ provided with the uniform norm $\|f - g\| = \sup_{x \in [0,1]} | f(x) - g(x) |$. Assume that $\Lambda = \Gamma = \mathbb{R}$, $W = \mathbb{R}^2$ and for all $f \in X$, $\lambda, \; \gamma \in \mathbb{R}$,

$$K_1(f, \lambda) = \begin{cases} A & \text{iff is continuous}, \\ A^c & o.w., \end{cases}$$

where A is the set of all continuous functions in X and $A^c = X \setminus A$. Suppose

$$K_2(f, \lambda) = K_3(f, \lambda, \gamma) = \begin{cases} \{f\} & \text{iff is continuous}, \\ A^c & o.w., \end{cases}$$

and $\Phi : X \times Y \times Z \times \mathbb{R} \rightrightarrows \mathbb{R}^2$ is defined by

$$\Phi(f, g, h, \gamma) = \begin{cases} \mathbb{R}^2{}_+ & \text{if } \; f = g = h \\ B_1(0) & o.w., \end{cases}$$

where $B_1(0)$ is the open ball with radius 1 around $(0, 0)$. Assume for all $f \in X$, $\lambda, \gamma \in \mathbb{R}$, $C(f, \lambda, \gamma) = \mathbb{R}^2{}_+$ and $e(f, \lambda, \gamma) = (1, 1)$. Now for $\beta = \beta_1$, $r = r_1$ and $J = 3$, one has

$$\begin{aligned} S_{3r_1\beta_1} &= \{f \in clK_1(f,\lambda) : \ g \in K_2(f,\lambda),\ h \in K_3(f,\lambda,\gamma)\ \Phi(f,g,h,\gamma) \subseteq \mathbb{R}^2{}_+\} \\ &= \{f \in X : \ g \in K_2(f,\lambda),\ h \in K_3(f,\lambda,\gamma)\ \Phi(f,g,h,\gamma) = \mathbb{R}^2{}_+\} \\ &= A. \end{aligned}$$

Select the sequence $\{f_n\}$ as $f_n(x) = nx$, for all $x \in [0, 1]$, then $\{f_n\}$ is an LP of type I and type II asymptotically solving sequence corresponding to $(\lambda_n, \gamma_n) \to (\lambda, \gamma)$, but $\{f_n\}$ is not convergent. Then, this problem is neither a generalized LP of type I nor a generalized LP well-posed of type II.

Remark 1 If Problem $(P_{Jr\beta}(\bar{\lambda}, \bar{\gamma}))$ is a generalized LP well-posed (resp. LP well-posed of type II), then Problem $(P_{Jr\beta}(\bar{\lambda}, \bar{\gamma}))$, is a generalized LP well-posed (resp. LP well-posed of type I), but in general, they are different. The next example confirms our claim.

Example 5 Assume that $X = \mathbb{R}^2$, $Y = Z = W = \Gamma = \Lambda = \mathbb{R}$, $K_1(x,\lambda) = [0,1] \times \{0\}$, $K_2(x,\lambda) = K_3(x,\lambda,\gamma) = [0,1]$, and $\Phi : \mathbb{R}\times\mathbb{R}\times\mathbb{R}\times\mathbb{R} \longrightarrow \mathbb{R}$ is defined by

$$\Phi((x_1,x_2),y,z,\gamma) = -{x_1}^2 + ({x_1}^4 + x_1){x_2}^2,$$

for all $x \in \mathbb{R}^2$, $y, z, \lambda, \gamma \in \mathbb{R}$. Let $C(x,\lambda,\gamma) = \mathbb{R}_+$, and $e(x,\lambda,\gamma) = 1$. Now for $\beta = \beta_1$, $r = r_1$ and $J = 3$,

$$\begin{aligned} S_{3r_1\beta_1} &= \{x \in clK_1(x,\lambda) : \ y \in K_2(x,\lambda),\ z \in K_3(x,\lambda,\gamma),\ \ \Phi(x,y,z,\gamma) \subseteq \mathbb{R}_+\} \\ &= \{x \in \mathbb{R}^2\ y, z \in \mathbb{R},\ \ \Phi(x,y,z,\gamma) \subseteq \mathbb{R}_+\} \\ &= \{(0,0)\}. \end{aligned}$$

If $\{x_n\} = \{(x_{n_1}, x_{n_2})\} \subseteq clK_1(x_n, \lambda_n)$ is an LP of type I asymptotically solving sequence corresponding to $(\lambda_n, \gamma_n) \to (\lambda, \gamma)$, then there is a positive sequence $\{\varepsilon_n\}$ that converges to 0 and

$$y \in K_2(x_n,\lambda_n),\ z \in K_3(x_n,\lambda_n,\gamma_n),\ \ \Phi(x_n,y,z,\gamma_n) + \varepsilon_n e(x_n,\lambda_n,\gamma_n) \subseteq \mathbb{R}_+,$$

which means that

$$y, z \in [0,1],\ \ -{x_{n1}}^2 + ({x_{n1}}^4 + x_{n1}){x_{n2}}^2 + \varepsilon_n e(x_n,\lambda_n,\gamma_n) \subseteq \mathbb{R}_+.$$

Since $\{x_n\} = \{(x_{n1}, x_{n2})\} \subseteq clK_1(x_n, \lambda_n)$, then $x_{n2} = 0$ and therefore we obtain $\lim_n x_n = (0,0)$, thus Problem $(P_{3r_1\beta_1}(\lambda,\gamma))$ is an LP well-posed of type I.

But, if $\{x_n\} = \{(n, \frac{1}{n})\}$, then $\{x_n\}$ is an LP of type II asymptotically solving sequence corresponding to $(\lambda_n, \gamma_n) \to (\lambda, \gamma)$, but $\{x_n\}$ is not convergent. Hence, this problem is not LP well-posed of type II.

Remark 2 If Problem $(P_{Jr\beta}(\bar{\lambda}, \bar{\gamma}))$ is a generalized LP well-posed of type I (resp. type II), then the solution set for Problem $(P_{Jr\beta}(\bar{\lambda}, \bar{\gamma}))$ is a nonempty compact set. In fact, if $\{x_n\}$ is a sequence in $S_{Jr\beta}(\bar{\lambda}, \bar{\gamma})$, consider a constant asymptotically solving sequence corresponding to $\{(\lambda_n, \gamma_n)\}$, that is for all n: $\lambda_n = \bar{\lambda}$ and $\gamma_n = \bar{\gamma}$. Therefore, there exists $x_0 \in S_{Jr\beta}(\bar{\lambda}, \bar{\gamma})$ and a subsequence x_{n_k} of x_n such that $x_{n_k} \to x_0$ and hence, $S_{Jr\beta}(\bar{\lambda}, \bar{\gamma})$ is compact.

In the next two results, we obtain sufficient conditions for generalized LP well-posedness of type I and type II of Problem $(P_{Jr\beta}(\bar{\lambda}, \gamma))$.

Theorem 1 *Assume that $\bar{E}(\lambda)$ is the set of all fixed points of $cl K_1(., \lambda)$ and*

(i) $S_{Jr_1\beta}(\lambda, \gamma)$ is nonempty;
(ii) $\bar{E}$ is compact valued and upper semi continuous;
(iii) K_2 and K_3 are lower semi continuous and Φ is single-valued continuous map;
(iv) $C(x, \lambda, \gamma)$ is closed.

Then, Problem $(P_{Jr_1\beta}(\lambda, \gamma))$ is a generalized LP well-posed of type I.

Proof It suffices to consider Problem $(P_{3r_1\beta}(\lambda, \gamma))$, the proof for the case $J = 4$, is similar. Assume that $\{(\lambda_n, \gamma_n)\} \subseteq \Lambda \times \Gamma$ is a sequence converging to (λ, γ) and sequence $\{x_n\}$ is LP of type I asymptotically solving sequence corresponding to $\{(\lambda_n, \gamma_n)\}$ for Problem $(P_{3r_1\beta}(\lambda, \gamma))$, then there is a sequence of positive numbers $\{\varepsilon_n\}$ which converges to 0 such that for all $y \in K_2(x_n, \lambda_n)$ and for all $z \in K_3(x_n, \lambda_n, \gamma_n)$, Eq. (1) holds. Since, for all n, $x_n \in \bar{E}(\lambda_n)$, then by the upper semi continuity of $\bar{E}$ and compactness of $\bar{E}(\lambda)$, we can deduce that $x_n \to x_0$ for some $x_0 \in \bar{E}(\lambda)$. Suppose that $x_0 \notin S_{3r_1\beta}(\lambda, \gamma)$, that is

$$\exists y_0 \in K_2(x_0, \lambda) : \exists z_0 \in K_3(x_0, \lambda, \gamma), \ \Phi(x_0, y_0, z_0, \gamma) \notin C(x_0, \lambda, \gamma). \quad (4)$$

Since K_2 and K_3 are lower semi continuous, then there exists a sequence $(y_n, z_n) \in K_2(x_n, \lambda_n) \times K_3(x_n, \lambda_n, \gamma_n)$, such that $y_n \to y_0$ and $z_n \to z_0$ and from Eq. (2), we deduce

$$\Phi(x_n, y_n, z_n, \gamma_n) + \varepsilon_n e(x_n, \lambda_n, \gamma_n) \in C(x_n, \lambda_n, \gamma_n).$$

Since Φ is continuous, then the sequence $w_n = \Phi(x_n, y_n, z_n, \gamma)$ tends to $w_0 = \Phi(x_0, y_0, z_0, \gamma)$ and

$$w_n + \varepsilon_n e(x_n, \lambda_n, \gamma_n) \in C(x_n, \lambda_n, \gamma_n).$$

But C is a closed map and $w_0 = \lim_n (w_n + \varepsilon_n e(x_n, \lambda_n, \gamma_n))$, therefore,

$$\Phi(x_0, y_0, z_0, \gamma) \in C(x_0, \lambda, \gamma),$$

which contradicts (4). □

Theorem 2 *Suppose that*

(i) $S_{Jr_1\beta}(\lambda, \gamma)$ is nonempty;
(ii) K_1 is closed with a compact range;

(iii) *K_2 and K_3 are lower semi continuous and Φ is a single-valued continuous map;*

(iv) *$C(x, \lambda, \gamma)$ is closed.*

Then, Problem $(P_{Jr_1\beta}(\lambda, \gamma))$ is a generalized LP well-posed of type II.

Proof It suffices to consider Problem $(P_{3r_1\beta}(\lambda, \gamma))$, the proof for the case $J = 4$, is similar. Assume that $\{(\lambda_n, \gamma_n)\} \subseteq \Lambda \times \Gamma$ is a sequence converging to (λ, γ) and sequence $\{x_n\}$ is an LP of type II asymptotically solving sequence corresponding to $\{(\lambda_n, \gamma_n)\}$ for Problem $(P_{3r_1\beta}(\lambda, \gamma))$, then there is a sequence of positive numbers $\{\varepsilon_n\}$ which converges to 0 such that for all $y \in K_2(x_n, \lambda_n)$ and for all $z \in K_3(x_n, \lambda_n, \gamma_n)$, Eqs. (1) and (2) hold. Since $d(x_n, K_1(x_n, \lambda_n)) \leq \varepsilon_n$, then there is a sequence $u_n \in K_1(x_n, \lambda_n)$ such that $d(x_n, u_n) \leq \varepsilon_n$. Then by the compactness of $cl K_1$, we can assume that $\{u_n\}$ has a convergent subsequence $\{u_m\}$ which converges to x_0. From $d(x_n, u_n) \leq \varepsilon_n$, we conclude that $x_m \to x_0$. By a similar argument as that of the last part of the proof of Theorem 1, we can deduce that $x_0 \in S_{3r_1\beta}(\lambda, \gamma)$ and the proof is complete. □

Remark 3 (a) In the two previous theorems, one can replace conditions (i) and (iii) with the following conditions:

> $(i)'$ $S_{Jr_2\beta}(\lambda, \gamma)$ is nonempty;
> $(iii)'$ K_2 is lower semicontinuous, Φ is a single-valued continuous map, and K_J is upper semicontinuous and compact valued.

Then, with minor modifications in the proofs, one can obtain the generalized LP well-posedness of type I (resp. type II) of the Problem $(P_{Jr_2\beta}(\lambda, \gamma))$.

(b) In the two previous theorems, if we replace conditions (i) and (iii) by the following conditions:

> $(i)''$ $S_{Jr_3\beta}(\lambda, \gamma)$ is nonempty;
> $(iii)''$ K_J are lower semi continuous, Φ is a single-valued continuous map and K_2 is compact valued and upper semicontinuous.

Then, with minor modifications in the proofs, one can deduce the generalized LP well-posedness of type I (resp. type II) of the Problem $(P_{Jr_3\beta}(\lambda, \gamma))$.

Many researchers show that LP well-posedness can be investigated and characterized via approximate of solutions, for example, for optimization problems, see [10, 29]; for equilibrium problems, see [4, 17, 20, 21]; and for variational inequality problems, see [11, 13, 18, 27, 28]. In order to consider this case, let us introduce the set-valued maps:

$$\Pi_{IJr\beta}, \Pi_{IIJr\beta} : \mathbb{R}^+ \times \Lambda \times \Gamma \rightrightarrows A$$

as follows:

$$\Pi_{I3r\beta}(\varepsilon, \bar{\lambda}, \bar{\gamma}) =: \{x \in A : x \in cl K_1(x, \bar{\lambda})),\ (y, z)\ r\ K_2(x, \bar{\lambda}) \times K_3(x, \bar{\lambda}, \bar{\gamma})$$
$$\beta(\Phi(x, y, z, \bar{\gamma}) + \varepsilon e(x, \bar{\lambda}, \bar{\gamma}), C(x, \bar{\lambda}, \bar{\gamma}))\}.$$

$$\Pi_{II3r\beta}(\varepsilon, \bar{\lambda}, \bar{\gamma}) =: \{x \in A : d(x, clK_1(x, \bar{\lambda})) \leq \varepsilon,\ (y, z)\ r\ K_2(x, \bar{\lambda}) \times K_3(x, \bar{\lambda}, \bar{\gamma})$$
$$\beta(\Phi(x, y, z, \gamma) + \varepsilon e(x, \bar{\lambda}, \bar{\gamma}), C(x, \bar{\lambda}, \bar{\gamma}))\}.$$

Similarly, by replacing K_3 by K_4, we can define:

$$\Pi_{I4r\beta}(\varepsilon, \bar{\lambda}, \bar{\gamma}) \text{ and } \Pi_{II4r\beta}(\varepsilon, \bar{\lambda}, \bar{\gamma}).$$

The above mentioned sets are increasing, i.e., for all $\lambda \in \Lambda$ and $\gamma \in \Gamma$ if $\varepsilon_1 \leq \varepsilon_2$, then

$$\Pi_{IJr\beta}(\varepsilon_1, \lambda, \gamma) \subseteq \Pi_{IJr\beta}(\varepsilon_2, \lambda, \gamma), \quad J \in \{3, 4\};$$

$$\Pi_{IIJr\beta}(\varepsilon_1, \lambda, \gamma) \subseteq \Pi_{IIJr\beta}(\varepsilon_2, \lambda, \gamma), \quad J \in \{3, 4\},$$

and furthermore, we obtain the following equalities

$$S_{Jr\beta}(\bar{\lambda}, \bar{\gamma}) = \bigcap_{\epsilon > 0} \Pi_{IJr\beta}(\varepsilon, \bar{\lambda}, \bar{\gamma}) = \Pi_{IJr\beta}(0, \bar{\lambda}, \bar{\gamma}), \quad J \in \{3, 4\}.$$

The next two results provide some alternative characterization of LP well-posedness of problem $(P_{Jr\beta}(\bar{\lambda}, \bar{\gamma}))$. In particular, they extend Theorems 4.1–4.4 in [4] and the corresponding ones in [11, 13, 17, 18, 20, 21, 27, 28].

Theorem 3 *If Problem $(P_{Jr\beta}(\bar{\lambda}, \bar{\gamma}))$ is a generalized LP well-posed of type I* (resp. *type II), then there is a nonempty compact subset G of $S_{Jr\beta}(\bar{\lambda}, \bar{\gamma})$ such that for every neighborhood V of* 0*, there is $\delta > 0$ such that*

$$x \in \Pi_{IJr\beta}(\delta, \bar{\lambda}, \bar{\gamma})(\text{resp. } x \in \Pi_{IIJr\beta}(\delta, \bar{\lambda}, \bar{\gamma})) + V \ \Rightarrow\ x \in G + V.$$

Conversely, if

(i) there is a nonempty compact subset G of $S_{Jr\beta}(\bar{\lambda}, \bar{\gamma})$ such that for every neighborhood V of 0*, there is $\delta > 0$ such that*

$$x \in \Pi_{IJr\beta}(\delta, \bar{\lambda}, \bar{\gamma})(\text{resp. } x \in \Pi_{IIJr\beta}(\delta, \bar{\lambda}, \bar{\gamma})) + V \ \Rightarrow\ x \in G + V,$$

(ii) $\Pi_{IJr\beta}$ (resp. $\Pi_{IIJr\beta}$) *is simultaneously upper semi continuous in its second and third arguments.*

Then, Problem $(P_{Jr\beta}(\bar{\lambda}, \bar{\gamma}))$ is a generalized LP well-posed of type I (resp. *type II).*

Proof The proof with some minor modifications is similar to the proof of Theorem 3.1 in [5] therefore, we omit it here. □

Corollary 1 *If Problem $(P_{Jr\beta}(\bar{\lambda}, \bar{\gamma}))$ is LP well-posed of type I* (resp. *type II), then there is $x_0 \in X$ such that for every neighborhood V of* 0*, there is $\delta > 0$ such that*

$$x \in \Pi_{IJr\beta}(\delta, \bar{\lambda}, \bar{\gamma})(\text{resp. } x \in \Pi_{IIJr\beta}(\delta, \bar{\lambda}, \bar{\gamma})) + V \quad \Rightarrow \quad x \in \{x_0\} + V.$$

Conversely, if

(i) *there exists* $x_0 \in S_{Jr\beta}(\bar{\lambda}, \bar{\gamma})$ *such that for every neighborhood* V *of* 0, *there is* $\delta > 0$ *such that*

$$x \in \Pi_{IJr\beta}(\delta, \bar{\lambda}, \bar{\gamma})(\text{resp.} x \in \Pi_{IIJr\beta}(\delta, \bar{\lambda}, \bar{\gamma})) + V \quad \Rightarrow \quad x \in \{x_0\} + V,$$

(ii) $\Pi_{IJr\beta}$ (resp. $\Pi_{IIJr\beta}$) *is simultaneously upper semi continuous in its second and third arguments.*

Then Problem $(P_{Jr\beta}(\bar{\lambda}, \bar{\gamma}))$, *is an LP well-posed of type I* (resp. *type II*).

Remark 4 The above corollary improves Theorems 3.1 and 3.2 in [15].

Example 6 Suppose that $X = Y = Z = \Lambda = \Gamma = W = \mathbb{R}$. Define $K_1(x, \lambda) = K_2(x, \lambda) = K_3(x, \lambda, \gamma) = [1, +\infty[$ and $\Phi : \mathbb{R} \times \mathbb{R} \times \mathbb{R} \times \mathbb{R} \rightrightarrows \mathbb{R}$ as $\Phi(x, y, z, \gamma) = [y^2 - x, +\infty[$, for all $x, \lambda, \gamma \in \mathbb{R}$. Let $C(x, \lambda, \gamma) = \mathbb{R}_+$ and $e(x, \lambda, \gamma) = 1$. Now for $\beta = \beta_1$, $r = r_1$ and $J = 3$, one has

$$\begin{aligned} S_{3r_1\beta_1} &= \{x \in cl K_1(x, \lambda) : \ y \in K_2(x, \lambda), \ z \in K_3(x, \lambda, \gamma), \ \Phi(x, y, z, \gamma) \subseteq \mathbb{R}^+\} \\ &= \{x \in [1, +\infty[: \ y \in [1, +\infty[, \ z \in [1, +\infty[, \ \Phi(x, y, z, \gamma) \subseteq \mathbb{R}^+\} \\ &= \{1\}. \end{aligned}$$

Hence,

$$\begin{aligned} &\Pi_{I3r_1\beta_1}(\delta, \lambda, p) \\ &= \{x \in cl K_1(x, \lambda) : y \in K_2(x, \lambda), z \in K_3(x, \lambda, \gamma), \ \Phi(x, y, z, \gamma) + \delta e(x, \lambda, \gamma) \subseteq \mathbb{R}^+\} \\ &= \{x \in [1, +\infty[: \ y \in [1, +\infty[, \ z \in [1, +\infty[, \ \Phi(x, y, z, \gamma) + \delta e(x, \lambda, \gamma) \subseteq \mathbb{R}^+\} \\ &= [1, 1 + \delta]. \end{aligned}$$

Thus, $\Pi_{I3r_1\beta_1}$ fulfills conditions (i) and (ii) of Corollary 1 and therefore, this problem is LP well-posed of type I.

The following result provides a sufficient condition for existence of condition (ii) of Theorem 3 and its corollary.

Theorem 4 *Assume* $(\bar{\lambda}, \bar{\gamma}) \in \Lambda \times \Gamma$, $\Pi_{IJr_1\beta_1}$ (resp.$\Pi_{IIJr_1\beta_1}$) *is nonempty and the following assumptions hold:*

(i) Φ *is lower semicontinuous;*
(ii) K_2 *and* K_3 *are lower semicontinuous on* $A \times \{\bar{\lambda}\}$ *(resp.* $A \times \{\bar{\lambda}\} \times \{\bar{\gamma}\}$*);*
(iii) $\bar{E}(\bar{\lambda})$ *is compact and* $\bar{E}$ *is upper semicontinuous;*
(iv) $C(x, \lambda, \gamma)$ *is closed.*

Then, $\Pi_{IJr_1\beta_1}$ (resp. $\Pi_{IIJr_1\beta_1}$) *is simultaneously upper semicontinuous in its second and third arguments.*

Proof For $J = 3$, it suffices to consider $\Pi_{IJr_1\beta_1}$, the proofs for the other cases are similar. By contradiction, for a fix $\varepsilon > 0$, suppose there is an open superset U of $\Pi_{I3r_1\beta_1}(\varepsilon, \bar{\lambda}, \bar{\gamma})$, a sequence $\{(\lambda_n, \gamma_n)\} \subseteq \Lambda \times \Gamma$ with $(\lambda_n, \gamma_n) \longrightarrow (\bar{\lambda}, \bar{\gamma})$ and a sequence $x_n \in \Pi_{I3r_1\beta_1}(\varepsilon, \lambda_n, \gamma_n)$, such that $x_n \notin U$ for all n. Since $\bar{E}$ is Upper semicontinuous at $\bar{\lambda}$ and by compactness of $\bar{E}(\bar{\lambda})$, we can deduce that $x_n \to x_0$ for some $x_0 \in \bar{E}(\bar{\lambda})$. If $x_0 \notin \Pi_{I3r_1\beta_1}(\varepsilon, \bar{\lambda}, \bar{\gamma})$, then there exist $y_0 \in K_2(x_0, \bar{\lambda})$ and $z_0 \in K_3(y_0, \bar{\lambda}, \bar{\gamma})$ such that

$$\Phi(x_0, y_0, z_0, \bar{\gamma}) + \varepsilon e(x_0, \bar{\lambda}, \bar{\gamma}) \nsubseteq C(x_0, \bar{\lambda}, \bar{\gamma}).$$

Therefore, there exists $w_0 \in \Phi(x_0, y_0, z_0, \bar{\gamma})$ such that $w_0 + \varepsilon e(x_0, \bar{\lambda}, \bar{\gamma}) \notin C(x_0, \bar{\lambda}, \bar{\gamma})$. On the other hand, by the lower semicontinuity of K_2 there is a sequence $y_n \in K_2(x_n, \lambda_n)$ such that $y_n \to y_0$ and the lower semicontinuity of K_3 implies the existence of a sequence $z_n \in K_3(y_n, \lambda_n, \gamma_n)$ such that $z_n \to z_0$.

As $x_n \in \Pi_{I3r\beta_1}(\varepsilon, \lambda_n, \gamma_n)$ thus, we have

$$\Phi(x_n, y_n, z_n, \gamma_n) + \varepsilon e(x_n, \lambda_n, \gamma_n) \subseteq C(x_n, \lambda_n, \gamma_n).$$

Since C is a closed map and e is continuous, we deduce

$$w_0 + e(x_0, \bar{\lambda}, \bar{\gamma}) \in C(x_0, \bar{\lambda}, \bar{\gamma}),$$

that contradicts our choice w_0. Therefore, $x_0 \in \Pi_{I3r_1\beta_1}(\varepsilon, \bar{\lambda}, \bar{\gamma}) \subseteq U$, so there is n_0 such that for all $n \geq n_0$, we have $x_n \in U$, which is a contradiction. □

Remark 5 (i) The above result holds for r_2 if K_2 is lower semi continuous and K_3 is upper semi continuous and compact valued.

(ii) The above result holds for r_3 if K_3 is lower semi continuous and K_2 is upper semi continuous and compact valued.

3 Scalarization Method for LP Well-Posedness

In this section by introducing two different classes of gap functions for our Problem $(P_{Jr\beta}(\bar{\lambda}, \bar{\gamma}))$, we obtain its LP well-posedness via LP well-posedness of these gap functions. Let us consider the following scalar optimization problem under perturbation:

$$(OP_{(\bar{\lambda},\bar{\gamma})}) \qquad \min \phi(x, \bar{\lambda}, \bar{\gamma}) \text{ s.t.} x \in cl K_1(x, \bar{\lambda}),$$

where $\phi : A \times \Lambda \times \Gamma \longrightarrow \mathbb{R}$. We designate the solution set and minimum value of Problem $(OP_{(\bar{\lambda},\bar{\gamma})})$ by $S(\bar{\lambda}, \bar{\gamma})$ and $\bar{v}$, respectively. Now, similar to Definitions 2 and 3, we define the notions of asymptotically solving sequence and LP-well-posedness in the scalar case.

Definition 4 ([4]) Suppose that $\{(\lambda_n, \gamma_n)\} \subseteq \Lambda \times \Gamma$ is a sequence converging to $(\bar{\lambda}, \bar{\gamma})$. A sequence $\{x_n\} \subset A$ is

(a) LP of type I asymptotically solving sequence corresponding to $\{(\lambda_n, \gamma_n)\}$, for Problem $(OP_{(\bar{\lambda},\bar{\gamma})})$, iff

 (i) $x_n \in clK_1(x_n, \lambda_n)$, for all n;
 (ii) there is a sequence of positive numbers $\{\varepsilon_n\}$ that converges to 0 and n_0 such that for all $n \geq n_0$, $\phi(x_n, \lambda_n, \gamma_n) \leq \varepsilon_n + \bar{v}$.

(b) LP of type II asymptotically solving sequence corresponding to $\{(\lambda_n, \gamma_n)\}$, for Problem $(OP_{(\bar{\lambda},\bar{\gamma})})$, condition (ii) of part (a) and $d(x_n, clK_1(x_n, \lambda_n)) \leq \varepsilon_n$, for all n.

Remark 6 Part (a) of the above definition is a generalization of LP of type I asymptotically solving sequence for optimization problem in [10, 12, 15, 26, 29]. While, part (b) is a generalization of Definition 3.1 of LP asymptotically solving sequence for optimization problem in [15].

Definition 5 (a) The Problem $(OP_{(\bar{\lambda},\bar{\gamma})})$ is LP well-posed of type I (resp. type II) if and only if,

 (i) there is only one solution for Problem $(OP_{(\bar{\lambda},\bar{\gamma})})$;
 (ii) for any sequence $\{(\lambda_n, \gamma_n)\} \subseteq \Lambda \times \Gamma$ converging to $(\bar{\lambda}, \bar{\gamma})$, every LP of type I (resp. type II,) asymptotically solving sequence for Problem $(OP_{(\bar{\lambda},\bar{\gamma})})$ corresponding to $\{(\lambda_n, \gamma_n)\}$ contains a subsequence that converges to $S(\bar{\lambda}, \bar{\gamma})$.

(b) The Problem $(OP_{(\bar{\lambda},\bar{\gamma})})$, is generalized LP well-posed of type I (resp. type II) if and only if,

 (i) there is a solution for Problem $(OP_{(\bar{\lambda},\bar{\gamma})})$;
 (ii) for any sequence $\{(\lambda_n, \gamma_n)\} \subseteq \Lambda \times \Gamma$ converging to $(\bar{\lambda}, \bar{\gamma})$, every LP of type I (resp. type II) asymptotically solving sequence for Problem $(OP_{(\bar{\lambda},\bar{\gamma})})$ corresponding to $\{(\lambda_n, \gamma_n)\}$ contains a subsequence that converges to some points of $S(\bar{\lambda}, \bar{\gamma})$.

Remark 7 The above definitions of LP well-posedness of type I extend the corresponding ones in [10, 12, 15, 26, 29] for Problem $(OP_{(\bar{\lambda},\bar{\gamma})})$.

Example 7 Suppose that $X = \Lambda = \Gamma = \mathbb{R}$. Define $K_1(x, \lambda) = [0, +\infty[$ and $\phi : [0, +\infty[\times\mathbb{R} \times \mathbb{R} \longrightarrow \mathbb{R}$ by $\phi(x, \lambda, \gamma) = |x - 1|$, for all $x \in X$, $\lambda \in \Lambda, \gamma \in \Gamma$. Then $S(\lambda, \gamma) = \{1\}$. If $\{x_n\} \subseteq clK_1(x_n, \lambda_n)$ is an asymptotically solving sequence corresponding to $(\lambda_n, \gamma_n) \longrightarrow (\lambda, \gamma)$, then there is a sequence of positive numbers $\{\varepsilon_n\}$ which converges to 0 and

$$\exists n_0 : \ \forall n \geq n_0 \ \ \phi(x_n, \lambda_n, \gamma_n) \leq \varepsilon_n,$$

which means that

$$\exists n_0 : \ \forall n \geq n_0 \ \ |x_n - 1| \leq \varepsilon_n,$$

since $\lim_n \varepsilon_n = 0$ and $1 - \varepsilon_n < x_n < 1 + \varepsilon_n$, then $\lim_n x_n = 1$. Hence, the sequence $\{1 - \frac{1}{n}\}$ is an example of an LP of type I asymptotically solving sequence for $\varepsilon_n = \frac{1}{n}$, for all n.

Example 8 Suppose that $X = \mathbb{R}^2$, $\Lambda = \Gamma = \mathbb{R}$, $A = \{(x_1, 0) : \; x_1 \in \mathbb{R}\}$. Define $K_1(x, \lambda) = [0, +\infty[\times[0, +\infty[$, $\phi : X \times \mathbb{R} \times \mathbb{R} \to \mathbb{R}$ by $\phi((x_1, x_2), \lambda, \gamma) = x_1^4 - (x_1^8 + x_1)x_2^4$, for $x \in X, \lambda, \gamma \in \mathbb{R}$. Then $S(\lambda, \gamma) = \{(0, 0)\}$. If $\{x_n\} = \{(n, \frac{1}{n})\}$, then $\{x_n\}$ is an LP of type I asymptotically solving sequence corresponding to $(\lambda_n, \gamma_n) \to (\lambda, \gamma)$, but $\{x_n\}$ is not convergent. Therefore, this problem is not an LP well-posed of type I.

Here, we use a modified version of a result of Sach [24] for obtaining a nonlinear scalarization function and define gap functions for problem $(P_{Jr\beta}(\bar{\lambda}, \bar{\gamma}))$.

Definition 6 Let $Q \subset Y$, C be a closed convex cone in Y with $\operatorname{int} C \neq \emptyset$ and $e \in \operatorname{int} C$, then

(i) $t_Q^{\beta} := \{t \geq 0 : \beta(Q + te, C)\}$.
(ii) Q is called C-bounded if for each neighborhood U of the origin of Y there is a real positive number t, such that $Q \subset C + tU$.
(iii) Q is called $-C$-closed if $Q - C$ is closed.
(iv) Q is called C-compact if any cover of Q of the form $\{Q + U_\alpha : U_\alpha \text{ open}\}$ has a finite subcover

Remark 8 One can show that when Q is C-compact, then Q is $-C$ closed and C-bounded. If the set valued function Φ satisfies condition (ii) (resp. (iii)) of Definition 6 at each point of $A \times B \times D \times \Gamma$, then we say that Φ is C-bounded (resp. $-C$-closed). It is evident that if Φ has bounded values in W, then it is C-bounded and furthermore, if Φ has C-compact values in W, then Φ is simultaneously C-bounded and $-C$-closed.

The proofs of the following results are similar to the corresponding ones in [24], with replacing C by $-C$, therefore it is omitted.

Lemma 2 *For a subset Q of Y, $e \in \operatorname{int} C$ and $\varepsilon > 0$, we have*

(i) If Q is C-bounded, then $s_Q^{\beta_1} := \min\{t; t \in t_Q^{\beta_1}\}$ is well-defined.
(ii) If Q is $-C$-closed, then $s_Q^{\beta_2} := \min\{t; t \in t_Q^{\beta_2}\}$ is well-defined.
(iii) If Q is C-bounded, then $s_Q^{\beta_1} = 0$ iff $\beta_1(Q, C)$, and when Q is compact, $s_Q^{\beta_1} < \varepsilon$ iff $\beta_1(Q, C - \varepsilon e)$.
(iv) If Q is $-C$-closed, then $s_Q^{\beta_2} = 0$ iff $\beta_2(Q, C)$ and $s_Q^{\beta_2} \leq \varepsilon$ iff $\beta_2(Q, C - \varepsilon e)$.

Remark 9 If we replace $\Phi(x, y, z, \gamma)$ for each $(x, , y, z, \gamma) \in A \times B \times D \times \Gamma$ by Q in the Definition 6, then by Remark 8, when Φ is compact valued, all of the consequences of Lemma 2 are valid for φ_3(resp. φ_4) $: X \times \Lambda \times \Gamma \longrightarrow \mathbb{R}$ defined as

$$\varphi_3(x,\lambda,\gamma) := \min\{t \in \mathbb{R}_+ : \ (y,z)\ r\ K_2(x,\lambda) \times K_3(x,\lambda,\gamma), \\ \beta(\Phi(x,y,z,\gamma) + te(x,\lambda,\gamma),\ C(x,\lambda,\gamma))\}. \tag{5}$$

$$(\text{resp. } \varphi_4(x,\lambda,\gamma) := \min\{t \in \mathbb{R}_+ : \ (y,z)\ r\ K_2(x,\lambda) \times K_4(y,\lambda,\gamma), \\ \beta(\Phi(x,y,z,\gamma) + te(x,\lambda,\gamma),\ C(x,\lambda,\gamma))\}). \tag{6}$$

Definition 7 A function $L : X \times \Lambda \times \Gamma \to \mathbb{R}$ is a gap function corresponding to function K_1, for Problem $(OP_{(\bar{\lambda},\bar{\gamma})})$ iff

(i) $L(x,\bar{\lambda},\bar{\gamma}) \geq 0,\ \forall x \in X$;
(ii) $\bar{x} \in S(\bar{\lambda},\bar{\gamma})$, iff $L(\bar{x},\bar{\lambda},\bar{\gamma}) = 0$ and $\bar{x} \in clK_1(\bar{x},\bar{\lambda})$.

Lemma 3 *If the map $C : X \times \Lambda \times \Gamma \to W$ is a closed map, then the function φ_3* (resp. φ_4) *is a gap function corresponding to function K_1 for Problem $(P_{3r\beta}(\bar{\lambda},\bar{\gamma}))$* (resp. $(P_{4r\beta}(\bar{\lambda},\bar{\gamma}))$).

Proof It suffices to consider $J = 3$, for $J = 4$ the proof is similar. Obviously, for all $(x,\bar{\lambda},\bar{\gamma}) \in X \times \Lambda \times \Gamma$, $\varphi_3(x,\bar{\lambda},\bar{\gamma}) \geq 0$ and if $x \in S_{3r\beta}(\bar{\lambda},\bar{\gamma})$, then we obtain $\varphi_3(x,\bar{\lambda},\bar{\gamma}) = 0$.

Conversely, if $\varphi_3(x,\bar{\lambda},\bar{\gamma}) = 0$ and $x \in clK_1(x,\bar{\lambda})$, then there is a sequence of positive numbers $\{\varepsilon_n\}$ which converges to 0 and

$$(y,z)\ r\ K_2(x,\bar{\lambda}) \times K_3(x,\bar{\lambda},\bar{\gamma}),\ \beta(\Phi(x,y,z,\bar{\gamma}) + \varepsilon_n e(x,\bar{\lambda},\bar{\gamma}),\ C(x,\bar{\lambda},\bar{\gamma})).$$

Since C is a closed map and $\lim_n \varepsilon_n = 0$, we obtain

$$(y,z)\ r\ K_2(x,\bar{\lambda}) \times K_3(x,\bar{\lambda},\bar{\gamma}),\ \beta(\Phi(x,y,z,\bar{\gamma}),\ C(x,\bar{\lambda},\bar{\gamma})),$$

then $x \in S_{3r\beta}(\bar{\lambda},\bar{\gamma})$. □

In the following theorem, we obtain an equivalence relation between the generalized LP well-posedness of Problem $(P_{Jr\beta}(\bar{\lambda},\bar{\gamma}))$ and LP well-posedness of Problem $(OP_{(\bar{\lambda},\bar{\gamma})})$ via its gap function.

Theorem 5 *Suppose that $C : X \times \Lambda \times \Gamma \rightrightarrows W$ is a closed map and Φ has C-compact values. Then, Problem $(P_{Jr\beta}(\bar{\lambda},\bar{\gamma}))$ is a generalized LP well-posed of type I* (resp. *type II*) *if and only if Problem $(OP_{(\bar{\lambda},\bar{\gamma})})$ for $\varphi_J(x,\bar{\lambda},\bar{\gamma})$ is a generalized LP well-posed of type I* (resp. *type II*).

Proof It suffices to consider $J = 3$; for $J = 4$ the proof is similar. Since solution sets of two problems are nonempty, then it suffices to show that every LP of type I (resp. type II) asymptotically solving sequence for Problem $(OP_{(\bar{\lambda},\bar{\gamma})})$ corresponding to $\{(\lambda_n,\gamma_n)\}$, is an LP of type I (resp. type II) asymptotically solving sequence for Problem $(P_{3r\beta}(\bar{\lambda},\bar{\gamma}))$ and vice versa. Suppose that a sequence $\{x_n\}$ is an LP of type

I (resp. type II) asymptotically solving sequence for Problem $(OP_{(\bar{\lambda},\bar{\gamma})})$, then th ere is a sequence of positive numbers $\{\varepsilon_n\}$ which converges to 0 and there is n_0 such that for all $n \geq n_0$, $\varphi_3(x_n, \lambda_n, \gamma_n) \leq \varepsilon_n$, since φ_3 is a gap function and $\bar{v} = 0$. Then, by definition of φ_3 and Remark 9 we have: $(y, z)\ r\ K_2(x_n, \lambda_n) \times K_3(x_n, \lambda_n, \gamma_n)$,

$$\beta(\Phi(x_n, y, z, \gamma_n) + \varepsilon_n e(x_n, \lambda_n, \gamma_n),\ C(x_n, \lambda_n, \gamma_n)),$$

Furthermore, $x_n \in cl K_1(x_n, \lambda_n)$ (resp. $d(x_n, cl K_1(x_n, \lambda_n) \leq \varepsilon_n)$. Therefore $\{x_n\}$ is an LP of type I (resp. type II) asymptotically solving sequence for Problem $(P_{3r\beta}(\bar{\lambda}, \bar{\gamma}))$. Conversely, suppose that a sequence $\{x_n\}$ is an LP of type I (resp. type II) asymptotically solving sequence corresponding to $\{(\lambda_n, \gamma_n)\}$ for Problem $(P_{3r\beta}(\bar{\lambda}, \bar{\gamma}))$. Then there is a sequence of positive numbers $\{\varepsilon_n\}$ which converges to 0 and

$$x_n \in cl K_1(x_n, \lambda_n) \text{ (resp. } d(x_n, cl K_1(x_n, \lambda_n) \leq \varepsilon_n)$$

such that

$$(y, z)\, r\, K_2(x_n, \lambda_n) \times K_3(x_n, \lambda_n, \gamma_n), \beta(\Phi(x_n, y, z, \gamma_n) + \varepsilon_n e(x_n, \lambda_n, \gamma_n), C(x_n, \lambda_n, \gamma_n)).$$

Then, from definition of $\varphi_3(x_n, \lambda_n, \gamma_n)$, we obtain $\varphi_3(x_n, \lambda_n, \gamma_n) \leq \varepsilon_n$. So, $\{x_n\}$ is an LP of type I (resp. type II) asymptotically solving sequence for Problem $(OP_{(\bar{\lambda},\bar{\gamma})})$. □

Example 9 In Example 3,

$$\begin{aligned}
&\varphi_{I3}(x, \lambda, \gamma) \\
&= \min\{t \in \mathbb{R}_+ :\ y \in K_2(x, \lambda),\ z \in K_3(x, \lambda, \gamma)\ \Phi(x, y, z, \gamma) + te(x, \lambda, \gamma) \subseteq \mathbb{R}^2{}_+\} \\
&= \min\{t \in \mathbb{R}_+ : y \in \mathbb{R},\ z \in \mathbb{R}\ \Phi(x, y, z, \gamma) + te(x, \lambda, \gamma) \subseteq \mathbb{R}^2{}_+\} \\
&= \begin{cases} 0 & \text{if } x \geq 0, \\ -x & o.w., \end{cases}
\end{aligned}$$

If $\{x_n\}$ is an LP of type I asymptotically solving sequence corresponding to $(\lambda_n, \gamma_n) \longrightarrow (\lambda, \gamma)$, then $\lim_n x_n = 0$. Therefore, Problem $(OP_{(\bar{\lambda},\bar{\gamma})})$ for φ_{I3} is LP well-posed of type I.

Remark 10 As it was shown in [5], one can deduce similarly the Levitin–Polyak well-posedness of generalized vector variational inequalities and generalized set-valued optimization problems from our results in this paper.

Acknowledgements The second author was partially supported by the Center of Excellence for Mathematics, University of Isfahan, Iran.

References

1. Agarwal, R., Balaj, M., O'Regan, A.: A unifying approach to variational relation problems. J. Optim. Theory Appl. **155**, 417–429 (2012)
2. Aliprantis, A., Border, K.: Infinite Dimensional Analysis. Springer, Berlin (2005)
3. Ansari, Q.H., Yao, J.C.: On vector quasi-equilibrium problems. Nonconvex Optim. Appl. **68**. Kluwer Academic Publishers, Norwell, MA (2003)
4. Chen, J., Wan, Z., Cho, Y.: Levitin-Polyak well-posedness by perturbations for systems of set-valued vector quasi-equilibrium problems. Math. Methods Oper. Res. **77**, 33–64 (2013)
5. Darabi, M., Zafarani, J.: Tykhonov well-posedness for quasi-equilibrium problems. J. Optim. Theory Appl. **165**, 458–479 (2015)
6. Darabi, M., Zafarani, J.: M-well-posedness and B-well-posedness for vector quasi-equilibrium problems, to appear in J. Nonlinear Convex Anal. (2016)
7. Dontchev, A., Zolezzi, T.: Well-Posed Optimization Problems. Lecture Notes in Mathematics. Springer, Berlin (1993)
8. Fu, J., Wang, S.: Generalized strong vector quasi-equilibrium problem with domination structure. J. Global Optim. **55**, 839–847 (2013)
9. Hu, R., Fang, Y.: Levitin-Polyak well-posedness of variational inequalities. Nonlinear Anal. **72**, 373–381 (2010)
10. Huang, X., Yang, X.: Generalized Levitin-Polyak well-posedness in constrained optimization. SIAM J. Optim. **17**, 243–258 (2006)
11. Huang, X., Yang, X.: Levitin-Polyak well-posedness of vector variational inequality problems with functional constraints. Numer. Funct. Anal. Optim. **31**, 440–459 (2010)
12. Huang, X., Yang, X.: Further study on the Levitin-Polyak well-posedness of constrained convex vector optimization problems. Nonlinear Anal. **75**, 1341–1347 (2012)
13. Jiang, B., Zhang, J., Huang, X.: Levitin-Polyak well-posedness of generalized quasivariational inequalities with functional constraints. Nonlinear. Anal. **70**, 1492–1530 (2009)
14. Lalitha, C., Bhatia, G.: Levitin-Polyak well-posedness for parametric quasi-variational inequality problem of the minty type. Positivity **16**, 527–541 (2012)
15. Lalitha, C., Chatterjee, P.: Levitin-Polyak well-posedness for constrained quasiconvex vector optimization problems. J. Global Optim. **59**, 191–205 (2014)
16. Levitin, E., Polyak, B.: Convergence of minimizing sequences in conditional extremum problems. Sov. Math. Dokl. **7**, 764–767 (1966)
17. Li, S., Li, M.: Levitin-Polyak well-posedness of vector equilibrium problems. Math. Methods Oper. Res. **69**, 125–140 (2009)
18. Li, X., Xia, F.: Levitin-Polyak well-posedness of a generalized mixed variational inequality in Banach spaces. Nonlinear Anal. **75**, 2139–2153 (2012)
19. Lin, J., Huang, Y.: Generalized vector quasi-equilibrium problems with applications to common fixed point theorems and optimization problems. Nonlinear Anal. **66**, 1275–1289 (2007)
20. Peng, J., Wang, Y., Wu, S.: Levitin-Polyak well-posedness of generalized vector equilibrium problems. Taiwan. J. Math. **15**, 2311–2330 (2011)
21. Peng, J., Wu, S., Wang, Y.: Levitin-Polyak well-posedness of generalized vector quasi-equilibrium problems with functional constraints. J. Global Optim. **52**, 779–795 (2012)
22. Peng, J., Wang, Y., Wu, S.: Levitin-Polyak well-posedness for vector quasi-equilibrium problems with functional constraints. Taiwan. J. Math. **16**, 635–649 (2012)
23. Peng, Y.W., Yang, X.M.: Levitin-Polyak well-posedness of a system of generalized vector variational inequality problems. J. Ind. Manag. Optim. **11**, 701–714 (2015)
24. Sach, P.: New nonlinear scalarization functions and applications. Nonlinear Anal. **75**, 2281–2292 (2012)
25. Tykhonov, A.: On the stability of the functional optimization problem. J. Comput. Math. Phys. **6**, 631–634 (1966)
26. Wang, S.H., Huang, N.J.: Levitin-Polyak well-posedness for generalized quasi-variational inclusion and disclusion problems and optimization problems with constraints. Taiwan. J. Math. **16**, 237–257 (2012)

27. Wang, S., Huang, N., Wong, M.: Strong Levitin-Polyak well-posedness for generalized quasi-variational inclusion problems with applications. Taiwan. J. Math. **16**, 665–690 (2012)
28. Xu, Z., Zhu, D., Huang, X.: Levitin-Polyak well-posedness in generalized vector variational inequality problem with functional constraints. Math. Methods Oper. Res. **67**, 505–524 (2008)
29. Zhang, L., Xia, F.: Scalarization method for Levitin-Polyak well-posedness of vectorial optimization problems. Math. Methods Oper. Res. **76**, 361–375 (2012)
30. Zolezzi, T.: Well-posedness criteria in optimization with application to the calculus of variations. Nonlinear Anal. **25**, 437–453 (1995)
31. Zolezzi, T.: Extended well-posedness of optimization problems. J. Optim. Theory Appl. **91**, 257–266 (1996)

Best Simultaneous Approximation in Quotient Spaces

T.D. Narang and Sahil Gupta

Abstract We discuss the problem of best simultaneous approximation in quotient spaces when the underlying spaces are metric linear spaces. We characterize simultaneous proximinality, simultaneous Chebyshevity, simultaneous pseudo-Chebyshevity and simultaneous quasi-Chebyshevity and see how these are transmitted to and from quotient spaces. The results proved in the paper generalize and extend several known results on the subject.

Keywords Best simultaneous approximation · Simultaneous proximinal set · Simultaneous Chebyshev set · Simultaneous pseudo-Chebyshev set · Simultaneous quasi-Chebyshev set

2010 Mathematics Subject Classification 41A28 · 41A52 · 41A65

1 Introduction and Preliminaries

C.B. Dunham [3] was the first to generalize the classical problem of approximating a continuous function to the problem of simultaneously approximating two continuous functions by a family of functions defined on a closed interval. Ahuja and Narang [1], Goel et al. [4], Holland et al. [5], Sastry and Naidu [12] and others studied this problem in normed linear spaces. Many researchers have discussed results on the existence, uniqueness and characterization of elements of best simultaneous approximation in normed linear spaces (see [1, 9] and references cited therein). Iranmanesh and Mohebi [6], Mohebi and Rezapour [7] have discussed these problems in quotient spaces when the underlying spaces are normed linear spaces. The situation in case of metric linear spaces and metric spaces is somewhat different. Although, some

T.D. Narang · S. Gupta (✉)
Guru Nanak Dev University, Amritsar, India
e-mail: sahilmath@yahoo.in

T.D. Narang
e-mail: tdnarang1948@yahoo.co.in

J.M. Cushing et al. (eds.), *Applied Analysis in Biological and Physical Sciences*, Springer Proceedings in Mathematics & Statistics 186,
DOI 10.1007/978-81-322-3640-5_21

attempts have been made to develop the theory of best simultaneous approximation in such spaces (see [10]) but this theory has not reached a satisfactory level as in the case of normed linear spaces. The present paper is also a step in this direction. In this paper, we discuss the problem of best simultaneous approximation in quotient spaces when the underlying spaces are metric linear spaces. We characterize simultaneous proximinality, simultaneous Chebyshevity, simultaneous pseudo-Chebyshevity and simultaneous quasi-Chebyshevity in metric linear spaces and see how these are transmitted to and from quotient spaces. The results proved in the paper generalize and extend several known results on best approximation and also results of [6, 7]. We start with a few definitions.

Let W be a non-empty subset of a metric space (X, d). A point $w_0 \in W$ is called a **best approximation** for $x \in X$, if

$$d(x, w_0) = d(x, W) \equiv \inf_{w \in W} d(x, w).$$

We denote the set of all best approximations to x in W by $P_W(x)$ i.e. $P_W(x) = \{w \in W : d(x, w) = d(x, W)\}$.

We say that the set W is **proximinal (Chebyshev)** if $P_W(x) \neq \phi$ (exactly a singleton) for each $x \in X$.

For a non-empty bounded subset S of X, we define

$$\delta(S, W) = \inf_{w \in W} \sup_{s \in S} d(s, w).$$

An element $w_0 \in W$ is called a **best simultaneous approximation** to S from W if $\sup_{s \in S} d(s, w_0) = \delta(S, W)$.

We denote the set of all best simultaneous approximations to S from W by $L_W(S)$ i.e. $L_W(S) = \{w \in W : \sup_{s \in S} d(s, w) = \delta(S, W)\}$.

The set W is called **simultaneous proximinal (simultaneous Chebyshev)** if for each bounded subset S of X there exist at least one (exactly one) best simultaneous approximation to S from W.

Clearly, every simultaneously proximinal set is proximinal.

A closed subset W of X is called **simultaneous quasi-Chebyshev** (see [7]) if the set $L_W(S)$ is non-empty and compact in X for every bounded subset S of X.

A metric space (X, d) is said to be **totally complete** [8] if its closed and bounded subsets are compact.

A linear space X with a translation invariant metric d (i.e. $d(x+z, y+z) = d(x, y)$ *for all* $x, y, z \in X$) such that addition and scalar multiplication are continuous on (X, d) is called a **metric linear space**.

Every normed linear space is a metric linear space but converse is not true.

The space X of all entire functions i.e. $X = \{f : f(z) = \sum_{n=0}^{\infty} a_n z^n,\ |a_n|^{\frac{1}{n}} \to 0$ *as* $n \to \infty\}$ with the metric d defined by $d(f, g) = max\{|a_0 - b_0|, |a_n - b_n|^{\frac{1}{n}}, n \geq 1\}$, *where* $f(z) = \sum_{n=0}^{\infty} a_n z^n$ *and* $g(z) = \sum_{n=0}^{\infty} b_n z^n$ is a non-normable metric linear space (see [2], p. 238).

For a non-empty convex set A of a metric linear space (X, d), we denote by $l(A)$ the linear manifold spanned by A i.e.

$$l(A) = \{\lambda x + (1-\lambda)y : x, y \in A, \lambda \ a \ scalar\}$$

For a fixed $z \in A$, the set $l(A) - z = \{x - z : x \in l(A)\}$ is a linear subspace of X, satisfying

$$l(A) - z = l(A - z)$$

The dimension of an arbitrary convex set $A \subseteq X$ is defined by

$$dim\ A = \begin{cases} dim\ l(A) & \text{if } A \neq \phi \\ -1 & \text{if } A = \phi \end{cases}$$

For every $z \in A$ we have

$$dim\ A = dim\ l(A) = dim\ [l(A) - z] = dim\ l(A - z) = dim\ (A - z).$$

A closed subset W of a metric linear spaces (X, d) is called **simultaneous pseudo-Chebyshev** (see [7]) if the set $L_W(S)$ is non-empty and finite dimensional in X for every bounded subset S of X.

It is easy to see that for any non-empty subset W of a metric linear space (X, d), $L_{W+x}(S + x) = L_W(S) + x$ for every $x \in X$.

If (X, d) is a metric linear space over a field F and M is a closed linear subspace of X. Then the quotient space

$$X/M = \{x + M : x \in X\}$$

with linear operations:
(i) $(x + M) + (y + M) = (x + y) + M$ for every $x, y \in X$.
(ii) $\lambda(x + M) = \lambda x + M$ for every $x \in X, \lambda \in F$.
is a metric linear space endowed with the metric

$$d(x + M, y + M) = \inf_{m \in M} d(x + m, y + m).$$

The canonical mapping $\pi : X \to X/M$ defined by $\pi(x) = x + M, x \in X$ is linear, continuous, open and satisfies $l(\pi(A)) = \pi(l(A))$ (see [11]).

2 Main Results

In this section, we prove how simultaneous proximinality and simultaneous Chebyshevity is transmitted to and from quotient spaces when the underlying spaces are metric linear spaces. We start with proving a few lemmas. For normed linear spaces, these lemmas were proved in [6].

Lemma 1 *Let* (X, d) *be a metric space and* M *a proximinal subset of* X*, then for each non-empty bounded set* S *in* X*, we have*

$$\delta(S, M) = \sup_{s\in S} \inf_{m\in M} d(s, m).$$

Proof Since $\inf_{m\in M} d(s, m) \leq d(s, m')$ for every m', we have

$$\sup_{s\in S} \inf_{m\in M} d(s, m) \leq \sup_{s\in S} d(s, m') \textit{ for every } m'$$

and so

$$\sup_{s\in S} \inf_{m\in M} d(s, m) \leq \inf_{m'\in M} \sup_{s\in S} d(s, m').$$

By the proximinality of M it follows that for each $s \in S$, there exist $m_s \in P_M(s)$ such that

$$d(s, m_s) = \inf_{m\in M} d(s, m)$$

Therefore, $\delta(S, M) = \inf_{m\in M} \sup_{s\in S} d(s, m) \leq \sup_{s\in S} d(s, m_s) = \sup_{s\in S} \inf_{m\in M} d(s, m) \leq \inf_{m\in M} \sup_{s\in S} d(s, m) = \delta(S, M)$. This gives $\delta(S, M) = \sup_{s\in S} \inf_{m\in M} d(s, m)$.

Lemma 2 *Let* (X, d) *be a metric linear space,* M *a proximinal subspace of* X*, and* S *a subset of* X*, then the following are equivalent:*
(i) S *is a bounded subset of* X*.*
(ii) S/M *is a bounded subset of* X/M*.*

Proof (i) $\Rightarrow$ (ii) Let $s \in S$ be arbitrary. Since $d(s + M, 0) \leq d(s, 0)$ *as* $0 \in M$, we get $\sup_{s\in S} d(s + M, 0) \leq \sup_{s\in S} d(s, 0) < \infty$ as S is bounded. Therefore, S/M is a bounded subset of X/M.
(ii) $\Rightarrow$ (i) Suppose S/M is bounded subset of X/M i.e. $\sup_{s\in S} d(s + M, 0) < \infty$.

Since M is proximinal, for each $s \in S$ there exist $m_s \in M$ such that $m_s \in P_M(s)$ and so $d(s, m_s) = \inf_{m\in M} d(s, m)$.

Using Lemma 1, we get

$$\sup_{s\in S} d(s, m_s) = \sup_{s\in S} \inf_{m\in M} d(s, m) = \inf_{m\in M} \sup_{s\in S} d(s, m).$$

Therefore for each $\varepsilon > 0$, there exist $m_\varepsilon \in M$ such that

$$\sup_{s\in S} d(s, m_\varepsilon) \leq \sup_{s\in S} d(s, m_s) + \varepsilon$$

Now, $d(s, 0) \leq d(s, m_\varepsilon) + d(m_\varepsilon, 0) \Rightarrow \sup_{s\in S} d(s, 0) \leq \sup_{s\in S} d(s, m_\varepsilon) + d(m_\varepsilon, 0) \leq \sup_{s\in S} d(s, m_s) + \varepsilon + d(m_\varepsilon, 0) = \sup_{s\in S} \inf_{m\in M} d(s, m) + d(m_\varepsilon, 0) + \varepsilon =$

$\sup_{s\in S} d(s+M,0)+d(m_\varepsilon,0)+\varepsilon<\infty$ as S/M is bounded. Therefore, $\sup_{s\in S} d(s,0)<\infty$ and hence S is bounded.

Concerning the simultaneous proximinality in quotient spaces, we have

Theorem 1 *Let* W *and* M *be subspaces of a metric linear space* (X,d). *If* $M\subseteq W$ *is simultaneous proximinal in* X*, then the following are equivalent:*
(i) W *is simultaneous proximinal in* X.
(ii) W/M *is simultaneous proximinal in* X/M.

Proof (i) ⇒ (ii) Let S/M be any bounded set in X/M, it follows from Lemma 2 that S is a bounded set in X.

Assume $w_0\in L_W(S)$ and $w_0+M\notin L_{W/M}(S/M)$. Now

$$w_0\in L_W(S)\Rightarrow \sup_{s\in S} d(s,w_0)=\delta(S,W)\equiv \inf_{w\in W}\sup_{s\in S} d(s,w).$$

If $w_0+M\notin L_{W/M}(S/M)$ then $\sup_{s+M\in S/M} d(s+M,w_0+M)\neq \inf_{w+M\in W/M}\sup_{s+M\in S/M} d(s+M,w+M)$ and so there exist $w'\in W$ such that

$$\sup_{s\in S} d(s-w'+M,0)<\sup_{s\in S} d(s-w_0+M,0)\leq \sup_{s\in S} d(s,w_0)=\delta(S,W) \quad (1)$$

On the other hand, for each $s\in S$, we have

$$d(s-w'+M,0)=\inf_{m\in M} d(s-w'+m,0)$$

and so for each $\varepsilon>0$ and each $s\in S$ there exist $m_s\in M$ such that

$$d(s-w'-m_s,0)\leq d(s-w'+M,0)+\varepsilon \quad (2)$$

As $w'+m_s\in W$, we conclude that $\delta(S,W)\leq \sup_{s\in S} d(s-(w'+m_s),0)=\sup_{s\in S} d(s,w'+m_s)\leq \sup_{s\in S} d(s-w'+M,0)+\varepsilon$ *by* (2).

Since $\varepsilon>0$ is arbitrary,

$$\delta(S,W)\leq \sup_{s\in S} d(s-w'+M,0) \quad (3)$$

From (1) and (3), we obtain

$$\delta(S,W)\leq \sup_{s\in S} d(s-w'+M,0)<\delta(S,W)$$

which is not possible and hence $w_0+M\in L_{W/M}(S/M)$.

(ii) ⇒ (i) As $w_0+M\in L_{W/M}(S/M)$, we have

$$\sup_{s+M\in S/M} d(s+M, w_0+M) = \inf_{w+M\in W/M} \sup_{s+M\in S/M} d(s+M, w+M) \tag{4}$$

Since M is proximinal, there exist $m_0 \in M$ such that

$$\sup_{s-w_0\in S-w_0} d(s-w_0, m_0) = \inf_{m\in M} \sup_{s-w_0\in S-w_0} d(s-w_0, m) \tag{5}$$

Using (5) and Lemma 1, we obtain

$$\sup_{s\in S} d(s-w_0-m_0, 0) = \inf_{m\in M} \sup_{s\in S} d(s-w_0-m, 0) = \sup_{s\in S} \inf_{m\in M} d(s-w_0-m, 0) =$$

$$\sup_{s\in S} d(s-w_0+M, 0) \leq \sup_{s\in S} d(s-w+M, 0) \textit{ for every } w \in W \textit{ by } (4).$$

This implies that $\sup_{s\in S} d(s-w_0-m_0, 0) \leq \sup_{s\in S} d(s, w)$ for every $w \in W$ *as* $0 \in M$. Hence $\sup_{s\in S} d(s, w_0+m_0) \leq \sup_{s\in S} d(s, w)$ for every $w \in W$. Since $w_0+m_0 \in W$, we obtain $w_0+m_0 \in L_W(S)$.

Corollary 1 *Let* M *be a simultaneous proximinal subspace of a metric linear space* (X, d) *and* $W \supseteq M$ *a simultaneous proximinal subspace of* X. *Then for each bounded set* S *in* X, *we have* $\pi(L_W(S)) = L_{W/M}(S/M)$.

Using the continuity of π, we obtain

Corollary 2 *Let* M *be a simultaneous proximinal subspace of a metric linear space* (X, d) *and* $W \supseteq M$ *a simultaneous quasi-Chebyshev subspace of* X. *Then* W/M *is simultaneous quasi-Chebyshev in* X/M.

Concerning the simultaneous Chebyshevity in quotient spaces, we have

Theorem 2 *Let* W *and* M *be subspaces of a metric linear space* (X, d). *If* $M \subseteq W$ *is simultaneous Chebyshev, then the following are equivalent:*
(i) W/M *is simultaneous Chebyshev in* X/M.
(ii) W *is simultaneous Chebyshev in* X.

Proof (i) $\Rightarrow$ (ii) Assume (ii) is false. Then some bounded subset S of X has two distinct best simultaneous approximations say l_0 and l_1 in W i.e. l_0 *and* $l_1 \in L_W(S)$.

Since $W \supseteq M$, it follows from Theorem 1 that l_0+M and $l_1+M \in L_{W/M}(S/M)$. By hypothesis, W/M is simultaneous Chebyshev and therefore $l_0+M = l_1+M$. So, there exist $m_0 \in M\backslash\{0\}$ such that $l_1 = l_0+m_0$. Since $l_0, l_1 \in L_W(S)$, we have

$$\sup_{s\in S} d(s-l_0-m_0, 0) = \sup_{s\in S} d(s-l_1, 0) = \sup_{s\in S} d(s, l_1) = \inf_{l\in W} \sup_{s\in S} d(s-l, 0) =$$

$\delta(S, W) = \delta(S-l_0, W) \leq \delta(S-l_0, M)$. This shows that both m_0 and 0 are best simultaneous approximation to $S-l_0$ from M i.e M is not simultaneous Chebyshev, a contradiction. Hence W is simultaneous Chebyshev in X.

(ii)$\Rightarrow$ (i) Assume (i) does not hold. Then for some bounded subset S of X, S/M has two distinct best simultaneous approximations say $w+M$ and $w'+M$ belonging to W/M. Thus $w-w' \notin M$.

Since M is simultaneously proximinal, there exist best simultaneous approximations m and m' to $S-w, S-w'$ respectively. Therefore, we have $m \in L_M(S-w)$, $m' \in L_M(S-w')$.

Since $W \supseteq M$ and $w+M, w'+M \in L_{W/M}(S/M)$ it follows from Theorem 1 that $w+m$ and $w'+m' \in L_W(S)$. But W is simultaneous Chebyshev in X, therefore $w+m=w'+m'$. This gives $w-w' \in M$ which is not true and hence (i) holds true.

Remark 1 For normed linear spaces, Theorems 1 and 2 were proved in [6].

We shall be needing the following result for characterizing simultaneous pseudo-Chebyshevity and simultaneous quasi-Chebyshevity in quotient spaces.

Lemma 3 *If* K *is a boundedly compact, closed subset of a metric space* (X,d), *then there exist a best simultaneous approximation to any bounded subset* F *of* X *i.e.* K *is simultaneous proximinal in* X.

Proof Let $g(k)=\sup_{f\in F} d(f,k)$ and $\delta=\delta(F,K)\equiv \inf_{k\in K}\sup_{f\in F} d(f,k)=\inf_{k\in K} g(k)$. By the definition of δ, there exist a sequence $\{k_n\}$ in K such that

$$g(k_n)\to \delta(F,K)\equiv\delta \tag{6}$$

Since $\{g(k_n)\}$ is a convergent sequence of scalars, it is bounded and therefore there exist a positive real number p such that

$$g(k_n)\le p \textit{ for all } n \textit{ i.e. } \sup_{f\in F} d(f,k_n)\le p \textit{ for all } n.$$

Now, $d(k_n,k_m)\le d(k_n,f)+d(f,k_m)$ for all f implies

$$\sup_{f\in F} d(k_n,k_m)\le \sup_{f\in F} d(k_n,f)+\sup_{f\in F} d(f,k_m)\Rightarrow \sup_{k_n,k_m\in K} d(k_n,k_m)\le 2p$$

This implies that $\{k_n\}$ is a bounded sequence in K. Since K is boundedly compact, there exist a subsequence $\{k_{n_i}\}$ of $\{k_n\}$ such that $k_{n_i}\to k_0\in K$. From (6), we have

$$\lim_{n\to\infty} g(k_n)=\delta \textit{ i.e. } \lim_{n\to\infty}\sup_{f\in F} d(f,k_n)=\delta.$$

Therefore, for a given $\varepsilon>0$, there exist a positive integer m such that

$$\left|\sup_{f\in F} d(f,k_n)-\delta\right|<\varepsilon \textit{ for all } n\ge m$$

This gives $\sup_{f\in F} d(f,k_n)<\delta+\varepsilon$ *for all* $n\ge m$. Consequently, $\sup_{f\in F} d(f,k_{n_i}) < \delta+\varepsilon$ *for all* $n_i\ge m$.

Since $\varepsilon > 0$ is arbitrary, $\sup_{f \in F} d(f, k_{n_i}) \leq \delta$ for all $n_i \geq m \Rightarrow d(f, k_{n_i}) \leq \delta$ for all $n_i \geq m$ and for every f.

Letting $n_i \to \infty$, we get

$$d(f, k_0) \leq \delta \textit{ for every } f \in F \Rightarrow \sup_{f \in F} d(f, k_0) \leq \delta$$

Also,

$$\delta = \inf_{k \in K} \sup_{f \in F} d(f, k) \leq \sup_{f \in F} d(f, k_0)$$

Hence $\sup_{f \in F} d(f, k_0) = \delta = \inf_{k \in K} \sup_{f \in F} d(f, k)$ i.e. K is simultaneous proximinal in X.

Concerning simultaneous pseudo Chebyshevity, we have

Theorem 3 *Let (X, d) be a totally complete metric linear space,* M *a finite dimensional subspace of X and $W \supseteq M$ a subspace of X. Then the following are equivalent:*
(i) W/M is simultaneous pseudo Chebyshev in X/M.
(ii) W is simultaneous pseudo Chebyshev in X.

Proof Since a finite dimensional subspace of a metric linear space is closed [11], M is a closed subspace of the totally complete metric linear space X and so it is boundedly compact [8]. Lemma 3 implies that M is simultaneously proximinal in X.
(i) $\Rightarrow$ (ii) Suppose W/M is simultaneous pseudo Chebyshev in X/M then W/M is simultaneous proximinal in X/M. Since M is simultaneous proximinal in X. It follows from Theorem 1 that W is simultaneous proximinal in X. Let S be a bounded set in X and $k_0 \in L_W(S)$ be arbitrary. Since $W \supseteq M$, it follows from Corollary 1 that

$$\pi[l(L_W(S) - k_0)] = l[\pi(L_W(S) - k_0)] = l[L_{W/M}(S/M) - (k_0 + M)] =$$
$$l[L_{W/M}(S/M) - (k_0 + M)] \tag{7}$$

Since W/M is simultaneous pseudo-Chebyshev in X/M,

$$dim\ l[L_{W/M}(S/M) - (k_0 + M)] = dim\ l[L_{W/M}(S/M)] = dim\ L_{W/M}(S/M) < \infty \tag{8}$$

From (7) and (8), we have

$$dim\ \pi[l(L_W(S) - k_0)] < \infty.$$

Since $\pi[l(L_W(S) - k_0)] = l[L_W(S) - k_0)]/M$, $dim\ \ \pi[l(L_W(S) - k_0)] = dim\ \ l$ $[L_W(S) - k_0)] - dim\ M$. As $dim\ M$ is finite, it follows that $dim\ l[L_W(S) - k_0)]$ is finite. Therefore, $dim\ l[L_W(S) - k_0)] = dim\ [L_W(S)]$ is finite. Hence, W is simultaneous pseudo-Chebyshev in X.
(ii) $\Rightarrow$ (i) Let S be a bounded set in X. Since W is simultaneous pseudo-Chebyshev

in X, $L_W(S)$ is a non-empty and finite dimensional set in X. In view of Theorem 1 and the fact that M is simultaneous proximinal in X, we get that W/M is simultaneous proximinal in X/M.

Thus, we have $dim\ [L_{W/M}(S/M)] = dim\ l[L_{W/M}(S/M)] = dim\ l[\pi(L_W(S)]$. But $dim\ l[\pi(L_W(S)] = dim\ \pi[l(L_W(S)] = dim\ l[L_W(S)]/M = dim\ [l(L_W(S)] - dim\ M$ implies $dim\ \pi[l(L_W(S)] < dim\ [l(L_W(S)] < \infty$ and so $dim\ L_{W/M}(S/M) < \infty$. Consequently, W/M is simultaneous pseudo-Chebyshev in X/M.

Remark 2 For normed linear spaces, Theorem 3 is given in [6].

The above proof shows that the condition of total completeness of the space in the above theorem can be removed if we take M to be a simultaneously proximinal subspace in X, so we have

Theorem 4 *Let (X, d) be a metric linear space,* M *a finite dimensional simultaneously proximinal subspace of X and $W \supseteq M$ a subspace of X. Then the following are equivalent:*
(i) W/M is simultaneous pseudo Chebyshev in X/M.
(ii) W is simultaneous pseudo Chebyshev in X.

Concerning simultaneous quasi Chebyshevity, we have

Theorem 5 *Let* M *be a closed subspace and $W \supseteq M$ a subspace of a totally complete metric linear space (X, d), then the following are equivalent:*
(i) W/M is simultaneous quasi-Chebyshev in X/M.
(ii) W is simultaneous quasi-Chebyshev in X.

Proof As M is a closed subset of totally complete metric linear space X, it is boundedly compact [8]. Lemma 3 implies that M simultaneously proximinal in X.
(i) $\Rightarrow$ (ii) Suppose W/M is simultaneous proximinal in X/M. Since M is simultaneous proximinal in X, it follows from Theorem 1 that W is simultaneous proximinal in X. Let S be a bounded set in X, then $L_W(S) \neq \phi$.

We now show that $L_W(S)$ is compact. Let $\{l_n\}$ be a sequence in $L_W(S)$. Then it follows from Theorem 1 that $\{l_n + M\}$ is a sequence in $L_{W/M}(S/M)$.

Since $L_{W/M}(S/M)$ is compact, there exist a subsequence $\{l_{n_k} + M\}$ of $\{l_n + M\}$ such that $\{l_{n_k} + M\} \to l_0 + M \in L_{W/M}(S/M)$. As M is proximinal in X, there exist $m_{n_k} \in M$ such that $m_{n_k} \in P_M(l_0 - l_{n_k})$ for every $k \geq 1$.

Hence $d(l_0 - l_{n_k} - m_{n_k}) = d(l_0 - l_{n_k}, M)$ for every $k \geq 1$. Taking limit as $k \to \infty$ and using the fact $l_{n_k} + M \to l_0 + M$, we get

$$\lim_{n\to\infty} d(l_0 - l_{n_k} - m_{n_k}, 0) = 0. \tag{9}$$

Moreover, M is simultaneous proximinal and hence closed. Without loss of generality we may assume that $\{m_{n_k}\}$ converges to an element say $m_0 \in M$. (otherwise consider a suitable subsequence of $\{m_{n_k}\}$ and use the fact that every bounded sequence has a convergent subsequence)

Let $l' = l_0 - m_0$. Then $l' \in W + M$, and we have

$$d(l' - l_{n_k}, 0) = d(l_0 - m_0 - l_{n_k}, 0) = d(l_0 - l_{n_k}, m_0) \leq d(l_0 - l_{n_k}, m_{n_k}) + d(m_{n_k}, m_0)$$

for every $k \geq 1$.
Taking limit as $k \to \infty$, we get

$$d(l' - l_{n_k}, 0) \to 0 \; as \; k \to \infty$$

Since $l_{n_k} \in L_{W+M}(S)$ for all $k \geq 1$ and $L_W(S)$ is closed, we conclude that $l' \in L_W(S)$. Therefore, $L_W(S)$ is compact.
(ii) $\Rightarrow$ (i) follows from Corollary 2.

The above proof shows that the condition of total completeness of the space in the above theorem can be removed if we take M to be a simultaneously proximinal subspace in X, so we have

Theorem 6 *Let* (X, d) *be a metric linear space,* M *a finite dimensional simultaneously proximinal subspace of* X *and* $W \supseteq M$ *a subspace of* X. *Then the following are equivalent:*
(i) W/M *is simultaneous quasi-Chebyshev in* X/M.
(ii) W *is simultaneous quasi-Chebyshev in* X.

Remark 3 For finite dimensional subspaces of normed linear spaces, Theorem 6 is given in [6].

Acknowledgements The research work of the first author has been supported by U.G.C. under Emeritus Fellowship and that of the second author under U.G.C., Senior Research Fellowship.

References

1. Ahuja, G.C., Narang, T.D.: On best simultaneous approximation. Nieuw Archief Voor Wiskunde **27**, 255–261 (1979)
2. Chandrasekhara Rao, K.: Functional Analysis. Narosa Publishing House, New Delhi (2002)
3. Dunham, C.B.: Simultaneous Chebyshev approximation of functions on an interval. Proc. Am. Math. Soc. **18**, 472–477 (1967)
4. Goel, D.S., Holland, A.S.B., Nasim, C., Sahney, B.N.: On best simultaneous approximation in normed linear spaces. Can. Math. Bull. **17**, 523–527 (1974)
5. Holland, A.S.B., Sahney, B.N., Tzimbalario, J.: On best simultaneous approximation. J. Indian Math. Soc. **40**, 69–73 (1976)
6. Iranmanesh, M., Mohebi, H.: On best simultaneous approximation in quotient spaces. Anal. Theory Appl. **23**, 35–49 (2007)

7. Mohebi, H., Rezapour, Sh.: On sum and quotient of quasi-Chebyshev subspaces in Banach spaces. Anal. Theory Appl. **19**, 266–270 (2003)
8. Narang, T.D.: On totally complete spaces. Math. Educ. **16**, 4–5 (1982)
9. Narang, T.D.: On best simultaneous approximation. J. Approx. Theory **39**, 87–90 (1983)
10. Narang, T.D.: Simultaneous approximation and Chebyshev centres in metric spaces. Mat. Vesnik **51**, 61–68 (1999)
11. Rudin, W.: Functional Analysis. McGraw-Hill, Inc., New York (1973)
12. Sastry, K.P.R., Naidu, S.V.R.: On best simultaneous approximation in normed linear spaces. Proc. Natl. Acad. Sci. India **43**, 249–250 (1973)

$H(.,.,.)$-η-Proximal-Point Mapping with an Application

Shamshad Husain, Huma Sahper and Sanjeev Gupta

Abstract In this article, we introduce the new notion of accretive mapping known as $H(.,.,.)$-η-mixed accretive mapping in q-uniformly smooth Banach spaces. It is the generalization of the $H(.,.)$-accretive mapping, introduced and studied by Zou and Huang [25]. Then, we will introduce the proximal-point mapping related with $H(.,.,.)$-η-mixed accretive mapping and discuss its Lipschitz continuity. We design an iterative algorithm for solving the system of variational inclusions by utilizing the proximal-point method and prove the convergence of iterative sequences generated by the algorithm. Few examples are considered to illustrate the introduced proximal-point mapping.

Keywords $H(.,.,.)$-η-mixed accretive mapping · Proximal-point mapping · Iterative algorithm · Convergence · Variational inclusions

2010 Mathematics Subject Classification Primary 47H19 · Secondary 49J40

1 Introduction

The study of the variational inequalities and variational inclusions is very crucial that allow mathematical models to some problems arising in mechanics, engineering science and economics etc. Firstly, Huang and Fang [8] studied the generalization of

S. Husain (✉) · H. Sahper
Department of Applied Mathematics, Aligarh Muslim University, Aligarh 202002, India
e-mail: s_husain68@yahoo.com

H. Sahper
e-mail: humashpr19@gmail.com

S. Gupta
Department of Humanities and Social Sciences, Indian Institute of Technology, Kanpur 208016, India
e-mail: guptasanmp@gmail.com

J.M. Cushing et al. (eds.), *Applied Analysis in Biological and Physical Sciences*, Springer Proceedings in Mathematics & Statistics 186,
DOI 10.1007/978-81-322-3640-5_22

m-accretive mappings and defined its proximal-point mapping in Banach space. Since then, a number of mathematicians introduced and studied the several generalized mappings such as H-accretive, A-η-accretive and G-η-monotone mappings etc. For example, see [2–21, 24] and references therein.

In the last few years, Zou and Huang [25] introduced and studied $H(.,.)$-accretive mapping, Wang et al. [22] introduced and studied $H(.,.)$-η-accretive mappings in the setting of Banach spaces and their proximal-point mappings, respectively. Very recently, Husain et. al. [11] introduced and studied $H(.,.)$-mixed mapping in the setting of Banach spaces and Ahmad et al. [1] introduced and studied $H(.,.)$-cocoercive mappings in the setting of Hilbert spaces.

This work is inspired by the research going in this direction. In this article we investigate a new notion of generalized accretive mapping known as $H(.,..,.)$-η-mixed accretive mapping in q-uniformly smooth Banach spaces and discuss the properties of its proximal-point mapping like single-valuedness and Lipschitz continuity. As an application, we consider the system of set-valued variational inclusion involving $H(.,..,.)$-η-mixed accretive mappings. Then, we design an iterative algorithm for solving the considered system of set-valued variational inclusions and show that the iterative sequences generated through the algorithm are strongly convergent to the solution of the system.

2 Preliminaries

Let E be a real Banach space equipped with norm $\|.\|$ and E^* be the topological dual space of E. Let $\langle .,. \rangle$ be the dual pair between E and E^*.

Definition 1 ([23]) A mapping $J_q : E \multimap E^*$, where $q > 1$, is said to be *generalized duality mapping*, if it is defined as

$$J_q(u) = \{f^* \in E^* : \langle u, f^* \rangle = \|u\|^q, \|f^*\| = \|u\|^{q-1}\}, \forall u \in E.$$

If J_2 is the *usual normalized duality mapping* on E, given as

$$J_q(u) = \|u\|^{q-1} J_2(u) \forall u (\neq 0) \in E.$$

If $E \equiv X$, a real Hilbert space, then J_2 becomes *identity mapping* on X.

Definition 2 ([23]) A Banach space E is called *smooth* if for every $u \in E$ with $\|u\| = 1$, there exists a unique $f \in E^*$ such that $\|f\| = f(u) = 1$.

The *modulus of smoothness* of E is a function $\rho_E : [0, \infty) \to [0, \infty)$, defined by

$$\rho_E(t) = \sup \left\{ \frac{1}{2}(\|u + v\| + \|u - v\|) - 1 : \|u\| \leq 1, \|v\| \leq t \right\}.$$

Definition 3 ([23]) A Banach space E is called
(i) *uniformly smooth* if

$$\lim_{t\to 0}\frac{\rho_E(t)}{t}=0;$$

(ii) *q-uniformly smooth*, for $q>1$, if there exists $c>0$ such that

$$\rho_E(t)\le ct^q,\ t\in[0,\infty).$$

Note that J_q is single-valued if E is uniformly smooth.

Lemma 1 ([23]) *Let E be a real uniformly smooth Banach space. Then E is q-uniformly smooth if and only if there exists $c_q>0$ such that, for all $u,v\in E$,*

$$\|u+v\|^q\le\|u\|^q+q\langle v,J_q(u)\rangle+c_q\|v\|^q.$$

Definition 4 Let us consider the mappings $P:E\to E$ and $\eta:E\times E\to E$. Then P is said to be
(i) η-accretive if

$$\langle P(u)-P(v),J_q(\eta(u,v))\rangle\ge 0,\ \forall u,v\in E;$$

(ii) strictly η-accretive, if P is η-accretive and equality holds if and only if $u=v$;
(iii) δ_1-η-strongly accretive with constant $\delta_1>0$ if

$$\langle P(u)-P(v),J_q(\eta(u,v))\rangle\ge\delta_1\|u-v\|^q,\ \forall u,v\in E;$$

(iv) Lipschitz continuous with $\lambda_P>0$, if

$$\|P(u)-P(v)\|\le\lambda_P\|u-v\|,\ \forall u,v\in E;$$

(v) μ-expansive with $\mu>0$, if that

$$\|P(u)-P(v)\|\ge\mu\|u-v\|,\ \forall u,v\in E;$$

if $\mu=1$, then it is expansive;
(vi) η is said to be Lipschitz with $\tau>0$, if

$$\|\eta(u,v)\|\le\tau\|u-v\|,\ \forall u,v\in E.$$

Definition 5 Let us consider the single-valued mappings $A,B,C:E\to E$, $\eta:E\times E\to E$ and $H:E\times E\times E\to E$. Then
(i) $H(A,.,.)$ is said to be α-η-cocoercive regarding A with $\alpha>0$, if

$$\langle H(Au, w, w) - H(Bv, w, w), J_q(\eta(u, v))\rangle \geq \alpha \|Au - Av\|^q, \forall u, v, w \in E;$$

(ii) $H(., B, .)$ is said to be β-η-relaxed cocoercive regarding B with $\beta > 0$, if

$$\langle H(w, Bu, w) - H(w, Bv, w), J_q(\eta(u, v))\rangle \geq (-\beta)\|Bu - Bv\|^q, \forall u, v, w \in E;$$

(iii) $H(., ., C)$ is said to be δ-η-strongly accretive regarding C with $\delta > 0$, if

$$\langle H(w, w, Cu) - H(w, w, Cv), J_q(\eta(u, v))\rangle \geq \delta \|u - v\|^q, \forall u, v, w \in E;$$

(iv) $H(A, ., .)$ is said to be Lipschitz continuous regarding A with $r_1 > 0$, if

$$\|H(Au, w, w) - H(Av, w, w)\| \leq r_1 \|u - v\|, \forall u, v, w \in E;$$

Similarly, we can give the Lipshitz continuity of H regarding B and C by taking constants $r_2 > 0$ and $r_3 > 0$ respectively.

Definition 6 Let $M : E \multimap E$ be a set-valued mapping, then M is said to be κ-η-relaxed accretive with $\kappa > 0$, if

$$\langle x - y, J_q(\eta(u, v))\rangle \geq (-\kappa)\|u - v\|^q, \forall u, v \in E, x \in M(u), y \in M(v).$$

Definition 7 A set-valued mapping $T : E \multimap \mathrm{CB}(E)$ is said to be D-Lipschitz continuous with $l > 0$ such that

$$D(T(u), T(v)) \leq l\|u - v\|, \forall u, v \in E,$$

where $D(., .)$ is the Hausdorff metric on $CB(E)$.

3 $H(., ., .)$-η-Mixed Accretive Mapping

First we consider the following assumptions, then we will introduce $H(., ., .)$-η-mixed accretive mapping and its proximal-point mapping.
Assumption c_1: Let $H : E \times E \times E \to E$ and $\eta : E \times E \to E$ be the single-valued mappings and assume that

1. $H(., ., .)$ is α-η-cocoercive regarding A,
2. $H(., ., .)$ is β-η-relaxed cocoercive regarding B,
3. $H(., ., .)$ is δ-η-strongly accretive regarding C.

Assumption c_2:

1. A is μ-expansive,
2. B is γ-Lipschitz continuous.

Definition 8 Let assumption c_1 holds, then the set-valued mapping $M : E \multimap E$ is said to be $H(.,.,.)$-η-mixed accretive regarding A, B and C, if
(i) M is κ-η-relaxed accretive;
(ii) $(H(A, B, C) + \rho M)(E) = E$, for all $\rho > 0$.

Example 1 Let $q = 2$ and $E = \mathbb{R}^2$ with the usual inner product. Let $A, B, C : \mathbb{R}^2 \to \mathbb{R}^2$ be defined by

$$
\begin{aligned}
Au &= \left(\frac{1}{2}u_1 - \frac{1}{2}u_2, -\frac{1}{2}u_1 + u_2\right),\\
Bv &= \left(-\frac{1}{2}v_1 - \frac{1}{2}v_2, \frac{1}{2}v_1 - \frac{1}{2}v_2\right),\\
Cw &= \left(\frac{1}{2}w_1 - \frac{1}{4}w_2, \frac{1}{4}w_1 + \frac{1}{3}w_2\right),
\end{aligned}
$$

for all $u = (u_1, u_2)$, $v = (v_1, v_2)$ and $w = (w_1, w_2) \in \mathbb{R}^2$.
Suppose that $H : \mathbb{R}^2 \times \mathbb{R}^2 \times \mathbb{R}^2 \to \mathbb{R}^2$ and $\eta : \mathbb{R}^2 \times \mathbb{R}^2 \to \mathbb{R}^2$ be defined by

$$
\begin{aligned}
H(Au, Bv, Cw) &= Au + Bv + Cw,\\
\eta(u, v) &= u - v, \qquad \forall u, v, w \in \mathbb{R}^2.
\end{aligned}
$$

Then, we can check easily the constants in Definition 8 having values $\alpha = \frac{2}{3}$, $\beta = 1$ and $\delta = \frac{1}{3}$.

Example 2 Let $E, A, B, C, H(A, B, C)$ and η are defined as in Example 1. Suppose that $M : E \multimap E$ be defined by $M(u) = (-3\pi, -3u_2), \forall u = (u_1, u_2) \in \mathbb{R}^2$.

Then, we can easily obtained the constant $\kappa = 3$ in Definition 8 and M is $H(.,.,.)$-η-mixed accretive mapping due to $(H(A, B, C) + \rho M)(E) = E, \forall \rho > 0$.

Theorem 1 *Let $M : E \multimap E$ be a $H(.,.,.)$-η-mixed accretive mapping regarding A, B and C. Let assumptions (c_1) and (c_2) hold with $\alpha > \beta$, $\mu > \gamma$ and $\ell = \alpha\mu^q - \beta\gamma^q + \delta > \kappa$. If the inequality given below*

$$\langle u - v, J_q(\eta(x, y))\rangle \geq 0, \tag{1}$$

satisfied for each $(y, v) \in \mathrm{Gph}(M)$, then $u \in Mx$, *where*

$$\mathrm{Gph}(M) = \{(p, q) \in E \times E : q \in M(p)\}. \tag{2}$$

Proof Assume on contrary that $\exists\ (x_0, u_0) \notin \mathrm{Gph}(M)$

$$\langle u_0 - v, J_q(\eta(x_0, y))\rangle \geq 0, \forall (y, v) \in \mathrm{Gph}(M). \tag{3}$$

By definition of $H(.,.,.)$-η-mixed accretive mapping, we have $(H(A, B, C) + \rho M)$ $(E) = E$, holds for all $\rho > 0$, then $\exists (x_1, u_1) \in \mathrm{Gph}(M)$ s.t.

$$H(Ax_0, Bx_0, Cx_0) + \rho u_0 = H(Ax_1, Bx_1, Cx_1) + \rho u_1 \in E. \tag{4}$$

Now,

$$\begin{aligned}\rho u_0 - \rho u_1 &= H(Ax_1, Bx_1, Cx_1) - H(Ax_0, Bx_0, Cx_0) \in E\\ \langle \rho u_0 - \rho u_1, J_q(\eta(x_0, x_1))\rangle &\\ &= -\langle H(Ax_0, Bx_0, Cx_0) - H(Ax_1, Bx_1, Cx_1), J_q(\eta(x_0, x_1))\rangle. \end{aligned}\tag{5}$$

Setting $(y, v) = (x_1, u_1)$ in (3) and then from the resultant, (4) and κ-η-relaxed accretivity of M, we obtain

$$\begin{aligned}-\kappa\|x_0 - x_1\|^q &\le \rho\langle u_0 - u_1, J_q(\eta(x_0, x_1))\rangle\\ &= -\langle H(Ax_0, Bx_0, Cx_0) - H(Ax_1, Bx_1, Cx_1), J_q(\eta(x_0, x_1))\rangle\\ &= -\langle H(Ax_0, Bx_0, Cx_0) - H(Ax_1, Bx_0, Cx_0), J_q(\eta(x_0, x_1))\rangle\\ &\quad - \langle H(Ax_1, Bx_0, Cx_0) - H(Ax_1, Bx_1, Cx_0), J_q(\eta(x_0, x_1))\rangle\\ &\quad - \langle H(Ax_1, Bx_1, Cx_0) - H(Ax_1, Bx_1, Cx_1), J_q(\eta(x_0, x_1))\rangle. \end{aligned}\tag{6}$$

By assumptions (c_1) and (c_2), we have

$$\begin{aligned}-\kappa\|x_0 - x_1\|^q &\le -\alpha\|Ax_0 - Ax_1\|^q + \beta\|Bx_0 - Bx_1\|^q - \delta\|x_0 - x_1\|^q\\ &\le -(\alpha\mu^q - \beta\gamma^q + \delta)\|x_0 - x_1\|^q\\ &= -\ell\|x_0 - x_1\|^q \le 0, \text{ where } \ell = \alpha\mu^q - \beta\gamma^q + \delta\\ &\le -(\ell - \kappa)\|x_0 - x_1\|^q \le 0,\end{aligned}$$

which gives $x_0 = x_1$, since $\ell > \kappa$. By (3), we have $u_0 = u_1$, a contradiction.

Hence $(x_0, u_0) = (x_1, u_1) \in \text{Gph}(M)$ and so $x_0 \in Mu_0$. This complete the proof.

Theorem 2 *Let $M : E \multimap E$ be a $H(.,.,.)$-η-mixed accretive mapping regarding A, B and C. If assumptions (c_1) and (c_2) hold with $\alpha > \beta$, $\mu > \gamma$ and $\ell = \alpha\mu^q - \beta\gamma^q + \delta > \rho\kappa$, then $(H(A, B, C) + \rho M)^{-1}$ is single-valued.*

Proof For any given $z \in E$, let $p, q \in (H(A, B, C) + \rho M)^{-1}(z)$. It follows that

$$\begin{cases} -H(Ap, Bp, Cp) + z \in \rho Mp,\\ -H(Aq, Bq, Cq) + z \in \rho Mq.\end{cases}$$

By using the κ-η-relaxed accretivity of M, we have

$$-\kappa\|p - q\|^q \le \frac{1}{\rho}\langle -H(Ap, Bp, Cp) + z - (-H(Aq, Bq, Cq) + z), J_q(\eta(p, q))\rangle$$

$$\begin{aligned}-\rho\kappa\|p-q\|^q &\leq \langle -H(Ap,Bp,Cp)+z-(-H(Aq,Bq,Cq)+z), J_q(\eta(p,q))\rangle \\ &= -\ \langle H(Ap,Bp,Cp)-H(Aq,Bq,Cq), J_q(\eta(p,q))\rangle \\ &= -\ \langle H(Ap,Bp,Cp)-H(Aq,Bp,Cp), J_q(\eta(p,q))\rangle \\ &\quad -\ \langle H(Aq,Bp,Cp)-H(Aq,Bq,Cp), J_q(\eta(p,q))\rangle \\ &\quad -\ \langle H(Aq,Bq,Cp)-H(Aq,Bq,Cq), J_q(\eta(p,q))\rangle. \end{aligned} \tag{7}$$

By assumptions (c_1) and (c_2), we have

$$\begin{aligned}-\rho\kappa\|p-q\|^q &\leq -\alpha\|Ap-Aq\|^q+\beta\|Bp-Bq\|^q-\delta\|p-q\|^q \\ &\leq -(\alpha\mu^q-\beta\gamma^q+\delta)\|p-q\|^q \\ &= -\ell\|p-q\|^q \leq 0, \text{ where } \ell=\alpha\mu^q-\beta\gamma^q+\delta \\ &\leq -(\ell-\rho\kappa)\|p-q\|^q \leq 0,\end{aligned}$$

since $\ell > \rho\kappa$. Hence, $\|p-q\| \leq 0$. Thus, we have $p=q$, and so $(H(A,B,C)+\rho M)^{-1}$ is single-valued.

Definition 9 Let $M : E \multimap E$ be a $H(.,.,.)$-η-mixed accretive mapping regarding A, B and C. If assumptions (c_1) and (c_2) hold with $\alpha > \beta$, $\mu > \gamma$ and $\ell = \alpha\mu^q - \beta\gamma^q + \delta > \rho\kappa$, then the proximal-point mapping $R^{H(.,.,.)-\eta}_{\rho,M} : E \to E$ is defined by

$$R^{H(.,.,.)-\eta}_{\rho,M}(a) = (H(A,B,C)+\rho M)^{-1}(a), \forall a \in E. \tag{8}$$

Theorem 3 *Let $M : E \multimap E$ be a $H(.,.,.)$-η-mixed accretive mapping regarding A, B and C. If assumptions (c_1) and (c_2) hold with $\alpha > \beta$, $\mu > \gamma$ and $\ell = \alpha\mu^q - \beta\gamma^q + \delta > \rho\kappa$, then the proximal-point mapping $R^{H(.,.,.)-\eta}_{\rho,M} : E \to E$ is $\frac{\tau^{q-1}}{\ell-\rho\kappa}$-Lipschitz continuous, i.e. satisfy the following inequality*

$$\|R^{H(.,.,.)-\eta}_{\rho,M}(a) - R^{H(.,.,.)-\eta}_{\rho,M}(b)\| \leq \frac{\tau^{q-1}}{\ell-\rho\kappa}\|a-b\|, \quad \forall a, b \in E.$$

Proof Let us consider the given point $a, b \in E$. It follows from (8) that

$$\begin{aligned}R^{H(.,.,.)-\eta}_{\rho,M}(a) &= (H(A,B,C)+\rho M)^{-1}(a),\\ R^{H(.,.,.)-\eta}_{\rho,M}(b) &= (H(A,B,C)+\rho M)^{-1}(b).\end{aligned}$$

$$\begin{aligned}&\frac{1}{\rho}(a - H(A(R^{H(.,.,.)-\eta}_{\rho,M}(a)), B(R^{H(.,.,.)-\eta}_{\rho,M}(a)), C(R^{H(.,.,.)-\eta}_{\rho,M}(a))) \in M(R^{H(.,.,.)-\eta}_{\rho,M}(a)),\\ &\frac{1}{\rho}(b - H(A(R^{H(.,.,.)-\eta}_{\rho,M}(b)), B(R^{H(.,.,.)-\eta}_{\rho,M}(b)), C(R^{H(.,.,.)-\eta}_{\rho,M}(b))) \in M(R^{H(.,.,.)-\eta}_{\rho,M}(b)).\end{aligned}$$

Let $p_1 = R^{H(.,.,.)-\eta}_{\rho,M}(a)$ and $p_2 = R^{H(.,.,.)-\eta}_{\rho,M}(b)$.

By using the κ-η-relaxed accretivity of M, we have

$$\frac{1}{\rho}\langle (a - H(A(p_1), B(p_1), C(p_1))) - (b - H(A(p_2), B(p_2), C(p_2))), J_q(\eta(p_1, p_2))\rangle$$
$$\geq -\kappa \|p_1 - p_2\|^q,$$
$$\langle a - b - (H(A(p_1), B(p_1), C(p_1)) - H(A(p_2), B(p_2), C(p_2))), J_q(\eta(p_1, p_2))\rangle$$
$$\geq -\rho\kappa \|p_1 - p_2\|^q,$$

thus we have

$$\begin{aligned}\langle a - b, J_q(\eta(p_1, p_2))\rangle \geq\ & \langle H(A(p_1), B(p_1), C(p_1)) - H(A(p_2), B(p_2), C(p_2)), J_q(\eta(p_1, p_2))\rangle \\ & - \rho\kappa \|p_1 - p_2\|^q.\end{aligned}$$

Further, we have

$$\begin{aligned}
&\|a - b\|\tau^{q-1}\|p_1 - p_2\|^{q-1} \\
&\quad \geq \|a - b\|\|\eta(p_1, p_2)\|^{q-1} \\
&\quad \geq \langle a - b, J_q(\eta(p_1, p_2))\rangle \\
&\quad \geq \langle H(A(p_1), B(p_1), C(p_1)) - H(A(p_2), B(p_2), C(p_2)), J_q(\eta(p_1, p_2))\rangle \\
&\qquad - \rho\kappa\|p_1 - p_2\|^q \\
&\quad \geq \langle H(A(p_1), B(p_1), C(p_1)) - H(A(p_2), B(p_1), C(p_1)), J_q(\eta(p_1, p_2))\rangle \\
&\qquad - \langle H(A(p_2), B(p_1), C(p_1)) - H(A(p_2), B(p_2), C(p_1)), J_q(\eta(p_1, p_2))\rangle \\
&\qquad - \langle H(A(p_2), B(p_2), C(p_1)) - H(A(p_2), B(p_2), C(p_2)), J_q(\eta(p_1, p_2))\rangle \\
&\qquad - \rho\kappa\|p_1 - p_2\|^q \\
&\quad \geq \alpha\|A(p_1) - A(p_2)\|^q - \beta\|B(p_1) - B(p_2)\|^q + \delta\|C(p_1) - C(p_2)\|^q - \rho\kappa\|p_1 - p_2\|^q \\
&\quad \geq \alpha\mu^q\|p_1 - p_2\|^q - \beta\gamma^q\|p_1 - p_2\|^q + \delta\|p_1 - p_2\|^q - \rho\kappa\|p_1 - p_2\|^q \\
&\quad = (\alpha\mu^q - \beta\gamma^q + \delta - \rho\kappa)\|p_1 - p_2\|^q \\
&\quad = (\ell - \rho\kappa)\|p_1 - p_2\|^q, \text{ where } \ell = \alpha\mu^q - \beta\gamma^q + \delta.
\end{aligned}$$

Hence

$$\|a - b\|\tau^{q-1}\|p_1 - p_2\|^{q-1} \geq (\ell - \rho\kappa)\|p_1 - p_2\|^q,$$

that is

$$\|R_{\rho,M}^{H(.,.,.)-\eta}(a) - R_{\rho,M}^{H(.,.,.)-\eta}(b)\| \leq \frac{\tau^{q-1}}{\ell - \rho\kappa}\|a - b\|, \forall u, v \in E.$$

4 System of Generalized Set-Valued Mixed Quasi-variational-like Inclusions

In this section, we formulate a system of generalized set-valued mixed quasi-variational-like inclusions involving $H(.,.,.)$-η-mixed accretive mapping in q-uniformly smooth Banach spaces.

Throughout the rest of the paper, unless otherwise stated, we consider that for each $i \in \{1, 2\}$, E_i be q_i-uniformly smooth Banach spaces with norm $\|.\|_i$, $CB(E_i)$ the family of all bounded and closed subsets of E_i. Let $H_i : E_i \times E_i \times E_i \to E_i$, $\eta_i : E_i \times E_i \to E_i$, $A_i, B_i, C_i, g_i : E_i \to E_i$, $N_i, P_i : E_1 \times E_2 \to E_i$, are all single-valued mappings and $S_i : E_1 \multimap \mathrm{CB}(E_1)$, $T_i : E_2 \multimap \mathrm{CB}(E_2)$ are four set-valued mappings. For each $i \in \{1, 2\}$, let $M_i : E_i \multimap E_i$, be a set-valued mapping such that M_i is a $H_i(A_i, B_i, C_i)$-η_i-mixed accretive mapping regarding A_i, B_i, C_i, and range$(g_i)\cap$dom$(M_i) \neq \emptyset$. Then the problem (9) is to find (x, y, u, v, w, z) such that $(x, y) \in E_1 \times E_2, u \in S_1(x), v \in T_1(y), w \in S_2(x), z \in T_2(y)$ and

$$\begin{cases} 0 \in N_1(x, y) + P_1(u, v) + M_1(g_1(x)), \\ 0 \in N_2(x, y) + P_2(w, z) + M_2(g_2(y)). \end{cases} \tag{9}$$

The system (9) is called the system of generalized set-valued mixed quasi-variational-like inclusions.

Lemma 2 *An element (x, y, u, v, w, z), where $(x, y) \in E_1 \times E_2, u \in S_1(x), v \in T_1(y), w \in S_2(x), z \in T_2(y)$, is a solution of the system (9), if and only if (x, y, u, v, w, z), satisfies the following relation:*

$$\begin{cases} g_1(x) = R^{H_1(.,.,.)-\eta_1}_{\rho_1,M_1}[H_1(A_1(g_1(x)), B_1(g_1(x)), C_1(g_1(x))) - \rho_1(N_1(x, y) + P_1(u, v))], \\ g_2(y) = R^{H_2(.,.,.)-\eta_2}_{\rho_2,M_2}[H_2(A_2(g_2(y)), B_2(g_2(y)), C_2(g_2(y))) - \rho_2(N_2(x, y) + P_2(w, z))], \end{cases} \tag{10}$$

where $R^{H_1(.,.,.)-\eta_1}_{\rho_1,M_1} = (H_1(A_1, B_1, C_1) + \rho_1 M_1)^{-1}$, $R^{H_2(.,.,.)-\eta_2}_{\rho_2,M_2} = (H_2(A_2, B_2, C_2) + \rho_2 M_2)^{-1}$, and $\rho_1, \rho_2 > 0$ are two constants.

Proof Consider first that an element (x, y, u, v, w, z) is a solution to system (9). Then it follows that

$$\begin{aligned} &0 \in N_1(x, y) + P_1(u, v) + M_1(g_1(x)), \\ &\Longrightarrow H_1(A_1(g_1(x)), B_1(g_1(x)), C_1(g_1(x))) \in H_1(A_1(g_1(x)), B_1(g_1(x)), C_1(g_1(x))) \\ &\qquad\qquad + \rho_1 N_1(x, y) + \rho_1 P_1(u, v) + \rho_1 M_1(g_1(x)) \\ &\Longrightarrow H_1(A_1(g_1(x)), B_1(g_1(x)), C_1(g_1(x))) - \rho_1 N_1(x, y) - \rho_1 P_1(u, v) \\ &\qquad\qquad \in (H_1(A_1, B_1, C_1) + \rho_1 M_1)(g_1(x)) \\ &\Longrightarrow g_1(x) = (H_1(A_1, B_1, C_1) + \rho_1 M_1)^{-1}[H_1(A_1(g_1(x)), B_1(g_1(x)), C_1(g_1(x))) \\ &\qquad\qquad - \rho_1 N_1(x, y) - \rho_1 P_1(u, v)] \\ &\Longrightarrow g_1(x) = R^{H_1(.,.,.)-\eta_1}_{\rho_1,M_1}[H_1(A_1(g_1(x)), B_1(g_1(x)), C_1(g_1(x))) - \rho_1 N_1(x, y) - \rho_1 P_1(u, v)]. \end{aligned}$$

Similarly, we can obtain

$$g_2(y) = R^{H_2(.,.,.)-\eta_2}_{\rho_2,M_2}[H_2(A_2(g_2(y)), B_2(g_2(y)), C_2(g_2(y))) + \rho_2 N_2(x, y) + \rho_2 P_2(w, z)].$$

Similarly the converse of lemma as follow:

$$\begin{aligned}
& g_1(x) = R^{H_1(.,.,.)-\eta_1}_{\rho_1,M_1}[H_1(A_1(g_1(x)), B_1(g_1(x)), C_1(g_1(x))) - \rho_1 N_1(x, y) - \rho_1 P_1(u, v)], \\
& \Longrightarrow g_1(x) = (H_1(A_1, B_1, C_1) + \rho_1 M_1)^{-1}[H_1(A_1(g_1(x)), B_1(g_1(x)), C_1(g_1(x))) \\
& \qquad\qquad - \rho_1 N_1(x, y) - \rho_1 P_1(u, v)] \\
& \Longrightarrow H_1(A_1(g_1(x)), B_1(g_1(x)), C_1(g_1(x))) - \rho_1 N_1(x, y) - \rho_1 P_1(u, v) \\
& \qquad\qquad \in (H_1(A_1, B_1, C_1) + \rho_1 M_1)(g_1(x)) \\
& \Longrightarrow H_1(A_1(g_1(x)), B_1(g_1(x)), C_1(g_1(x))) \in H_1(A_1(g_1(x)), B_1(g_1(x)), C_1(g_1(x))) \\
& \qquad\qquad + \rho_1 N_1(x, y) + \rho_1 P_1(u, v) + \rho_1 M_1(g_1(x)) \\
& \Longrightarrow 0 \in N_1(x, y) + P_1(u, v) + M_1(g_1(x)).
\end{aligned}$$

Similarly, we can obtain

$$0 \in N_2(x, y) + P_2(w, z) + M_2(g_2(y)).$$

Algorithm 1 For each $x, y \in E_1 \times E_2$, $G_1(x) \subseteq g_1(E_1)$, $G_2(y) \subseteq g_2(E_2)$, where set-valued mappings $G_1 : E_1 \multimap E_1$ and $G_2 : E_2 \multimap E_2$ defined by

$$G_1(x) = \bigcup_{u \in S_1(x)} \bigcup_{v \in T_1(y)} (R^{H_1(.,.,.)-\eta_1}_{\rho_1,M_1}(H_1(A_1(g_1(x)), B_1(g_1(x)), C_1(g_1(x))) - \rho_1 N_1(x, y) - \rho_1 P_1(u, v))), \tag{11}$$

$$G_2(y) = \bigcup_{w \in S_2(x)} \bigcup_{z \in T_2(y)} (R^{H_2(.,.,.)-\eta_2}_{\rho_2,M_2}(H_2(A_2(g_2(y)), B_2(g_2(y)), C_2(g_2(y))) - \rho_2 N_2(x, y) - \rho_2 P_2(w, z))), \tag{12}$$

where for each $i = 1, 2$, $M_i : E_i \multimap E_i$ is $H_i(A_i, B_i, C_i)$-η_i-mixed accretive mapping regarding A_i, B_i, C_i, respectively.

For given $(x_0, y_0) \in E_1 \times E_2$, $u_0 \in S_1(x_0)$, $v_0 \in T_1(y_0)$, $w_0 \in S_2(x_0)$, $z_0 \in T_2(y_0)$, let

$$\begin{aligned}
a_0 = & R^{H_1(.,.,.)-\eta_1}_{\rho_1,M_1}(H_1(A_1(g_1(x_0)), B_1(g_1(x_0)), C_1(g_1(x_0))) \\
& - \rho_1 N_1(x_0, y_0) - \rho_1 P_1(u_0, v_0)) \in G_1(x_0) \subseteq g_1(E_1),
\end{aligned}$$

$$
\begin{aligned}
b_0 =& R^{H_2(.,.,.)-\eta_2}_{\rho_2,M_2}(H_2(A_2(g_2(y_0)), B_2(g_2(y_0)), C_2(g_2(y_0)))\\
&- \rho_2 N_2(x_0, y_0) - \rho_2 P_2(w_0, z_0)) \in G_2(y_0) \subseteq g_2(E_2).
\end{aligned}
$$

Hence, there exists $(x_1, y_1) \in E_1 \times E_2$ such that $a_0 = g_1(x_1), b_0 = g_2(y_1)$. Since $u_0 \in S_1(x_0) \in CB(E_1), v_0 \in T_1(y_0) \in CB(E_2), w_0 \in S_2(x_0) \in CB(E_1)$ and $z_0 \in T_2(y_0) \in CB(E_2)$, then by Nadler's result [17], $\exists\ u_1 \in S_1(x_1), v_1 \in T_1(y_1), w_1 \in S_2(x_1)$ and $z_1 \in T_2(y_1)$ such that

$$
\begin{aligned}
\|u_0 - u_1\|_1 &\le \left(1 + 1^{-1}\right) D_1(S_1(x_0), S_1(x_1)),\\
\|v_0 - v_1\|_2 &\le \left(1 + 1^{-1}\right) D_2(T_1(y_0), T_1(y_1)),\\
\|w_0 - w_1\|_1 &\le \left(1 + 1^{-1}\right) D_1(S_2(x_0), S_2(x_1)),\\
\|z_0 - z_1\|_2 &\le \left(1 + 1^{-1}\right) D_2(T_2(y_0), T_2(y_1)),
\end{aligned}
$$

where $D_i(.,.)$ is the Hausdorff metric on $CB(E_i)$, for $i = 1, 2$. Let

$$
\begin{aligned}
a_1 =& R^{H_1(.,.,.)-\eta_1}_{\rho_1,M_1}(H_1(A_1(g_1(x_1)), B_1(g_1(x_1)), C_1(g_1(x_1)))\\
&- \rho_1 N_1(x_1, y_1) - \rho_1 P_1(u_1, v_1)) \in G_1(x_1) \subseteq g_1(E_1),
\end{aligned}
$$

$$
\begin{aligned}
b_1 =& R^{H_2(.,.,.)-\eta_2}_{\rho_2,M_2}(H_2(A_2(g_2(y_1)), B_2(g_2(y_1)), C_2(g_2(y_1)))\\
&- \rho_2 N_2(x_1, y_1) - \rho_2 P_2(w_1, z_1)) \in G_2(y_1) \subseteq g_2(E_2).
\end{aligned}
$$

Hence, there exists $(x_2, y_2) \in E_1 \times E_2$ such that $a_1 = g_1(x_2), b_1 = g_2(y_2)$. By induction, we can define sequences $\{x_n\}, \{g_1(x_n)\}, \{y_n\}, \{g_2(y_n)\}, \{u_n\}, \{v_n\}, \{w_n\}$ and $\{z_n\}$ as follows:

$$
\begin{aligned}
g_1(x_{n+1}) =& R^{H_1(.,.,.)-\eta_1}_{\rho_1,M_1}(H_1(A_1(g_1(x_n)), B_1(g_1(x_n)), C_1(g_1(x_n)))\\
&- \rho_1 N_1(x_n, y_n) - \rho_1 P_1(u_n, v_n)), \qquad (13)
\end{aligned}
$$

$$
\begin{aligned}
g_2(y_{n+1}) =& R^{H_2(.,.,.)-\eta_2}_{\rho_2,M_2}(H_2(A_2(g_2(y_n)), B_2(g_2(y_n)), C_2(g_2(y_n)))\\
&- \rho_2 N_2(x_n, y_n) - \rho_2 P_2(w_n, z_n)), \qquad (14)
\end{aligned}
$$

$$
\begin{aligned}
u_n \in S_1(x_n), &\quad \|u_n - u_{n+1}\|_1 \le \left(1 + (n+1)^{-1}\right) D_1(S_1(x_n), S_1(x_{n+1})), \quad (15)\\
v_n \in T_1(y_n), &\quad \|v_n - v_{n+1}\|_2 \le \left(1 + (n+1)^{-1}\right) D_2(T_1(y_n), T_1(y_{n+1})), \quad (16)\\
w_n \in S_2(x_n), &\quad \|w_n - w_{n+1}\|_1 \le \left(1 + (n+1)^{-1}\right) D_1(S_2(x_n), S_2(x_{n+1})), \quad (17)\\
z_n \in T_2(y_n), &\quad \|z_n - z_{n+1}\|_2 \le \left(1 + (n+1)^{-1}\right) D_2(T_2(y_n), T_2(y_{n+1})), \quad (18)
\end{aligned}
$$

for all $n = 0, 1, 2, \ldots,$ and $\rho_1, \rho_2 > 0$ are constants.

5 Existence of Solutions for the System (9)

In this section, first we recall some definitions and then we discuss about the existence of solution of system (9) and convergence analysis of iterative sequences generated by Algorithm 1.

Definition 10 For $i \in \{1, 2\}$, the single-valued mapping $N_i : E_1 \times E_2 \to E_i$ is said to be

(i) Lipschitz continuous with constant $\zeta_i > 0$ in the first component, if

$$\|N_i(x_1, .) - N_i(x_2, .)\|_i \leq \zeta_i \|x_1 - x_2\|_1, \forall x_1, x_2 \in E_1;$$

(ii) Lipschitz continuous with constant $\nu_i > 0$ in the second component, if

$$\|N_i(., y_1) - N_i(., y_2)\|_i \leq \nu_i \|y_1 - y_2\|_2, \forall y_1, y_2 \in E_2;$$

(iii) strongly accretive with constant $n_i > 0$ in the first component, if

$$\langle N_i(x_1, .) - N_i(x_2, .), J_{q_i}(\eta_i(x_1, x_2))\rangle_i \leq n_i \|x_1 - x_2\|_1^{q_1}, \forall x_1, x_2 \in E_1;$$

(iv) strongly accretive with constant $\vartheta_i > 0$ in the second component, if

$$\langle N_i(., y_1) - N_i(., y_2), J_{q_i}(\eta_i(y_1, y_2))\rangle_i \leq \vartheta_i \|y_1 - y_2\|_2^{q_2}, \forall y_1, y_2 \in E_2.$$

Definition 11 For $i \in \{1, 2\}$, let $P_i : E_1 \times E_2 \to E_i$ be a single-valued mapping, $S_i : E_1 \multimap CB(E_1)$ and $T_i : E_2 \multimap CB(E_2)$ be the set-valued mappings. Then P_i is said to be

(i) Lipschitz continuous with constant $\epsilon_i > 0$ in the first component regarding S_i, if

$$\|P_i(u_1, .) - P_i(u_2, .)\|_i \leq \epsilon_i \|u_1 - u_2\|_1, \forall (x, y) \in E_1 \times E_2, u_1 \in S_1(x), u_2 \in S_2(x);$$

(ii) Lipschitz continuous with constant $\sigma_i > 0$ in the second component regarding T_i, if

$$\|P_i(., v_1) - P_i(., v_2)\|_i \leq \sigma_i \|v_1 - v_2\|_2, \forall (x, y) \in E_1 \times E_2, v_1 \in T_1(y), v_2 \in T_2(y).$$

Theorem 4 *In problem (9), for each $x, y \in E_1 \times E_2$, $G_1(x) \subseteq g_1(E_1)$, $G_2(y) \subseteq g_2(E_2)$, where $G_1 : E_1 \multimap E_1$ and $G_2 : E_2 \multimap E_2$ defined by (11) and (12). For each $i \in \{1, 2\}$, assume that*
(i) S_i is D_1-Lipschitz continuous with constant l_{S_i} and T_i is D_2-Lipschitz continuous with constant l_{T_i};
(ii) A_i is α_i-expansive, and B_i, η_i are β_i, τ_i-Lipschitz continuous, respectively;
(iii) g_i is λ_{g_i}-Lipschitz continuous and ξ_{g_i}-strongly accretive;

(iv) $H_i(A_i, B_i, C_i)$ is r_i, s_i and t_i-Lipschitz continuous regarding A_i, B_i and C_i, respectively;
(v) N_i is ξ_i-strongly accretive in the ith component, ζ_i-Lipschitz continuous in the first component and ν_i-Lipschitz continuous in the second component;
(vi) P_i is ϵ_i-Lipschitz continuous in the first component regarding S_i and σ_i- Lipschitz continuous in the second component regarding T_i;
In addition

$$0 < \frac{\tau_1^{q_1-1}k_1}{\xi_{g_1}}[L_1 + \rho_1\epsilon_1 l_{S_1}] + \frac{\tau_2^{q_2-1}k_2\rho_2}{\xi_{g_2}}[\zeta_2 + \epsilon_2 l_{S_2}] < 1,$$

$$0 < \frac{\tau_1^{q_1-1}k_1\rho_1}{\xi_{g_1}}[\nu_1 + \sigma_1 l_{T_1}] + \frac{\tau_2^{q_2-1}k_2}{\xi_{g_2}}[L_2 + \rho_2\sigma_2 l_{T_2}] < 1, \tag{19}$$

where

$$\begin{aligned} L_1 &= \big[(r_1+s_1+t_1)^{q_1}\lambda_{g_1}^{q_1} - q_1\rho_1\xi_1 + q_1\rho_1\zeta_1\{(r_1+s_1+t_1)^{q_1-1}\lambda_{g_1}^{q_1-1} + \tau_1^{q_1-1}\} \\ &\qquad + \rho_1^{q_1}C_{q_1}\zeta_1^{q_1}\big]^{\frac{1}{q_1}}, \\ L_2 &= \big[(r_2+s_2+t_2)^{q_2}\lambda_{g_2}^{q_2} - q_2\rho_2\xi_2 + q_2\rho_2\nu_2\{(r_2+s_2+t_2)^{q_2-1}\lambda_{g_2}^{q_2-1} + \tau_2^{q_2-1}\} \\ &\qquad + \rho_2^{q_2}C_{q_2}\nu_2^{q_2}\big]^{\frac{1}{q_2}}, \\ k_1 &= \frac{\tau_1^{q_1-1}}{\alpha_1\mu_1^{q_1} - \beta_1\gamma_1^{q_1} + \delta_1 - \rho_1\kappa_1}, \alpha_1\mu_1^{q_1} - \beta_1\gamma_1^{q_1} + \delta_1 > \rho_1\kappa_1, \alpha_1 > \beta_1, \mu_1 > \gamma_1, \text{ and} \\ k_2 &= \frac{\tau_2^{q_2-1}}{\alpha_2\mu_2^{q_2} - \beta_2\gamma_2^{q_2} + \delta_2 - \rho_2\kappa_2}, \alpha_2\mu_2^{q_2} - \beta_2\gamma_2^{q_2} + \delta_2 > \rho_2\kappa_2, \alpha_2 > \beta_2, \mu_2 > \gamma_2. \end{aligned}$$

Then, the system (9) has a solution $(x, y, u, v, w, z) \in E$, and the iterative sequences $\{x_n\}$, $\{y_n\}$, $\{u_n\}$, $\{v_n\}$, $\{w_n\}$ and $\{z_n\}$ generated by Algorithm 1 strongly converge to x, y, u, v, w and z, respectively.

Proof For each $i = 1, 2$, S_i is D_1-Lipschitz continuous with constant l_{S_i} and T_i is D_2-Lipschitz continuous with constant l_{T_i}. It follows from (15)–(18)

$$\begin{aligned} \|u_{n+1} - u_n\|_1 &\le \left(1 + (n+1)^{-1}\right) D_1(S_1(x_{n+1}), S_1(x_n)) \\ &\le \left(1 + (n+1)^{-1}\right) l_{S_1}\|x_{n+1} - x_n\|_1, \end{aligned} \tag{20}$$

$$\|v_{n+1} - v_n\|_2 \le \left(1 + (n+1)^{-1}\right) D_2(T_1(y_{n+1}), T_1(y_n))$$

$$\leq \left(1 + (n+1)^{-1}\right) l_{T_1} \|y_{n+1} - y_n\|_2, \tag{21}$$

$$\|w_{n+1} - w_n\|_1 \leq \left(1 + (n+1)^{-1}\right) D_1(S_2(x_{n+1}), S_2(x_n))$$

$$\leq \left(1 + (n+1)^{-1}\right) l_{S_2} \|x_{n+1} - x_n\|_1, \tag{22}$$

$$\|z_{n+1} - z_n\|_2 \leq \left(1 + (n+1)^{-1}\right) D_2(T_2(y_{n+1}), T_2(y_n))$$

$$\leq \left(1 + (n+1)^{-1}\right) l_{T_2} \|y_{n+1} - y_n\|_2, \tag{23}$$

for $n = 0, 1, 2, \ldots$

Since, for each $i = 1, 2$, g_i is ξ_{g_i}-strongly accretive and η_i is τ_i Lipschitz continuous, we have

$$\begin{aligned}
\|g_1(x_{n+1}) - g_1(x_n)\|_1 \tau_1^{q_1-1} \|x_{n+1} - x_n\|_1^{q_1-1} &\geq \|g_1(x_{n+1}) - g_1(x_n)\|_1 \|\eta_1(x_{n+1}, x_n)\|_1^{q_1-1} \\
&\geq \langle g_1(x_{n+1}) - g_1(x_n), J_{q_1}(\eta_1(x_{n+1}, x_n))\rangle_1 \\
&\geq \xi_{g_1} \|x_{n+1} - x_n\|_1^{q_1},
\end{aligned}$$

implies that

$$\|x_{n+1} - x_n\|_1 \leq \frac{\tau_1^{q_1-1}}{\xi_{g_1}} \|g_1(x_{n+1}) - g_1(x_n)\|_1. \tag{24}$$

In the view of (24), we can obtain

$$\|y_{n+1} - y_n\|_2 \leq \frac{\tau_2^{q_2-1}}{\xi_{g_2}} \|g_2(y_{n+1}) - g_2(y_n)\|_2. \tag{25}$$

For each $i \in \{1, 2\}$ and $(x, y) \in E_1 \times E_2$, the proximal-point mappings $R^{H_1(.,.,.)-\eta_1}_{\rho_1, M_1}$ and $R^{H_2(.,.,.)-\eta_2}_{\rho_2, M_2}$ are $k_1 = \frac{\tau_1^{q_1-1}}{\alpha_1\mu_1^{q_1} - \beta_1\gamma_1^{q_1} + \delta_1 - \rho_1\kappa_1}$ and $k_2 = \frac{\tau_2^{q_2-1}}{\alpha_2\mu_2^{q_2} - \beta_2\gamma_2^{q_2} + \delta_2 - \rho_2\kappa_2}$-Lipschitz continuous, respectively. It follows from (13) and Theorem 3 that

$$\begin{aligned}
\|g_1(x_{n+1}) - g_1(x_n)\|_1 &= \|R^{H_1(.,.,.)-\eta_1}_{\rho_1, M_1}\{H_1(A_1(g_1(x_n)), B_1(g_1(x_n)), C_1(g_1(x_n))) \\
&\quad - \rho_1 N_1(x_n, y_n) - \rho_1 P_1(u_n, v_n)\} \\
&\quad - [R^{H_1(.,.,.)-\eta_1}_{\rho_1, M_1}\{H_1(A_1(g_1(x_{n-1})), B_1(g_1(x_{n-1})), C_1(g_1(x_{n-1}))) \\
&\quad - \rho_1 N_1(x_{n-1}, y_{n-1}) - \rho_1 P_1(u_{n-1}, v_{n-1})\}]\|_1 \\
&\leq k_1 \|H_1(A_1(g_1(x_n)), B_1(g_1(x_n)), C_1(g_1(x_n))) \\
&\quad - \rho_1 N_1(x_n, y_n) - \rho_1 P_1(u_n, v_n) \\
&\quad - \{H_1(A_1(g_1(x_{n-1})), B_1(g_1(x_{n-1})), C_1(g_1(x_{n-1}))) \\
&\quad - \rho_1 N_1(x_{n-1}, y_{n-1}) - \rho_1 P_1(u_{n-1}, v_{n-1})\}\|_1
\end{aligned}$$

$$
\begin{aligned}
&= k_1 \|H_1(A_1(g_1(x_n)), B_1(g_1(x_n)), C_1(g_1(x_n))) \\
&\quad - H_1(A_1(g_1(x_{n-1})), B_1(g_1(x_{n-1})), C_1(g_1(x_{n-1}))) \\
&\quad - \rho_1(N_1(x_n, y_n) - N_1(x_{n-1}, y_n)) \\
&\quad - \rho_1(N_1(x_{n-1}, y_n) - N_1(x_{n-1}, y_{n-1})) \\
&\quad - \rho_1(P_1(u_n, v_n) - P_1(u_{n-1}, v_{n-1}))\|_1 \\
&\le k_1 \|H_1(A_1(g_1(x_n)), B_1(g_1(x_n)), C_1(g_1(x_n))) \\
&\quad - H_1(A_1(g_1(x_{n-1})), B_1(g_1(x_{n-1})), C_1(g_1(x_{n-1}))) \\
&\quad - \rho_1(N_1(x_n, y_n) - N_1(x_{n-1}, y_n))\|_1 \\
&\quad + k_1\rho_1 \|N_1(x_{n-1}, y_n) - N_1(x_{n-1}, y_{n-1})\|_1 \\
&\quad + k_1\rho_1 \|P_1(u_n, v_n) - P_1(u_{n-1}, v_{n-1})\|_1.
\end{aligned} \tag{26}
$$

Using the r_i, s_i and t_i-Lipschitz continuity of $H_i(A_i, B_i, C_i)$ regarding A_i, B_i and C_i for each $i = 1, 2$, respectively and λ_{g_i}-Lipschitz continuity of g_i, we have

$$
\begin{aligned}
&\|H_1(A_1(g_1(x_n)), B_1(g_1(x_n)), C_1(g_1(x_n))) - H_1(A_1(g_1(x_{n-1})), B_1(g_1(x_{n-1})), C_1(g_1(x_{n-1})))\|_1 \\
&\le \|H_1(A_1(g_1(x_n)), B_1(g_1(x_n)), C_1(g_1(x_n))) - H_1(A_1(g_1(x_{n-1})), B_1(g_1(x_n)), C_1(g_1(x_n)))\|_1 \\
&\quad + \|H_1(A_1(g_1(x_{n-1})), B_1(g_1(x_n)), C_1(g_1(x_n))) \\
&\quad - H_1(A_1(g_1(x_{n-1})), B_1(g_1(x_{n-1})), C_1(g_1(x_n)))\|_1 \\
&\quad + \|H_1(A_1(g_1(x_{n-1})), B_1(g_1(x_{n-1})), C_1(g_1(x_n))) \\
&\quad - H_1(A_1(g_1(x_{n-1})), B_1(g_1(x_{n-1})), C_1(g_1(x_{n-1})))\|_1 \\
&\le (r_1 + s_1 + t_1)\|g_1(x_n) - g_1(x_{n-1})\|_1 \\
&\le (r_1 + s_1 + t_1)\lambda_{g_1} \|x_n - x_{n-1}\|_1.
\end{aligned} \tag{27}
$$

In view of (27), we have the following

$$
\begin{aligned}
&\|H_2(A_2(g_2(y_n)), B_2(g_2(y_n)), C_2(g_2(y_n))) - H_2(A_2(g_2(y_{n-1})), B_2(g_2(y_{n-1})), C_2(g_2(y_{n-1})))\|_2 \\
&\le (r_2 + s_2 + t_2)\lambda_{g_2} \|y_n - y_{n-1}\|_2.
\end{aligned} \tag{28}
$$

For each $i = 1, 2$, using the ζ_i and ν_i-Lipschitz continuity of N_i in their first and second component, respectively, we have

$$\|N_1(x_n, y_n) - N_1(x_{n-1}, y_n)\|_1 \le \zeta_1 \|x_n - x_{n-1}\|_1, \tag{29}$$

$$\|N_2(x_n, y_n) - N_2(x_{n-1}, y_n)\|_2 \le \zeta_2 \|x_n - x_{n-1}\|_1, \tag{30}$$

$$\|N_1(x_n, y_n) - N_1(x_n, y_{n-1})\|_1 \le \nu_1 \|y_n - y_{n-1}\|_2, \tag{31}$$

and

$$\|N_2(x_n, y_n) - N_2(x_n, y_{n-1})\|_2 \le \nu_2 \|y_n - y_{n-1}\|_2. \tag{32}$$

By using Lemma 1, (27), (29), ξ_1-strongly accretiveness of N_1 in the first component and τ_1-Lipschitz continuity of η_1, we have

$$\begin{aligned}
&\|H_1(A_1(g_1(x_n)), B_1(g_1(x_n)), C_1(g_1(x_n))) - H_1(A_1(g_1(x_{n-1})), B_1(g_1(x_{n-1})), C_1(g_1(x_{n-1})))\\
&\qquad - \rho_1(N_1(x_n, y_n) - N_1(x_{n-1}, y_n))\|_1^{q_1}\\
&\le \|H_1(A_1(g_1(x_n)), B_1(g_1(x_n)), C_1(g_1(x_n))) - H_1(A_1(g_1(x_{n-1})), B_1(g_1(x_{n-1})), C_1(g_1(x_{n-1})))\|_1^{q_1}\\
&\quad - q_1\rho_1\langle N_1(x_n, y_n) - N_1(x_{n-1}, y_n), J_{q_1}(\eta_1(x_n, x_{n-1}))\rangle_1\\
&\quad - q_1\rho_1\langle N_1(x_n, y_n) - N_1(x_{n-1}, y_n), J_{q_1}[H_1(A_1(g_1(x_n)), B_1(g_1(x_n)), C_1(g_1(x_n)))\\
&\quad - H_1(A_1(g_1(x_{n-1})), B_1(g_1(x_{n-1})), C_1(g_1(x_{n-1})))] - J_{q_1}(\eta_1(x_n, x_{n-1}))\rangle_1\\
&\quad + \rho_1^{q_1} C_{q_1}\|N_1(x_n, y_n) - N_1(x_{n-1}, y_n))\|_1^{q_1}\\
&\le (r_1 + s_1 + t_1)^{q_1}\lambda_{g_1}^{q_1}\|x_n - x_{n-1}\|_1^{q_1} - q_1\rho_1\xi_1\|x_n - x_{n-1}\|_1^{q_1}\\
&\quad + q_1\rho_1\|N_1(x_n, y_n) - N_1(x_{n-1}, y_n)\|_1\big[\|H_1(A_1(g_1(x_n)), B_1(g_1(x_n)), C_1(g_1(x_n)))\\
&\quad - H_1(A_1(g_1(x_{n-1})), B_1(g_1(x_{n-1})), C_1(g_1(x_{n-1})))\|_1^{q_1-1} + \|\eta_1(x_n, x_{n-1})\|_1^{q_1-1}\big]\\
&\quad + \rho_1^{q_1} C_{q_1}\zeta_1^{q_1}\|x_n - x_{n-1}\|_1^{q_1}\\
&\le (r_1 + s_1 + t_1)^{q_1}\lambda_{g_1}^{q_1}\|x_n - x_{n-1}\|_1^{q_1} - q_1\rho_1\xi_1\|x_n - x_{n-1}\|_1^{q_1}\\
&\quad + q_1\rho_1\zeta_1\|x_n - x_{n-1}\|_1\big[(r_1 + s_1 + t_1)^{q_1-1}\lambda_{g_1}^{q_1-1}\|x_n - x_{n-1}\|_1^{q_1-1}\\
&\quad + \tau_1^{q_1-1}\|x_n - x_{n-1}\|_1^{q_1-1}\big] + \rho_1^{q_1} C_{q_1}\zeta_1^{q_1}\|x_n - x_{n-1}\|_1^{q_1}\\
&= (r_1 + s_1 + t_1)^{q_1}\lambda_{g_1}^{q_1}\|x_n - x_{n-1}\|_1^{q_1} - q_1\rho_1\xi_1\|x_n - x_{n-1}\|_1^{q_1}\\
&\quad + q_1\rho_1\zeta_1\big[(r_1 + s_1 + t_1)^{q_1-1}\lambda_{g_1}^{q_1-1} + \tau_1^{q_1-1}\big]\|x_n - x_{n-1}\|_1^{q_1}\\
&\quad + \rho_1^{q_1} C_{q_1}\zeta_1^{q_1}\|x_n - x_{n-1}\|_1^{q_1}\\
&= \big[(r_1 + s_1 + t_1)^{q_1}\lambda_{g_1}^{q_1} - q_1\rho_1\xi_1 + q_1\rho_1\zeta_1\{(r_1 + s_1 + t_1)^{q_1-1}\lambda_{g_1}^{q_1-1} + \tau_1^{q_1-1}\}\\
&\quad + \rho_1^{q_1} C_{q_1}\zeta_1^{q_1}\big]\|x_n - x_{n-1}\|_1^{q_1}.
\end{aligned}$$

This implies that

$$\begin{aligned}
&\|H_1(A_1(g_1(x_n)), B_1(g_1(x_n)), C_1(g_1(x_n))) - H_1(A_1(g_1(x_{n-1})), B_1(g_1(x_{n-1})), C_1(g_1(x_{n-1})))\\
&\qquad - \rho_1(N_1(x_n, y_n) - N_1(x_{n-1}, y_n))\|_1\\
&\le \big[(r_1 + s_1 + t_1)^{q_1}\lambda_{g_1}^{q_1} - q_1\rho_1\xi_1 + q_1\rho_1\zeta_1\{(r_1 + s_1 + t_1)^{q_1-1}\lambda_{g_1}^{q_1-1} + \tau_1^{q_1-1}\}\\
&\qquad + \rho_1^{q_1} C_{q_1}\zeta_1^{q_1}\big]^{\frac{1}{q_1}}\|x_n - x_{n-1}\|_1 = L_1\|x_n - x_{n-1}\|_1,
\end{aligned} \tag{33}$$

where

$$\begin{aligned}
L_1 = &\big[(r_1 + s_1 + t_1)^{q_1}\lambda_{g_1}^{q_1} - q_1\rho_1\xi_1 + q_1\rho_1\zeta_1\{(r_1 + s_1 + t_1)^{q_1-1}\lambda_{g_1}^{q_1-1} + \tau_1^{q_1-1}\}\\
&+ \rho_1^{q_1} C_{q_1}\zeta_1^{q_1}\big]^{\frac{1}{q_1}}.
\end{aligned}$$

For each $i = 1, 2$, using the ϵ_i and σ_i-Lipschitz continuity of P_i in their first second component, respectively, we have

$$\begin{aligned}
&\|P_1(u_n, v_n) - P_1(u_{n-1}, v_{n-1})\|_1 \\
&\quad \leq \|P_1(u_n, v_n) - P_1(u_{n-1}, v_n)\|_1 + \|P_1(u_{n-1}, v_n) - P_1(u_{n-1}, v_{n-1})\|_1 \\
&\quad \leq \epsilon_1 \|u_n - u_{n-1}\|_1 + \sigma_1 \|v_n - v_{n-1}\|_2 \\
&\quad \leq \epsilon_1 \left(1 + n^{-1}\right) l_{S_1} \|x_n - x_{n-1}\|_1 + \sigma_1 \left(1 + n^{-1}\right) l_{T_1} \|y_n - y_{n-1}\|_2
\end{aligned}$$

$$= \epsilon_1 l_{S_1} \left(1 + n^{-1}\right) \|x_n - x_{n-1}\|_1 + \sigma_1 l_{T_1} \left(1 + n^{-1}\right) \|y_n - y_{n-1}\|_2. \tag{34}$$

Similarly,

$$\begin{aligned}
\|P_2(w_n, z_n) - P_2(w_{n-1}, z_{n-1})\|_2 = \epsilon_2 l_{S_2} \left(1 + n^{-1}\right) \|x_n - x_{n-1}\|_1 \\
+ \sigma_2 l_{T_2} \left(1 + n^{-1}\right) \|y_n - y_{n-1}\|_2.
\end{aligned} \tag{35}$$

It follows from (26), (31), (33) and (34) that

$$\begin{aligned}
\|g_1(x_{n+1}) - g_1(x_n)\|_1 \leq k_1 \left[L_1 + \rho_1 \left(1 + n^{-1}\right) \epsilon_1 l_{S_1}\right] \|x_n - x_{n-1}\|_1 \\
+ k_1 \rho_1 \left[\nu_1 + \left(1 + n^{-1}\right) \sigma_1 l_{T_1}\right] \|y_n - y_{n-1}\|_2.
\end{aligned} \tag{36}$$

Using (24) and (36), we have

$$\begin{aligned}
\|x_{n+1} - x_n\|_1 \leq \frac{\tau_1^{q_1 - 1} k_1}{\xi_{g_1}} \left[L_1 + \rho_1 \left(1 + n^{-1}\right) \epsilon_1 l_{S_1}\right] \|x_n - x_{n-1}\|_1 \\
+ \frac{\tau_1^{q_1 - 1} k_1 \rho_1}{\xi_{g_1}} \left[\nu_1 + \left(1 + n^{-1}\right) \sigma_1 l_{T_1}\right] \|y_n - y_{n-1}\|_2.
\end{aligned} \tag{37}$$

Let

$$\|x_{n+1} - x_n\|_1 \leq \theta_{1n} \|x_n - x_{n-1}\|_1 + \theta_{2n} \|y_n - y_{n-1}\|_2, \tag{38}$$

where

$$\theta_{1n} = \frac{\tau_1^{q_1 - 1} k_1}{\xi_{g_1}} \left[L_1 + \rho_1 \left(1 + n^{-1}\right) \epsilon_1 l_{S_1}\right],$$

and

$$\theta_{2n} = \frac{\tau_1^{q_1-1} k_1 \rho_1}{\xi_{g_1}} \left[\nu_1 + \left(1 + n^{-1}\right) \sigma_1 l_{T_1} \right].$$

By (14) and Theorem 3, we have

$$\begin{aligned}
\|g_2(y_{n+1}) - g_2(y_n)\|_2 = \|R^{H_2(.,.,.)-\eta_2}_{\rho_2,M_2}&\{H_2(A_2(g_2(y_n)), B_2(g_2(y_n)), C_2(g_2(y_n)))\\
&- \rho_2 N_2(x_n, y_n) - \rho_2 P_2(w_n, z_n)\}\\
&- [R^{H_2(.,.,.)-\eta_2}_{\rho_2,M_2}\{H_2(A_2(g_2(y_{n-1})), B_2(g_2(y_{n-1})), C_2(g_2(y_{n-1})))\\
&- \rho_2 N_2(x_{n-1}, y_{n-1}) - \rho_2 P_2(w_{n-1}, z_{n-1})\}]\|_2
\end{aligned}$$

$$\begin{aligned}
&\le k_2 \|H_2(A_2(g_2(y_n)), B_2(g_2(y_n)), C_2(g_2(y_n)))\\
&\quad - \rho_2 N_2(x_n, y_n) - \rho_2 P_2(w_n, z_n)\\
&\quad - \{H_2(A_2(g_2(y_{n-1})), B_2(g_2(y_{n-1})), C_2(g_2(y_{n-1})))\\
&\quad - \rho_2 N_2(x_{n-1}, y_{n-1}) - \rho_2 P_2(w_{n-1}, z_{n-1})\}\|_2\\
&= k_2 \|H_2(A_2(g_2(y_n)), B_2(g_2(y_n)), C_2(g_2(y_n)))\\
&\quad - H_2(A_2(g_2(y_{n-1})), B_2(g_2(y_{n-1})), C_2(g_2(y_{n-1})))\\
&\quad - \rho_2(N_2(x_n, y_n) - N_2(x_n, y_{n-1}))\\
&\quad - \rho_2(N_2(x_n, y_{n-1}) - N_2(x_{n-1}, y_{n-1}))\\
&\quad - \rho_2(P_2(w_n, z_n) - P_2(w_{n-1}, z_{n-1}))\|_2\\
&\le k_2 \|H_2(A_2(g_2(y_n)), B_2(g_2(y_n)), C_2(g_2(y_n)))\\
&\quad - H_2(A_2(g_2(y_{n-1})), B_2(g_2(y_{n-1})), C_2(g_2(y_{n-1})))\\
&\quad - \rho_2(N_2(x_n, y_n) - N_2(x_n, y_{n-1}))\|_2\\
&\quad + k_2 \rho_2 \|N_2(x_n, y_{n-1}) - N_2(x_{n-1}, y_{n-1})\|_2\\
&\quad + k_2 \rho_2 \|P_2(w_n, z_n) - P_2(w_{n-1}, z_{n-1})\|_2. \qquad (39)
\end{aligned}$$

By using Lemma 1, (28), (32), ξ_2-strongly accretivity of N_2 in the second component and τ_2-Lipschitz continuity of η_2, we have

$$\begin{aligned}
\|H_2(A_2(g_2(y_n)), B_2(g_2(y_n)), C_2(g_2(y_n))) &- H_2(A_2(g_2(y_{n-1})), B_2(g_2(y_{n-1})), C_2(g_2(y_{n-1})))\\
&- \rho_2(N_2(x_n, y_n) - N_2(x_n, y_{n-1}))\|_2^{q_2}\\
\le \|H_2(A_2(g_2(y_n)), &B_2(g_2(y_n)), C_2(g_2(y_n))) - H_2(A_2(g_2(y_{n-1})), B_2(g_2(y_{n-1})), C_2(g_2(y_{n-1})))\|_2^{q_2}\\
&- q_2 \rho_2 \langle N_2(x_n, y_n) - N_2(x_n, y_{n-1}), J_{q_2}(\eta_2(y_n, y_{n-1}))\rangle_2\\
&- q_2 \rho_2 \langle N_2(x_n, y_n) - N_2(x_n, y_{n-1}), J_{q_2}[H_2(A_2(g_2(y_n)), B_2(g_2(y_n)), C_2(g_2(y_n)))\\
&- H_2(A_2(g_2(y_{n-1})), B_2(g_2(y_{n-1})), C_2(g_2(y_{n-1})))] - J_{q_2}(\eta_2(y_n, y_{n-1}))\rangle_2\\
&+ \rho_2^{q_2} C_{q_2} \|N_2(x_n, y_n) - N_2(x_n, y_{n-1}))\|_2^{q_2}\\
\le (r_2 + s_2 + t_2)^{q_2} &\lambda_{g_2}^{q_2} \|y_n - y_{n-1}\|_2^{q_2} - q_2 \rho_2 \xi_2 \|y_n - y_{n-1}\|_2^{q_2}\\
&+ q_2 \rho_2 \|N_2(x_n, y_n) - N_2(x_n, y_{n-1})\|_2 \big[\|H_2(A_2(g_2(y_n)), B_2(g_2(y_n)), C_2(g_2(y_n)))\\
&- H_2(A_2(g_2(y_{n-1})), B_2(g_2(y_{n-1})), C_2(g_2(y_{n-1})))\|_2^{q_2-1} + \|\eta_2(y_n, y_{n-1})\|_2^{q_2-1}\big]\\
&+ \rho_2^{q_2} C_{q_2} \nu_2^{q_2} \|y_n - y_{n-1}\|_2^{q_2}
\end{aligned}$$

$$\begin{aligned}
&\leq (r_2+s_2+t_2)^{q_2}\lambda_{g_2}^{q_2}\|y_n-y_{n-1}\|_2^{q_2} - q_2\rho_2\xi_2\|y_n-y_{n-1}\|_2^{q_2} \\
&\quad + q_2\rho_2\nu_2\|y_n-y_{n-1}\|_2\big[(r_2+s_2+t_2)^{q_2-1}\lambda_{g_2}^{q_2-1}\|y_n-y_{n-1}\|_2^{q_2-1} \\
&\quad + \tau_2^{q_2-1}\|y_n-y_{n-1}\|_2^{q_2-1}\big] + \rho_2^{q_2}C_{q_2}\nu_2^{q_2}\|y_n-y_{n-1}\|_2^{q_2} \\
&= (r_2+s_2+t_2)^{q_2}\lambda_{g_2}^{q_2}\|y_n-y_{n-1}\|_2^{q_2} - q_2\rho_2\xi_2\|y_n-y_{n-1}\|_2^{q_2} \\
&\quad + q_2\rho_2\nu_2\big[(r_2+s_2+t_2)^{q_2-1}\lambda_{g_2}^{q_2-1} + \tau_2^{q_2-1}\big]\|y_n-y_{n-1}\|_2^{q_2} \\
&\quad + \rho_2^{q_2}C_{q_2}\nu_2^{q_2}\|y_n-y_{n-1}\|_2^{q_2} \\
&= \big[(r_2+s_2+t_2)^{q_2}\lambda_{g_2}^{q_2} - q_2\rho_2\xi_2 + q_2\rho_2\nu_2\{(r_2+s_2+t_2)^{q_2-1}\lambda_{g_2}^{q_2-1} + \tau_2^{q_2-1}\} \\
&\quad + \rho_2^{q_2}C_{q_2}\nu_2^{q_2}\big]\|y_n-y_{n-1}\|_2^{q_2}.
\end{aligned}$$

This implies that

$$\begin{aligned}
&\|H_2(A_2(g_2(y_n)),B_2(g_2(y_n)),C_2(g_2(y_n))) - H_2(A_2(g_2(y_{n-1})),B_2(g_2(y_{n-1})),C_2(g_2(y_{n-1}))) \\
&\qquad - \rho_2(N_2(x_n,y_n)-N_2(x_n,y_{n-1}))\|_2 \\
&\quad \leq \big[(r_2+s_2+t_2)^{q_2}\lambda_{g_2}^{q_2} - q_2\rho_2\xi_2 + q_2\rho_2\nu_2\{(r_2+s_2+t_2)^{q_2-1}\lambda_{g_2}^{q_2-1} + \tau_2^{q_2-1}\} \\
&\qquad + \rho_2^{q_2}C_{q_2}\nu_2^{q_2}\big]^{\frac{1}{q_2}}\|y_n-y_{n-1}\|_2 = L_2\|y_n-y_{n-1}\|_2,
\end{aligned} \tag{40}$$

where

$$\begin{aligned}
L_2 = \big[&(r_2+s_2+t_2)^{q_2}\lambda_{g_2}^{q_2} - q_2\rho_2\xi_2 + q_2\rho_2\nu_2\{(r_2+s_2+t_2)^{q_2-1}\lambda_{g_2}^{q_2-1} + \tau_2^{q_2-1}\} \\
&+ \rho_2^{q_2}C_{q_2}\nu_2^{q_2}\big]^{\frac{1}{q_2}}.
\end{aligned}$$

It follows from (30), (35), (39) and (40) that

$$\begin{aligned}
\|g_2(y_{n+1})-g_2(y_n)\|_2 &\leq k_2\rho_2\left[\zeta_2 + \left(1+n^{-1}\right)\epsilon_2 l_{S_2}\right]\|x_n-x_{n-1}\|_1 \\
&\quad + k_2\left[L_2 + \rho_2\left(1+n^{-1}\right)\sigma_2 l_{T_2}\right]\|y_n-y_{n-1}\|_2.
\end{aligned} \tag{41}$$

Using (25) and (41), we have

$$\begin{aligned}
\|y_{n+1}-y_n\|_2 &\leq \frac{\tau_2^{q_2-1}k_2\rho_2}{\xi_{g_2}}\left[\zeta_2 + \left(1+n^{-1}\right)\epsilon_2 l_{S_2}\right]\|x_n-x_{n-1}\|_1 \\
&\quad + \frac{\tau_2^{q_2-1}k_2}{\xi_{g_2}}\left[L_2 + \rho_2\left(1+n^{-1}\right)\sigma_2 l_{T_2}\right]\|y_n-y_{n-1}\|_2.
\end{aligned} \tag{42}$$

Let

$$\|y_{n+1}-y_n\|_2 \leq \theta_{3n}\|x_n-x_{n-1}\| + \theta_{4n}\|y_n-y_{n-1}\|, \tag{43}$$

where

$$\theta_{3n} = \frac{\tau_2^{q_2-1} k_2 \rho_2}{\xi_{g_2}} \left[\zeta_2 + \left(1 + n^{-1}\right) \epsilon_2 l_{S_2} \right],$$

and

$$\theta_{4n} = \frac{\tau_2^{q_2-1} k_2}{\xi_{g_2}} \left[L_2 + \rho_2 \left(1 + n^{-1}\right) \sigma_2 l_{T_2} \right].$$

On adding (38) and(43), we get

$$\begin{aligned} \|x_{n+1} - x_n\|_1 + \|y_{n+1} - y_n\|_2 &\leq (\theta_{1n} + \theta_{3n})\|x_n - x_{n-1}\|_1 + (\theta_{2n} + \theta_{4n})\|y_n - y_{n-1}\|_2, \\ &\leq \theta_n(\|x_n - x_{n-1}\|_1 + \|y_n - y_{n-1}\|_2), \end{aligned} \tag{44}$$

where

$$\theta_n = \max\{(\theta_{1n} + \theta_{3n}), (\theta_{2n} + \theta_{4n})\}.$$

Let

$$\theta = \max\{(\theta_1 + \theta_3), (\theta_2 + \theta_4)\},$$

where

$$\theta_1 = \frac{\tau_1^{q_1-1} k_1}{\xi_{g_1}} \left[L_1 + \rho_1 \epsilon_1 l_{S_1} \right], \theta_2 = \frac{\tau_1^{q_1-1} k_1 \rho_1}{\xi_{g_1}} \left[\nu_1 + \sigma_1 l_{T_1} \right],$$

and

$$\theta_3 = \frac{\tau_2^{q_2-1} k_2 \rho_2}{\xi_{g_2}} \left[\zeta_2 + \epsilon_2 l_{S_2} \right], \theta_4 = \frac{\tau_2^{q_2-1} k_2}{\xi_{g_2}} \left[L_2 + \rho_2 \sigma_2 l_{T_2} \right].$$

Here θ_n tend to θ, as n goes to ∞. By (19), $0 < \theta < 1$ and so (44) implies that $\{x_n\}$ and $\{y_n\}$ are Cauchy sequences. Then, $\exists x \in E_1$ and $y \in E_2$ such that x_n, y_n converges to x, y, respectively, as n goes ∞.

Now, we show that u_n converges to $u \in S_1(x)$, v_n converges to $v \in S_2(y)$, w_n converges to $w \in T_1(x)$ and z_n converges to $z \in T_2(y)$. From (20)–(23), the sequences $\{u_n\}$, $\{v_n\}$, $\{w_n\}$ and $\{z_n\}$ are Cauchy sequences. Hence, there $\exists\ u \in E_1, v \in E_2, w \in E_1, z \in E_2$ such that u_n converges to u, v_n converges to v, w_n converges to w and z_n converges to z as n goes to ∞.

Furthermore,

$$\begin{aligned} d(u, S_1(x)) &\leq \|u - u_n\|_1 + d(u_n, S_1(x)) \\ &\leq \|u - u_n\|_1 + D_1(S_1(x_n), S_1(x)) \\ &\leq \|u - u_n\|_1 + l_{S_1}\|x_n - x\|_1 \to 0, \text{ as } n \to \infty. \end{aligned}$$

Which shows that $d(u, S_1(x)) = 0$. Since $S_1(x) \in CB(E_1)$, we have $u \in S_1(x)$. Similarly, it is easy to observe that $v \in S_2(y), w \in T_1(x), z \in T_2(y)$. By continuity of $g_1, g_2, H_1, H_2, A_1, A_2, B_1, B_2, C_1, C_2, \eta_1, \eta_2, N_1,$ $N_2, P_1, P_2, R^{H_1(.,.,.)-\eta_1}_{\rho_1,M_1}$, $R^{H_2(.,.,.)-\eta_2}_{\rho_2,M_2}$ and Algorithm 1, the following relation satisfied by x, y, u, v, w, z that is

$$g_1(x) = R^{H_1(.,.,.)-\eta_1}_{\rho_1,M_1}(H_1(A_1(g_1(x)), B_1(g_1(x)), C_1(g_1(x))) - \rho_1 N_1(x, y) - \rho_1 P_1(u, v)),$$

$$g_2(y) = R^{H_2(.,.,.)-\eta_2}_{\rho_2,M_2}(H_2(A_2(g_2(y)), B_2(g_2(y)), C_2(g_2(y))) - \rho_2 N_2(x, y) - \rho_2 P_2(w, z)).$$

By Lemma 2, (x, y, u, v, w, z) is the solution of the system (9).

References

1. Ahmad, R., Dilshad, M., Wong, M. M., Yao, J. C.: $H(.,.)$-cocoercive operator and an application for solving generalized variational inclusions. Abstr. Appl. Anal. vol. 2011, Article ID 261534, 12 p (2011)
2. Aubin, J.P., Cellina, A.: Differential Inclusions. Springer, Berlin, Germany (1984)
3. Fang, Y.P., Huang, N.J.: H-monotone operator and resolvent operator technique for variational inclusions. Appl. Math. Comput. **145**(2–3), 795–803 (2003)
4. Fang, Y.P., Huang, N.J.: H-monotone operators and system of variational inclusions. Commun. Appl. Nonlinear Anal. **11**(1), 93–101 (2004)
5. Fang, Y.P., Huang, N.J.: H-accretive operators and resolvent operator technique for solving variational inclusions in Banach spaces. Appl. Math. Lett. **17**, 647–653 (2004)
6. Fang, Y.P., Huang, N.J., Thompson, H.B.: A new system of variational inclusions with (H, η)-monotone operators in Hilbert spaces. Comput. Math. Appl. **49**(2–3), 365–374 (2005)
7. Huang, N.J.: Nonlinear implicit quasi-variational inclusions involving generalized m-accretive mappings. Arch. Inequal. Appl. **2**(4), 413–425 (2004)
8. Huang, N.J., Fang, Y.P.: Generalized m-accretive mappings in Banach spaces. J. Sichuan Univ. **38**(4), 591–592 (2001)
9. Huang, N.J., Fang, Y.P.: A new class of generalized variational inclusions involving maximal η-monotone mappings. Publ. Math. Debr. **62**(1–2), 83–98 (2003)
10. Husain, S., Gupta, S.: Algorithm for solving a new system of generalized variational inclusions in Hilbert spaces. J. Calc. Var. vol. 2013, Article ID 461371, 8 p. (2013)
11. Husain, S., Gupta, S., Mishra, V.N.: Graph convergence for the $H(.,)$-mixed mapping with an application for solving the system of generalized variational inclusions. Fixed Point Theory Appl. **2013**, 304 (2013)
12. Husain, S., Gupta, S., Mishra, V. N.: Generalized $H(.,..,.)$-η- cocoercive operators and generalized set-valued variational- like inclusions. J. Math. vol. 2013, Article ID 738491, 10 p. (2013)
13. Husain, S., Gupta, S., Sahper, H.: Algorithm for solving a new system of generalized nonlinear quasi-variational-like inclusions in Hilbert spaces. Chin. J. Math. vol. 2014, Article ID 957482, 7 p. (2014)
14. Kazmi, K.R., Khan, F.A.: Iterative approximation of a unique solution of a system of variational-like inclusions in real q-uniformly smooth Banach spaces. Nonlinear Anal. **67**, 917–929 (2007)
15. Lan, H.Y., Cho, Y.J., Verma, R.U.: Nonlinear relaxed cocoercive variational inclusions involving (A, η)-accretive mappings in banach spaces. Comput. Math. Appl. **51**, 1529–1538 (2006)

16. Lan, H.Y., Kim, J.H., Cho, Y.J.: On a new system of nonlinear A-monotone multivalued variational inclusions. J. Math. Anal. Appl. **327**, 481–493 (2007)
17. Nadler, S.B.: Multivalued contraction mapping. Pac. J. Math. **3**(3), 457–488 (1969)
18. Peng, J., Zhu, D.: A new system of generalized mixed-quasi-variational inclusions with (H, η)-monotone operators. J. Math. Anal. Appl. **327**, 175–187 (2007)
19. Verma, R.U.: Generalized nonlinear variational inclusion problems involving A-monotone mappings. Appl. Math. Lett. **19**(9), 960–963 (2006)
20. Verma, R.U.: Sensitivity analysis for generalized strongly monotone variational inclusions based on the (A, η)-resolvent operator technique. Appl. Math. Lett. **19**(12), 1409–1413 (2006)
21. Verma, R.U.: General system of (A, η)-monotone variational inclusion problems based on generalized hybrid iterative algorithm. Nonlinear Anal.: Hybrid Syst. **1**(3), 326–335 (2007)
22. Wang, Z.B., Ding X.P.: $(H(.,.)\text{-}\eta)$-accretive operators with an application for solving set-valued variational inclusions in Banach spaces. Comput. Math. Appl. **59**, 1559–1567 (2010)
23. Xu, H.K.: Inequalities in Banach spaces with applications. Nonlinear Anal. **16**(12), 1127–1138 (1991)
24. Zhang, Q.B.: Generalized implicit variational-like inclusion problems involving G-η-monotone mappings. Appl. Math. Lett. **20**(2), 216–221 (2007)
25. Zou, Y.Z., Huang, N.J.: $H(.,)$-accretive operator with an application for solving variational inclusions in Banach spaces. Appl. Math. Comput. **204**(2), 809–816 (2008)

Some Integral Inequalities for Log-Preinvex Functions

Akhlad Iqbal and V. Samhita

Abstract In this paper, a new type of Hermite–Hadamard inequalities is established for log-preinvex functions. Some natural applications to special means of real numbers are also discussed.

Keywords Invex sets · Preinvex functions · Log-preinvex functions · Hermite–Hadamard inequality

Mathematics Subject Classification 26D10 · 26D15 · 26D99

1 Introduction

Log-preinvex functions are nonconvex functions and are the generalized class of log-convex functions. Log-convex functions play an important role in the theory of special functions and mathematical statistics [9, 17, 18, 21]. Let I be a closed interval. A real valued function $f : I \to R$ is said to be convex on I if $f(tx + (1 - t)y) \leq tf(x) + (1 - t)f(y)$ for all $x, y \in I$ and $t \in [0, 1]$. The well-known classical Hermite–Hadamard inequality

$$f\left(\frac{a+b}{2}\right) \leq \frac{1}{b-a}\int_a^b f(x)dx \leq \frac{f(a)+f(b)}{2}$$

gives us an estimate of the mean value of a convex function $f : I \to R$.

A. Iqbal (✉)
Department of Mathematics, Aligarh Muslim University, Aligarh 202002, India
e-mail: akhlad6star@gmail.com; akhlad.mm@amu.ac.in

V. Samhita
Department of Mathematics, BITS Pilani Hyderabad Campus,
Hyderabad, Telangana, India
e-mail: vadrevu.samhita@gmail.com

J.M. Cushing et al. (eds.), *Applied Analysis in Biological and Physical Sciences*,
Springer Proceedings in Mathematics & Statistics 186,
DOI 10.1007/978-81-322-3640-5_23

In recent years, some refinements and several generalizations of the Hermite–Hadamard Inequality have been extensively investigated by the many authors, see [4–8, 10–13, 15, 22, 24]. Dragomir and Aggarwal [7] proved two inequalities for differentiable convex mappings.

An important and significant generalization of convexity is invexity, which was introduced by Hanson [14] in 1981. Weir and Mond [23] introduced the concept of preinvex functions while Jeyakumar [16] studied the properties of preinvex functions and their role in optimization and mathematical programming. Ahmad et al. [2] introduced the concept of geodesic η-preinvex functions on Riemannian manifolds.

The main purpose of this paper is to establish some new refined inequalities of Hermite–Hadamard's type for log-preinvex functions. Applications to special means have also been considered.

2 Preliminaries

Definition 2.1 ([14]) A set $S \subseteq R^n$ is said to be invex with respect to $\eta : S \times S \to R^n$ if for every $x, y \in S$ and $t \in [0, 1]$

$$y + t\eta(x, y) \in S \tag{1}$$

Every convex set is invex for $\eta(x, y) = x - y$ but converse need not be true. Let $S \subseteq R^n$ be an invex set with respect to $\eta : S \times S \to R^n$. For every $x, y \in S$ in the η-path P_{xv} joining the points x and $v := x + \eta(y, x)$ is defined as follows

$$P_{xv} := \{z : z = x + t\eta(y, x) : t \in [0, 1]\}.$$

Definition 2.2 ([23]) Let $S \subseteq R^n$ be an invex set with respect to $\eta : S \times S \to R^n$. Then, the function $f : S \to R$ is said to be preinvex with respect to η, if for every $x, y \in S$ and $t \in [0, 1]$,

$$f(y + t\eta(x, y)) \leq tf(x) + (1 - t)f(y). \tag{2}$$

Every convex function is a preinvex function but the converse is not true. For example, the function $f(x) = - \mid x \mid$ is not a convex function but it is a preinvex function with respect to η, where

$$\eta(x, y) := \begin{cases} x - y, & \text{if } x \leq 0, y \leq 0 \text{ and } x \geq 0, y \geq 0, \\ y - x, & \text{otherwise.} \end{cases}$$

The concept of invex and preinvex functions have played a very important role in the development of generalized convexities.

Definition 2.3 ([20]) A function $f : I \to (0, \infty)$ on the invex set S is said to be logarithmic preinvex with respect to η, if

$$f(y + t\eta(x, y)) \leq (f(x))^t (f(y))^{(1-t)}, \quad \text{for every } x, y \in S,\ t \in [0, 1] \tag{3}$$

Note that for $\eta(x, y) = x - y$, the invex set S reduces to convex set and consequently log-preinvex function to log-convex function.

3 Main Results

In this section, we generalize Hermite–Hadamard inequality for log-preinvex functions. Barani et al. [3] generalized the result of Dragomir and Aggarwal [7] as follows:

Lemma 3.1 ([3]) *Let $A \subset R$ be an open invex subset with respect to $\theta : AXA \to R$ and $a, b \in A$ with $\theta(a, b) \neq 0$. Suppose that $f : A \to R$ is a differentiable function. If f' is integrable on the θ path P_{bc}, $c = b + \theta(a, b)$, then the following equality holds*

$$-\frac{f(b) + f(b + \theta(a, b))}{2} + \frac{1}{\theta(a, b)} \int_b^{b+\theta(a,b)} f(x)dx = \frac{\theta(a, b)}{2} \int_0^1 (1 - 2t) f'(b + t\theta(a, b))dt$$

Using Lemma 3.1, we prove the following theorem.

Theorem 3.1 *Let $A \subset R$ be an open invex subset with respect to $\theta : AXA \to R$ and $f : A \to R$ be a differentiable function. If $|f'|$ is log-preinvex on A then for every $a, b \in A$ with $\theta(a, b) \neq 0$ the following inequality holds*

$$\left| \frac{f(b) + f(b + \theta(a, b))}{2} - \frac{1}{\theta(a, b)} \int_b^{b+\theta(a,b)} f(x)dx \right|$$

$$\leq \frac{\theta(a, b)}{2} \left[\frac{|f'(a)| - |f'(b)|}{log|f'(a)| - log|f'(b)|} - 2\left(\frac{\sqrt{|f'(a)|} - \sqrt{|f'(b)|}}{log|f'(a)| - log|f'(b)|} \right)^2 \right]$$

Proof Using Lemma 3.1 and log preinvexity of $|f'|$, we get

$$\left| \frac{f(b)+f(b+\theta(a,b))}{2} - \frac{1}{\theta(a,b)} \int_b^{b+\theta(a,b)} f(x)dx \right| = \left| \frac{\theta(a,b)}{2} \int_0^1 (1 - 2t) f'(b + t\theta(a, b))dt \right|$$

$$\leq \frac{\theta(a,b)}{2} \int_0^1 |1 - 2t|\, |f'(b)|^{1-t} |f'(a)|^t dt$$

Now integrating by parts, the above result is obtained.

Theorem 3.2 *Let $A \subset R$ be an open invex subset with respect to $\theta : AXA \to R$ and $f : A \to R$ be a differentiable function. Suppose that $p \in R$ with $p > 1$. If $|f'|^{p/(p-1)}$ is log-preinvex on A then for every $a, b \in A$ with $\theta(a, b) \neq 0$ the following inequality holds*

$$\left| \frac{f(b) + f(b + \theta(a, b))}{2} - \frac{1}{\theta(a, b)} \int_b^{b+\theta(a,b)} f(x)dx \right| \leq \frac{|\theta(a, b)|}{2(p + 1)^{1/p}} \left[\frac{|f'(a)^q - |f'(b)|^q}{q(log|f'(a)| - log|f'(b)|)} \right]^{1/q}$$

Proof Using Lemma 3.1 and Holders' Inequality, we get

$$\begin{aligned} \left| \tfrac{f(b)+f(b+\theta(a,b))}{2} - \tfrac{1}{\theta(a,b)} \int_b^{b+\theta(a,b)} f(x)dx \right| &= \left| \tfrac{\theta(a,b)}{2} \int_0^1 (1-2t) f'(b + t\theta(a, b))dt \right| \\ &\leq \tfrac{\theta(a,b)|}{2} \left[\int_0^1 |1 - 2t|^p dt \right]^{1/p} \\ &\quad \times \left[\int_0^1 |f'(b + t\theta(a, b)|^q dt \right]^{1/q} \\ &= \tfrac{|\theta(a,b)|}{2(p+1)^{1/p}} \left[\int_0^1 |f'(b + t\theta(a, b))|^q dt \right]^{1/q} \\ &\leq \tfrac{|\theta(a,b)|}{2(p+1)^{1/p}} \left[\int_0^1 |f'(b)|^{q(1-t)} |f'(a)|^{qt} dt \right]^{1/q} \\ &= \tfrac{|\theta(a,b)|}{2(p+1)^{1/p}} \left[\tfrac{|f'(a)|^q - |f'(b)|^q}{q(log|f'(a)| - log|f'(b)|)} \right]^{1/q}, \end{aligned}$$

where $\frac{1}{p} + \frac{1}{q} = 1$. This completes the proof.

It is to be noted that if $A = [a, b]$ and $\theta(x, y) = x - y$ for every $x, y \in A$ then, these results of log-preinvex will be converted to that of log-convex.

Mohan and Neogy [19] gives the following condition C.

Condition C The mapping $\eta : S \times S \to R^n$ is said to satisfy the condition C if for every $x, y \in S$ and $t \in [0, 1]$,

$$\eta(y, y + t\eta(x, y)) = -t\eta(x, y),$$

$$\eta(x, y + t\eta(x, y)) = (1 - t)\eta(x, y)$$

Now, from condition C, for every $x, y \in S$ and every $t_1, t_2 \in [0, 1]$, we have

$$\eta(y + t_2\eta(x, y), y + t_1\eta(x, y)) = (t_2 - t_1)\eta(x, y).$$

Lemma 3.2 *Let $S \subseteq R^n$ be an invex set with respect to $\eta : SXS \to R^n$ and $f : S \to R$ be a function. Suppose that η satisfies the condition C on S. Then for every $x, y \in S$*

the function f is log-preinvex with respect to η on the η path P_{xv} if and only if the function $\phi : [0, 1] \to R$ defined by $\phi(t) := f(x + t\eta(y, x))$ is log convex on [0, 1].

Proof Suppose that ϕ is log convex on [0, 1] and $z_1 := x + t_1\eta(y, x) \in P_{xv}$, $z_2 := x + t_2\eta(y, x) \in P_{xv}$. Fix $\lambda \in [0, 1]$. Since η satisfies condition C,

$$\begin{aligned} f(z_1 + \lambda\eta(z_2, z_1)) &= f(x + ((1-\lambda)t_1 + \lambda t_2)\eta(y, x)) \\ &= \phi((1 - \lambda t_1) + \lambda t_2) \\ &= (\phi(t_1)^{1-\lambda})(\phi(t_2)^{\lambda}) \\ &= f(z_1)^{1-\lambda} f(z_2)^{\lambda}. \end{aligned}$$

Hence, f is log preinvex with respect to η on the η path P_{xv}.

Conversely, let $x, y \in S$ and the function f is log preinvex with respect to η on the η path P_{xv}. Suppose that η satisfies the condition C and $t_1, t_2 \in [0, 1]$. Then for every $\lambda \in [0, 1]$ we have,

$$\begin{aligned} \phi((1-\lambda)t_1 + \lambda t_2) &= f(x + ((1-\lambda)t_1 + \lambda t_2)\eta(y, x)) \\ &\le f(x + t_1\eta(y, x) + \lambda\eta(x + t_2\eta(y, x), x + t_1\eta(y, x)) \\ &\le f(x + t_1\eta(y, x))^{1-\lambda} f(x + t_2\eta(y, x))^{\lambda} \\ &= \phi(t_1)^{1-\lambda}\phi(t_2)^{\lambda}. \end{aligned}$$

Therefore, ϕ is log convex on [0, 1].

Lemma 3.3 *Assume that $a, b \in R$ with $a < b$ and $f : [a, b] \to R$ is a differentiable function on (a, b). If $|f'|$ is log convex on [a, b] then the following inequality holds true*

$$\left|\frac{f(a) + f(b)}{2} - \frac{1}{b-a}\int_a^b f(x)dx\right| \le \frac{b-a}{2}\left[\frac{|f'(a)| - |f'(b)|}{log|f'(a)| - log|f'(b)|} - 2\left(\frac{\sqrt{|f'(a)|} - \sqrt{|f'(b)|}}{log|f'(a)| - log|f'(b)|}\right)^2\right]$$

Proof Since $|f'|$ is log convex, using Lemma 3.1, we get

$$\begin{aligned} \left|\tfrac{f(b)+f(a)}{2} - \tfrac{1}{b-a}\int_b^{b+\theta(a,b)} f(x)dx\right| &= \left|\tfrac{\theta(a,b)}{2}\int_0^1 (1-2t)f'(ta + (1-t)b)dt\right| \\ &\le \tfrac{b-a}{2}\int_0^1 |1-2t|\,|f'(b)|^{1-t}|f'(a)|^t dt. \end{aligned}$$

Now integrating by parts,the above result is obtained.

Theorem 3.3 *Let $S \subseteq R^n$ be an invex set with respect to $\eta : SXS \to R^n$. Suppose that η satisfies the condition C on S. Suppose that every $x, y \in S$ the function $f : S \to R^+$ is log preinvex with respect to η on the η path P_{xv}. Then for every $(a, b) \in (0, 1)$ with $a < b$ the following inequality holds*

$$\left|1/2\int_0^a f(x+s\eta(y,x))ds + 1/2\int_0^b f(x+s\eta(y,x))ds - \frac{1}{b-a}\int_a^b\int_0^s f(x+s\eta(y,x))dtds\right|$$
$$\leq (b-a)\left[\frac{f(x+b\eta(y,x)) - f(x+a\eta(y,x)) - f(x+b\eta(y,x))f(x+a\eta(y,x))}{log(f(x+b\eta(y,x))) - log(f(x+a\eta(y,x)))}\right]^{1/2}$$
$$\leq \frac{b-a}{2}\left[\frac{|f(x+a\eta(y,x))| - |f(x+b\eta(y,x))|}{log|f(x+a\eta(y,x))| - log|f(x+b\eta(y,x))|}\right.$$
$$\left. -2\left(\frac{\sqrt{|f(x+a\eta(y,x))|} - \sqrt{|f(x+b\eta(y,x))|}}{log|f(x+a\eta(y,x))| - log|f(x+b\eta(y,x))|}\right)^2\right]$$

Proof Let $x, y \in$ S and a,b $\in$ (0,1) with $a < b$. Since f is log-preinvex with respect to η on the η path P_{xv} by Lemma 3.3 the function $\phi : [0, 1] \to R^+$ defined by

$$\phi(t) := f(x+t\eta(y,x))$$

is log convex on [0, 1]. Now we define the function $\varphi : [0, 1] \to R^+$ as follows

$$\varphi(t) := \int_0^t \phi(t)ds = \int_0^t f(x+s\eta(y,x))ds$$

Obviously for every $t \in (0, 1)$ we have

$$\varphi'(t) = \phi(t) = f(x+t\eta(y,x)) \geq 0.$$

Applying Lemma 3.3 to the function φ implies that

$$\left|\frac{\varphi(a)+\varphi(b)}{2} - \frac{1}{b-a}\int_a^b \varphi(s)ds\right|$$

$$\leq \frac{b-a}{2}\left[\frac{|\varphi'(a)| - |\varphi'(b)|}{log|\varphi'(a)| - log|\varphi'(b)|} - 2\left(\frac{\sqrt{|\varphi'(a)|} - \sqrt{|\varphi'(b)|}}{log|\varphi'(a)| - log|\varphi'(b)|}\right)^2\right]$$

and hence we deduce that the above theorem holds.

4 Inequalities for Second Order Differentiable Functions

Now, we derive some results for functions whose second order derivative absolute values are log-preinvex. The following Lemma was given by Barani et al. [3].

Lemma 4.1 ([3]) *Let $A \subset R$ be an open invex subset with respect to $\theta : AXA \to R$ and $a, b \in A$ with $\theta(a, b) \neq 0$. Suppose that $f : A \to R$ is a differentiable function. If f'' is integrable on the θ path P_{bc}, $c = b + \theta(a, b)$, then the following equality holds*

$$\frac{f(b) + f(b + \theta(a, b))}{2} - \frac{1}{\theta(a, b)} \int_b^{b+\theta(a,b)} f(x)dx = \frac{\theta(a, b)^2}{2} \int_0^1 t(1 - t) f''(b + t\theta(a, b))dt$$

Using Lemma 4.1, we prove the following result.

Theorem 4.1 *Let $A \subset R$ be an open invex subset with respect to $\theta : AXA \to R$ and $a, b \in A$ with $\theta(a, b) \neq 0$. Suppose that $f : A \to R$ is a twice differentiable function on A. If $|f''|$ is log-preinvex and f'' integrable on the θ path P_{bc}, $c = b + \theta(a, b)$, then the following inequality holds*

$$\left| \frac{f(b)+f(b+\theta(a,b))}{2} - \frac{1}{\theta(a,b)} \int_b^{b+\theta(a,b)} f(x)dx \right|$$

$$\leq \frac{\theta(a, b)^2}{2} \left[\frac{|f''(a)| + |f''(b)|}{(log|f''(a)| - log|f''(b)|)^2} + \frac{2(|f''(b)| - |f''(a)|)}{(log|f''(a)| - log|f''(b)|)^3} \right]$$

Proof From Lemma 4.1, we have

$$\begin{aligned}
\left| \frac{f(b)+f(b+\theta(a,b))}{2} - \frac{1}{\theta(a,b)} \int_b^{b+\theta(a,b)} f(x)dx \right| &= \left| \frac{\theta(a,b)}{2}^2 \int_0^1 t(1-t) f''(b + t\theta(a, b))dt \right| \\
&\leq \frac{\theta(a,b)}{2}^2 \int_0^1 t(1-t)|f''(b)|^{1-t}|f''(a)|^t dt \\
&= \frac{\theta(a,b)^2}{2} \left[\frac{|f''(a)|+|f''(b)|}{(log|f''(a)|-log|f''(b)|)^2} \right. \\
&\quad \left. + \frac{2(|f''(b)|-|f''(a)|)}{(log|f''(a)|-log|f''(b)|)^3} \right]
\end{aligned}$$

which completes the proof.

Theorem 4.2 *Let $A \subset R$ be an open invex subset with respect to $\theta : AXA \to R$ and $a, b \in A$ with $\theta(a, b) \neq 0$. Suppose that $f : A \to R$ is a twice differentiable function on A. If $|f''|^{p/p-1}$ is log-preinvex on A for $p > 1$. If f'' integrable on the θ path P_{bc}, $c = b + \theta(a, b)$, then the following inequality holds*

$$\left| \frac{f(b) + f(b + \theta(a, b))}{2} - \frac{1}{\theta(a, b)} \int_b^{b+\theta(a,b)} f(x)dx \right|$$

$$\leq \frac{\theta(a,b)^2}{16}\sqrt{\pi}^{-1/p}\frac{\Gamma(p+1)}{\Gamma(p+3/2)}^{1/p}\left(\frac{|f''(a)|^q-|f''(b)|^q}{q(log|f''(a)|-log|f''(b)|)}\right)^{1/q}$$

Proof By log-preinvexity of $|f''|^q$, Lemma 4.1 and Holder's inequality, we get

$$\left|\frac{f(b)+f(b+\theta(a,b))}{2}-\frac{1}{\theta(a,b)}\int_b^{b+\theta(a,b)}f(x)dx\right| \leq \frac{\theta(a,b)}{2}^2\int_0^1 t(1-t)f''(b+t\theta(a,b))dt$$

$$\leq \frac{\theta(a,b)^2}{2}\left(\int_0^1(t-t^2)^p dt\right)^{1/p}$$

$$\times\left(\int_0^1|f''(b+\theta(a,b)|^q dt\right)^{1/q}$$

$$\leq \frac{\theta(a,b)}{2}^2\left[\frac{2^{-1-2p}\sqrt{\pi}\Gamma(1+p)}{\Gamma(3/2+p)}\right]^{1/p}$$

$$\times\left[\int_0^1|f''(b)|^{q(1-t)}|f''(a)|^{qt}dt\right]^{1/q}$$

$$\leq \frac{\theta(a,b)^2}{16}\sqrt{\pi}^{-1/p}\frac{\Gamma(p+1)}{\Gamma(p+3/2)}^{1/p}$$

$$\times\left(\frac{|f''(a)|^q-|f''(b)|^q}{q(log|f''(a)|-log|f''(b)|)}\right)^{1/q}$$

which completes the proof.

Theorem 4.3 *Let $A \subset R$ be an open invex subset with respect to $\theta : AXA \to R$ and $a, b \in A$ with $\theta(a,b) \neq 0$. Suppose that $f : A \to R$ is a twice differentiable function on A. If $|f''|^q$ is log-preinvex on A for $q > 1$ and f'' integrable on the θ path P_{bc}, $c = b + \theta(a,b)$,then the following inequality holds*

$$\frac{f(b)+f(b+\theta(a,b))}{2}-\frac{1}{\theta(a,b)}\int_b^{b+\theta(a,b)}f(x)dx$$

$$\leq \frac{\theta(a,b)^2}{12}(6)^{1/q}\left[\frac{|f''(a)|^q+|f''(b)|^q}{(q(log|f''(a)|-log|f''(b)|))^2}+\frac{2(|f''(b)|^q-|f''(a)|^q)}{(q(log|f''(a)|-log|f''(b)|))^3}\right]^{1/q}$$

Proof By Lemma 4.1 and using the well known weighted power mean inequality, we get

$$\left|\frac{f(b)+f(b+\theta(a,b))}{2}-\frac{1}{\theta(a,b)}\int_b^{b+\theta(a,b)}f(x)dx\right|$$

$$\leq \frac{\theta(a,b)}{2}^2 \int_0^1 t(1-t)|f''(b+t\theta(a,b))|dt$$

$$\leq \frac{\theta(a,b)^2}{2}\left(\int_0^1 (t-t^2)dt\right)^{1-1/q}\left(\int_0^1 (t-t^2)|f''(b+\theta(a,b)|^q dt\right)^{1/q}$$

$$\leq \frac{\theta(a,b)^2}{2}(1/6)^{1-1/q}\left[\int_0^1 (t-t^2)|f''(b)|^{q(1-t)}|f''(a)|^{qt}dt\right]^{1/q}$$

$$\leq \frac{\theta(a,b)^2}{12}(6)^{1/q}\left[\frac{|f''(a)|^q+|f''(b)|^q}{(q(log|f''(a)|-log|f''(b)|))^2}+\frac{2(|f''(b)|^q-|f''(a)|^q)}{(q(log|f''(a)|-log|f''(b)|))^3}\right]^{1/q}$$

which completes the proof.

5 Applications to Special Means

Let a and b be two positive numbers. We have:
Arithmetic mean

$$A(a,b)=\frac{a+b}{2},$$

Logarithmic mean

$$L_p(a,b)=\frac{a-b}{lna-lnb},$$

and generalized logarithmic mean

$$L_p(a,b)=\left[\frac{a^{p+1}-b^{p+1}}{(p+1)(a-b)}\right]^{1/p}, \quad p\neq -1,0;$$

There are several results connecting these means, see [1] for some new relations; however very few results are known for arbitrary real numbers. For this, it is clear that we can extend some of the above means as follows:

$$A(\alpha,\beta)=\frac{\alpha+\beta}{2}, \qquad \alpha,\beta\in R$$

$$\bar{L}(\alpha,\beta)=\frac{\beta-\alpha}{ln|\beta|-ln|\alpha|}, \qquad \alpha,\beta\in R\backslash 0$$

$$L_n(\alpha,\beta)=\left[\frac{\beta^{n+1}-\alpha^{n+1}}{(n+1)(\beta-\alpha)}\right]^{\frac{1}{n}} \qquad n\in N,\ n\geq 1,\ \alpha,\beta\in R,\ \alpha<\beta$$

Now using the results of Sects. 3 and 4, we prove the following inequalities connecting the above means for arbitrary real numbers.

Proposition 5.1 *Let $a, b \in R$, $a < b$ and $n \in N$, $n \geq 2$. Then the following inequality holds:*

$$|A(a^n, b^n) - (L_n(a, b))^n| \leq \frac{n|a-b|}{2}\left[\bar{L}(|b^{n-1}|, |a^{n-1}|) - \frac{1}{2}(\bar{L}(\sqrt{|b^{n-1}|}, \sqrt{|a^{n-1}|}))^2\right]$$

Proof The proof is immediate from Theorem 3.1 applied for $f(x) = x^n$, $x \in R$, $\theta(a, b) = a - b$.

Proposition 5.2 *Let $a, b \in R$, $a < b$ and $n \in N$, $n \geq 2$. Then, for all $p > 1$, the following inequality holds:*

$$|A(a^n, b^n) - L_n^n(a, b)| \leq \frac{n|a-b|}{2(p+1)^{1/p}}\left[\bar{L}(|b^{n-1}|^q, |a^{n-1}|^q)\right]^{\frac{1}{q}}$$

Proof The proof is immediate from Theorem 3.2 applied for $f(x) = x^n$, $x \in R$, $\theta(a, b) = a - b$.

Proposition 5.3 *Let $a, b \in R\backslash 0$, $a < b$. Then the following inequality holds:*

$$|A\left(\frac{1}{a}, \frac{1}{b}\right) - \bar{L}^{-1}(a, b)| \leq \frac{|a-b|}{2}\left[L\left(\frac{1}{b^2}, \frac{1}{a^2}\right) - \frac{1}{2}(\bar{L}(|\frac{1}{b}|, |\frac{1}{a}|))^2\right]$$

Proof The proof is obvious from Theorem 3.1 applied for $f(x) = \frac{1}{x}$, $x \in [a, b]$, $\theta(a, b) = a - b$.

Proposition 5.4 *Let $a, b \in R\backslash 0$, $a < b$. Then for all $p > 1$, the following inequality holds:*

$$|A\left(\frac{1}{a}, \frac{1}{b}\right) - \bar{L}^{-1}(a, b)| \leq \frac{|a-b|}{2(p+1)^{1/p}}\left[L\left(\left(\frac{1}{b^2}\right)^q, \left(\frac{1}{a^2}\right)^q\right)\right]^{\frac{1}{q}}$$

Proof The proof is obvious from Theorem 3.2 applied for $f(x) = \frac{1}{x}$, $x \in [a, b]$, $\theta(a, b) = a - b$.

Proposition 5.5 *Let $a, b \in R$, $a < b$ and $n \in N$, $n \geq 2$. Then the following inequality holds*

$$\frac{1}{(n+1)(n+2)}\left|A(a^{n+2}, b^{n+2}) - L_{n+2}^{n+2}(a, b)\right| \leq \frac{1}{n^2}\bar{L}(a, b)^2\left[A(|a^n|, |b^n|) - L(|a^n| - |b^n|)\right]$$

Proof The proof is immediate from Theorem 4.1 applied for $f(x) = \frac{x^{n+2}}{(n+1)(n+2)}$, $x \in R$, $\theta(a, b) = a - b$.

Proposition 5.6 *Let $a, b \in R$, $a < b$ and $n \in N$, $n \geq 2$. Then, for all $p > 1$, the following inequality holds*

$$\frac{1}{(n+1)(n+2)}\left|A(a^{n+2}, b^{n+2}) - L_{n+2}^{n+2}(a,b)\right| \leq \frac{\theta(a,b)^2}{16}\sqrt{\pi}^{1/p}\frac{\Gamma(p+1)}{\Gamma(p+3/2)}^{1/p}\left(L(|a^n|^q, |b^n|^q)\right)^{\frac{1}{q}}$$

Proof The proof is immediate from Theorem 4.2 applied for $f(x) = \frac{x^{n+2}}{(n+1)(n+2)}$, $\in R$, $\theta(a,b) = a - b$.

Proposition 5.7 *Let $a, b \in R\backslash 0$, $a < b$. Then the following inequality holds*

$$\left|A(alog|a|, blog|b|) + \frac{1}{2}A(a,b) - \frac{b^2log|b| - a^2log|a|}{2(b-a)}\right| \leq L^2(a,b)\left[A(|\frac{1}{a}|, |\frac{1}{b}|) - L(|\frac{1}{a}|, |\frac{1}{b}|)\right]$$

Proof The proof is obvious from Theorem 4.1 applied for $f(x) = xlogx - x$, $x \in [a,b]$, $\theta(a,b) = a - b$.

Proposition 5.8 *Let $a, b \in R\backslash 0$, $a < b$. Then for all $p > 1$, the following inequality holds*

$$\left|A(alog|a|, blog|b|) + \frac{1}{2}A(a,b) - \frac{b^2log|b| - a^2log|a|}{2(b-a)}\right| \leq \frac{\theta(a,b)^2}{16}\sqrt{\pi}^{1/p}\frac{\Gamma(p+1)}{\Gamma(p+3/2)}^{1/p}$$
$$\times\left[L\left(\left|\frac{1}{b^q}\right|^q, \left|\frac{1}{a^q}\right|^q\right)\right]^{\frac{1}{q}}$$

Proof The proof is obvious from Theorem 4.2 applied for $f(x) = xlogx - x$, $x \in [a,b]$, $\theta(a,b) = a - b$.

Acknowledgements The first author acknowledges that this work is supported by University Grant Commission under grant No. F.20-23/2013 (BSR).

References

1. Agarwal, R.P., Dragomir, S.S.: An application of Hayashi's inequality for differentiable functions. Comput. Math. Appl. **32**(6), 95–99 (1996)
2. Ahmad, I., Iqbal, A., Ali, S.: On properties of Geodesic η-preinvex functions. Adv. Oper. Res. Article ID 381831, 10 p. (2009)
3. Barani, A., Ghazanfari, A.G., Dragomir, S.S.: Hermite-Hadamard inequality for functions whose derivatives absolute values are preinvex. J. Inequal. Appl. **2012**, 247 (2012)
4. Bombardelli, M., Varosanec, S.: Properties of h-convex functions related to the Hermite-Hadamard-Fejer inequalities. Comput. Math. Appl. **58**, 1869–1877 (2010)
5. Dragomir, S.S.: Two mappings in connection to Hadamard's inequalities. J. Math. Anal. Appl. **167**, 49–56 (1992)
6. Dragomir, S.S.: On Hadamards inequalities for convex functions. Math. Balk. **6**, 215–222 (1992)
7. Dragomir, S.S., Agarwal, R.P.: Two inequalities for differentiable mappings and applications to special means of real numbers and to trapezoidal formula. Appl. Math. Lett. **11**, 91–95 (1998)

8. Dragomir, S.S., Pearce, C.E.M.: Quasi-convex functions and Hadamard's inequality. Bull. Aust. Math. Soc. **57**(3), 377–385 (1998)
9. Dragomir, S.S., Pearce, C.E.M.: Selected Topics on Hermite-Hadamard Inequalities and Applications. RGMIA Monographs, Victoria University (2000)
10. Dragomir, S.S., Pecaric, J.E., Sandor, J.: A note on the Jensen-Hadamard inequality. Anal. Numr. Thor. Approx. **19**, 29–34 (1990)
11. Dragomir, S.S., Pecaric, J.E., Persson, L.E.: Some inequalities of Hadamard type. Soochow J. Math. **21**, 335–341 (1995)
12. Dragomir, S.S., Cho, Y.J., Kim, S.S.: Inequalities of Hadamard's type for Lipschitzian mappings and their applications. J. Math. Anal. Appl. **245**, 489–501 (2000)
13. Hadjisavvas, N.: Hadamard-type inequalities for quasi-convex functions. J. Inequal. Pure Appl. Math. **4**(1) article 13 (2003)
14. Hanson, M.A.: On sufficiency of the Kuhn-Tucker conditions. J. Math. Anal. Appl. **80**, 545–550 (1981)
15. Ion, D.A.: Some estimates on the Hermite-Hadamard inequality through quasi-convex functions, Annals of University of Craiova, Math. Comput. Sci. Ser. **34**, 19–22 (2007)
16. Jeyakumar, V.: Strong and weak invexity in mathematical programming. Methods Oper. Res. **55**, 109–125 (1985)
17. Manfrino, R.B., Delgado, R.V., Ortega, J.A.G.: Inequalities a Mathematical Olympiad Approach. Birkhauser, Basel (2009)
18. Mitrinovic, D.S., Pecaric, J.E., Fink, A.M.: Classical and New Inequalities in Analysis. Kluwer Academic Publishers, Dordrecht (1993)
19. Mohan, S.R., Neogy, S.K.: On invex sets and preinvex function. J. Math. Anal. Appl. **189**, 901–908 (1995)
20. Noor, M.A.: Variational-like inequalities. Optimization **30**, 323–330 (1994)
21. Peciaric, J.E., Proschan, F., Tong, Y.L.: Convex Functions, Partial Orderings and Statistical Applications. Academic Press, New York (1991)
22. Qaisar, S., Hussain, S., Chuanjiang, He.: On new inequalities of Hermite-Hadamard type for functions whose third derivatives absolute values are quasi-convex with applications. J. Egypt. Math. Soc. **22**, 137–146 (2014)
23. Weir, T., Mond, B.: Preinvex functions in multiobjective optimization. J. Math. Anal. Appl. **136**, 29–38 (1988)
24. Yang, G.S., Hwang, D.Y., Tseng, K.L.: Some inequalities for differentiable convex and concave mappings. Comput. Math. Appl. **47**, 207–216 (2004)

Estimates for Initial Coefficients of Certain Starlike Functions with Respect to Symmetric Points

Kanika Khatter, V. Ravichandran and S. Sivaprasad Kumar

Abstract It is well known that the class of all analytic functions f defined on the unit disk satisfying $\Re(zf'(z)/(f(z)-f(-z)))>0$ is a subclass of close-to-convex functions and the nth Taylor coefficient of these functions are bounded by one. However, no bounds for the nth coefficients of functions f satisfying $2zf'(z)/(f(z)-f(-z)) \prec \varphi(z)$ are known except for $n=2,3$. The sharp bounds for the fourth coefficient of analytic univalent functions f satisfying the subordination $2zf'(z)/(f(z)-f(-z)) \prec \varphi(z)$ is obtained. The bound for the fifth coefficients is also obtained in certain special cases including φ is e^z and $\sqrt{1+z}$.

Keywords Starlike function · Coefficient estimates · Subordination

2010 Mathematics Subject Classification 30C45 · 30C80

1 Introduction

Let $\mathscr{A}$ be the class of all normalized analytic functions of the form $f(z)=z+\sum_{n=2}^{\infty}a_nz^n$ in the open unit disk $\mathbb{D}:=\{z\in\mathbb{C}:|z|<1\}$ and $\mathscr{S}$ be the subclass of $\mathscr{A}$ consisting of univalent functions. A function $f\in\mathscr{A}$ is starlike with respect to symmetric points in $\mathbb{D}$ if for every r less than and sufficiently close to one and every ζ on $|z|=r$, the angular velocity of $f(z)$ about the point $f(-\zeta)$ is positive at $z=\zeta$ as z traverses the circle $|z|=r$ in the positive direction. Analytically, a function $f\in\mathscr{A}$ is starlike with respect to symmetric points if

K. Khatter (✉) · S. Sivaprasad Kumar
Department of Applied Mathematics, Delhi Technological University, Delhi 110 085, India
e-mail: kanika.khatter@yahoo.com

S. Sivaprasad Kumar
e-mail: spkumar@dce.ac.in

V. Ravichandran
Department of Mathematics, University of Delhi, Delhi 110 007, India
e-mail: vravi68@gmail.com

J.M. Cushing et al. (eds.), *Applied Analysis in Biological and Physical Sciences*, Springer Proceedings in Mathematics & Statistics 186,
DOI 10.1007/978-81-322-3640-5_24

$$\operatorname{Re}\left(\frac{zf'(z)}{f(z)-f(-z)}\right)>0 \quad (z\in\mathbb{D}).$$

The class $\mathcal{S}_s^*$ of all functions starlike with respect to symmetric points was introduced and studied by Sakaguchi [11]. The functions belonging to this class are close-to-convex and hence univalent. It is well known that this class $\mathcal{S}_s^*$ includes the class of convex functions CV and the class of odd starlike functions [11]; the functions in $\mathcal{S}_s^*$ also satisfy the sharp coefficient inequality $|a_n|\le 1$, see [3] and [9] for other related classes. An analytic function f is said to be subordinate to F, written $f\prec F$ or $f(z)\prec F(z)\quad(z\in\mathbb{D})$ if there exists an analytic function $w:\mathbb{D}\to\mathbb{D}$ satisfying $w(0)=0$ and $f(z)=F(w(z))$ in $|z|<1$. Let φ be a univalent function with positive real part which maps $\mathbb{D}$ onto a domain which is symmetric with respect to the real line and starlike with respect to $\varphi(0)=1$ and $\varphi'(0)>0$. Let $\mathcal{S}^*(\varphi)$ be the class of functions $f\in\mathcal{S}$ for which $zf'(z)/f(z)\prec\varphi(z)$ and $\mathcal{C}(\varphi)$ be the class of functions $f\in\mathcal{S}$ for which $1+zf''(z)/f'(z)\prec\varphi(z)$. The above classes were introduced and studied by Ma and Minda [4]. Also, for such φ, Ravichandran [9] introduced the following class:

$$S_s^*(\varphi)=\left\{f\in\mathcal{S}:\frac{2zf'(z)}{f(z)-f(-z)}\prec\varphi(z)\right\}$$

and later in [12], the sharp bound for the Fekete–Szegö coefficient functional $|a_3-\mu a_2^2|$ were obtained. This immediately gives the bound for the first two coefficients of functions in the above classes.

In this paper, our aim is to determine the bound for the fourth coefficient of functions belonging to the class $S_s^*(\varphi)$. This is done by first expressing the coefficients of f in terms of the coefficients B_n of φ and the coefficient c_n of a function with a positive real part. The coefficient estimate for a_4 follows from a result of Prokhorov and Szynal [8]. The bound for the fifth coefficient of functions in $S_s^*(\varphi)$ is highly non-linear. We are able to estimate a_5 in certain important special cases of φ:

$$\mathcal{S}_{s,e}^*:=S_s^*(e^z),\quad \mathcal{S}_{s,L}^*:=S_s^*(\sqrt{1+z}),\quad\text{and}$$
$$\mathcal{S}_{s,RL}^*:=S_s^*\left(\sqrt{2}-(\sqrt{2}-1)\sqrt{\frac{1-z}{1+2(\sqrt{2}-1)z}}\right).$$

These classes are analogues of the corresponding classes of starlike functions introduced and studied respectively in [6, 7, 13].

Let $\mathcal{P}$ be the calss of all analytic functions $p(z)=1+\sum_{n=0}^{\infty}c_nz^n$ with Re $p(z)>0$ for $z\in\mathbb{D}$ and Ω be the class of all analytic functions $w:\mathbb{D}\to\mathbb{D}$ of the form $w(z)=w_1z+w_2z^2+\cdots$. To prove our results, we need the following results; the results in (a)–(c) of Lemma 1 are respectively in [1, 4, 10] while we refer the reader to [8] for the very lengthy expression of $H(q_1,q_2)$.

Lemma 1 *Let $p(z) = 1 + \sum_{n=0}^{\infty} c_n z^n \in \mathscr{P}$. Then,*

(a) $|c_2 - \upsilon c_1^2| \leq 2\max\{1, |2\upsilon - 1|\}$,
(b) $|c_3 - 2\beta c_1 c_2 + \delta c_1^3| \leq 2$ *if* $0 \leq \beta \leq 1$ *and* $\beta(2\beta - 1) \leq \delta \leq \beta$.
(c) $|\gamma c_1^4 + a c_2^2 + 2\alpha c_1 c_3 - (3/2)\beta c_1^2 c_2 - c_4| \leq 2,$ *when* $0 < \alpha < 1,\ 0 < a < 1$ *and* $8a(1-a)((\alpha\beta - 2\gamma)^2 + (\alpha(a+\alpha) - \beta)^2) + \alpha(1-\alpha)(\beta - 2a\alpha)^2 \leq 4\alpha^2 (1-\alpha)^2 a(1-a)$.

Lemma 2 ([8]) *Let $H(q_1, q_2)$ be as in [8]. If $w \in \Omega$, then for any real numbers q_1 and q_2, the following sharp estimate $|w_3 + q_1 w_1 w_2 + q_2 w_1^3| \leq H(q_1, q_2)$ holds, where*

$$H(q_1, q_2) = \begin{cases} 1 & \text{for } (q_1, q_2) \in D_1 \cup D_2 \\ |q_2| & \text{for } (q_1, q_2) \in \cup_{k=3}^{7} D_k \\ \frac{2}{3}(|q_1| + 1)\left(\frac{|q_1|+1}{3(|q_1|+1+q_2)}\right)^{\frac{1}{2}} & \text{for } (q_1, q_2) \in D_8 \cup D_9 \\ \frac{1}{3} q_2 \left(\frac{q_1^2-4}{q_1^2-4q_2}\right)\left(\frac{q_1^2-4}{3(q_2-1)}\right)^{\frac{1}{2}} & \text{for } (q_1, q_2) \in D_{10} \cup D_{11} - \{\pm 2, 1\} \\ \frac{2}{3}(|q_1| - 1)\left(\frac{|q_1|-1}{3(|q_1|-1-q_2)}\right)^{\frac{1}{2}} & \text{for } (q_1, q_2) \in D_{12} \end{cases} \tag{1}$$

The extremal functions, up to rotations, are of the form

$$w(z) = z^3, \quad w(z) = z, \quad w(z) = w_0(z) = \frac{z([(1-\lambda)\varepsilon_2 + \lambda\varepsilon_1] - \varepsilon_1\varepsilon_2 z)}{1 - [(1-\lambda)\varepsilon_1 + \lambda\varepsilon_2]z},$$

$$w(z) = w_1(z) = \frac{z(t_1 - z)}{1 - t_1 z}, \quad w(z) = w_2(z) = \frac{z(t_2 + z)}{1 + t_2 z}$$

$$|\varepsilon_1| = |\varepsilon_2| = 1, \quad \varepsilon_1 = t_0 - e^{\frac{-i\theta_0}{2}}(a \mp b), \quad \varepsilon_2 = -e^{\frac{-i\theta_0}{2}}(ia \pm b),$$

$$a = t_0 \cos\frac{\theta_0}{2}, \quad b = \sqrt{1 - t_0^2 \sin^2\frac{\theta_0}{2}}, \quad \lambda = \frac{b \pm a}{2b}$$

$$t_0 = \left[\frac{2q_2(q_1^2+2) - 3q_1^2}{3(q_2-1)(q_1^2-4q_2)}\right]^{\frac{1}{2}}, \quad t_1 = \left(\frac{|q_1|+1}{3(|q_1|+1+q_2)}\right)^{\frac{1}{2}},$$

$$t_2 = \left(\frac{|q_1|-1}{3(|q_1|-1-q_2)}\right)^{\frac{1}{2}}, \quad \cos\frac{\theta_0}{2} = \frac{q_1}{2}\left[\frac{q_2(q_1^2+8) - 2(q_1^2+2)}{2q_2(q_1^2+2) - 3q_1^2}\right].$$

The sets $D_k,\ k = 1, 2, \ldots, 12$, are defined as follows

$$
\begin{aligned}
D_1 &= \{(q_1, q_2) : |q_1| \le \frac{1}{2}, |q_2| \le 1\}, \\
D_2 &= \{(q_1, q_2) : \frac{1}{2} \le |q_1| \le 2, \frac{4}{27}(|q_1|+1)^3 - (|q_1|+1) \le q_2 \le 1\}, \\
D_3 &= \{(q_1, q_2) : |q_1| \le \frac{1}{2}, q_2 \le -1\}, \\
D_4 &= \{(q_1, q_2) : |q_1| \ge \frac{1}{2}, q_2 \le -\frac{2}{3}(|q_1|+1)\}, \\
D_5 &= \{(q_1, q_2) : |q_1| \le 2, q_2 \ge 1\}, \\
D_6 &= \{(q_1, q_2) : 2 \le |q_1| \le 4, q_2 \ge \frac{1}{12}(q_1^2+8)\}, \\
D_7 &= \{(q_1, q_2) : |q_1| \ge 4, q_2 \ge \frac{2}{3}(|q_1|-1)\}, \\
D_8 &= \{(q_1, q_2) : \frac{1}{2} \le |q_1| \le 2, -\frac{2}{3}(|q_1|+1) \le q_2 \le \frac{4}{27}(|q_1|+1)^3 - (|q_1|+1)\}, \\
D_9 &= \{(q_1, q_2) : |q_1| \ge 2, -\frac{2}{3}(|q_1|+1) \le q_2 \le \frac{2|q_1|(|q_1|+1)}{q_1^2+2|q_1|+4}\}, \\
D_{10} &= \{(q_1, q_2) : 2 \le |q_1| \le 4, \frac{2|q_1|(|q_1|+1)}{q_1^2+2|q_1|+4} \le q_2 \le \frac{1}{12}(q_1^2+8)\}, \\
D_{11} &= \{(q_1, q_2) : |q_1| \ge 4, \frac{2|q_1|(|q_1|+1)}{q_1^2+2|q_1|+4} \le q_2 \le \frac{2|q_1|(|q_1|-1)}{q_1^2-2|q_1|+4}\}, \\
D_{12} &= \{(q_1, q_2) : |q_1| \ge 4, \frac{2|q_1|(|q_1|-1)}{q_1^2-2|q_1|+4} \le q_2 \le \frac{2}{3}(|q_1|-1)\}.
\end{aligned}
$$

2 Main Results

By making use of Lemma 2, we prove the following bound for the fourth coefficient for functions in the class $S_s^*(\varphi)$.

Theorem 1 *Let the function* $f(z) = z + a_2z^2 + a_3z^3 + \cdots \in \mathcal{S}_s^*(\varphi)$ *where* $\varphi(z) = 1 + B_1z + B_2z^2 + B_3z^3 + \cdots$. *Then*

$$|a_4| \le \frac{B_1}{4} H(q_1, q_2),$$

where $H(q_1, q_2)$ *is as defined in* (1),

$$q_1 := \frac{4B_2 + B_1^2}{2B_1}; \qquad q_2 := \frac{2B_3 + B_1B_2}{2B_1}. \tag{2}$$

Proof If $f \in S_s^*(\varphi)$, then there is an analytic function $w(z) = w_1 z + w_2 z^2 + \cdots \in \Omega$ such that

$$\frac{2zf'(z)}{f(z) - f(-z)} = \varphi(w(z)). \tag{3}$$

Since

$$\frac{2zf'(z)}{f(z) - f(-z)} = 1 + 2a_2 z + 2a_3 z^2 + (-2a_2 a_3 + 4a_4)z^3 + (-2a_3^2 + 4a_5)z^4 + \cdots$$

and

$$\varphi(w(z)) = 1 + B_1 w_1 z + (B_2 w_1^2 + B_1 w_2)z^2 + (B_3 w_1^3 + 2B_2 w_1 w_2 + B_1 w_3)z^3 + \cdots$$

we get, from (3),

$$\begin{aligned}
a_2 &= \frac{1}{2} B_1 w_1, \\
a_3 &= \frac{1}{2}(B_2 w_1^2 + B_1 w_2), \\
a_4 &= \frac{1}{4}\left(\left(B_3 + \frac{1}{2} B_1 B_2\right) w_1^3 + \left(2B_2 + \frac{1}{2} B_1^2\right) w_1 w_2 + B_1 w_3\right).
\end{aligned}$$

The coefficient a_4 can be rewritten as

$$a_4 = \frac{B_1}{4}(w_3 + q_1 w_1 w_2 + q_2 w_1^3)$$

where q_1 and q_2 are given by 2. Lemma 2 immediately yields the desired estimate $|a_4| \le B_1 H(q_1, q_2)/4$.

Our next theorem provides sharp bound on $|a_5|$ for three different choices of φ. The bounds for a_2, a_3, a_4 are also included here for completeness.

Theorem 2 *Let* $f(z) = z + a_2 z^2 + a_3 z^3 + \cdots$.

(a) If $f \in \mathscr{S}_{s,L}^*$, *then*

$$|a_2| \le \frac{1}{4}, \quad |a_3| \le \frac{1}{4}, \quad |a_4| \le \frac{1}{8} \quad \textit{and} \quad |a_5| \le \frac{1}{8}.$$

(b) If $f \in \mathscr{S}_{s,RL}^*$, *then*

$$|a_2| \le \frac{5}{4} - \frac{3}{2\sqrt{2}}, \quad |a_3| \le \frac{5}{4} - \frac{3}{2\sqrt{2}}, \quad |a_4| \le \frac{5}{8} - \frac{3}{4\sqrt{2}} \quad \textit{and} \quad |a_5| \le \frac{5}{8} - \frac{3}{4\sqrt{2}}.$$

(c) If $f \in \mathcal{S}^*_{s,e}$, then

$$|a_2| \leq \frac{1}{2}, \quad |a_3| \leq \frac{1}{2}, \quad |a_4| \leq \frac{1}{4} \quad \text{and} \quad |a_5| \leq \frac{1}{4}.$$

All the bounds here are sharp.

Proof For the function $f(z) = z + a_2z^2 + a_3z^3 + \cdots \in \mathcal{S}_s^*(\varphi)$, we first express a_n in terms of the coefficients of the functions $\varphi(z) = 1 + B_1z + B_2z^2 + \cdots$ and $p_1(z) = 1 + c_1z + c_2z^2 + \cdots \in \mathcal{P}$. In terms of the coefficients b_n of the function p defined by

$$p(z) := \frac{2zf'(z)}{f(z) - f(-z)} = 1 + b_1z + b_2z^2 + \cdots,$$

the coefficients a_n are expressed by

$$na_n = \sum_{k=1}^{\lceil n/2 \rceil} b_{n+1-2k}a_{2k-1}. \tag{4}$$

From Eq. (4), we have

$$2a_2 = b_1, \quad 2a_3 = b_2, \quad -2a_2a_3 + 4a_4 = b_3, \quad -2a_3^2 + 4a_5 = b_4. \tag{5}$$

Since φ is univalent and $p \prec \varphi$, the function

$$p_1(z) = \frac{1 + \varphi^{-1}(p(z))}{1 - \varphi^{-1}(p(z))} = 1 + c_1z + c_2z^2 + \cdots$$

belongs to $\mathcal{P}$. Equivalently,

$$p(z) = \varphi\left(\frac{p_1(z) - 1}{p_1(z) + 1}\right).$$

Using the last equation, we obtain each b_i's in terms of c_i's and B_i's as follows:

$$\begin{aligned}
b_1 &= \frac{1}{2}B_1c_1, \\
b_2 &= \frac{1}{4}\left((B_2 - B_1)c_1^2 + 2B_1c_2\right), \\
b_3 &= \frac{1}{8}\left((B_1 - 2B_2 + B_3)c_1^3 + 4(B_2 - B_1)c_1c_2 + 4B_1c_3\right), \\
b_4 &= \frac{1}{16}\Big((-B_1 + 3B_2 - 3B_3 + B_4)c_1^4 + 6(B_3 - 2B_2 + B_1)c_1^2c_2 \\
&\quad + 4(B_2 - B_1)c_2^2 + 8(B_2 - B_1)c_1c_3 + 8B_1c_4\Big).
\end{aligned} \tag{6}$$

Thus, from (5) and (6), we get

$$a_2 = \frac{1}{4} B_1 c_1, \tag{7}$$
$$a_3 = \frac{1}{8} \left((B_2 - B_1) c_1^2 + 2 B_1 c_2 \right), \tag{8}$$
$$a_4 = \frac{1}{64} \left((2B_1 - B_1^2 - 4B_2 + B_1 B_2 + 2B_3) c_1^3 + (2B_1^2 + 8B_2 - 8B_1) c_1 c_2 + 8 B_1 c_3 \right) \tag{9}$$

and

$$\begin{aligned} a_5 = \frac{1}{128} \Big(& (-2B_1 + B_1^2 + 6B_2 - 2B_1 B_2 + B_2^2 - 6B_3 + 2B_4) c_1^4 \\ & + (12B_1 - 4B_1^2 - 24B_2 + 4B_1 B_2) c_1^2 c_2 + (4B_1^2 + 8B_2 - 8B_1) c_2^2 \\ & + (16B_2 - 16B_1) c_1 c_3 + 16 B_1 c_4 \Big). \end{aligned} \tag{10}$$

(a) Let $f \in \mathscr{S}^*_{s,L}$. Then,

$$\varphi(z) = \sqrt{1+z} = 1 + \frac{1}{2} z - \frac{1}{8} z^2 + \frac{1}{16} z^3 - \frac{5}{128} z^4 + \cdots .$$

Thus $B_1 = 1/2$, $B_2 = -1/8$, $B_3 = 1/16$ and $B_4 = -5/128$. On substituting these values in (7), (8), (9) and (10), we get

$$\begin{aligned} a_2 &= \frac{1}{8} c_1, \\ a_3 &= \frac{1}{64} (-5c_1^2 + 8c_2), \\ a_4 &= \frac{1}{1024} (21 c_1^3 - 72 c_1 c_2 + 64 c_3), \\ a_5 &= \frac{1}{8192} (-116 c_1^4 + 544 c_1^2 c_2 - 256 c_2^2 - 640 c_1 c_3 + 512 c_4). \end{aligned}$$

Since $|c_n| \leq 2$ for $n \geq 1$, we have $|a_2| \leq 1/4$. By using Lemma 1(a) we obtain $|a_3| \leq 1/4$. Since

$$a_4 = \frac{1}{16} (c_3 - 2\beta c_1 c_2 + \delta c_1^3)$$

where $\beta = 9/16$ and $\delta = 21/64$, Lemma 1(b) shows that $|a_4| \leq 1/8$. Similarly,

$$a_5 = \frac{1}{16} (\gamma c_1^4 + a c_2^2 + 2\alpha c_1 c_3 - (3/2) \beta c_1^2 c_2 - c_4),$$

where $\gamma = 29/128$, $a = 1/2$, $\alpha = 5/8$, $\beta = 17/24$. Lemma 1(c) shows that $|a_5| \leq 1/8$. Define the functions f_k $(k = 1, 2, \ldots)$ by

$$\frac{2zf_k'(z)}{f_k(z) - f_k(-z)} = \sqrt{1+z^k} = 1 + \frac{z^k}{2} - \frac{z^{2k}}{8} + \frac{z^{3k}}{16} + \cdots, \quad (f_k(0) = 0,\ f_k'(0) = 1).$$

Then

$$f_1(z) = z + \frac{1}{4}z^2 + \cdots, \qquad f_2(z) = z + \frac{1}{4}z^3 + \cdots,$$

$$f_3(z) = z + \frac{1}{8}z^4 + \cdots, \text{ and } \quad f_4(z) = z + \frac{1}{8}z^5 + \cdots.$$

Clearly the functions $f_k \in \mathcal{S}^*_{s,L}$. Moreover the kth coefficient is sharp for f_{k-1} where $k = 2, 3, 4, 5$.

(b) Let $f \in \mathcal{S}^*_{s,RL}$. Then,

$$\begin{aligned}\varphi(z) &= \sqrt{2} - (\sqrt{2}-1)\sqrt{\frac{1-z}{1+2(\sqrt{2}-1)z}} \\ &= 1 + \frac{5-3\sqrt{2}}{2}z + \frac{71-51\sqrt{2}}{8}z^2 + \frac{589-415\sqrt{2}}{16}z^3 + \frac{20043-14179\sqrt{2}}{128}z^4 + \cdots.\end{aligned}$$

Thus $B_1 = (5-3\sqrt{2})/2$, $B_2 = (71-51\sqrt{2})/8$, $B_3 = (589-415\sqrt{2})/16$ and $B_4 = (20043-14179\sqrt{2})/128$. Using these values in (7), (8), (9) and (10), we get

$$\begin{aligned}a_2 &= \frac{1}{8}(-1+\sqrt{2})(-c_1 + 2\sqrt{2}c_1), \\ a_3 &= \frac{1}{64}(-1+\sqrt{2})\left((-27+12\sqrt{2})c_1^2 + (-8+16\sqrt{2})c_2\right), \\ a_4 &= \frac{1}{1024}\left((1179-818\sqrt{2})c_1^3 + 8(145-108\sqrt{2})c_1c_2 + 64(5-3\sqrt{2})c_3\right)\end{aligned}$$

and

$$\begin{aligned}a_5 = \frac{1}{8192}&((14638-10453\sqrt{2})c_1^4 - 48(-508+351\sqrt{2})c_1^2c_2 \\ &- 64(-94+69\sqrt{2})c_2^2 - 384(-17+13\sqrt{2})c_1c_3 - 512(-5+3\sqrt{2})c_4).\end{aligned}$$

Since $|c_n| \leq 2$ for $n \geq 1$, we have

$$|a_2| \leq \frac{5}{4} - \frac{3}{2\sqrt{2}}.$$

Use of Lemma 1(a) shows that

$$|a_3| \leq \frac{5}{4} - \frac{3}{2\sqrt{2}}.$$

Since

$$a_4 = \frac{5-3\sqrt{2}}{16}\left(c_3 - 2\beta c_1 c_2 + \delta c_1^3\right)$$

where

$$\beta = (108\sqrt{2}-145)/(16(5-3\sqrt{2})), \quad \delta = (1179 - 818\sqrt{2})/(64(5-3\sqrt{2})).$$

Lemma 1(b) shows that

$$|a_4| \leq \frac{5}{8} - \frac{3}{4\sqrt{2}}.$$

Similarly, a_5 can be rewritten as

$$a_5 = \frac{-5+3\sqrt{2}}{8}\left(\gamma c_1^4 + a c_2^2 + 2\alpha c_1 c_3 - (3/2)\beta c_1^2 c_2 - c_4\right)$$

where

$$\gamma = (14638 - 10453\sqrt{2})/(512(-5+3\sqrt{2})), \quad a = (94 - 69\sqrt{2})/(8(-5+3\sqrt{2})),$$
$$\alpha = (3(17-3\sqrt{2}))/(8(-5+3\sqrt{2})) \quad \text{and} \quad \beta = (-508+351\sqrt{2})/(16(-5+3\sqrt{2})).$$

Lemma 1(c) shows that

$$|a_5| \leq \frac{5}{8} - \frac{3}{4\sqrt{2}}.$$

Define the functions f_k $(k = 1, 2, \ldots)$ by

$$\frac{2zf_k'(z)}{f_k(z) - f_k(-z)} = \sqrt{2} - (\sqrt{2}-1)\sqrt{\frac{1-z^k}{1+2(\sqrt{2}-1)z^k}} \quad (f_k(0) = 0,\ f_k'(0) = 1).$$

Then

$$f_1(z) = z + \frac{5-3\sqrt{2}}{4}z^2 + \cdots, \qquad f_2(z) = z + \frac{5-3\sqrt{2}}{4}z^3 + \cdots,$$

$$f_3(z) = z + \frac{5-3\sqrt{2}}{8}z^4 + \cdots, \text{ and } \quad f_4(z) = z + \frac{5-3\sqrt{2}}{8}z^5 + \cdots.$$

Clearly, the functions $f_k \in \mathcal{S}^*_{s,RL}$ and the kth coefficient is sharp for f_{k-1} $(k = 2, 3, 4, 5)$
(c) Let $f \in \mathcal{S}^*_{s,e}$. Then,

$$q(z) = e^z = 1 + z + \frac{z^2}{2} + \frac{z^3}{6} + \frac{z^4}{24} + \frac{z^5}{120} + \cdots.$$

Thus, $B_1 = 1$, $B_2 = 1/2$, $B_3 = 1/6$ and $B_4 = 1/24$. Using these values in (7), (8), (9) and (10), we get

$$\begin{aligned} a_2 &= \frac{1}{4}c_1, \\ a_3 &= \frac{1}{16}(-c_1^2 + 4c_2), \\ a_4 &= \frac{1}{384}(-c_1^3 - 12c_1c_2 + 48c_3) \end{aligned}$$

and

$$a_5 = \frac{1}{384}(c_1^4 - 24c_1^2c_3 + 48c_4).$$

Since $|c_n| \leq 2$ for $n \geq 1$, therefore $|a_2| \leq 1/2$. Use of Lemma 1(a) shows that $|a_3| \leq 1/2$. Since

$$a_4 = \frac{1}{8}(c_3 - 2\beta c_1c_2 + \delta c_1^3)$$

where $\beta = 1/8$ and $\delta = -1/48$, Lemma 1(b) shows that $|a_4| \leq 1/4$. Similarly,

$$a_5 = \frac{1}{8}(\gamma c_1^4 + ac_2^2 + 2\alpha c_1c_3 - (3/2)\beta c_1^2c_2 - c_4)$$

where $\gamma = -1/48$, $a = 0$, $\alpha = 1/4$, $\beta = 0$. Note that Lemma 2 holds for $a = 0$. Applying Lemma 1(c) with $a = 0$, we have $|a_5| \leq 1/4$. Define the functions f_k $(k = 1, 2, \ldots)$ by

$$\frac{2zf_k'(z)}{f_k(z) - f_k(-z)} = e^{kz} = 1 + z^k + \frac{z^{2k}}{2!} + \frac{z^{3k}}{3!} + \cdots \quad (f_k(0) = 0,\ f_k'(0) = 1).$$

Then

$$f_1(z) = z + \frac{1}{2}z^2 + \cdots \qquad f_2(z) = z + \frac{1}{2}z^3 + \cdots,$$

$$f_3(z) = z + \frac{1}{4}z^4 + \cdots, \text{ and } \ f_4(z) = z + \frac{1}{4}z^5 + \cdots.$$

Clearly the functions $f_k \in \mathscr{S}^*_{s,e}$. Clearly the kth coefficient is sharp for f_{k-1} for $k = 2, 3, 4, 5$.

References

1. Ali, R.M.: Coefficients of the inverse of strongly starlike functions. Bull. Malays. Math. Sci. Soc. (2) **26**(1), 63–71 (2003)
2. Ali, R.M., Ravichandran, V., Seenivasagan, N.: Coefficient bounds for p-valent functions. Appl. Math. Comput. **187**(1), 35–46 (2007)
3. Das, R.N., Singh, P.: On subclasses of schlicht mapping. Indian J. Pure Appl. Math. **8**(8), 864–872 (1977)
4. Ma, W.C., Minda, D.: A unified treatment of some special classes of univalent functions. In: Proceedings of the Conference on Complex Analysis, pp. 157–169. Tianjin (1992). (Conference Proceedings and Lecture Notes in Analysis, I). International Press, Cambridge, MA
5. Ma, W.C., Minda, D.: Uniformly convex functions. II. Ann. Polon. Math. **58**(3), 275–285 (1993)
6. Mendiratta, R., Nagpal, S., Ravichandran, V.: A subclass of starlike functions associated with left-half of the lemniscate of Bernoulli. Int. J. Math. **25**(9) 1450090 (2014)
7. Mendiratta, R., Nagpal, S., Ravichandran, V.: On a subclass of strongly starlike functions associated with exponential function. Bull. Malays. Math. Sci. Soc. **38**(1), 365–386 (2015)
8. Prokhorov, D.V., Szynal, J.: Inverse coefficients for (α, β)-convex functions. Ann. Univ. Mariae Curie-Skłodowska Sect. A **35**(1981), 125–143 (1984)
9. Ravichandran, V.: Starlike and convex functions with respect to conjugate points. Acta Math. Acad. Paedagog. Nyházi. (N.S.) **20**(1), 31–37 (2004)
10. Ravichandran, V., Verma, S.: Bound for the fifth coefficient of starlike functions. C. R. Acad. Sci. Paris. Ser. **I**(353), 505–510 (2015)
11. Sakaguchi, K.: On a certain univalent mapping. J. Math. Soc. Jpn. **11**, 72–75 (1959)
12. Shanmugam, T.N., Ramachandran, C., Ravichandran, V.: Fekete-Szegő problem for subclasses of starlike functions with respect to symmetric points. Bull. Korean Math. Soc. **43**(3), 589–598 (2006)
13. Sokół, J., Stankiewicz, J.: Radius of convexity of some subclasses of strongly starlike functions. Zeszyty Nauk. Politech. Rzeszowskiej Mat. No. **19**, 101–105 (1996)

Note on Convex Functionals in the Dual Spaces of Nonreflexive Banach Spaces

Yuqing Chen, Yeol Je Cho and Jong Kyu Kim

Abstract Let E be a real nonreflexive Banach space, E^* be the dual space of E and $\phi : D(\phi) \subseteq E^* \to R \cup \{+\infty\}$ be a convex functional. Comparing with the classical Legendre-Fenchel conjugate, we define the weak* conjugate of ϕ as

$$\phi_*(x) = \sup_{p \in D(\phi)} \{(p, x) - \phi(p)\}$$

and also, comparing with the classical sub-differential, we define the weak* sub-differential of ϕ as

$$\partial^* \phi(p) = \{x \in E^* : \phi(q) - \phi(p) \geq (q - p, x), \ \ \forall \, q \in D(\phi)\}.$$

In this paper, we study their properties and relationships with the classical conjugate and the sub-differential.

Keywords Legendre-Fenchel conjugate · Weak* conjugate · Classical sub-differential · Weak* sub-differential

2010 Mathematics Subject Classification 55U30 · 39B62

Y.Q. Chen
College of Applied Mathematics Guangdong University of Technology, Guangzhou 510006, Guangdong, People's Republic of China
e-mail: ychen64@163.com

Y.J. Cho
Department of Mathematics Education, The Research Institute of Natural Sciences, Gyongsang National University, Jinju 660-701, Korea
e-mail: yjcho@gnu.ac.kr

J.K. Kim (✉)
Department of Mathematics Education, Kyungnam University, Changwon 51767, Korea
e-mail: jongkyuk@kyungnam.ac.kr

J.M. Cushing et al. (eds.), *Applied Analysis in Biological and Physical Sciences*, Springer Proceedings in Mathematics & Statistics 186,
DOI 10.1007/978-81-322-3640-5_25

1 Introduction

Convex functionals play important roles in variational problems, the existence problems of partial differential equations, control and optimize problems and variational inequality problems, etc., and it has been extensively studied by many authors, especially, the theory of convex functionals has been well developed (see [4–6, 8, 13–15, 21, 22, 25, 26, 28]). In recent years, abstract convex functional has also been studied by Bartz and Reich [7] and Jeyakumar et al. [16].

Let E be a nonreflexive Banach space, E^* be the dual space of E and E^{**} be the second dual space of E. In this paper, we further study the convex functional defined on the dual space of E. To be precise, let $\phi : D(\phi) \subseteq E^* \to R \cup \{+\infty\}$ be a convex functional, $\phi^*(f) = \sup_{p \in D(\phi)}\{(p, f) - \phi(p)\}$ be the classical Legendre–Fenchel conjugate, and

$$\partial\phi(p) = \{f \in E^{**} : \phi(q) - \phi(p) \geq (p - q, f), \ \ \forall\, q \in D(\phi)\}$$

be the classical sub-differential. Then we define the *weak* conjugate* of ϕ as

$$\phi_*(x) = \sup_{p \in D(\phi)} \{(p, x) - \phi(p)\}$$

and the *weak* sub-differential* of ϕ as

$$\partial^*\phi(p) = \{x \in E^* : \phi(q) - \phi(p) \geq (q - p, x), \ \ \forall\, q \in D(\phi)\}.$$

Then these two concepts coincide with their classical ones when E is reflexive, but different when E is not reflexive. Also, we study their properties and relationships with the classical conjugate and the sub-differential.

Throughout this paper, we use $\rightharpoonup^*$ to represent the convergence in the weak* topology and $\to$ represent the convergence in the norm topology.

2 The Weak* Conjugate and Weak* Sub-differential of Convex Functionals

In this section, let E be a real Banach space, E^* be the dual space of E and E^{**} be the dual space of E^*. Under the canonical embedding mapping, E can be embedded as a subspace of E^{**} and so we always regard E as a subspace of E^{**}.

Now, we introduce the weak* conjugate and weak* sub-differential of a convex functional and study their properties.

Let $\phi : D(\phi) \subseteq E^* \to R \cup \{+\infty\}$ be a functional. We recall that the classical sub-differential of a convex function $\phi : E \to R \cup \{+\infty\}$, *Legendre–Fenchel conjugate* of ϕ is defined by

$$\phi^*(f) = \sup_{p \in D(\phi)} \{(p, f) - \phi(p)\}$$

for all $f \in E^{**}$ and the *second conjugate* of ϕ is defined by

$$\phi^{**}(p) = \sup_{f \in D(\phi^*)} \{(p, f) - \phi^*(p)\}$$

for all $p \in E^*$, respectively. Similarly, we define the *weak* Legendre–Fenchel conjugate* of ϕ by

$$\phi_*(x) = \sup_{p \in D(\phi)} \{(p, x) - \phi(p)\}$$

for all $x \in E$ and define

$$\phi_*^*(p) = \sup_{x \in D(\phi_*)} \{(p, x) - \phi_*(x)\}$$

for all $p \in E^*$, respectively.

Remark 2.1 If E is reflexive, then $\phi_* = \phi^*$ and $\phi_*^* = \phi^{**}$, but, in the nonreflexive case, these equalities don't necessarily hold. See the following example.

Example 2.1 Let $E = c_0 = \{(a_i) : a_i \in R, i = 1, 2, \ldots, \lim_{i \to \infty} a_i = 0\}$ with the sup norm. Then we have

$$E^* = l^1 = \{(a_i) : a_i \in R, i = 1, 2, \ldots, \Sigma_{i=1}^{\infty} |a_i| < +\infty\}$$

and

$$E^{**} = l^\infty = \{(a_i) : a_i \in R, i = 1, 2, \ldots, \sup_{i \geq 1} |a_i| < +\infty\}.$$

Take $v = (v_i) \in l^\infty \setminus c_0$ and $w = (w_i) \in c_0$, set $\phi(p) = \Sigma_{i=1}^{\infty} p_i v_i$ and $\psi(p) = \Sigma_{i=1}^{\infty} p_i w_i$ for all $p \in l^1$. Then, we get

(a) $\phi_*(x) = +\infty$ for all $x \in c_0$;

(b) $\phi_*^*(p) = +\infty$ for all $p \in l^1$;

(c)

$$\phi^*(x) = \begin{cases} 0, & if \ \ x = v \\ +\infty, & otherwise \end{cases}$$

for all $x \in l^\infty$;

(d) $\phi^{**}(p) = \phi(p)$ for all $p \in l^1$;

(e)

$$\psi_*(x) = \begin{cases} 0, & if \;\; x = w \\ +\infty, & otherwise \end{cases}$$

for all $x \in c_0$;

(f)

$$\psi^*(x) = \begin{cases} 0, & if \;\; x = w \\ +\infty, & otherwise \end{cases}$$

for all $x \in l^\infty$;

(g) $\psi_*^*(p) = \psi(p)$ for all $p \in l^1$.

Proposition 2.2 *Let $\phi : E^* \to R \cup \{+\infty\}$ be a proper convex functional. Then we have the following:*

(1) $\phi_*(x) + \phi(p) \geq (p, x)$ for all $x \in E$ and $p \in E^*$;

(2) $\phi_*^*(p) \leq \phi^{**}(p) \leq \phi(p)$ for all $p \in E^*$;

(3) if ϕ is lower semi-continuous in the weak topology of E^*, then $\phi_*^*(p) = \phi(p)$.*

Proof By the definition, (1) and (2) are trivial. We only prove (3).

By the assumption that ϕ is lower semi-continuous in the weak* topology of E^*, we know that $epi\phi = \{(p, \alpha) \in E^* \times R : \phi(p) \leq \alpha\}$ is a weak* closed convex subset.

Suppose that $\phi_*^*(p_0) < \phi(p_0)$ for some $p_0 \in E^*$. Then

$$(p_0, \phi_*^*(p_0)) \notin epi(\phi).$$

By Hahn–Banach's separation theorem (see [27]), there exists $x_0 \in E$, and $\lambda_0 \in R$ such that

$$(p_0, x_0) + \lambda_0 \phi_*^*(p_0) < \beta \leq \inf_{(p,\alpha) \in epi(\phi)} \{(p, x_0) + \lambda_0 \alpha\}$$

for some $\beta \in R$. Since $\lambda_0 \phi_*^*(p_0) < \lambda_0 \phi(p_0)$, we have $\lambda_0 > 0$. Therefore, we have

$$(p_0, -\frac{x_0}{\lambda_0}) - \phi_*^*(p_0) > -\frac{\beta}{\lambda_0} \geq (p, -\frac{x_0}{\lambda_0}) - \phi(p)$$

for all $p \in D(\phi)$ and so we have

$$(p_0, -\frac{x_0}{\lambda_0}) - \phi_*^*(p_0) > -\frac{\beta}{\lambda_0} \geq \phi_*(-\frac{x_0}{\lambda_0}).$$

Thus we have $\phi_*^*(p_0) > \phi_*^*(p_0)$, which is a contradiction. Therefore, we have $\phi_*^*(p_0) \geq \phi(p_0)$. From (2), we have $\phi_*^*(p) = \phi(p)$ for all $p \in E^*$. This completes the proof. □

Let $\phi : D(\phi) \subseteq E^* \to R \cup \{+\infty\}$ be a mapping. We recall that the *classical sub-differential* of ϕ at p is defined by

$$\partial\phi(p) = \{f \in E^{**} : \phi(q) - \phi(p) \geq (p - q, f), \ \ \forall\, q \in D(\phi)\}.$$

It is a well-known result of Rockfellar [25] that $\partial\phi$ is a maximal monotone mapping. Similarly, we define the *weak* sub-differential* of ϕ at p as follows:

$$\partial^*\phi(p) = \{x \in E : \phi(q) - \phi(p) \geq (q - p, x), \ \ \forall\, q \in D(\phi)\}$$

for all $p \in D(\phi)$ and $\partial^*\phi(p) = \emptyset$ for all $p \notin D(\phi)$.

It is obvious that $\partial^*\phi(p) \subseteq \partial\phi(p)$, and those two concepts coincides when E is reflexive. In the nonreflexive case, they are not necessarily equal as the following example can be seen.

Example 2.3 Let $E = c_0$, $E^* = l^1$, $E^{**} = l^\infty$ and $\phi(p) = \|p\|$ for all $p \in l^1$. Then, by the direct computation, we get $\partial^*\phi(0) = \{x \in c_0 : \|x\| \leq 1\}$ and $\partial\phi(0) = \{x \in l^\infty : \|x\| \leq 1\}$.

The following result is obvious:

Proposition 2.4 *Let $\phi : E^* \to R \cup \{+\infty\}$ be a convex functional. Then we have the following:*

(1) $\partial^*\phi(p)$ is a weak closed convex subset of E;
(2) $0 \in \partial^*\phi(p_0)$ if and only if $\phi(p_0) = \inf_{p \in D(\phi)} \phi(p)$;
(3) $\partial^*\phi : E^* \to E$ is monotone.

Theorem 2.5 *Let $\phi : E^* \to R \cup \{+\infty\}$ be a convex functional which is lower semi-continuous in the weak* topology, and $p \in E^*$, $x \in E$. Then the following statements are equivalent:*

(1) $\phi(p) + \phi_*(x) = (p, x)$;
(2) $x \in \partial^*\phi(p)$;
(3) $p \in \partial\phi_*(x)$;
(4) $x \in \partial\phi_*^*(p)$;
(5) $x \in \partial\phi(p)$.

Proof (1) $\Rightarrow$ (2) is obvious.

(2) $\Rightarrow$ (3) If $x \in \partial^*\phi(p)$, then $\phi(q) - \phi(p) \geq (q - p, x)$ for all $q \in E^*$ and so we have $(p, x) - \phi(p) \geq (q, x) - \phi(q)$ and it implies $(p, x) - \phi(p) \geq \phi_*(x)$. Consequently, we have

$$\phi_*(y) + (p, x - y) \geq \phi_*(x)$$

for all $y \in E$. Therefore, $p \in \partial\phi_*(x)$.

(3) ⇒ (4) If $p \in \partial\phi_*(x)$, then $\phi_*(y) - \phi_*(x) \geq (p, y - x)$ for all $y \in E^*$ and so we have $(p, x) - \phi_*(x) \geq \phi_*^*(p)$. Thus it follows that $\phi_*^*(q) + (p - q, x) \geq \phi_*^*(p)$ for all $q \in E^*$ and so $x \in \phi_*^*(p)$.

(4) ⇒ (5) Since ϕ is lower semi-continuous in the weak* topology, by Proposition 2.2, we have $\phi(p) = \phi_*^*(p)$ and so (5) is true.

(5) ⇒ (1) If $x \in \partial\phi(p)$, then $\phi(q) - \phi(p) \geq (q - p, x)$ for all $q \in E^*$ and so we have $(p, x) - \phi(p) \geq \phi_*(x)$, i.e., $(p, x) \geq \phi(p) + \phi_*(x)$. Thus, from Proposition 2.2-(1), it follows that (1) holds. This completes the proof. □

Let $L(x, p) : E \times E^* \to R \cup \{+\infty\}$ be a Lagrange convex functional and $L^*(p, x) = \sup_{(y,q)\in E\times E^*}\{(p, y) + (q, x) - L(y, q)\}$, which plays important role in variational inequality problems (see Ghoussoub [15] and the references therein).

Proposition 2.6 *Let $\phi : E^* \to R \cup \{+\infty\}$ be a convex functional and $L(x, p) = \phi_*(x) + \phi(p)$ for all $(x, p) \in E \times E^*$. Then $L^*(p, x) = \phi_*(x) + \phi_*^*(p)$. Moreover, if ϕ is lower semi-continuous in the weak* topology of E^*, then $L^*(p, x) = L(x, p)$.*

Proof The proof is straightforward and so we omit it. □

Definition 2.7 (see [12]) Let X be a topological space. A function $f : X \to R$ is said to be *sequentially lower semi-continuous from above at* x_0 if, for any $x_n \to x_0$, $f(x_{n+1}) \leq f(x_n)$ implies that $f(x_0) \leq \lim_{n\to\infty} f(x_n)$.

Remark 2.2 It is known that a lower semi-continuous function is a lower semi-continuous function from above, but the reverse is not true. A lower semi-continuous from above and convex function with the coercive condition in a reflexive Banach space attains its minimum (see [12]). It is well known that, for a convex function in a reflexive Banach space, the lower semi-continuity in strong topology is equivalent to the lower semi-continuity in weak topology, but this is not true for lower semi-continuity from above (see [2]). For more on lower semi-continuous functions from above with its generalizations and applications in nonconvex equilibrium problems, variational inequality problems and fixed point problems, see [1, 3, 10, 11, 17–20, 23, 24].

Proposition 2.8 *Let $\phi : E^* \to R \cup \{+\infty\}$ be a convex functional which is sequentially lower semi-continuous from above in the weak* topology and $\lim_{\|p\|\to+\infty} \phi(p) = +\infty$. Then there exists $p_0 \in E^*$ such that $\phi(p_0) = \inf_{p\in D(\phi)} \phi(p)$.*

Proof We take a sequence $\{p_n\}$ in E^* such that

$$\phi(p_1) \geq \phi(p_2) \geq \cdots \geq \phi(p_n) \geq \cdots$$

and

$$\phi(p_n) \to \inf_{p\in D(\phi)} \phi(p).$$

Since $\lim_{\|p\|\to+\infty}\phi(p)=+\infty$, $\{p_n\}$ is a bounded sequence in E^* and so $\{p_n\}$ has a convergent subsequence of $\{p_n\}$ with $p_{n_k} \rightharpoonup^* p_0$ in E^*. By the assumption, ϕ is sequentially lower semi-continuous from above and so $\phi(p_0)\leq \lim_{n\to\infty}\phi(p_n)$. Therefore, it follows that

$$\phi(p_0)=\inf_{p\in D(\phi)}\phi(p).$$

This completes the proof. □

Proposition 2.9 *The functional* $\|\cdot\|^2 : E^*\to R$ *is sequentially lower semi-continuous in the weak* topology.*

Proof Suppose that $p_n \rightharpoonup^* p_0$. Then $p_0(x)=\lim_{n\to\infty}p_n(x)$ for all $x\in E$ and so we have

$$|p_0(x)|\leq \liminf_{n\to\infty}\|p_n\|\|x\|$$

for all $x\in E$. Thus we have

$$\|p_0\|=\sup_{\|x\|=1}|p_0(x)|\leq \liminf_{n\to\infty}\|p_n\|$$

and so

$$\|p_0\|^2\leq \liminf_{n\to\infty}\|p_n\|^2.$$

This completes the proof. □

Theorem 2.10 *Let* $\phi : E^*\to R\cup\{+\infty\}$ *be a convex functional which is sequentially lower semi-continuous in the weak* topology. Then we have*

$$\partial^*(\phi+\epsilon\|\cdot\|^2)(E^*)=E$$

for all $\epsilon>0$.

Proof For any $x\in E$, we set $\psi(p)=\phi(p)+\epsilon\|p\|^2-p(x)$ for $x\in D(\phi)$. It is obvious that ψ is sequentially lower semi-continuous in the weak* topology and so ψ is sequentially lower semi-continuous from above in the weak* topology and

$$\lim_{\|p\|\to+\infty}\psi(p)=+\infty.$$

By Proposition 2.8, there exists $p_0\in E^*$ such that $\phi(p_0)=\inf_{p\in D(\psi)}\psi(p)$. By Proposition 2.4-(2),

$$0\in\partial^*(\phi+\epsilon\|\cdot\|^2-(\cdot)(x))(p_0),$$

which is equivalent to $x\in\partial^*(\phi+\epsilon\|\cdot\|^2)(p_0)$. This completes the proof. □

Acknowledgements This work was supported by the Basic Science Research Program through the National Research Foundation of Korea funded by the Ministry of Education, Science and Technology (NRF-2013053358 and 2013R1A1A2054617).

References

1. Al-Homidan, S., Ansari, Q.H., Yao, J.C.: Some generalizations of Ekeland-type variational principle with applications to equilibrium problems and fixed point theory. Nonlinear Anal. **69**, 126–139 (2008)
2. Aruffo, A., Bottaro, G.: Generalizations of sequential lower semicontinuity. Boll. Uni. Mat. Ital. Ser. **9**(1), 293–318 (2008)
3. Aruffo, A.B., Bottaro, G.: Some variational results using generalizations of sequential lower semicontinuity. Fixed Point Theory Appl. **323487**, 21 (2010)
4. Attouch, H., Brezis, H.: Duality for the sum of convex functions in general Banach spaces. In: Aspects of Mathematics and its Application. North-Holland, Amsterdam (1986)
5. Aubin, J.P., Ekeland, I.: Applied Nonlinear Analysis, Reprint of the 1984 Original. Dover Publications, Inc., New York (2006)
6. Barbu, V., Precupanu, T.: Convexity and Optimization in Banach Spaces, 2nd edn. D. Reidel Publishing Co., Dordrecht (1986)
7. Bartz, S., Reich, S.: Abstract convex optimal antiderivatives. Ann. I. H. Poincar **29**, 435–454 (2012)
8. Borwein, J.M., Vanderwerff, J.D.: Convex functions of Legendre type in general Banach spaces, Technical report, CECM, Simon Fraser University, Barnaby (2000)
9. Borwein, J.M., Lewis, A.S.: Convex Analysis and Nonlinear Optimization, Theory and Examples. Springer, New York (2000)
10. Bugajewskia, D., Kasprzak, P.: Fixed point theorems for weakly F-contractive and strongly F-expansive mappings. J. Math. Anal. Appl. **359**, 126–134 (2009)
11. Castellania, M., Pappalardob, M., Passacantandob, M.: Existence results for nonconvex equilibrium problems. Optim. Methods Softw. **25**, 49–58 (2010)
12. Chen, Y.Q., Cho, Y.J., Yang, L.: Note on the results with lower semi-continuity. Bull. Korean Math. Soc. **39**, 535–541 (2002)
13. Clarke, F.H.: Optimization and Nonsmooth Analysis. Wiley Inc. Publication, New York. A Wiley-Interscience (1983)
14. Ekeland, I., Temam, R.: Analyse Convexe et Problemes Variationnels. Dunod, Collection Etudes Mathematiques (1974)
15. Ghoussoub, N.: Self-dual Partial Differential Systems and Their Variational Principles. Springer, Heidelberg (2008)
16. Jeyakumar, V., Rubinov, A., Wu, Z.Y.: Generalized Fenchels conjugation formulas and duality for abstract convex functions. J. Optim. Theory Appl. **132**, 441–458 (2007)
17. Khanh, P.Q., Quy, D.N.: A generalized distance and enhanced Ekeland's variational principle for vector functions. Nonlinear Anal. **73**, 2245–2259 (2010)
18. Khanh, P.Q., Quy, D.N.: On Ekeland's variational principle for Pareto minima of set-valued mappings. J. Optim. Theory Appl. **153**, 280–297 (2012)
19. Lin, L.J., Du, W.S.: Ekeland's variational principle, minimax theorems and existence of nonconvex equilibria in complete metric spaces. J. Math. Aanal. Appl. **323**, 360–370 (2006)
20. Lin, L.J., Du, W.S.: On maximal element theorems, variants of Ekeland's variational principle and their applications. Nonlinear Anal. **68**, 1246–1262 (2008)
21. Minty, G.J.: On the monotonicity of the gradient of a convex function. Pacific J. Math. **14**, 243–247 (1964)
22. Phelps, R.R.: Convex Functions, Monotone Operators and Differentiability, Lecture Notes in Mathematics, vol. 1364. Springer, Heidelberg (1993)

23. Qiu, J.H.: On Ha's version of set-valued Ekelands variational principle. Acta Math. Sinica **28**, 717–726 (2012)
24. Qiu, J.H., He, F.: A general vectorial Ekeland's variational principle with a p-distance. Acta Math. Sinica **29**, 1655–1678 (2013)
25. Rockafellar, R.T.: On the maximal monotonicity of subdifferential mapping. Pacific J. Math. **33**, 209–216 (1970)
26. Rockafellar, R.T.: Convex Analysis. Princeton University Press, Princeton (1970)
27. Rudin, W.: Functional Analysis. MacGraw-Hill, New York (1973)
28. Zalinescu, C.: Convex Analysis in General Vector Spaces. World Scientific, Singapore (2002)

Nonlinear Aspects of Certain Linear Phenomena in Banach Spaces

M.A. Sofi

Abstract As opposed to the linear theory of Banach spaces which deals with the description of Banch spaces in terms of their linear topological properties involving the use of linear subspaces, linear maps and their relatives, the nonlinear theory seeks to achieve the same goal using nonlinear objects/quantities attached to a Banach space. These latter objects include Lipschitz maps, bilinear/multilinear maps and polynomials with domains of definition being replaced by subsets of the given space. It is remarkable that the linear structure of a Banach spaces is determined to a large extent by Lipschitz maps and that in a number of interesting cases, the linear structure is also captured by the larger class of uniformly continuous mappings acting between Banach spaces. In the present work, this will be illustrated by a number of results, both old and new, involving the extendability of Lipschitz maps and certain phenomena associated with it.

Keywords Banach space · Hilbert space · Lipschitz map · Extension operator

2010 Mathematics Subject Classication Primary: 46B03 · 46B20 · Secondary: 45C15

1 Introduction

A Banach space—and more generally, a Frechet space—comes equipped with a metric topology (derived from its norm in the former case) which is tied up to its linear structure by a set of compatibility conditions required of the norm function (linear topology in the latter case) on the given space. Thus, apart from the structure which is endowed on a Banach space by virtue of its norm, two of the many other underlying structures derive from the uniformity and the (associated) topology induced on it by the metric. The question regarding the extent to which the linear topological structure

---

M.A. Sofi (✉)
Department of Mathematics, University of Kashmir, Srinagar 190006, India
e-mail: aminsofi@gmail.com

J.M. Cushing et al. (eds.), *Applied Analysis in Biological and Physical Sciences*, Springer Proceedings in Mathematics & Statistics 186,
DOI 10.1007/978-81-322-3640-5_26

of a Banach space is determined by these weaker structures assumes significance in view of the symbiotic relationship between these structures as has been unraveled over a period of time. The phenomenal progress having been witnessed in recent years on 'nonlinear geometry' of Banach spaces has led to a better understanding of certain deeper aspects of Banach space theory on the one hand and to far reaching implications of it in theoretical computer science on the other.

To illustrate this point, let us recall the famous Maur–Ulam theorem [23] which says that a bijective isometry between (real) Banach spaces which preserves the origin is linear! In effect, it says that two (real) Banach spaces which can be identified as metric spaces are already equivalent as linear topological spaces. In other words, the metric topology of a Banach space is strong enough to capture its linear structure. On the other hand, the norm-topology of a Banach space is too weak to distinguish a Banach soace from another. In fact, an old theorem of M.I. Kadec [19] says that any two separable infinite dimensional Banach spaces are (topologically) homeomorphic. However, as opposed to these extreme possibilities, it turns out that the uniform classification of Banach spaces belongs somewhere in the middle. In this classification, what we consider are uniform homeomorphisms, and more importantly, bi-Lipschitz mappings in place of isometries and topological homeomorphisms. To put this point into perspective, a well known theorem of Enflo [12] says that a Banach space is linearly homeomorphic to a Hilbert space as soon as it is uniformly homeomorphic to it. On the other hand, it remains an important open problem whether two (infinite dimensional) separable Banach spaces are linearly homeomorphic if they are Lipschitz homeomorphic.

The present work is devoted to a discussion of these issues in the context of extendability of Lipschitz maps between Banach spaces. As will be shown in what follows, in most of the cases—as in the linear theory—the phenomenon of extendability of nonlinear maps determines to a large extent the linear structure of the underlying Banach as is illustrated in the following examples which may be looked upon as nonlinear analogues of results already known from the linear theory of Banach spaces.

Before we do that, let us describe a few examples involving some nonlinear characterisations of Hilbert spaces which, however, do not have a counterpart in the linear theory.

2 Nonlinear Characterisations of Hilbert Spaces

Example 1 Given a real Banach space X, then each of the following conditions is equivalent to X being Hilbertian.

(a) Reference [6] $\|f\|\|g\| \leq 2\|fg\|, \forall f, g \in X^*$.
(b) Reference [5] $\|L\| = \|L^\wedge\|$ for each symmetric bilinear from L on X. (Here $L^\wedge$ denote the polynomial associated to L).

(c) Reference [5] $\forall x(\neq 0), y(\neq 0) \in X$, $\exists$ a symmetric bilinear form L on X such that $L(x, y) = \|x\|\,\|y\|$.

Comments:

(i) The norm of the product in the RHS of (a) is meant in the sense of fg acting as a bilinear form on $X : fg(x, y) = f(x)g(y),\ x, y \in X$. In general the norm of an n-linear L on X is defined by:

$$\|L\| = \sup\left\{|L(x_1, x_2, \ldots, x_n)|;\ x_1, x_2 \ldots, x_n \in B_X\right\}.$$

(ii) A complex analogue of (c) is also valid whereby L is required to be symmetric Hermitian: $L(x, y) = \overline{L(y, x)}$, $\forall x, y \in X$.

(iii) Without requiring symmetry of L, the assertion in Example 1(c) always holds in an arbitrary normed space by choosing $L(x, y) = f(x)g(y)$. This follows from Hahn Banach extension theorem.

Example 2 A Banach space X is a Hilbert space if and only if given $x, y \in S_X$ (unit sphere of X), there exists a linear isometry T on X such that $T(\{x, y\}) = \{x, y\}$, i.e., X is 2-transitive.

Example 3 ([22, Theorem 1.9]) Given a (complex) Banach space X such that for each contraction T on X and a polynomial $p(z)$, it holds that

$$\|p(T)\| \leq \sup_{|z|\leq 1} |p(z)|,$$

then X is a Hilbert space.

(iv) It is a celebrated open problem in functional analysis—known as Banach–Mazur rotation problem—whether a separable Banach space is isometric (resp. isomorphic) to a Hilbert space if X is transitive (=1-transitive), i.e. if T in Example 2 above can be chosen such that $T(x) = y$. It is well known that Hilbert spaces satisfy this property and that this property characterises Hilbertisability in finite dimensional spaces. The latter statement is a consequence of an old result of Auerbach asserting the existence of an inner product on each finite dimensional space X, which induces a larger group of isometries on X than the group of isometries with respect to the given norm. On the other hand, a transitive norm is always maximal in the sense that no equivalent norm on the space admits a larger group of isometries, a fact combined with the above result of Auerbach yields the desired assertion. See [16] for a comprehensive survey of state of art surrounding this problem upto 2002.

(v) The proof of the statement in Example 3 is a consequence of the 2-dimensional analogue of the above statement in (iv) combined with the well known fact (due to Kakutani [20]) that a Banach space is isometric to a Hilbert space if (and only if) each of its 2-dimensional subspaces is an inner product space.

(vi) The converse of the statement that the inequality in Example 3 above is valid for contractions T acting in a Hilbert space is already a well known result in operator theory popularly known as Von Neumann inequality (see [25, Chap. 1]).

Example 4 Lipschitz equivalent of an operator-valued Hahn Banach theorem:

An easily checked consequence of the projection theorem in Hilbert space X is the assertion that every bounded linear map defined on a subspace of X and taking values in an arbitrarily given Banach space can be extended to a bounded linear map on the whole space X. The fact that this extension property of bounded linear maps characterises X as being a Hilbert space follows from the celebrated Lindenstrauss–Tzafiriri complemented subspace theorem which says that a Banach space is Hilbertian exactly when all of its closed subspaces are complemented (see [1, Sect. 12.4]).

The problem involving the Lipschitz analogue of the above extension property pertains to the extension of Lipschitz maps, now defined on arbitrary subsets rather than merely on subspaces of X. Given metric spaces X and Y, we say that a map $f : X \to Y$ is c-Lipschitz if there exists $c > 0$ such that $d(f(x), f(y)) \leq cd(x, y)$, $\forall x, y \in X$.

In this connection, we have the following theorem:

Theorem 5 ([7, Chap. 2], see also [27]) *For a Banach space X, TFAE:*

(i) Given a subset A of X and a 1-Lipschitz map $f : A \to \ell_2$, there exists a 1-Lipschitz map $g : X \to \ell_2$ such that $f = g$ on A.
(ii) X is a Hilbert space.

Proof (*Outline*) The fact that (ii) implies (i) is (the infinite dimensional analogue of) an old result of Kirszbraun which says that $\mathbb{R}^m$-valued c-Lipschitz maps acting on subsets of Euclidean spaces can always be extended to a c-Lipschitz map on the whole space. The proof is straightforward for functions taking values in $\mathbb{R}$. Indeed, if A is a subset of $\mathbb{R}^n$ and $f : A \to \mathbb{R}$ is a c-Lipschitz map, then the map

$$F(x) = \inf\{f(a) + c|x - a| : a \in A\}$$

defines a c-Lipschitz map on $\mathbb{R}^n$ which extends f. (It is clear that the same argument works for a metric space in place of $\mathbb{R}^n$—an important fact which is due originally to McShane). Further, the special case of Kirszbraun's theorem noted above applied to the co-ordinate functions gives a cL-Lipschitz extension of a given c-Lipschitz function $f : A \to \mathbb{R}^m (A \subset \mathbb{R}^n)$ where L is a constant depending upon m. (Precisely, $L = \sqrt{m}$). In effect, Kirszbraun's original theorem is exactly the last statement modulo the assertion that the constant L appearing there is redundant! However, that is much harder to prove! Here, let us also point out that it is relatively easier to prove Kirszbraun's theorem((ii)⇒(i)) for the special case involving 1-Lipschitz mappings defined on convex domains in a Hilbert spaces and taking values in an arbitrary Banach space. The key idea involved in the proof of this statement is the existence of a Chebyscheff projection induced by a convex subset of a Hilbert space where it also turns out to be a contraction, i.e. a 1-Lipschitz map.

Regarding the implication (i)⇒(ii), it turns out that a stronger statement is true, namely that if (i) holds for some Banach space Z in place of ℓ_2 which is merely assumed to be strictly convex, then both X and Z are Hilbert spaces!

The conclusion in (ii) is achieved under the weaker assumption—which trivially follows from (i)—that 1-Lipchitz maps defined on three-point subsets of the domain can be extended to a 1-Lipschiz map on any four-point superset. Here, it is important to point out that the extension is always possible if it is sought to be effected from a 2-point set to a larger set having three points. Indeed, if $A = \{x, y\}$ and if $u : A \to Y$ is a 1-Lipschitz map, then for $z \notin \{x, y\}$ and for t defined by

$$t = \min\left\{1, \frac{\|z - y\|}{\|x - y\|}\right\},$$

the formula $v(z) = tu(x) + (1 - t)u(y)$ gives a 1-Lipschitz mapping from $\{x, y, z\}$ into Y, which obviously extends the given u. □

Example 6 The Hahn–Banach extension property for subspaces Y of a Hilbert space X yields the existence of a bounded linear map $\psi : Y^* \to X^*$ such that $\psi(g)|_Y = g$ for each $g \in Y^*$ (For each $g \in Y^*$, simply choose f to be the composite of g with the orthogonal projection of X onto Y). Further, the existence of ψ for each choice of Y characterises X as a Hilbert space. A nonlinear analogue of this property could be one of the following types:

(a) Polynomial analogue: Existence of ψ from the space of polynomials on Y to the space of polynomials on X.
(b) Lipschitz analogue (for subspaces): Existence of ψ from the space of Lipschitz functions on the subspace Y to the space of Lipschitz functions on X.
(c) Lipschitz analogue (for subsets): Existence of ψ from the space of Lipschitz functions on subsets Y of X to the space of Lipschitz functions on X.

Comments:

(a′) As opposed to continuous linear functionals, polynomials defined on subspaces do not always extend to larger spaces. For example, the scalar product on ℓ_2 cannot be extended to a bilinear functional on ℓ_∞. This is because bilinear forms on ℓ_∞ are weakly sequentially continuous (because ℓ_∞ has the Dunford–Pettis property) whereas the inner product is not! However, as in the case of linear functionals defined on subspaces of a Hilbert space, polynomials also admit extensions from subspaces of Hilbert spaces to the whole space. Further, there are non-Hilbertian Banach spaces which admit extension of polynomials from subspaces to the larger space. Indeed, Maurey's extension theorem (See [11, Chap. 12]) guarantees the extendability of bilinear forms (and therefore of scalar polynomials of degree 2) for type 2 Banach spaces. This raises the question of the description of Banach spaces X such that each polynomial on each subspace of X admit an extension to a polynomial of the same degree on X. Whereas it remains to be known whether there are non-Hilbertain Banach spaces witnessing the aforementioned extension property, it turns out that the existence and boundedness of the linear 'extension' operator between appropriate spaces of polynomials (of a fixed degree n) involving the given Banach space X does indeed characterise X as a Hilbert space. In what follows, the symbol $P^n(Z)$ shall denote the space of polynomials of degree at most n on a Banach space Z.

Theorem 7 *For a Banach space X, TFAE:*

(i) *X is a Hilbert space.*
(ii) *For each subspace Y of X, there exists a bounded linear (extension) map $\psi : P^n(Y) \to P^n(X)$, i.e., such that $\psi(g)|_Y = g$ for each $g \in P^n(Y)$.*
(iii) *For each subspace Y of X, there exists a bounded linear (extension) map $\psi : Y^* \to X^*$, i.e., such that $\psi(g)|_Y = g$ for each $g \in Y^*$.*

Proof (*Sketch*) As noted earlier, the implications (i)⇒(ii) and (i)⇒(iii) are straightforward consequences of the existence of the orthogonal projection from X onto Y. To prove that (ii)⇒(iii), let $\psi : P^n(Y) \to P^n(X)$ be as given in (ii). Fix $e \in Y$ and choose $g \in Y^*$ such that $\|g\| = 1$ and $g(e) = 1$. Let $\eta : Y^* \to P^n(Y)$ and $\varphi : P^n(X) \to X^*$ be defined by:

$$\eta(f) = f.g^{n-1}, \quad \varphi(P) = DP(e) - (n-1)P(e)q$$

where $DP(e)$ denotes the Frechet derivative of P at e. We also note that if A is the symmetric n-linear form associated to P, then $DP(e)(x) = nA(e, e, \ldots, x)$. In case of $P = f.g^{(n-1)}$, we see for the associated symmetric n-linear form A that

$$\begin{aligned} A(e, e, \ldots, x) &= \frac{1}{n}\left[A(x, e, \ldots, e) + A(e, x, \ldots, x) + \cdots + A(e, e \ldots, x)\right] \\ &= \frac{1}{n}\left[f(x) + f(e)g(x) + \cdots + f(e)g(x)\right] \\ &= \frac{1}{n}\left[f(x) + (n-1)f(e)g(x)\right]. \end{aligned}$$

The desired map $\rho : Y^* \to X^*$ is given by: $\rho = \varphi \circ \psi \circ \eta$. Indeed, given $f \in Y^*$ and $x \in Y$, we have

$$\begin{aligned} \rho(f)(x) &= \varphi \circ \psi \circ \eta(f)(x) \\ &= D\left[\psi \circ \eta(f)\right](x) - (n-1)\eta(f)(e)g(x) \\ &= D\left[\eta(f)\right](e)(x) - (n-1)\eta(f)(e)g(x) \\ &= nA(e, e, \ldots, x) - (n-1)f(e)g(x) \\ &= f(x). \end{aligned}$$

Finally, whereas the proof of (iii)⇒(i) is well known from the linear theory, we include a neat proof of this implication, using the fact that condition (iii) can be equivalently described as a local property in the following sense:

($*$) There exists $c > 0$ such that for each finite dimensional subspace Y of X, there exists a continuous linear extension map $\psi : Y^* \to X^*$ with $\|\psi\| \leq c$ (see [12] for details).

Indeed, consider finite dimensional subspaces $Y \subset Z \subset X$ and let $\psi : Y^* \to X^*$ be a continuous linear extension map as in (i) with $\|\psi\| \leq c$. Taking conjugates gives a map $\psi^* : X^{**} \to Y^{**}$ which when restricted to Z produces a projection $P : Z \to Y$ with

$$\|P\| \leq \|\psi^*\| = \|\psi\| \leq c.$$

From the proof of the Lindenstrauus–Tzafiriri theorem on the complemented subspace theorem (see [1, Theorem 12.4.2]), it follows that there exists $f(c) > 0$, independent of $Z \subset X$, such that $d(Z, \ell_2^{\dim Z}) \leq f(c)$. Since Z was chosen arbitrarily, by virtue of [18], we conclude that X is isomorphic to a Hilbert space. □

3 Existence of Continuous Linear 'Extension' Maps

Here we discuss the Lipschitz analogue of the above theorem and ask if in the setting of (a′), it is possible to choose for each subspace Y of X a bounded linear 'extension' map $\psi : \mathrm{Lip}(Y) \to \mathrm{Lip}(X)$ such that $\psi(g)|_Y = g$ for each $g \in \mathrm{Lip}(Y)$. Here $\mathrm{Lip}(Y)$ stands for the space of all *Lipschitz* functions $f : Y \to \mathbb{R}$ which is easily checked to be a Banach space when equipped with the norm:

$$\|f\| = \sup_{x \neq y} \frac{\|f(x) - f(y)\|}{\|x - y\|}.$$

We shall see below that for subspaces Y of a Banach space X, the existence of a bounded linear 'extension' map $\psi : \mathrm{Lip}(Y) \to \mathrm{Lip}(X)$ imposes rather severe restrictions on Y. We shall need the following definitions.

Definition 8 A subspace Y of a Banach space X is said to be locally complemented if there exists $c > 0$ such that for each finite dimensional subspace M of X, there exists a continuous linear map $f : M \to Y$ with $\|f\| \leq c$ and $f(x) = x$ for all $x \in M \cap Y$. Equivalently, the annihilator of Y is complemented as a subspace of X^*.

Of course, a complemented subspace is trivially locally complemented. On the other hand, it is well known that c_0 is not complemented in ℓ_∞. However, the fact that c_0 is locally complemented in ℓ_∞ is a consequence of the next theorem combined with the fact that c_0 is a Lipschitz retract of ℓ_∞ ([7, Example 1.5]). It also follows from the more general statement that every Banach space is locally complemented in its bidual.

Theorem 9 *Let X be a Banach space and Y a subspace of X such that there exists a bounded linear 'extension' map $L : Lip(Y) \to Lip(X)$, i.e., $L(g)|_Y = g$ for each $g \in Lip(Y)$. Then there exists a bounded linear extension map $\psi : Y^* \to X^*$. Equivalently, Y is locally complemented in X.*

Proof We begin by using L to define another bounded linear map $\psi : \text{Lip}(Y) \to \text{Lip}(X)$ which actually takes values in X^* and when restricted to Y^* yields a bounded linear extension map from Y^* into X^*. The local complementedness of Y will then follow from the theorem of Kalton mentioned above. The map ψ is constructed using the existence of a Banach limit $\int .. dx$ which is a continuous linear functional on $\ell_\infty(X)$ satisfying the following conditions:

(a) $\| \int .. dx \| = 1$.
(b) $\int 1 dx = 1$.
(c) $\int f(x + x')dx = \int f(x)dx, \forall f \in \ell_\infty(X)$ and $\forall x' \in X$.

The new map $\psi : \text{Lip}(Y) \to \text{Lip}(X)$ is now defined by the formula $\psi(f)(z) = \int\{\int[(Lf)(x + y + z) - (Lf)(x + y)]dy\}dx,\ f \in Lip(Y),\ z \in X$.

The boundedness of L together with (a) and the Lipschitz property of f yield ψ as a well defined bounded linear map such that for $z_1, z_2 \in X$, we have, by virtue of (c),

$$\begin{aligned}
\psi(f)(z_1 + z_2) &= \int \left\{ \int [(Lf)(x + y + z_1 + z_2) - (Lf)(x + y)]\, dy \right\} dx \\
&= \int \left\{ \int [(Lf)(x + y + z_1 + z_2) - (Lf)(x + y + z_2)]\, dy \right\} dx \\
&\qquad + \int \left\{ \int [(Lf)(x + y + z_2) - (Lf)(x + y)]\, dy \right\} dx \\
&= \int \left\{ \int [(Lf)(x + y + z_1) - (Lf)(x + y)]\, dy \right\} dx \\
&\qquad + \int \left\{ \int [(Lf)(x + y + z_2) - (Lf)(x + y)]\, dy \right\} dx \\
&= \psi(f)(z_1) + \psi(f)(z_2).
\end{aligned}$$

Restricting ψ to Y^* gives a map $\psi : Y^* \to X^*$ such that for $g \in Y^*$ and $z \in Y$, an application of the extension property of L combined with (b) and (c) gives

$$\begin{aligned}
\psi(g)(z) &= \int \left\{ \int [(Lg)(x + y + z) - (Lg)(x + y)]\, dy \right\} dx \\
&= \int \left\{ \int [(Lg)(x + y + z) - (Lg)(y + z)]\, dy \right\} dx \\
&\qquad + \int \left\{ \int [(Lg)(y + z) - (Lg)(x + y)]\, dy \right\} dx
\end{aligned}$$

$$
\begin{aligned}
&= \int \left\{ \int [(Lg)(x+y) - (Lg)(y)]\, dy \right\} dx \\
&\qquad + \int \left\{ \int [(Lg)(y+z) - (Lg)(x+y)]\, dy \right\} dx \\
&= \int \left\{ \int [(Lg)(y+z) - (Lg)(y)]\, dy \right\} dx \\
&= \int \left\{ \int [g(y+z) - g(y)]\, dy \right\} dx \\
&= \int \left\{ \int g(z) dy \right\} dx = g(z).
\end{aligned}
$$

This completes the proof of the theorem. □

Combined with Theorem 7 ((iii)⇒(i)), the above theorem yields the following characterisation of Hilbert space.

Corollary 10 *For a Banach space X, TFAE:*

(i) X is a Hilbert space.
(ii) For each subspace Y of X, there exists a bounded linear 'extension' map ψ : $Lip(Y) \to Lip(X)$.

As a first important consequence of the above corollary, the following Lipschitz analogue of an isomorphic characterisation of Hilbert spaces follows directly from it (see [23, Theorem 4.12]).

Corollary 11 *For a Banach space X, TFAE:*

(i) X is a Hilbert space.
(ii) For each subspace Y of X and each Banach space Z, each Lipschitz map on Y and taking values in Z can be extended to a Lipschitz map on X.

Indeed, the given condition applied to each closed subspace Y of X yields Y as Lipschitz retract of X and this gives rise to a bounded linear extension map $\psi : \text{Lip}(Y) \to \text{Lip}(X)$ which, by virtue of Corollary 10, yields that X is a Hilbert space.

Remark 12 The question regarding the description of Banach spaces X resulting from (ii) in the above corollary with $Z = \ell_2$ is an important open problem belonging to this circle of ideas. It is known that whereas this property holds for a class of Banach spaces that includes L_p spaces for $2 < p < \infty$, a Banach space enjoying the latter property, namely property (ii) with $Z = \ell_2$ has type p for $p < 2$. However, it is conjectured that such spaces are type 2 Banach spaces. To the best of our knowledge, the linear analogue of this problem also remains open. On the other hand, (as was seen in Theorem 5), using nonexpansive mappings on arbitrary subsets in place of Lipschitz maps on subspaces of X (as in the above corollary) with the choice of $Z = \ell_2$ yields X (isometrically) as a Hilbert space.

The next corollary deals with the Lipschitz analogue of a well known result involving the description of Banach spaces X such that every closed convex subset of X is a nonexpansive retract of X. That this property holds for Hilbert spaces follows from the easily-checked observation that the 'nearest point projection' P_C corresponding to a given closed subset C of X is indeed a nonexpansive (1-Lipschitz) map which is the identity map on C. Here P_C is defined to be the map given by:

$$P_C(x) = \inf\{\|x - y\| : y \in C\}, \quad x \in X.$$

The fact that the stated property holds precisely for Hilbert spaces is a famous theorem due originally to Reich [26]. The next result provides an isomorphic analogue of the latter statement, which follows exactly on similar lines as in the previous corollary.

Corollary 13 *A Banach space has the property that each closed and convex subset of X is a Lipschitz retract of X if and only if X is isomorphic to a Hilbert space.*

(c′) We now investigate the question involving the existence of an 'extension' operator $\psi : \text{Lip}(A) \to \text{Lip}(X)$ where A is now chosen to be an *arbitrary subset* of the metric space X. A metric space with this property is said to satisfy the simultaneous Lipschitz extension property (SLEP). An answer to this question turns out to be highly nontrivial, more so when X is not a Banach space. In this general setting, this question has been a subject of intensive research spanning different areas of mathematics including, in particular, geometry and group theory. There are interesting examples of situations where such extensions always exist which include, for example, $\mathbb{R}^n$ (with respect to any norm), the Heisenberg group, metric trees of arbitrary cardinality, groups of polynomial growth and certain classes of Riemannian manifolds of bounded geometry. However, there are examples of 2-dimensional Riemannian manifolds of bounded geometry and their subsets for which such an extension does not exist. The best source of information on these and related issues is the ground breaking work of A. Brudnyi and Y. Brudnyi [9]. Making use of some deep but well known facts from the local theory of Banach spaces, we shall see below that the extension procedure breaks down for all infinite dimensional Banach spaces, including of course, infinite dimensional Hilbert spaces where the situation was shown to be extremely pleasant from the viewpoint of the extension procedure involving linear subspaces.

Theorem 14 *A Banach space X has (SLEP) if and only if X is finite dimensional.*

Proof (*Sketch*) Let us introduce certain numerical parameters that will be used to quantify the existence of the extension operator mentioned above. To this end, for each subset A of a metric space M, denote by $\text{Ext}(A, M)$ the set of 'extension' operators $\psi : \text{Lip}(A) \to \text{Lip}(M)$, i.e., $\psi(g)|_Y = g$ for each $g \in \text{Lip}(Y)$. We set

$$\lambda(M) = \sup\{\inf \|T\| : T \in Ext(A, M), A \subset M\},$$

and call it the 'global Lipschitz extension constant' of M.

We claim that $\lambda(X) = \infty$ for each infinite dimensional Banach space X. In particular, this would imply that each infinite dimensional Banach space admits a subset A without the simultaneous extension property. We shall need the following tools to prove our statement.

(a) Dvoretzky's spherical sections theorem (see [11, Chap. 19]): Each infinite dimensional Banach space X contains for each n, an n-dimensional subspace X_n which is 2-isometric with ℓ_2^n.

(b) (Reference [8]) There exists $c > 0$ (independent of n) such that $\lambda(\ell_2^n) \geq cn^{1/8}$.

Whereas (a) is already well known as a highly nontrivial fact from the local theory of Banach spaces, the proof of (b) hinges on another deep result involving an alternative description of $\lambda(M)$ given by the following formula:

$$\lambda(M) = \sup\{\nu(M, Z); Z \in FD\}.$$

Here FD denotes the class of all finite dimensional Banach spaces and $\nu(X, Z)$ is given by

$$\nu(M, Z) = \sup\{\nu(A, M, Z); A \subset M\}.$$

where $\nu(A, M, Z)$ is defined to be the infimum of all constants $c > 0$ such that each $g \in \mathrm{Lip}(A, Z)$ admits an extension to $f \in \mathrm{Lip}(M, Z)$ such that

$$\|f\|_{\mathrm{Lip}(M,Z)} \leq c\|g\|_{\mathrm{Lip}(A,Z)}.$$

To prove the desired assertion, for each $n > 0$, choose a subspace X_n of X according to (a). Thus there exists a linear isomorphism $\varphi : \ell_2^n \to X_n$ such that $\|\varphi\|\|\varphi^{-1}\| < 2$. Fix $\epsilon > 0$ and let $A \subset \ell_2^n$. There exists $E \in \mathrm{Ext}(\varphi(A), X_n)$ such that $\|E\| < \lambda(\varphi(A), X_n) + \epsilon$. Consider the mappings $\psi : \mathrm{Lip}(X_n) \to \mathrm{Lip}(\ell_2^n)$ and $L : \mathrm{Lip}(A) \to \mathrm{Lip}(\varphi(A))$ given by

$$\Psi(f) = f \circ \varphi \quad \text{and} \quad L(g) = g \circ \varphi^{-1}|_{\varphi(A)},\ f \in \mathrm{Lip}(X_n),\ g \in \mathrm{Lip}(A).$$

Then F given by $F = \Psi \circ E \circ L$ defines a linear map $F : \mathrm{Lip}(A) \to \mathrm{Lip}(\ell_2^n)$ such that

$$\|F\| \leq \|\Psi\|\|E\|\|L\| \leq \|\varphi\|\|E\|\|\varphi^{-1}\| \leq 2\|E\|.$$

Using the above estimates, we get

$$\lambda(A, \ell_2^n) \leq \|F\| \leq 2\|E\| < 2(\lambda(\varphi(A), X_n) + \epsilon) \leq 2(\lambda(X_n) + \epsilon)$$

Letting $\epsilon \to 0$ gives $\lambda(A, \ell_2^n) \leq 2(\lambda(X_n)$. Now it is easy to see that $\lambda(A) \leq c\lambda(B)$ whenever A is a c-Lipschitz retract of B. Applying the latter assertion to (the 2-Lipschitz retract) X_n(in X), it follows that $\lambda(X_n) \leq 2\lambda(X)$. Combined with the above estimate, this yields

$$\lambda(X) \geq \frac{1}{2}\lambda(X_n) \geq \frac{1}{4}\lambda(A, \ell_2^n).$$

Taking supremum over $A \subset \ell_2^n$ and using (b), we get

$$\lambda(X) \geq \frac{1}{4}\lambda(\ell_2^n) \geq \frac{c}{4}n^{1/8}.$$

Since n is arbitrary, it follows that $\lambda(X) = \infty$. Finally, the assertion that $\lambda(X) < \infty$ for finite dimensional Banach spaces is a famous result due to Whitney (see [29, Chap. 6.2.3]). □

The following definition will be found useful in what follows. Let us use the symbol $\text{Lip}_0(X)$ for the subspace of $\text{Lip}(X)$ consisting of those functions f such that $f(\theta) = 0$ for some distinguished point $\theta \in X$. It turns out that $\text{Lip}_0(X)$ is a dual space and so, there exists a Banach space, denoted $\mathfrak{I}(X)$, such that $\mathfrak{I}(X)^* = \text{Lip}_0(X)$. The space $\mathfrak{I}(X)$ shall be called the *free Banach space* over X. For notational convenience and the fact that in what follows, we shall be dealing exclusively with the space $\text{Lip}_0(X)$, we shall continue to use the symbol $\text{Lip}(X)$ for the space $\text{Lip}_0(X)$ defined above.

Remark 15 It turns out that for a Banach space X, we have $\lambda(X) = \lambda(B_X)$. In fact, using some well known techniques involving the so-called Gromov Hausdorff distance between metric spaces, it can be shown [8] that for a subset $S \subset M$, the equality: $\lambda(S) = \lambda(M)$ holds provided there exists a bi-Lipschitz map $\delta : M \to M$, with the following properties:

(i) $S \subset \delta(S)$
(ii) $\cup_{i=0}^{\infty}\delta^i(S)$ is dense in M.
(iii) The linear map $\Delta : \text{Lip}(M) \to \text{Lip}(M)$ defined by $\Delta(f)(x) = f(\delta(x)), x \in M$ satisfies

$$\|\Delta\|\|\Delta^{-1}\| = 1.$$

Indeed, for the case of M being a Banach space X, the choice of $S = B_X$ and of δ given by $\delta(x) = 2x$ satisfies (i)–(iii) above and so yields the equality $\lambda(X) = \lambda(B_X)$ as desired.

Combined the last assertion with Theorem 14, it follows that for an infinite dimensional Banach space X, we have $\lambda(B_{X^*}) = \infty$. Now choosing X to be separable yields a compact metrizable spaces $M = B_{X^*}$ (in the weak*-topology) for which, however, it is not clear if $\lambda(M) = \infty$. On the other hand, there are examples of compact metric spaces $\mathfrak{I}(M)$ for which the Lipschitz free space (M) lacks the λ-bounded approximation property (λ-BAP) (See [14]). Interestingly, it turns out that these two properties are closely related. In fact, for compact metric spaces M, the condition $\lambda(M) < \infty$ leads to the Lipschitz free space $\mathfrak{I}(M)$ having BAP. The latter statement is a consequence of the following uniform extension phenomenon characterising λ-BAP. (See Godefroy [15]).

Theorem 16 *For a compact metric space* M*, TFAE:*

(i) The space $\mathfrak{I}(M)$ *has* λ*-BAP.*
(ii) There exists a sequence $\epsilon_n > 0$ *with* $\epsilon_n \to 0$ *such that for some sequence* $(M_n)_n$ *of finite* ϵ_n*-dense subsets of* M*, there exist operators* $\psi_n : Lip(M_n) \to Lip(M)$ *with* $\|\psi_n\| \leq \lambda$ *and* $\|R_n\psi_n - I\|_{L,\infty} \to 0$*, where* R_n *is the restriction operator to* M_n*,* $R_n : Lip(M) \to Lip(M_n)$ *and* $\|T\|_{L,\infty}$ *denotes the norm of an operator* T *acting on the domain space being equipped with the Lipschitz norm and the range space with the uniform norm.*

Noting that the finiteness of $\lambda(M)$ implies condition (ii) of the above theorem, the following question is natural:

Question: For a compact metric space M such that $\mathfrak{I}(M)$ has λ-BAP, does it follow that $\lambda \leq \lambda(M) < \infty$?

The above question was explicitly posed by Godefroy in [15]. We feel the answer to this question would turn out to be negative if the question mentioned in the paragraph preceding Theorem 16 had an affirmative answer. In other words, we have

Question: Is it true that $\lambda(M) = \infty$, where M is the closed unit ball of the dual of an infinite dimensional Banach space equipped with the weak*-topology.

We conjecture that the answer is in the affirmative. Prof. A Brudnyi surmises the assertion to be valid in the more general case of infinite dimensional compact metric spaces.

(b) As opposed to the existence or otherwise of extension operators involving spaces of Lipschitz maps on (domains in) metric spaces, the famous Dugundji–Tietze theorem guarantees the existence of extension operators in the context of spaces of continuous functions defined on closed subspaces of metric spaces. However, for nonmetrizable spaces, such an extension operator may not exist. In a recent work, A. Aviles and W. Marciszweski [3] show that there exist subsets S of the unit ball B_H of a nonseparable Hilbert space H which do not admit an extension operator from $C(S)$ to $C(B_H)$.

4 Lipschitz Analogue of Injectivity

In the category of Banach spaces, the class of injective Banach spaces and the class of Hilbert spaces are in a certain sense dual to each other and intersect precisely in finite dimensional spaces. As seen in the previous section, many of the linear properties of Hilbert spaces lend themselves to a suitable nonlinear analogue. We shall soon discover that it is possible to push this analogy further in the context of injective Banach spaces and characterise them in terms of Lipschitz maps.

Definition 17 A Banach space X is said to be injective if for each Banach space Z, each X-valued bounded linear map from any subspace of Z extends to a bounded linear map on Z.

Comments:

(i) It is not difficult to see that injectivity of a Banach space X implies the existence of $\lambda \geq 1$ such that X is λ-injective, i.e., each X-valued continuous linear mapping f on any subspace of a given Banach space can be extended to a continuous linear map g on the whole space such that $\|g\| \leq \lambda \|f\|$. Thus, each injective Banach space is λ-injective for some $\lambda \geq 1$. A complete characterisation of 1-injective Banach spaces is given in the following theorem of Nachbin, Goodner, Hasumi and Kelley.

Theorem 18 (See [1, Chap. 4]) *A Banach space X is (isometrically) injective if and only if it is linearly isometric with $C(K)$ for some compact Hausdorff space K which is extremally disconnected (i.e., each open set in K has an open closure).*

(ii) It is easy to check that a Banach space X is injective if and only if it is complemented in any larger Banach space.

Following is the Lipschitz analogue of the above characterisation of 1-injectivity.

Theorem 19 *For a Banach space X, TFAE:*

(i) X is 1-injective.
(ii) X is hyperconvex.
(iii) X is a 1-absolute Lipschitz retract.

Hyperconvexity as referred to in the above theorem is to be understood in the following sense.

Definition 20 A metric space X is said to be *hyperconvex* (in the sense of Aronszajn and Panitchpakdi [2]) if $\cap_{i \in \Lambda} B(x_i, r_i) \neq \emptyset$ for any collection $\{x_i\}_{i \in \Lambda}$ in X and $\{r_i\}_{i \in \Lambda} \subset \mathbb{R}^+$ such that $d(x_i, x_j) \leq r_i + r_j$, $\forall i, j \in \Lambda$.

For normed spaces, the above condition is easily seen to be equivalent to the binary intersection property (BIP).

Remarkably, the above characterisation also holds for metric spaces X which are 1-injective in the sense of Banach spaces except that in the metric setting, what we are seeking is the extendability of X-valued 1-Lipschitz mappings on any metric space to a Lipschitz map on any larger metric space with the same Lipschitz norm. Equivalently, X is a 1-Lipschitz retract of any metric space containing it, or in other words, X is a 1-absolute Lipschitz retract. Also, the result no longer holds for λ—injectivity as long as $\lambda > 1$—the space c_0 is well known to be a 2-absolute Lipschitz retract which is not 2-injective. In any case, it turns out that X being a λ-absolute Lipschitz retract implies that X^{**} is λ-injective. A proof of this assertion follows from Theorem 9 of Section B as shown in a recent work of the author [28].

The following useful theorem guarantees the existence of an extension operator from the space of Lipschitz functions on a certain class of Banach spaces to that on their superspaces. More generally, we have

Theorem 21 *Let X be a Banach space which is an absolute Lipschitz retract, i.e. there exists a Lipschitz retract onto X of any Banach space Z containing X as a subspace. Then for each Banach space Y, there exists a continuous linear extension operator from $Lip(X, Y)$ to $Lip(Z, Y)$.*

The easy proof follows by mapping $S \in \text{Lip}(X, Y)$ to $ST \in \text{Lip}(Z, Y)$ where T is a Lipschitz retraction of Z onto X. Further, it follows from Theorem 9 that a subspace which is a Lipschitz retract of a larger Banach space is locally complemented in it. Also the question regarding the existence of a locally complemented subspace which is not a Lipschitz retract of the ambient space is intimately connected with the famous open problem whether a Banach space is a Lipschitz retract of its bidual (see [7, p. 183]).

With the definition of absolute Lipschitz retract being as given above, it would be interesting to explore the ℓ_2-analogue of this definition—i.e. $Y = \ell_2$—and ask for the description of Banach spaces Z such that the extendability of ℓ_2-valued Lipschitz mappings on Z to larger Banach spaces could be effected within a uniform bound on the Lipschitz norm of the extended map. The linear counterpart of this situation is provided by the following theorem which characterises such Banach spaces Z as those which are Hilbert–Schmidt, in the sense that each bounded linear map acting on a Hilbert space is a Hilbert–Schmidt map as soon as it factors over Z.

Theorem 22 ([23]) *For a Banach space Z, the following are equivalent:*

(i) A bounded linear map on Z into ℓ_2 extends to a bounded linear map on any superspace of Z.
(ii) Z is a Hilbert-Schmidt space.

The search for Banach spaces Z for which property (i) of Theorem 22 holds for Lipschitz maps remains elusive. We conclude this section with the following conjecture:

Conjecture: Given a Banach space Z, the Lipschitz free space $\mathfrak{I}(Z)$ is a Hilbert Schmidt space if (and only if) each ℓ_2-valued Lipschitz map on Z extends to a Lipschitz map on any Banach space containing Z.

Comments (iv): Part of the programme involving the search for a suitable Lipschitz analogue of a linear phenomenon in Banach space theory seeks to propose and devise a satisfactory theory of well known operator ideals of linear operators in Banach space theory, including in particular, the ideal of p-summing maps in the Lipschitz setting. The success story surrounding recent breakthroughs in Banach space theory has been due in no small measure to the decisive role played by the theory of p-summing maps in linear theory. In recent years, several approaches have been proposed to the idea of a polynomial/Lipschitz analogue of p-summing maps and their relatives including a Lipschitz variant of Grothendieck's theorem which has been proved for Lipschitz maps defined on the so called weighted metric trees. This has been made possible by extending the theory of p-summing maps to Lipschitz p-summing maps acting between metric spaces, in tune with the general philosophy that the numerical quantities associated with the linear structure of Banach spaces lend themselves to an appropriate analogue in the metric space setting. The theory of type and cotype which has witnessed such a far reaching influence even beyond the frontiers of Banach space theory has been successfully employed in the metric space setting—thanks to the seminal work of Mendelson and Naor [24] and Keith Ball [4] in this direction. This has led to a remarkable breakthrough involving the discovery of a nonlinear analogue of the famous Maurey Extension Theorem in the framework of metric spaces [4]. However, the picture is far from satisfactory in the nonlinear setting as can be testified by a whole lot of problems that continue to remain open in this area of research. We isolate but two of them as they are crucial to a possible resolution of the conjecture proposed above. In the following, the symbol Π_2^{Lip} shall denote the class of Lipschitz 2-summing maps.

Problem 1 Is it true that a Banach space X is Hilbert Schmidt if and only if

$$\text{Lip}(X, \ell_2) = \Pi_2^{\text{Lip}}(X, \ell_2)$$

An important ingredient of an approach to the proof of the above problem would be provided by an affirmative solution to the following problem.

Problem 2 To what extent do the Banach space properties of a Banach space X carry over to the free space $\mathfrak{I}(X)$ over X? In particular, does it follow that for X being a Hilbert Schmidt space X, it follows that the free space $\mathfrak{I}(X)$ is also a Hilbert Schmidt space?

C. Lipschitz compact maps: The study of compact linear maps acting between Banach spaces is an old and a thoroughly investigated theme in functional analysis. An important outcome of this line of research has been the extent to which it has become possible to capture the structure/geometry of a Banach space in terms of the coincidence of the class of compact maps acting on or into them with other classes of maps encountered in functional analysis. An important example of this phenomenon is furnished by a well known theorem of Davis, Figiel, Johnson and Pelczynski [10] which characterises weakly compact linear maps precisely as those that factor over a reflexive Banach space. The nonlinear counterpart of this assertion is a recent development in this area that characterises weakly compact Lipschitz maps exactly as in the linear setting. But first a definition.

Definition 23 (*see* [16]) Given a map $f : M \to X$ acting on a metric space M and taking values in a Banach space X, its Lipschitz image is defined to be the set $\text{Lim}(f)$ given by

$$\lim(f) = \left\{ \frac{f(x) - f(y)}{d(x, y)}, x, y \in X, x \neq y \right\}.$$

It is clear that for linear maps acting between normed spaces, $\lim(f)$ coincides with the image of the unit ball under f. This motivates the following definition of a Lipschitz (weakly) compact map.

Definition 24 (*see* [17]) With the setting of the above definition and a distinguished point $\theta \in M$, the map $f : M \to X$ with $f(\theta) = 0$ is said to be Lipschitz weakly(compact) if $\lim(f)$ is a (weakly) relatively compact subset of X.

Comments:

(i) It is clear that for linear maps acting between Banach spaces, the notions of Lipschitz (weak) compactness and (weak) compactness coincide.
(ii) There is a canonical isometric embedding $e_M : x \to \delta_x$ of M into $\Im(M)$, where δ_x is defined by: $\delta_x(f) = f(x),\ f \in \mathrm{Lip}_0(M, X)$. This makes it possible to define a (unique) map $T_f : \Im(M) \to X$—the linearisation of the given Lipschitz—such that $T_f \circ e_M = f$. Further, it can be shown that f is Lipschitz (weakly) compact if and only if T_f is a (weakly) compact linear map.

We have the following theorem.

Theorem 25 (see [17]) *With M and X as given above, a Lipschitz map $f : M \to X$ is weakly compact if and only if it factors over a reflexive Banach space: there exist a reflexive Banach space Z, a bounded linear map $h : Z \to X$ and a Lipschitz map $g : M \to Z$ such that $h \circ g = f$.*

Epilogue

The linear theory of Banach spaces which has witnessed spectacular success over the past several decades has, in recent years, lent itself to extensive research from a variety of new points of view that include (a) quantisation of the classical theory of Banach spaces (b) polynomial variants of linear theory of Banach spaces and most importantly (c) Lipschitz classification of Banach spaces. It turns out that, at least in most cases of interest, the linear theory of Banach spaces is captured to an appreciable extent by their Lipschitz structure as is testified by a whole lot of results discussed in the previous paragraphs. The well known but a highly nontrivial fact due to Enflo (see [7, Chap. 10]) already mentioned in the introduction - that a Banach space which is Lipschitz (or even uniformly) homeomorphic to a Hilbert space is also linearly homeomorphic to it shows how the linear structure of a Banach space is closely tied up its metric (uniform) structure. Let us also mention that in recent years, a great deal of effort has been invested in investigating Lipschitz analogues of various notions and results from the linear theory that include, in particular, the notions of numerical range, numerical index, Daugavet property etc. which coincide with their counterparts in the linear setting, as it indeed should be. An important open problem belonging to this circle of ideas concerns the equality of the (linear) numerical index of a Banach space with the Lipschitz numerical index, even as the equality of these numerical parameters has been verified for a wide class of Banach space that include finite dimensional spaces, Hilber spaces, and more generally for L_p-spaces on a σ-finite positive measure space (Ω, Σ) with Σ being countably generated and the class of separable Banach spaces with the Radon Nikodym property (See [30] for a comprehensive account on this line of research). Very recently (arXiv:1601.07821v1 [math.FA]), efforts have been initiated to investigate the Lipschitz variants of the Bishop-Phelps phenomena involving the structure of norm attaining maps on a given Banach space.

Acknowledgements This work was initiated during the author's visit at the Harish Chandra Institute, Allahabad during Jan.–Feb.2015. He would like to thank his hosts for their kind invitation and for the hospitality provided to him during the period of his stay at the institute. Thanks are also due to the CSIR, New Delhi for providing support under its Emeritus Scientist Scheme vide grant No. 21 (0969)/13/EMR-II

References

1. Albiac, F., Kalton, N.J.: Topics in Banach Space Theory. GTM, vol. 223. Springer, New York (2006)
2. Aronszajn, N., Panitchpakdi, P.: Extension of uniformly continuous transformations and hyper-convex metric spaces. Pac. J. Math. **6**, 405–439 (1956)
3. Aviles, A., Marciszweski, W.: Extension operators on balls and on spaces of finite sets. Stud. Math. **227**(2), 165–182 (2015)
4. Ball, K.: Markov chains, Riesz transforms and Lipschitz maps. Geom. Funct. Anal. **2**, 137–172 (1992)
5. Benitz, C., Sarantopoulos, Y.: Characterisation of real inner product spaces by means of symmetric bilinear forms. J. Math. Anal. Appl. **180**, 207–220 (1993)
6. Benitz, C., Sarantopoulos, Y.: Lower bounds for norms of products of polynomials. Math. Proc. Camb. Philos. Soc. **124**, 395–408 (1998)
7. Benjamini, Y., Lindenstrauss, J.: Geometric Nonlinear Functional Analysis, Volume 1, vol. 48. American Mathematical Society, Colloquium Publications, Providence (2000)
8. Brudnyi, A., Brudnyi, Yu.: Linear and nonlinear extensions of Lipschitz functions from subsets of metric spaces. Saint Petersberg Math. Jour. **19**(3), 397–406 (2008)
9. Brudnyi, A., Brudnyi, Yu.: Metric spaces with linear extensions preserving Lipschitz condition. Am. J. Math. **129**, 217–314 (2007)
10. Davis, W.J., Figiel, T., Johnson, W.B., Pelczynski, A.: Factoring weakly compact operstors. J. Func. Anal. **1**, 311–327 (1974)
11. Diestel, J., Jarchow, H., Tonge, A.: Absolutely Summing Operators, Cambridge Tracts in Mathematics. Camb. Univ. Press, Cambridge (1995)
12. Enflo, P.: Uniform structures and square roots in topological groups, I, II, Isr. J. Math. **8**, 23–252, 253–272 (1970)
13. Fernandez-Unzueta, M., Prieto, A.: Extension of polynomials defined on subspaces. Math. Proc. Camb. Philos. Soc. **148**, 505–518 (2010)
14. Godefroy, G., Ozawa, N.: Free Banach spaces and the approximation properties. Proc. Am. Math. Soc. **142**(5), 1681–1687 (2014)
15. Godefroy, G.: Extension of Lipschitz functions and Grothendiecks bounded approximation property. North West. Eur. J. Math. **1**, 1–6 (2015)
16. Guerrero, J.B., Rodriguez-Palacios, A.: Transitivity of the norm on Banach spaces. Extr. Math. **17**(1), 1–58 (2002)
17. Jimenez-Vargas, A., Sepulcre, J.M., Villegas-Vallecillos, M.: Lipschitz compact operators. J. Math. Anal. Appl. **415**, 889–901 (2014)
18. Joichi, J.T.: Normed linear spaces equivalent to inner product spaces. Proc. Am. Math. Soc. **17**, 423–426 (1966)
19. Kadec, M.I.: A proof of topological equivalence of all separable infinite dimensional Banach spaces. Funk. Anal. i. Priloz. **1**, 61–70 (1967)

20. Kakutani, S.: Some characterisations of Euclidean space. Jpn. J. Math. **16**, 93–97 (1939)
21. Kalton, N.J.: Locally complemented subspaces and L_p-spaces for $0 < p < 1$. Math. Nachr. 115, 71–94 (1984)
22. Kaufmann, P.L.: Products of Lipschitz-free spaces and applications. Stud. Math. **226**(3), 213–227 (2015)
23. Mazur, S., Ulam, S.: Sur les transformations isometriques despaces vectoriels norms. C.R. Acad. Sci. Paris **194**, 946–948 (1932)
24. Mendel, M., Naor, A.: Metric cotype. Ann. Math. **168**, 247–298 (2008)
25. Pisier, G.: Similarity problems and completely bounded maps. Springer Lecture Notes, vol. 1618 (1995)
26. Reich, S.: Extension problems for accretive subsets of Banach spaces. J. Func. Anal. **26**, 378–395 (1977)
27. Sofi, M.A.: Some problems in functional analysis inspired by Hahn-Banach type theorems. Ann. Func. Anal. **5**(2), 1–29 (2014)
28. Sofi, M.A.: On nonlinear structure of Banach spaces and characterization of Hilbert spaces (submitted)
29. Stein, E.M.: Singular integrals and differentiability properties of functions. Princeton University Press, Princeton (1970)
30. Wang, R., Huang, X., Tan, D.: On the numerical radius of Lipschitz operators in Banach spaces. J. Math. Anal. Appl. **411**(1), 118 (2014)

Some Results on Fixed Points of Weak Contractions for Non Compatible Mappings via (E.A)-Like Property

T. Som, A. Kundu and B.S. Choudhury

Abstract In the present work we have established common fixed point results for non compatible mappings satisfy some weak contractions in metric spaces extending the works of Pathak et al. FILOMAT 21(2):211–234, 2007, [31], Babu and Alemayehu, J. Adv. Res. Pure Math. 2(2), 89–106, 2010, [7] and others in turn. The results are based on the newly introduced (E.A) like properties and common (E.A)-like properties in metric spaces, the initial version of which was introduced by Aamri and Moutawakil in 2002.

Keywords R-weakly commuting · Occasionally weakly compatible · Weak contraction · Control functions · E.A. Like property

AMS Subject Classification 47H10 · 54H25

1 Introduction and Mathematical Preliminaries

In fixed point theory we are always concerned with finding a set of sufficient conditions for the mappings and the spaces concerned, which guarantee a fixed point or a common fixed point result. Common fixed point theorems for contractive type mappings necessarily require a commutativity or a variant condition, a condition on the ranges of the mappings, continuity of one or more mappings besides a contractive

T. Som (✉)
Department of Mathematical Sciences, IIT (BHU), Varanasi 221005, India
e-mail: tsom.apm@iitbhu.ac.in

A. Kundu
Department of Mathematics, Siliguri Institute of Technology, Darjeeling 734009, West Bengal, India
e-mail: kunduamaresh@yahoo.com

B.S. Choudhury
Department of Mathematics, IIEST, Shibpur, Howrah 711103, West Bengal, India
e-mail: binayak12@yahoo.co.in

J.M. Cushing et al. (eds.), *Applied Analysis in Biological and Physical Sciences*, Springer Proceedings in Mathematics & Statistics 186,
DOI 10.1007/978-81-322-3640-5_27

condition. And every significant fixed point or common fixed point theorem attempts to weaken or obtain a necessary version of one or more of these conditions [5, 6, 17, 18, 28, 29]. By weakening or adding different conditions, several authors have generalized Banach's contraction mapping principle in different directions in the context of metric spaces [1, 9, 26, 35]. The notion of weakly commuting maps was initiated by Sessa [34]. Jungck [19] gave the concept of of compatible maps and showed that weakly commuting mappings are compatible, but converse is not true. Junck [20] further weakened the notion of compatibility by introducing weak compatibility. Of course weakly compatible mapping need not be compatible and several fixed point theorem has been deduced with out appeal of continuity.

However, on the other hand the study of common fixed points of noncompatible mappings is also very interesting [3]. Work along these lines has recently been initiated by Pant [27] by introducing point wise R-weakly commuting mappings. There are examples of noncompatible maps among pairs of mappings which are discontinuous at their common fixed point [30].

The mappings f and g are said to be noncompatible if there exists a sequence $\{x_n\}$ in X such that for some t in X but $\lim_{n\to\infty} d(fgx_n, gfx_n)$ is either non-zero or non-existent.

The following are some essential concepts for our discussion in this paper.

Definition 1.1 Let X be a set and let f, g be two self mappings of X. A point x in X is called a coincidence point of f and g iff $fx = gx$. We shall call $w = fx = gx$ a point of coincidence of f and g.

Definition 1.2 (*Weakly compatible mappings*) [20] Two mappings $f, g : X \to X$, where (X, d) is a metric space, are said to be weakly compatible if they commute at their coincidence points, that is, if $ft = gt$ for some $t \in X$ implies that $fgt = gft$.

Definition 1.3 ([27]) Two selfmaps f and g on a metric space (X, d) are called point wise R-weakly commuting if given x in X there exists $R > 0$ such that

$$d(fgx, gfx) \leq Rd(fx, gx).$$

To generalise the notion of weakly compatible maps, in 2008, Al-Thagafi and Shahzad [4] defined the concept of occasionally weakly compatible (owc) as mentioned below.

Definition 1.4 ([4]) Two self-maps f and g of a set X are occasionally weakly compatible *(owc)* iff there is a point x in X which is a coincidence point of f and g (i.e., $fx = gx$) at which f and g commutate (i.e., $fgx = gfx$).

Lemma 1.1 ([21]) *Let X be a set, f, g are owc self-mappings of X. If f and g have a unique point of coincidence, $w = fx = gx$, then w is the unique common fixed point of f and g.*

Remark Every pair of weakly compatible maps is occasionally weakly compatible, but its converse need not be true [4].

Two weakly compatible mappings having coincidence points are occasionally weakly compatible. In [2], it was shown that the converse is not true.

In 2002, Aamri and Moutawakil [1] generalized the notion of non compatible mapping in metric space by introducing the concept of (E.A) property and imposing it they settle down the general commutativity condition to the commutativity condition at a point of coincidence without apple of continuity of a mappings. we also notice that (E.A) property minimizes the commutativity conditions of the maps to the commutativity at their points of coincidence. More-over, (E.A) property allows replacing the completeness requirement of the space with a more natural condition of closeness of the range. If f and g are both noncompatible then they do satisfy the (E.A) property. The beauty of this property lies in the fact that it allows the construction procedure to get Cauchy sequence in a natural way. On the other hand the (E.A) property enables us to study the existence of common fixed point of nonexpansive or Lipschitz type conditions in the setting of noncomplete metric spaces.

Definition 1.5 [1, 8] Let f and g be two self-maps of a metric space (X, d). We say that f and g satisfy the property (E.A) if there exists a sequence $\{x_n\}$ in X such that

$$\lim_{n\to\infty} fx_n = \lim_{n\to\infty} gx_n = z, \text{ for some } z \in X.$$

Definition 1.6 [31] (*Common (E.A) Property*) Let $A, B, S, T : X \to X$ where (X, d) is a metric space. Then the pair $\{A, S\}$ and $\{B, T\}$ are said to satisfy common (E.A) property if there exist two sequences $\{x_n\}$ and $\{y_n\}$ in X such that

$$\lim_{n\to\infty} Ax_n = \lim_{n\to\infty} Sx_n = \lim_{n\to\infty} Ty_n = \lim_{n\to\infty} By_n = z \text{ for some } z \in X.$$

Wadhwa et al. in [37] introduced the notion of (E.A)-like property in a fuzzy metric space, by relaxing the containment condition of range spaces. Here we define (E.A)-like property in a metric space.

Definition 1.7 Let f and g be two self-maps of a metric space (X, d). We say that f and g satisfy the property (E.A.) Like property if there exists a sequence $\{x_n\}$ such that

$$\lim_{n\to\infty} fx_n = \lim_{n\to\infty} gx_n = z, \text{ for some } z \in fX \text{ or } gX, \; i.e., z \in f(X) \cup g(X).$$

Definition 1.8 (*Common (E.A)-like Property*) Let A, B, S and T be self maps of a metric space (X, d).Then the pairs (A, S) and (B, T) said to satisfy common(E.A)-like property if there exists two sequences $\{x_n\}$ and $\{y_n\}$ in X such that

$$\lim_{n\to\infty} Ax_n = \lim_{n\to\infty} Sx_n = \lim_{n\to\infty} Ty_n = \lim_{n\to\infty} By_n = z \text{ for } z \in S(X) \cap T(X) \text{ or } z \in A(X) \cap B(X).$$

Pathak, Rodrigues-Lopez and Verma [31] observed that 'weak compatibility' and 'property (E.A)' are independent of each other. Also it was shown by Babu and Alemayehu [7] that ‘occasionally weak compatibility’ and ‘property (E.A)’ are independent to each other.

Role of (E.A)-like property in proving common fixed point theorems can be concluded by following,

(1) if two mappings satisfy (E.A)-like property then they satisfy (E.A) property also
(2) *(E.A)-like property* relaxes the condition of containment of ranges and closeness of the ranges which are necessary with (E.A) property.

Alber and Guerre-Delabriere in [5] suggested another way of generalization of the Banach contraction principle by introducing the concept of weak contraction in Hilbert spaces. Rhoades [33] has shown that the result which Alber et al. proved in [5] is also valid in complete metric spaces. Particularly, in [13], Dutta and Choudhury has proved a generalization different from that used by Rhoades [33]. Other weakly contractive mappings and their generalizations and extensions in various space have been discussed in several works some of which are noted in [10–19, 22–25, 32, 36, 38].

The purpose of this paper is to prove common fixed point theorems for generalised weak contractions via (E.A)-like property in metric space. Here we extend the three theorems to the case of non compatible mappings such that they mutually satisfy the(E.A)-like- property without using the continuity requirement or the containment of ranges and closeness of the ranges. So our work provides some new contributions to the field of metric fixed point theory.

2 Main Results

In this section, we prove some common fixed point theorems for non compatible in metric spaces where the functions mutually satisfy “(E.A)-like property”. Now we prove the following theorem.

Theorem 2.1 *Let f and g be two weakly compatible self mappings of a metric spaces (X, d) and satisfy for all $x, y \in X$.*

(a)

$$d(fx, fy) \leq k \max(d(gx, gy), d(fx, gx), d(fy, gy), d(fx, gy), d(fy, gx)) \text{ for } 0 \leq k < 1 \tag{2.1}$$

(b) if f and g have “(E.A)-like property” then f and g have a unique common fixed point.

Proof Since f and g satisfy “(E.A)-like property” there exists a sequence $\{x_n\}$ in X such that

$$\lim_{n\to\infty} fx_n = \lim_{n\to\infty} gx_n = z \text{ for some } z \in fX \text{ or } gX. \tag{2.2}$$

Suppose that $\lim_{n\to\infty} fx_n = z \in gX$, therefore, $z = gu$ for some $u \in X$. Now we prove that $fu = gu$. From (2.1) we get

$$d(fu, fx_n) \le k \max(d(gu, gx_n), d(fu, gu), d(fx_n, gx_n), d(fu, gx_n), d(fx_n, gu)) \quad (2.3)$$

Taking the limit as $n \to \infty$ in the above inequality and using (2.2) we get

$$d(fu, z) \le k \max(d(z, z), d(fu, z), d(z, z), d(fu, z), d(z, z))$$

$$d(fu, z) \le kd(fu, z).$$

Since $k < 1$, only we have $d(fu, z) = 0$, that is $fu = z$.

Therefore,

$$fu = z = gu. \quad (2.4)$$

Since f and g are weakly compatible therefore from definition it follows,

$$fz = fgu = gfu = gz. \quad (2.5)$$

Now we prove that $f(z) = z$. Suppose that $f(z) \neq z$.

Then, from (2.1), we obtain

$$d(fz, fx_n) \le k \max(d(gz, gx_n), d(fz, gz), d(fx_n, gx_n), d(fz, gx_n), d(fx_n, gz)) \quad (2.6)$$

Taking $n \to \infty$, in the above two inequalities, using (2.5) we get

$$d(fz, z) \le k \max(d(gz, z), d(fz, gz), d(z, z), d(fz, z), d(z, gz)).$$

In view of (2.5), i.e., $fz = gz$ the above inequality implies

$$d(fz, z) \le kd(fz, z).$$

Since $k < 1$, only we have $z = fz$.

Hence

$$z = fz = gz. \quad (2.7)$$

That is z is a common fixed point of f and g. The uniqueness is easily followed from the theorem.

Corollary *Let f and g be two R-weakly commuting self mappings of a metric spaces (X, d) and satisfy both the conditions (a) and (b) of Theorem 2.1 for all $x, y \in X$. Then f and g have a unique common fixed point.*

Theorem 2.2 *Let A, B, S and T be self mappings of a metric spaces (X, d) and satisfy for all $x, y \in X$*

(a)

$$\psi(d(Ax, By)) \le \psi(M(x, y)) - \beta(M(x, y)) + \theta(N(x, y)) \tag{2.8}$$

where

$$M(x, y) \in \{d(Sx, Ty), d(Ax, Sx), d(By, Ty), \tfrac{1}{2}[d(Sx, By) + d(Ax, Ty)]\} \tag{2.9}$$

and

$$N(x, y) = \min\{d(Sx, Ty), d(Ax, Sx), d(By, Ty), d(Sx, By), d(Ax, Ty)\} \tag{2.10}$$

(i) $\psi : [0, \infty) \to [0, \infty)$ is a monotone non-decreasing function such that $\psi(t) = 0$ if and only if $t = 0$, called Altering distance function.
(ii) $\beta : [0, \infty) \to [0, \infty)$ is a lower semi continuous function satisfy $\beta(t) = 0$ iff $t = 0$.
(iii) $\theta : [0, \infty) \to [0, \infty)$ is a continuous function such that $\theta(t) = 0$ if and only if $t = 0$.

Also,

(b1) if the pairs (A, S) and (B, T) satisfy common "(E.A)-like property"
(b2) both the pairs (A, S) and (B, T) are Occasionally Weakly Compatible on X, then A, B, S and T have a unique common fixed point in X.

Proof Since the pairs (A, S) and (B, T) satisfy common "(E.A)-like property" therefore there exists two sequences $\{x_n\}$ and $\{y_n\}$ in X such that

$$\lim_{n\to\infty} Ax_n = \lim_{n\to\infty} Sx_n = \lim_{n\to\infty} Ty_n = \lim_{n\to\infty} By_n = z \tag{2.11}$$

where,

$$z \in S(X) \cap T(X) \text{ or } z \in A(X) \cap B(X). \tag{2.12}$$

First we assume that $z \in S(X) \cap T(X)$, then

$$\lim_{n\to\infty} Ax_n = z \in SX \text{ that is, } z = Su \text{ for some } u \in X. \tag{2.13}$$

Putting $x = u$ and $y = y_n$ in (2.8) and using (2.9), (2.10) we have

$$\psi(d(Au, By_n)) \le \psi(M(u, y_n)) - \beta(M(u, y_n)) + \theta(N(u, y_n)) \tag{2.14}$$

where

$$M(u, y_n) \in \{d(Su, Ty_n), d(Au, Su), d(By_n, Ty_n), \tfrac{1}{2}[d(Su, By_n) + d(Au, Ty_n)]\} \tag{2.15}$$

and

$$N(u, y_n) = \min\{d(Su, Ty_n), d(Au, Su), d(By_n, Ty_n), d(Su, By_n), d(Au, Ty_n)\}. \tag{2.16}$$

Taking $n \to \infty$, in (2.15) and (2.16) respectively, and using (2.11), (2.13), we obtain

$$\lim_{n\to\infty} M(u, y_n) \in \{d(z, Au), \tfrac{1}{2}d(Au, z)\} \tag{2.17}$$

$$\text{and } \lim_{n\to\infty} N(u, y_n) = 0. \tag{2.18}$$

Now taking the limit infimum as $n \to \infty$, in (2.14), and using (2.17) and (2.18) and the properties of ψ, β and the fact that $\theta(0) = 0$ we have,

$$\liminf_{n\to\infty} \psi(d(Au, By_n)) \leq \liminf_{n\to\infty} \psi(M(u, y_n)) - \limsup_{n\to\infty} \beta(M(u, y_n))$$

From the above inequality, for two different values of $M(u, y_n)$, only we have $d(z, Au) = 0$, that is, $z = Au = 0$, hence

$$Au = z = Su. \tag{2.19}$$

On the other hand as (B, T) satisfy common "(E.A)-like property" then

$$\lim_{n\to\infty} By_n = z \in TX \text{ that is, } z = Tv \text{ for some } v \in X. \tag{2.20}$$

Putting $x = x_n$ and $y = v$ in (2.8) and using (2.9), (2.10) we have

$$\psi(d(Ax_n, Bv)) \leq \psi(M(x_n, v)) - \beta(M(x_n, v)) + \theta(N(x_n, v)) \tag{2.21}$$

where

$$M(x_n, v) \in \{d(Sx_n, Tv), d(Ax_n, Sx_n), d(Bv, Tv), \tfrac{1}{2}[d(Sx_n, Bv) + d(Ax_n, Tv)]\} \tag{2.22}$$

and

$$N(x_n, v) = \min\{d(Sx_n, Tv), d(Ax_n, Sx_n), d(Bv, Tv), d(Sx_n, Bv), d(Ax_n, Tv)\}. \tag{2.23}$$

Taking $n \to \infty$, in (2.22) and (2.23) respectively, and using (2.11), (2.16), we obtain

$$\lim_{n\to\infty} M(x_n, v) \in \{d(z, Bv)\} \tag{2.24}$$

$$\text{and } \lim_{n\to\infty} N(u, y_n) = 0. \tag{2.25}$$

Now taking the limit infimum as $n \to \infty$, in (2.21), and using (2.24), (2.25) and the properties of ψ, β and the fact that $\theta(0) = 0$ we have,

$$\liminf_{n\to\infty} \psi(d(Ax_n, Bv)) \leq \liminf_{n\to\infty} \psi(M(x_n, v)) - \limsup_{n\to\infty} \beta(M(x_n, v)).$$

From the above inequality we have $\beta(d(z, Bv)) = 0$, which implies $d(z, Bv) = 0$, that is

$$Bv = z = Tv. \tag{2.26}$$

Since the pair $(A; S)$ is owc, there exists u_0 such that

$$Au_0 = Su_0 = u_1 \tag{2.27}$$

and

$$\begin{aligned} ASu_0 &= SAu_0 = w_1 \\ i.e., Au_1 &= Su_1 = w_1. \end{aligned} \tag{2.28}$$

Again the pair $(B; T)$ is owc, then there exists v_0 such that

$$Bv_0 = Tv_0 = u_2. \tag{2.29}$$

And also satisfy

$$\begin{aligned} BTv_0 &= TBv_0 = w_2 \\ i.e., \ Bu_2 &= Tu_2 = w_2. \end{aligned} \tag{2.30}$$

We now show that $w_1 = z = w_2$. Suppose $w_1 \neq z$. Now using (2.8)–(2.10), (2.26) and (2.28) we get

$$\psi(d(w_1, z)) = \psi(d(Au_1, Bv)) \leq \psi(M(u_1, v)) - \beta(M(u_1, v)) + \theta(N(u_1, v)) \tag{2.31}$$

where

$$M(u_1, v) \in \{d(Su_1, Tv), d(Au_1, Su_1), d(Bv, Tv), \frac{1}{2}[d(Su_1, Bv) + d(Au_1, Tv)]\}$$

using (2.26) and (2.28) we have

$$M(u_1, v) \in \{d(w_1, z)\} \tag{2.32}$$

again from (2.10)

$$N(u_1, v) = \min\{d(Su_1, Tv), d(Au_1, Su_1), d(Bv, Tv), d(Su_1, Bv), d(Au_1, Tv)\}$$

In view of (2.26), (2.28)

$$N(u_1, v) = 0. \tag{2.33}$$

Thus from (2.31) and using (2.32), (2.33) and the fact that ψ, β are continuous and $\theta(0) = 0$, we obtain

$$\psi(d(w_1, z)) \leq \psi(d(w_1, z)) - \beta(d(w_1, z))$$

the above relation holds good only when $\beta(d(w_1, z)) = 0$, that is $w_1 = z$. Similarly we can show that $w_2 = z$.

Hence

$$w_1 = z = w_2. \tag{2.34}$$

Therefore from (2.28) and (2.30), we get

$$Au_1 = Su_1 = Bu_2 = Tu_2 = z. \tag{2.35}$$

Next we prove that $u_1 = u_2$. Suppose $u_1 \neq u_2$.

Using (2.8)–(2.10) we have

$$\psi(d(u_1, u_2)) = \psi(d(Au_0, Bv_0)) \leq \psi(M(u_0, v_0)) - \beta(M(u_0, v_0)) + \theta(N(u_0, v_0)) \tag{2.36}$$

where

$$M(u_0, v_0) \in \{d(Su_0, Tv_0), d(Au_0, Su_0), d(Bv_0, Tv_0), \tfrac{1}{2}[d(Su_0, Bv_0) + d(Au_0, Tv_0)]\}$$
$$\text{that is } M(u_0, v_0) \in \{d(u_1, u_2)\} \tag{2.37}$$

and

$$N(u_0, v_0) = \min\{d(Su_0, Tv_0), d(Au_0, Su_0), d(Bv_0, Tv_0), d(Su_0, Bv_0), d(Au_0, Tv_0)\}$$
$$i.e., \; N(u_0, v_0) = 0. \tag{2.38}$$

From (2.36), using (2.37) and (2.38) we have,

$$\psi(d(u_1, u_2)) \leq \psi(d(u_1, u_2)) - \beta(d(u_1, u_2)) + \theta(0).$$

Hence using the properties of ψ, β and the fact that $\theta(0) = 0$ we arrive at a contradiction. Thus $u_1 = u_2$. Therefore,

$$Au_1 = Su_1 = Bu_1 = Tu_1 = z. \tag{2.39}$$

That is z is a point of coincidence of A, B, S and T. To prove the uniqueness suppose z_1 and z_2 ($z_1 \neq z_2$) are two points of coincidences of A, B, S and T. Therefore there exists u_1 and $u_2$in X such that

$$Au_1 = Su_1 = Bu_1 = Tu_1 = z_1 \text{ and } Au_2 = Su_2 = Bu_2 = Tu_2 = z_2. \tag{2.40}$$

From (2.8),

$$\psi(d(z_1, z_2)) = \psi(d(Au_1, Bu_2)) \leq \psi(M(u_1, u_2)) - \beta(M(u_1, u_2)) + \theta(N(u_1, u_2)) \tag{2.41}$$

where

$$\begin{aligned} M(u_1, u_2) &\in \{d(Su_1, Tu_2), d(Au_1, Su_1), d(Bu_2, Tu_2), \tfrac{1}{2}[d(Su_1, Bu_2) + d(Au_1, Tu_2)]\} \\ &\in \{d(z_1, z_2)\} \end{aligned} \tag{2.42}$$

and

$$\begin{aligned} N(u_1, u_2) &= \min\{d(Su_1, Tu_2), d(Au_1, Su_1), d(Bu_2, Tu_2), d(Su_1, Bu_2), d(Au_1, Tu_2)\} \\ i.e., \ N(u_1, u_2) &= 0. \end{aligned} \tag{2.43}$$

Hence, from (2.41), using (2.42), (2.43) and the properties of ψ, β and the fact that $\theta(0) = 0$, we get

$$\begin{aligned} \psi(d(z_1, z_2)) = \psi(d(Au_1, Bu_2)) &\leq \psi(d(z_1, z_2)) - \beta(d(z_1, z_2)) + \theta(0) \\ &< \psi(d(z_1, z_2)) \end{aligned}$$

which is a contradiction that is, $z_1 = z_2$. Hence z is a unique point of coincidence of A, B, S and T.

Since z is a unique point of coincidence of A, B, S and T, then from (2.39) we have,

$$Au_1 = Su_1 = Bu_1 = Tu_1 = z.$$

As, each of the pairs (A, S) and (B, T) are occasionally weakly compatible, thus

$$Az = ASu_1 = SAu_1 = Sz = t_1 \tag{2.44}$$

and

$$Bz = BTu_1 = TBu_1 = Tz = t_2 \tag{2.45}$$

To prove the common fixed point of A, B, S and T we show that $t_1 = z = t_1$. Suppose $t_1 \neq z$. Then

$$\psi(d(t_1, z)) = \psi(d(Az, Bu_1)) \leq \psi(M(z, u_1)) - \beta(M(z, u_1)) + \theta(N(z, u_1)) \tag{2.46}$$

where

$$\begin{aligned} M(z, u_1) &\in \{d(Sz, Tu_1), d(Az, Sz), d(Bu_1, Tu_1), \tfrac{1}{2}[d(Sz, Bu_1) + d(Az, Tu_1)]\} \\ &\in \{d(t_1, z)\} \end{aligned} \tag{2.47}$$

and

$$N(z, u_1) = \min\{d(Sz, Tu_1), d(Az, Sz), d(Bu_1, Tu_1), d(Sz, Bu_1), d(Az, Tu_1)\}$$
$$i.e., \ N(z, u_1) = 0. \tag{2.48}$$

Hence, from (2.46), using (2.47), (2.48) and the properties of ψ, β and the fact that $\theta(0) = 0$, we get

$$\begin{aligned}\psi(d(t_1, z)) = \psi(d(Az, Bu_1)) &\leq \psi(d(t_1, z)) - \beta(d(t_1, z)) + \theta(0)\\ &< \psi(d(t_1, z))\end{aligned}$$

which is a contradiction. Hence we have $t_1 = z$. Similarly we can prove that $t_2 = z$. Thus $t_1 = t_2 = z$. Therefore from (2.44) and (2.45)

$$Az = Sz = Bz = Tz = z.$$

Then z is a common fixed point of A, B, S and T. As we prove earlier z is a unique point of coincidence of A, B, S and T, therefore z is a unique common fixed point of A, B, S and T. Hence the theorem.

References

1. Aamri, M., Moutawakil, D. El.: Some new common fixed point theorems under strict contractive conditions. J. Math. Anal. Appl. **27**, 181–188 (2002)
2. Abbas, M., Doric, D.: Common fixed point theorem for four selfmappings satisfying a generalized weak contractive condition. Filomat **24**, 1–10 (2010)
3. Agarwal, R.P, Bisht, R.K., Shahzad, N.: A comparison of various noncommuting conditions in metric fixed point theory and their applications. Fixed Point Theory Appl. (2014)
4. AL-Thagafi M.A., Shahzad, N.: A note on occasionally weakly compatible maps, Int. J. Math. Anal. **32**, 55–58 (2009)
5. Alber Ya. I., Guerre-Delabriere S.: Principles of weakly contractive maps in Hilbert spaces, New Results in Operator Theory. In: Gohberg, I., Lyubich, Y. (eds.) Advances and Applications, vol. 98, pp. 7–22. Birkhäuser, Basel (1997)
6. Ali J., Imdad M.: An implicit function implies several contraction conditions, Sarajevo J. Math. **4**(17), no. 2, 269–285 (2008)
7. Babu G.V.R., Alemayehu G.N.: Point of coincidence and common fixed points of a pair of generalized weakly contractive maps. J. Adv. Res. Pure Math. **2**(2), 89–106 (2010)
8. Babu G.V.R., Sailaja P.D.: Common fixed points of (ψ, φ) weak quasi contractions with property (E.A). Int. J. Math. Sci. Comput. **1**(2), 29–37 (2011)
9. Ciric, L.B.: A generalization of Banach's contraction principle. Proc. Am. Math. Soc. **45**(2), 267–273 (1974)
10. Choudhury B.S., Metiya N.: Fixed points of weak contractions in cone metric spaces. Nonlinear Anal. TMA **72**, 1589–1593 (2010)
11. Choudhury, B.S., Kundu, A.: (ψ, α, β)-weak contractions in partially ordered metric spaces. Appl. Math. Lett. **25**, 6–10 (2012)
12. Choudhury, B.S., Konar, P., Rhoades, B.E., Metiya, N.: Fixed point theorems for generalized weakly contractive mappings. Nonlinear Anal. TMA **74**, 2116–2126 (2011)

13. Dutta P.N., Choudhury B.S.: A generalisation of contraction principle in metric spaces, Fixed Point Theory Appl. Article ID 406368 (2010)
14. Doric, D.: Common fixed point for generalized (ψ, φ)-weak contractions. Appl. Math. Lett. **22**, 1896–1900 (2009)
15. Eslamian M., Abkar A.: A fixed point theorems for generalized weakly contractive mappings in complete metric space, Ital. J. Pure Appl. Math. (To appear)
16. Harjani, J., Sadarangani, K.: Fixed point theorems for weakly contractive mappings in partially ordered sets. Nonlinear Anal. **71**, 3403–3410 (2009)
17. Imdad, M., Sharma, A., Chauhan, S.: Unifying a multitude of common fixed point theorems in symmetric spaces. Filomat **28**(6), 1113–1132 (2014)
18. Imdad, M., Chauhan, S., Kumam, P.: Fixed point theorems for two hybrid pairs of non-self mappings under joint common limit range property in metric spaces. J. Nonlinear Convex Anal. **16**(2), 243–254 (2015)
19. Jungck, G.: Compatible mappings and common fixed points. Int. J. Math. Math. Sci. **9**(4), 771–779 (1986)
20. Jungck, G.: Common fixed points for noncontinuous nonself maps on nonmetric spaces. Far East J. Math. Sci. **4**, 199–215 (1996)
21. Jungck, G., Rhoades, B.E.: Fixed point theorems for occasionally weakly compatible mappings. Fixed Point Theory **7**, 286–296 (2006)
22. Liu, W., Wu, J., Li, Z.: Fixed points for (ψ, φ)-weak contractions. Appl. Math. Lett. **24**, 1–4 (2011)
23. Liu, Z., Zhang, X., Ume, J.S., Kang, S.M.: Common fixed point theorems for four mappings satisfying ψ-weakly contractive conditions. Fixed Point Theory Appl. **2015**, 20 (2015). doi:10.1186/s13663-015-0271-z
24. Nashine H.K., Altun I.: Fixed point theorems for generalized weakly contractive condition in ordered metric spaces. Fixed point Theory Appl. **2011**, Article ID 132367, 20 p. (2011)
25. Nashine, H.K., Samet, B.: Fixed point results for mappings satisfying (ψ, ϕ)-weakly contractive condition in partially ordered metric spaces. Nonlinear Anal. **74**, 2201–2209 (2011)
26. Pant, R.P.: Common fixed points of four maps. Bull. Calcutta Math. Soc. **90**, 281–286 (1998)
27. Pant, R.P.: R-weak commutativity and common fixed points. Soochow J. Math. **25**, 37–42 (1999)
28. Pant, R.P.: Common fixed points of contractive maps. J. Math. Anal. Appl. **226**, 251–258 (1998)
29. Pant, R.P.: A study on fixed points of noncommuting mappings. D.Sc. thesis, Kumaun University, Nainital, India (2000)
30. Pant, R.P.: Discontinuity and fixed points. J. Math. Anal. Appl. **240**, 284–289 (1999)
31. Pathak, H.K., Rodriguez-Lopez, R., Verma, R.K.: A common fixed point theorem using implicit relation and property (E.A) in metric spaces. FILOMAT **21**(2), 211–234 (2007)
32. Popescu, O.: Fixed points for (ψ, φ)- weak contractions. Appl. Math. Lett. **24**, 1–4 (2011)
33. Rhoades B. E.: Some theorems on weakly contractive maps, Nonlinear Anal. TMA, **47**(4), 2683–2693 (200l)
34. Seasa, S.: On a weak commutativity condition of mappings in fixed point considerations. Publ. Inst. Math. **32**, 149–153 (1982)
35. Suzuki, T.: A generalized Banach contraction principle that characterizes metric completeness. Proc. Am. Math. Soc. **136**(5), 1861–1869 (2008)
36. Verma, M., Chandel, R.S.: Common fixed point theorems using "(E.A)like" property and implicit relation in Fuzzy metric spaces. Adv. Fixed Point Theory **2**(4), 510–518 (2012)
37. Wadhwa, K., Dubey, H., Reena, J.: Impact of "E.A. Like" property on common fixed point theorems in fuzzy metric spaces. J. Adv. Stud. Topol. **3**(1), 52–59 (2012)
38. Zhang, Q., Song, Y.: Fixed point theory for generalized $\phi-$ weak contractions. Appl. Math. Lett. **22**(1), 75–78 (2009)